Hydrogen Storage Technology
Materials and Applications

About the Cover

The cover combines figures from the book with other images to convey the following message: Our concerns of energy security, political stability, and reducing global warming are planet-wide concerns. The work reported in *Hydrogen Storage Technology: Materials and Applications* represents upward-pointing progress (in both materials and engineering) in eliminating these concerns. The upward direction tracks the observed hydrogen release from $Mg(BH_4)_2$, shown as the blue line. Ultimately, through hydrogen technology, we will achieve a secure environment for all inhabitants of the earth, characterized by clean water, soil, and air.

Cover design by Lennie Klebanoff and Scott Shamblin of Taylor & Francis. Lennie Klebanoff acknowledges helpful conversations about the cover design with his sister, Susan Klebanoff.

Hydrogen Storage Technology
Materials and Applications

Edited by
Lennie Klebanoff

CRC Press
Taylor & Francis Group
Boca Raton London New York

CRC Press is an imprint of the
Taylor & Francis Group, an **informa** business

CRC Press
Taylor & Francis Group
6000 Broken Sound Parkway NW, Suite 300
Boca Raton, FL 33487-2742

First issued in paperback 2016

© 2012 by Taylor & Francis Group, LLC
CRC Press is an imprint of Taylor & Francis Group, an Informa business

No claim to original U.S. Government works

ISBN 13: 978-1-138-19929-3 (pbk)
ISBN 13: 978-1-4398-4107-5 (hbk)

Library of Congress Cataloging-in-Publication Data

Hydrogen storage technology : materials and applications / editor, Lennie Klebanoff.
    p. cm.
    "A CRC title."
    ISBN 978-1-4398-4107-5 (alk. paper)
    1. Hydrogen--Storage. 2. Energy storage. I. Klebanoff, Lennie.

    TP245.H9H93 2013
    621.31'26--dc23   2012024194

Visit the Taylor & Francis Web site at
http://www.taylorandfrancis.com

and the CRC Press Web site at
http://www.crcpress.com

Contents

Editor's Introduction... vii
The Editor.. xiii
The Contributors xv

Section I The Need for Hydrogen in the 21st Century and Devices for Converting It to Power

1 The Need for Hydrogen-Based Energy Technologies in the 21st Century.................3
   Jay Keller, Lennie Klebanoff, Susan Schoenung, and Mary Gillie

2 Hydrogen Conversion Technologies and Automotive Applications 31
   Lennie Klebanoff, Jay Keller, Matt Fronk, and Paul Scott

Section II Hydrogen Storage Materials and Technologies

3 Historical Perspectives on Hydrogen, Its Storage, and Its Applications 65
   Bob Bowman and Lennie Klebanoff

4 Hydrogen Storage in Pressure Vessels: Liquid, Cryogenic, and Compressed Gas ... 91
   Guillaume Petitpas and Salvador Aceves

5 Hydrogen Storage in Interstitial Metal Hydrides.. 109
   Ben Chao and Lennie Klebanoff

6 Development of On-Board Reversible Complex Metal Hydrides for
   Hydrogen Storage 133
   Vitalie Stavila, Lennie Klebanoff, John Vajo, and Ping Chen

7 Storage Materials Based on Hydrogen Physisorption... 213
   Channing Ahn and Justin Purewal

8 Development of Off-Board Reversible Hydrogen Storage Materials....................... 239
   *Jason Graetz, David Wolstenholme, Guido Pez, Lennie Klebanoff, Sean McGrady, and
   Alan Cooper*

Section III Engineered Hydrogen Storage Systems: Materials, Methods, and Codes and Standards

9 Engineering Properties of Hydrogen Storage Materials... 331
   Daniel Dedrick

10 Solid-State H₂ Storage System Engineering: Direct H₂ Refueling...........................347
Terry Johnson and Pierre Bénard

11 Engineering Assessments of Condensed-Phase Hydrogen Storage Systems........385
Bob Bowman, Don Anton, and Ned Stetson

12 Codes and Standards for Hydrogen Storage in Vehicles..405
Christine Sloane

Editor's Epilogue and Acknowledgments...427

Index...433

Editor's Introduction

This book describes the current state of the art in storing hydrogen gas so that it can be used to provide power. While all methods of hydrogen storage are discussed, a particular emphasis is given to solid-state hydrogen storage methods.

Hydrogen is ubiquitous in the universe, comprising about 75% of the visible mass of the known universe [1] and about 75% by mass of our Sun. Yet here on Earth, pure hydrogen (i.e., gaseous H_2) exists only in very small amounts in the Earth's atmosphere (5.0×10^{-5} mole fraction) [2]. This scarcity arises because at room temperature molecular H_2 has sufficient velocity (1.77 km/s) [3] that a significant fraction of its molecular velocity distribution is greater than the Earth's escape velocity of 11.0 km/s [4]. Thus, over long periods of time, any molecular hydrogen in the Earth's atmosphere eventually leaks into space.

The reactivity of hydrogen with other elements reduces further the natural occurrence of molecular H_2, with hydrogen being bound up as atoms in water (H_2O), methane (CH_4), as well as in coal, oil, in the organic chemistry of living systems and a multitude of man-made products. So, although H_2 is a flammable gas, it does not exist within or above the Earth in sufficient abundance to qualify it as a "fuel" that can simply be collected and burned to provide power. Rather, if one desires molecular H_2 for a particular purpose, one has to extract it from these various chemical compounds which contain it, and this requires effort and energy. For this reason, hydrogen is often referred to as an "energy carrier" or as a means of "energy storage."

How much energy does it take to retrieve H_2 from these materials? Hydrogen is more energy intensive to manufacture than a fossil fuel such as gasoline. Recently, the European Commission has conducted extensive energy analyses of many types of "fuels" [5]. One can consider the energy ratio $E_{expended}/E_{fuel}$ where $E_{expended}$ is the primary energy expended (some of it wasted) to create the fuel, and E_{fuel} is the combustion energy content (lower heating value, LHV) of the fuel produced. Hydrogen production via steam reforming of natural gas (NG), the most widely used production method, has an $E_{expended}/E_{fuel}$ value of 0.75 [5], indicating one can extract more energy out of the hydrogen (utilizing E_{fuel}) than it took energy to make it ($E_{expended}$). Hydrogen thus forms the basis for an energetically viable transportation energy infrastructure.

However, the largest benefits of using hydrogen stem from the high-efficiency energy conversion devices (such as fuel cells, and H_2 internal combustion engines) that hydrogen enables. This inclusion of the energy conversion device in the energy considerations is the basis for "Well-to-Wheels (WTW) analyses, and strongly favors hydrogen technology. In Section I (Chapters 1 and 2), the myriad societal benefits that are already coming from the initiation of a hydrogen infrastructure are examined (Chapter 1). Chapter 2 explains the high-efficiency hydrogen conversion devices that are making these benefits possible.

As examined in Chapter 1, there are three primary concerns motivating building an energy carrier infrastructure based on H_2:

1. **Fuel Resource Insecurity.** There will come a day when our fossil fuels, such as coal, NG, and propane, as well as our liquid fuels, gasoline and diesel derived from fossil-based crude oil, all eventually run out since they are finite resources. The fuel resources will be gone. This is *fuel resource insecurity*. Since hydrogen can be made from electrolysis of water using renewable (solar, wind—which are

inexhaustible) electricity, hydrogen can relieve fuel resource insecurity, particularly in the transportation sector. Chapter 1 examines the issues and timescales associated with the fuel resource insecurity of the current fossil fuels.

2. **Political Energy Insecurity.** If one country is dependent on another country for shipments of energy, then there is potential that the energy supply chain can be turned off or manipulated as an instrument of political pressure. This is political energy insecurity. Since hydrogen can be made from domestic energy resources that are available to almost all countries, using hydrogen can relieve political energy insecurity. This is also discussed in Chapter 1.

3. **Global Climate Change.** *Our climate is changing, and humans are causing it.* The combustion of fossil-fuel-based resources releases greenhouse gases (GHGs) such as CO_2 and N_2O into the atmosphere. Since H_2 does not contain carbon, there is no CO_2 released at the point of use. So, making hydrogen in a carbon-free way will establish a zero-carbon technology in every application where hydrogen is used. Hydrogen also enables technologies such as fuel cells that have zero NO_x as well. Chapter 1 describes how the global climate change problem has really been with us for a century, and how resolving this problem demands establishing a zero-carbon energy carrier system such as hydrogen.

I would like to share my own thinking on the problem of CO_2 emissions as it has developed over time. My initial investigations of this issue started from a position of not really believing man-made emissions of CO_2 were a problem. However, I learned some things that changed my attitude.

The first piece of learning was the fact that the atmosphere of Earth is remarkably thin. It is much thinner than is apparent while standing on Earth's surface. At an altitude of 10 miles, the atmospheric pressure is 0.1 atmospheres [6]. So, to within 90% accuracy, the atmosphere is gone if you travel 10 miles straight up. This is remarkable. When looking up from the surface, the sky looks endless. Even when viewed from commercial aircraft, looking up, the sky seems very thick. But if you think about how short a distance 10 miles is when traveling horizontally, it is not very far at all. Think about taking a 10-mile trip to the next town over. It's nothing!

When compared to the radius of Earth (3959 miles), the thickness of the atmosphere is about the same relative thickness as the skin of an apple. Figure I.1 is a photograph of the situation [7]. All of the beneficial warming caused by the existence of Earth's atmosphere in the first place, producing about 60°F of surface warming [8], is caused by this very thin gaseous layer on Earth shown in Figure I.1. It starts to become clear how modifying such a thin layer can start to have important consequences for global climate. It becomes even clearer when it is realized that this very thin gas layer is vulnerable to contamination across the entire vast surface area of the planet.

The second piece of understanding came from learning that CO_2 is the largest "driver" or "forcer" in the increases in surface warming during the past century [9], yet the absolute CO_2 concentration in the atmosphere is fairly low, about 395 ppm of the total atmosphere [10]. This CO_2 concentration is only about 0.04% of the total atmosphere, which is a really small number. So, not only is the atmosphere very thin, but within that very thin layer, small changes in atmospheric composition can have important effects. So, "small" increases in CO_2 matter, and it turns out that the first increases make the biggest impacts. For me, this elevated my level of concern.

The third piece of information for me was the realization that my intuition about the nature of the world can be *really wrong*. I think the best example of this is to consider the

FIGURE I.1 (See color insert.)
The Earth's atmosphere photographed from space. (Photo courtesy of NASA.)

example of a sheet of newspaper folding onto itself [11]. If I take a very large single sheet of newspaper and fold it in half onto itself, I get a paper "stack" with resulting thickness two times the original sheet thickness. Fold it again, and I get a resulting thickness four times the original thickness. How thick will the newspaper thickness be if I continue this process of folding 100 times? When I was first confronted with this question, my intuition alone told me the final stack height would be maybe 100 feet tall. The correct answer is the stack would be 3.4 billion light years tall—3.4 billion light years! For context, the size of the universe is thought to be about 78 billion light years [12]. My intuitive answer was really wrong. In fact, words cannot really describe how wrong my intuition was about the nature of this folding of a sheet of newspaper.

Rather than conclude that I have a mental problem, I prefer to believe that we as human beings are not readily able to understand the world around us. The information available to us through our eyes and our ears does not convey the true situation we find ourselves in. Figure I.2 shows our true circumstances [13]. We live on a ball, surrounded by a really thin blue gas layer, suspended in a dark, black vacuum. Our blue ball is traveling about 437,000 miles per hour as we and our solar system orbit the center of the Milky Way Galaxy [14]. Nothing in our daily experience would suggest this is the case. From the ground, the Earth looks flat. The sky looks deep and blue. We seem not to be moving at all. Our local experience does not convey the truth of our situation, and as a result, our capacity to see things as they really are is quite poor as far as understanding the physical world around us is concerned.

I believe this may have been why I initially thought man-made emissions of CO_2 were not a problem. As discussed, I did not have a good physical feel for how thin the atmosphere really is. Furthermore, I do not have a good physical feel for the vast area that constitutes the surface of the Earth, emissions from which can threaten the atmosphere. I have traveled extensively all over the world, and I think jet travel actually reduces my capacity to appreciate the size of the Earth's surface. It is immense, far bigger than my capacity to picture it. Furthermore, I do not have a good physical feel for how many people 7 billion is. The largest collection of people I have ever seen at one time is about 60,000. I cannot intuitively conceive of 7 billion people, almost all burning fossil fuel, putting unimaginable amounts of CO_2 into the atmosphere. I think these intuitive limitations hindered me initially from thinking CO_2 emissions were a problem. I suspect this may be why many

FIGURE I.2 (See color insert.)
The earth photographed from the moon, *Apollo 8*, 1968. (Photo courtesy of NASA.)

people resist the idea that humans can be causing global climate change through fossil fuel technology. It does not "feel right" to many people. I can relate.

But, progress via science is about being open to the evidence wherever it leads (especially when it feels uncomfortable), dealing with phenomena in most cases far removed from our everyday experiences, and trying not to hold tight to our preconceptions about the nature of the world. I now look at this problem of man-made emissions of GHGs, examine the published data, think about how thin the atmosphere is, examine how very small changes in atmospheric composition can have large effects, try to contemplate the vast quantities of fossil fuel being burned, realize the fact that my own a priori physical intuition about my circumstances can be really wrong, and now think, "How could global climate change not be attributable to humans?" This view is supported by powerful scientific analyses performed by well-respected technical bodies, such as the U.S. National Academy of Sciences [15] and the Intergovernmental Panel on Climate Change (IPCC) [16], as explained thoroughly in Chapter 1.

This is a good time to review the science and engineering of hydrogen storage technology. From 2006 to 2011, the U.S. Department of Energy (DOE) funded three major centers to advance hydrogen storage materials: the Metal Hydride Center of Excellence (MHCoE) [17], the Chemical Hydride Center of Excellence [18], and the Sorption Center of Excellence [19]. Results from all three centers' efforts, as well as recent results from individual investigators around the world, are reviewed.

Section I of this book (Chapters 1 and 2) is led by Associate Editor Dr. Jay Keller, formerly of Sandia National Laboratories. In Chapter 1, Jay and his coauthors Susan Schoenung, Mary Gillie, and myself review motivations for developing a hydrogen-based energy carrier and storage infrastructure and present in one place for anybody to read why hydrogen is not only desirable to address fuel resource insecurity and political energy insecurity but also, in the case of mitigating global climate change, *required*. Chapter 2 (myself, Jay Keller, Matt Fronk, and Paul Scott) describes the technologies that use hydrogen

(hydrogen internal combustion engines and fuel cells) to produce useful work and electricity. Reviewing these "hydrogen conversion devices" allows a better understanding of the total system energy efficiencies of hydrogen technology and clarifies what the hydrogen storage needs really are.

I am the Associate Editor for Section II of the book (Chapters 3–8). In Chapter 3, Bob Bowman and I give some historical accounts of hydrogen technology and the storage methods used in these past applications. In Chapter 4, Guillaume Petitpas and Salvador Aceves review gaseous and cryogenic storage of hydrogen from a thermodynamic perspective as a prelude to discussions of solid-state storage materials. For over 50 years, interstitial metal hydrides have been studied for their hydrogen storage properties. This history is reviewed by Ben Chao and me in Chapter 5.

The interstitial metal hydrides possess remarkable hydrogen storage properties but are too heavy for some applications. Chapter 6 (by Vitalie Stavila, myself, John Vajo, and Ping Chen) reviews lightweight (high gravimetric capacity) complex metal hydrides. Chapter 7 (Channing Ahn and Justin Purewal) examines physisorption-based materials. Both the complex metal hydrides and physisorption-based materials are considered "on-board reversible" hydrogen storage materials; the spent hydrogen material can be recharged with molecular H_2 without removing the material from the application. In Chapter 8, Jason Graetz, David Wolstenholme, Guido Pez, myself, Sean McGrady and Alan Cooper give a comprehensive description of "off-board reversible" hydrogen storage materials.

As someone once said, the devil is in the details, and no matter how good a storage material may be, unless it can be made into an actual working tank system, it is of little value. The Associate Editor for Section III of the book (Chapters 9–12) is Mr. Daniel Dedrick of Sandia National Laboratories. In Chapter 9, he gives an account of the fundamental thermal properties of materials that must be understood and managed to create a real solid-state storage system. In Chapter 10, Terry Johnson and Pierre Bénard describe the engineering designs of real hydrogen storage systems that have successfully managed the many issues that arise in hydrogen storage tank operation. In Chapter 11, Bob Bowman and coauthors Don Anton and Ned Stetson give a broad systems engineering assessment of condensed-phase hydrogen storage systems. Chris Sloane in Chapter 12 explains the "codes and standards" work that enables hydrogen technology to be used in a fully understood, certified, and safe manner. Finally, I share some closing thoughts and give acknowledgments in the Editor's Epilogue and Acknowledgments.

A number of the figures in these chapters are reproduced in color in a "color insert." Please check out the color insert when a figure caption refers to it. A periodic table is also provided on the inside back cover.

I end this Editor's Introduction with a disclaimer. The views expressed here are not necessarily those of Sandia National Laboratories, Lockheed Martin Corporation (which manages Sandia), or those of the U.S. DOE. These are my views, formed from a lifetime of having the privilege of working in science and technology, I hope for the betterment of all humanity and the precious planet we call home.

May you find this is an informative, clear, inspiring, and enjoyable experience. Now, turn the page.

Lennie Klebanoff
Sandia National Laboratories, Livermore, California

References

1. Technical data for hydrogen. http://periodictable.com/Elements/001/data.html.
2. R. C. Weast, *The Handbook of Chemistry and Physics*, CRC Press, Boca Raton, FL, 1976, p. F-205.
3. A. Roth, *Vacuum Technology*, Elsevier, Amsterdam, p. 36.
4. R. T. Weidner and R. L. Sells, *Elementary Physics*, Allyn and Bacon, Boston, 1975, p. 212.
5. Well-to-wheels analysis of future automotive fuels and powertrains in the European context, well-to-tank report, March 2007. http://ies.jrc.ec.europa.eu/WTW.
6. Roth, *Vacuum Technology*, p. 5.
7. NASA image. http://www.spaceflight.nasa.gov/gallery/images/shuttle/sts-123/hires/s123e007722.jpg.
8. J. Hansen, D. Johnson, A. Lacis, S. Lebedeff, P. Lee, D. Rind, and G. Russell, *Science* **213**, 957 (1981).
9. J. Hansen et al., *J. Geophys. Res.* **110**, D18104 (2005).
10. The website of the United States National Oceanic and Atmospheric Administration (NOAA) reports atmospheric CO_2 levels at http://www.climate.gov/#UnderstandingClimate.
11. S. Harris, *The End of Faith: Religion, Terror and the Future of Reason*, Norton, New York, 2004, p. 183.
12. J. S. Key et al., *Phys. Rev. D* **75**, 084034 (2007).
13. Earthrise, Apollo 8, NASA images, December 29, 1968. http://www.nasa.gov/images/content/136062main_bm4_3.jpg.
14. See http://solar-center.stanford.edu/FAQ/Qsolsysspeed.html.
15. *America's Climate Choices, Panel on Advancing the Science of Climate Change*, National Research Council, 2010, ISBN 0-309-14589-9.
16. S. Solomon, D. Qin, M. Manning, Z. Chen, M. Marquis, K. B. Averyt, M. Tignor, and H. L. Miller, eds., *IPCC (Intergovernmental Panel on Climate Change) Report: Climate Change 2007: The Physical Science Basis. Contribution of Working Group I to the Fourth Assessment Report of the IPCC*, Cambridge University Press, Cambridge, UK, 996 pp.
17. For a description of the Metal Hydride Center of Excellence, see http://www.sandia.gov/MHCoE/.
18. For a description of the Chemical Hydride Center of Excellence, see http://www.hydrogen.energy.gov/annual_progress10_storage.html.
19. For a description of the Sorption Center of Excellence, see http://www.nrel.gov/basic_sciences/carbon_based_hydrogen_center.cfm#hsce.

The Editor

Dr. Leonard E. Klebanoff was born in Washington, D.C., and raised in nearby Bethesda, Maryland. He received his BS in chemistry and MS in organic chemistry from Bucknell University (Lewisburg, PA). He received a PhD in physical chemistry from the University of California, Berkeley. There he worked with Prof. David Shirley to investigate the chemical, magnetic, and electronic properties of chromium surfaces. After Berkeley, Lennie did a postdoctoral stint at the National Bureau of Standards (now NIST) where he used spin-polarized electron scattering to study the surface magnetism of nickel.

From 1987 to 1997, Lennie was a chemistry professor at Lehigh University (Bethlehem, PA). There he invented the technique of spin-resolved x-ray photoelectron spectroscopy (SRXPS) and used it to examine the spin dependence of core-level photoelectric transitions from ferromagnetic surfaces and films. At Lehigh, he eventually attained the rank of Full Professor with tenure.

After 12 years away from his beloved San Francisco Bay area, he longed to return and in 1997 took a position at Sandia National Laboratories in Livermore, California. He was appointed the environmental team leader for the Extreme Ultraviolet Lithography (EUVL) program, a large collaboration between the three Bay Area national laboratories and a consortium of semiconductor companies including Intel, Motorola, and AMD. By 2003, his interests began to take him into the alternative energy arena, where he served with California State officials to develop the Governor's Blueprint for a Hydrogen Highway in California.

In 2006, Lennie was named Director of the U.S. Department of Energy (DOE) Metal Hydride Center of Excellence (MHCoE), a 21-institution center funded by the DOE Office of Energy Efficiency and Renewable Energy (EERE) to advance the science of solid-state storage of hydrogen using metal hydrides. Lennie also served as the Sandia technical lead in this MHCoE effort. After his 5-year tenure as the MHCoE Director, he continued his hydrogen-related work, leading the Sandia effort in fuel cell market transformation, including the development of a Fuel Cell Mobile Lighting system that has been used the past 3 years on the red carpet at the Academy Awards (Oscars) and Golden Globes award shows in Hollywood.

Lennie has authored over 95 scientific papers and holds 21 patents.

In his spare time, Lennie enjoys spending time with his wife, Nitcha. He is a scratch golfer (after he hits the ball, he scratches his head, wondering where the ball went) and still likes to play basketball with his college friends, Old and Sigs, with whom he played varsity basketball at Bucknell. He is particularly fond of the Transformers movies and eating vast quantities of home-cooked Thai food courtesy of Nitcha.

Lennie Klebanoff
Hydrogen and Combustion Technology Department
Sandia National Laboratories
Livermore, California
lekleba@sandia.gov

The Contributors

Salvador Aceves
Energy Conversion and Storage Group
Lawrence Livermore National Laboratory
Livermore, California
aceves6@llnl.gov

Dr. Salvador M. Aceves was born in Atlanta, Georgia and grew up in central Mexico. He received an undergraduate degree in mechanical engineering from the University of Guanajuato (Mexico) followed by MS and PhD degrees from Oregon State University. He worked for 2 years as an assistant professor at the University of Guanajuato followed by a year at the Daido Institute of Technology (Nagoya, Japan) as an NSF-JSPS postdoctoral fellow. He joined LLNL (Lawrence Livermore National Laboratory) in 1993 and is group leader for energy conversion and storage. As an applied thermodynamicist, he has researched energy utilization with a special focus on the transportation sector. Dr. Aceves's past and current research projects include electric vehicle climate control, electric and hybrid vehicle analysis, and homogeneous charge compression ignition engines. A strong believer in the future potential of hydrogen-based transportation, Salvador has for the last 18 years researched hydrogen delivery, hydrogen engines, and hydrogen safety, in addition to hydrogen storage in cryogenic vessels. Salvador Aceves is a fellow of the American Society of Mechanical Engineers and a distinguished alumnus from Oregon State University. An avid soccer player, he's managed to play continuously in local leagues for almost 20 years. His new team, the Jacobians, is mainly made up of the members of LLNL's energy conversion and storage group, and is a perennial title contender in the Livermore indoor soccer league.

Channing Ahn
Division of Engineering and Applied Science
California Institute of Technology
Pasadena, California
cca@caltech.edu

Dr. Channing C. Ahn has been on the faculty of Caltech since 1987. He was born in Boston, Massachusetts. He received his BS from the University of California, Berkeley, and his PhD in physics from the University of Bristol. He coauthored the *EELS Atlas* and edited *Electron Energy Loss Spectrometry in Materials Science.*

His work in hydrides started in 1995 in collaboration with R. C. Bowman, Jr., studying disproportionation reactions in AB_5 alloys. Since then, he has studied the microstructure of Sb-mediated growth during group IV MBE, the use of group IV elements as battery anode materials, and more recently, hydrogen adsorption phenomena in high-surface-area materials. He was principal investigator in both the Metal Hydride Center of Excellence and the Hydrogen Sorption Center of Excellence. He served as an expert member of the International Energy Agency Task 22, Fundamental and Applied Hydrogen Storage Materials Development effort. He presently serves as a member of the Scientific Advisory Committee for the Canadian Hydrogen Strategic Research Network (H2CAN). He has read physics texts for Recording for the Blind and played clarinet with the California Chamber Symphony.

Don Anton
Renewable Energy Directorate
Savannah River National Laboratory
Aiken, South Carolina
Donald.anton@srnl.doe.gov

Dr. Donald L. Anton was born in Berwyn, Illinois, a suburb of Chicago. He attended Purdue University in West Lafayette, Indiana, where he received a BS degree in materials science and engineering, working with Prof. J. F. Radavich on grain boundary precipitation in nickel base superalloys. Subsequently, Don attended Northwestern University in Evanston, Illinois, where he worked with Prof. Morris E. Fine under a National Science Foundation Fellowship, Low Cycle Fatigue of Precipitation Hardened Nickel Base Alloys, after which he was awarded a PhD in materials science and engineering. United Technologies Research Center in East Hartford, Connecticut, was his first position; he investigated the effects of precipitation hardening and alloy design on creep and fatigue properties of superalloy single crystals for gas turbine engine applications. Editing of the prestigious Seven Springs Superalloy Symposium proceedings in 1992 and 1996 ensued. Previous work in high-temperature alloys led to investigations of intermetallic compounds with very high melting points and editing of two more technical volumes on intermetallic composites in 1990 and 1992. With the end of the Cold War, efforts were redirected toward energy research. In coordination with the UTC fuel cell division, an initiative to store hydrogen more effectively in automobiles was developed. This led to the modeling, design, construction, and evaluation of the first operational complex hydride hydrogen storage system utilizing $NaAlH_4$. Savannah River National Laboratory, Aiken, South Carolina, is his current employer, where he works at the Center of Hydrogen Research. Dr. Anton is now the Director of the Hydrogen Storage Engineering Center of Excellence, funded by the Department of Energy, which is investigating materials-based hydrogen storage systems. Aiken is now his home; he lives with his wife, Mary, and son, Daniel. Don enjoys traveling and experiencing the human existence across the globe; his career in science has afforded ample opportunities. This has led to an interest in geocaching, which many of his colleagues have been exposed to through treks under waterfalls and along secluded seashores at some of the most beautiful locations around the world.

Pierre Bénard
Department of Physics at the Université du Québec à Trois-Rivières
Trois-Rivières Quebec, Canada
pierre.benard@uqtr.ca

Dr. Pierre Bénard was born in Montreal, Canada. Pierre is a professor of physics at the Université du Québec à Trois-Rivières and a member of the Hydrogen Research Institute. He holds a PhD in theoretical condensed matter physics from the Université de Sherbrooke and an MSc from the University of Toronto. His PhD thesis was on the magnetic properties of the normal phase of high-temperature superconductors. He has written over 100 refereed contributions. Pierre is adjunct director of the Hydrogen Research Institute (HRI). HRI, founded in 1994, is a research unit of the Université du Québec à Trois-Rivières dedicated to facilitating the use of hydrogen as an alternative energy vector. His research interests cover the study of microporous adsorbents for hydrogen storage and purification and the study of hydrogen safety issues using computational fluid dynamics. He is a member of the board of directors of CQMF, the Québec Centre on Functional Materials, and member of the scientific committee of the Hydrogen Canada NSERC (Natural Sciences and Engineering Research Council of Canada) Strategic Research Network (H2Can). He is also guest professor at the Technical University of Wuhan. In his spare time, Pierre enjoys reading about astronomy and history, used to practice fencing (not for a while though), and is a fan of sci-fi shows such as *Dr. Who* and *Eureka*.

Bob Bowman
Energy and Engineering Sciences Directorate
Oak Ridge National Laboratory
Oak Ridge, Tennessee
rcbjr1967@gmail.com

Dr. Robert C. Bowman, Jr., is currently a senior research-and-development (R&D) staff member at Oak Ridge National Laboratory (ORNL, Oak Ridge, TN). He was born in Dayton, Ohio. He holds a BS degree from Miami University (Oxford, OH); a master's degree from the Massachusetts Institute of Technology (Cambridge, MA); and a PhD from the California Institute of Technology (Pasadena, CA). All of his degrees are in chemistry. Bob has more than 40 years of research experience on the properties, behavior, and applications of metal hydrogen systems. He is the author/coauthor of over 300 published papers. In addition, he was coeditor of *Hydrogen in Disordered and Amorphous Solids* (1986), *Hydrogen in Semiconductors and Metals* (*MRS Symposium Proceedings*, 1998), and the proceedings of MH2006 published in the *Journal of Alloys and Compounds* (2007). He has been awarded three patents on using metal hydrides. Bob is an honorary member of the Steering Committee for the International Symposium on Metal-Hydrogen Systems, having served as its secretary-general between 2006 and 2010. He has been cochair for the 1995 Gordon Research Conference on Hydrogen Metals Systems and the cochair for MH2006. Prior to joining ORNL, Bob was a Principal Member-of-Technical Staff at the Jet Propulsion Laboratory (Pasadena, CA), where he worked on the development of hydrogen sorption cryogenic refrigerators for space flight applications and characterizations of advanced metal hydrides for hydrogen storage and energy conversion. He had been previously employed at Aerojet Electronic Systems (Azusa, CA), the Aerospace Corporation (El Segundo, CA), and the Monsanto-Mound Laboratory (Miamisburg, OH), where he performed basic and applied studies of numerous metal-hydrogen systems for nuclear and energy technologies. He was also a visiting associate in materials science at the California Institute of Technology.

Bob has a keen interest in history, especially topics relating to the expansion of the American frontier and the military during the 19th century. One of his favorite activities is being a historical interpreter and re-enactor as a U.S. dragoon from the Mexican-American War. He also enjoys traveling with his wife, Judy by planes, trains, and pickup trucks.

Ben Chao
BASF Battery Materials–Ovonic
BASF Corporation
Rochester Hills, Michigan
benjamin.chao@basf.com

Dr. Benjamin S. Chao was born and grew up in Taiwan. He came to the United States for advanced study after completing his BS in physics. He received his MS in physics from the University of Cincinnati, Ohio, and PhD in materials science from Syracuse University (Syracuse, NY). In 1980, Ben joined Energy Conversion Devices (ECD) as a research scientist working in the analytical group, structure lab. Later, he served as lab director. At ECD, he has participated in various materials development and characterization, including amorphous and nanocrystalline-silicon-based photovoltaic materials, phase change chalcogenide alloy materials for optical and electric memory media, thermoelectric materials, silicon oxide/nitride corrosion resistance coating materials, multiplayer devices, and yttrium barium copper oxide (YBCO) superconducting materials. When the United States Advanced Battery Consortium (USABC) awarded the first contract to ECD and its wholly owned subsidiary Ovonic Battery Company (OBC) in 1992, Ben joined the R&D group to advance the metal hydride hydrogen storage materials in nickel metal hydride (NiMH) battery for electric vehicle application. In 2000, he continued his R&D work in transition-metal-based alloy materials in Texaco Ovonic Hydrogen Systems, LLC, a joint venture between ECD and Texaco, and Ovonic Hydrogen Systems, LLC, a wholly owned subsidiary of ECD, to develop a family of reversible solid hydrogen storage systems for portable power, transportation, and stationary applications. He is author/coauthor of over 80 publications and 11 U.S. patents. Ben enjoys many activities in his leisure time: reading, including comics books such Charlie Brown, Calvin and Hobbes; traveling; movies; and sports. He is a diehard college basketball fan.

Ping Chen
Hydrogen and Advanced Materials Division
Dalian National Laboratories for Clean Energy
Dalian Institute of Chemical Physics, CAS
Dalian, China
pchen@dicp.ac.cn

Dr. Ping Chen was born in Shandong Province, China, and is a professor at the Dalian Institute of Chemical Physics (DICP, China). Ping received a BS degree and a PhD degree in chemistry in 1991 and 1997, respectively, from Xiamen University, China. She was a faculty member in the faculty of science at National University of Singapore (NUS) before she joined DICP in 2008. Her primary research interest includes the development of chemical and complex hydrides for hydrogen storage, catalysis, and organic synthesis. She pioneered the research in hydrogen storage over amide-hydride composite materials (2002) and amidoborane system (2008). Ping has over 80 peer-reviewed journal articles with more than 4000 citations.

Alan Cooper
Global Technology Center
Air Products and Chemicals, Inc.
Allentown, Pennsylvania
cooperac@airproducts.com

Dr. Alan C. Cooper was born in Iowa City, Iowa. He received his BS in chemistry from Alma College (Alma, MI) and a PhD in inorganic chemistry from Indiana University (Bloomington, IN). After postdoctoral research at Yale University (New Haven, CT), he joined the Corporate Science and Technology Center at Air Products and Chemicals, Inc. (Allentown, PA), where he is now a lead research chemist in the Global Technology Center. During his time at Air Products, he has been a principal investigator for three projects funded by the U.S. Department of Energy on new technologies for hydrogen storage and delivery. The projects included the investigation of advanced solid-state hydrogen adsorbents and the development of new liquid organic hydrogen carrier technology. Within the topic of hydrogen storage materials, Alan has authored 18 publications in peer-reviewed journals and has been awarded five U.S. patents. In his spare time, he enjoys listening to live music and participating in outdoor activities, such as mountain biking and hiking. Most of all, he enjoys spending time with his very active daughter and wife.

Daniel Dedrick
Hydrogen and Combustion Technology Department
Sandia National Laboratories
Livermore, California
dededri@sandia.gov

Mr. Daniel E. Dedrick, Associate Editor of Section III, was born in Port Angeles, Washington. Daniel manages the Hydrogen Program and the Hydrogen and Combustion Technologies

Department at Sandia National Laboratories in Livermore, California. Daniel has 10 years of experience in the research, development, and demonstration of the hydrogen energy systems, with specific expertise in storage systems. Daniel earned his MS in mechanical engineering from the University of California, Berkeley, with a specific focus on heat and mass transfer. Daniel was the first to measure and optimize the engineering properties of complex metal hydrides to enable the development of the world's only complex metal hydride system that was demonstrated to refuel and provide fuel at rates appropriate for high-demand fuel cell electric vehicles. Daniel is active in various critical hydrogen and fuel cell organizations, including the Fuel Cells and Hydrogen Energy Association, the International Energy Agency Hydrogen Implementing Agreement, IA-HySafe, and the California Fuel Cell Partnership. Today, Daniel continues his work to remove technical barriers to the deployment and widespread utilization of hydrogen for energy applications. Outside work, Daniel enjoys riding bicycles and spending time with his wife and son.

Matt Fronk
Director (retired), GM Fuel Cell R&D
Honeoye Falls, New York
mfronk@frontiernet.net

Mr. Matthew H. Fronk brings over 32 years of industry experience from General Motors in the areas of advanced fuel and emission controls as well as alternative energy development. For 20 years, he led GM's PEM fuel cell R&D program, which is currently in Honeoye Falls, New York. The result of this work was a 100-vehicle fuel cell electric vehicle fleet— the largest prototype fleet of its kind in the world. After his tenure at GM, he held the position of director of the Center for Sustainable Mobility at the Golisano Institute for Sustainability at Rochester Institute of Technology (RIT, Rochester, NY). At RIT, he led activities examining a full spectrum of emerging alternative fuel and propulsion technologies and their applications: biodiesel, ethanol, batteries, and fuel cells. Currently, Matt is a consultant in the alternative energy field. Matt was also instrumental in the creation of the NY Battery and Energy Storage Technology Consortium (NY BEST) and is currently supporting hydrogen infrastructure development activities in the northeast part of the U.S. In support of hydrogen infrastructure deployment, he is working closely with the automotive original equipment manufacturers, hydrogen suppliers, as well as government leaders, officials, and other involved organizations and individuals.

Matt holds a BSME degree from Union College in Schenectady, New York, which is also his birthplace. He and his wife, Donna, have three adult children. Outside his family activities, and in his spare time, he enjoys basketball, old cars, and fishing.

Mary Gillie
EnergeticUK Limited
Chester, United Kingdom
mary.gillie1@yahoo.co.uk

Dr. Mary Gillie was born in Leamington Spa, United Kingdom. She graduated in applied physics from the University of Edinburgh before modeling the control of wind farms for the PhD in electrical engineering at the University of Strathclyde in Glasgow. She is a senior consultant at EA Technology and technical director of EnergeticUK. She has worked on numerous projects to decarbonise and improve the efficiency of electricity systems. She has managed field trials of innovative voltage control techniques for the distribution network and is part of the technical team for the largest smartgrid project in the UK. Working with communities, she designed and implemented small scale generation and low carbon buildings for a village aiming to become carbon neutral. Internationally, she was the UK national expert for Task 18 of the IEA Hydrogen Implementing Agreement. Mary sees hydrogen playing a major role in managing low carbon energy networks as heat, power, and transport become more inter-dependent. She wants to see hydrogen's potential developed as a long term storage medium as the amount of renewables increases and as a fuel for long-range transport.

When taking time off from solving the world's energy crisis, Mary is an avid mountian runner, enjoys organic gardening, traveling, and swimming. She also likes a bit of culture such as going to the theatre and literature.

Jason Graetz
Sustainable Energy Technologies Department
Brookhaven National Laboratory
Upton, New York
graetz@bnl.com

Dr. Jason Graetz received his undergraduate degree in physics from Occidental College in Los Angeles, California in 1998. He received his master's and PhD degrees in materials science from the California Institute of Technology (Pasadena) in 2000 and 2003 and later joined the Hydrogen Storage Group at Brookhaven National Lab as a postdoctoral research associate. Jason is now a tenured scientist at Brookhaven and the leader of the Energy Storage Group; he oversees a number of projects focused on hydrogen storage and lithium batteries. His hydrogen storage work has involved the characterization of a variety of hydrides, investigating reaction kinetics, thermodynamics, and the role of catalysts. More recently, his research has focused on the synthesis of high-capacity aluminum-based hydrides and the development of low-energy chemical pathways to regenerate "irreversible" hydrides. Jason is a U.S. expert on hydrogen storage for the International Energy Agency and was the recipient of the 2006 Ewald Wicke Award and the 2008 Presidential Early Career Award for Scientists and Engineers (PECASE) for his work on hydrogen storage. Jason was born in Grosse Pointe, Michigan, and currently resides with his wife and son on the east end of Long Island, where he is an avid canoe paddler and surfer.

Terry Johnson
Energy Systems Engineering and Analysis
    Department
Sandia National Laboratories
Livermore, California
tajohns@sandia.gov

Mr. Terry A. Johnson is a Distinguished Member of the Technical Staff in the Energy Systems Engineering and Analysis Department at Sandia National Laboratories in Livermore, California. Terry was born in Tacoma, Washington, and hails from Winthrop, Washington. Prior to joining Sandia, he received his BS and MS degrees in mechanical engineering at Washington State University (Pullman). Terry is an energy systems engineer and is currently involved in several efforts related to hydrogen storage. He is helping to develop fueling protocols for high-pressure hydrogen storage tanks for fuel cell vehicles as well as developing a metal hydride storage system for a fuel cell forklift. Prior to this work, he led a 6-year effort to successfully develop and demonstrate an advanced hydrogen storage system for General Motors that was based on the complex metal hydride sodium alanate. Terry is an avid cyclist and spends many of his lunch hours chasing other cyclists around the Livermore hills.

Jay Keller
Zero Carbon Energy Solutions
Oakland, California
jay.o.keller@gmail.com

Dr. Jay O. Keller, Associate Editor of Section I, was born in San Diego, California. In 1979, he received his MS degree and in 1983 his PhD, both from the University of California at

Berkeley, in mechanical engineering and with a focus on energy. He has over 175 publications focusing on combustion and hydrogen energy in the refereed literature, proceedings of meetings and symposia, book chapters, and patents.

Jay has devoted the past 18 years to building a globally recognized program at Sandia National Labs that advances the science and technologies needed to enable a net-zero green house gas (GHG) emitting energy system. Hydrogen programs and projects he has directed include activities in hydrogen-fueled conventional technologies (reciprocating and turbine internal combustion engines), fuel cell systems development for utility and transportation applications, materials science for hydrogen storage, hydrogen effects on materials, molecular separation, fuel cell membranes, systems analysis and engineering, thermochemical solar and nuclear hydrogen production, and the science and engineering behind the development of international regulation codes and standards, such as understanding unintended hydrogen release behavior and hydrogen effects in materials.

Jay was the deputy director for the Metal Hydride Center of Excellence (MHCoE), which was the largest of three U.S. Centers of Excellence (CoE), with 21 partners (industry, university, and national laboratory). In 2010, the three CoEs were awarded the Department of Energy's (DOE) meritorious achievement award for their "outstanding contribution to advancing the state-of-the-art in hydrogen storage." In addition to the MHCoE, his program is responsible for providing the science and technology behind the development of hydrogen codes and standards (C&S) for the implementation of the hydrogen infrastructure. He is a technical advisor to the DOE and supports all C&S international engagements, including being the alternate to the co-chair for the Regulation Codes and Standards Working Group (RCSWG) of the International Partnership for Hydrogen and Fuel Cells in the Economy (IPHE). In addition, his program is active in five annexes of the Fuel Cell Implementing Agreement (FCIA) and Hydrogen Implementing Agreement (HIA) of the International Energy Agency (IEA); these annexes are Fuel Cell, Storage, Analysis, Production, and Safety.

As recognition for his leadership abilities, in 2009 Jay was awarded the DOE Hydrogen Program Meritorious Award (with special recognition for outstanding technical contributions). In his spare time, Jay enjoys bike riding in the beautiful Oakland hills.

Sean McGrady
Department of Chemistry
University of New Brunswick
Fredericton, New Brunswick, Canada
smcgrady@unb.ca

Dr. G. Sean McGrady obtained his BA (1986) and DPhil (1990) degrees in chemistry at the University of Oxford, England, under the supervision of Prof. Tony Downs, and was awarded a Royal Society Postdoctoral Fellowship to work at the University of Hanover in Germany. Returning to Oxford after a year, he took up a position with Oxford Lasers Limited before moving back to the university in 1993 as a junior research fellow and lecturer in chemistry. In 1998, he accepted a tenured university lecturership at King's College, University of London. On the closure of King's Chemistry Department in 2003, Sean immigrated to Canada, becoming an associate professor of chemistry at the University of New Brunswick (UNB, Fredericton, NB) and a full professor and university research scholar in 2006. In his time at UNB, Sean has raised over $5 million in research grants and contracts to support projects spanning a wide range of contemporary chemical problems and drawing on an unusually broad range of techniques; these include supercritical fluid chemistry, microwave activation of reactions, and synthesis of reactive and marginally stable materials under conditions of unusually high pressures and low temperatures. His current team consists of around a dozen researchers—a mix of undergraduate and graduate students, research assistants, and postdocs—working on a portfolio of projects in the areas of structural and materials chemistry, catalysis, and energy storage and conversion. In particular, his research into hydrogen storage materials and techniques has received international recognition: Sean has been running the largest university-based program in this field in Canada since 2008, and his group comprised the only non-U.S. member of the U.S. Department of Energy's hydrogen storage Metal Hydride Center of Excellence. Sean has published some 90 articles in high-profile international science journals and is the inventor of 12 U.S. and international patents, with several applications pending.

Guillaume Petitpas
Energy Conversion and Storage Group
Lawrence Livermore National Laboratory
Livermore, California
petitpas1@llnl.gov

Dr. Guillaume Petitpas was born in France, near Saint-Malo, in 1982. He successively earned his BS and MS degrees in mechanical and aeronautical engineering at ENSMA (Ecole Nationale Superieure de Mecanique et d'Aerotechnique, Poitiers, France). During this program, he studied at Queen's University of Belfast (Northern Ireland, 2004) and was a postgraduate exchange student at the University of Washington (Seattle, 2005). He received his PhD degree in chemical and process engineering from Ecole des Mines de Paris in 2008. His thesis on plasma-assisted production of hydrogen out of liquid fuels was partially sponsored by Renault SA (a leading French car manufacturer) and ADEME (French Environment and Energy Management Agency). Guillaume then moved to California for a position as postdoctoral research associate at Lawrence Livermore National Laboratory, under the supervision of Salvador Aceves, where he remains. While he mainly works on cryocompressed storage for automotive applications, his proficiencies in hydrogen also include delivery, safety, and education. He recently participated in an online course for laboratory personnel handling hydrogen, available at http://www.h2labsafety.org. When not "intensely" working at the lab, Guillaume enjoys aikido (almost fifth Kyu) and going to Santa Cruz to surf.

Guido Pez
3705 Vale View Drive
Allentown, Pennsylvania
pezgp@ptd.net

Dr. Guido P. Pez was born in Italy but acquired most of his education in Australia, where his family moved after World War II. He received his BSc degree from the University of New South Wales and in 1967 a PhD in chemistry from Monash University, which was followed by postdoctoral research at McMaster University (Hamilton, ON, Canada). In the United States, he first worked at Allied Chemical Corporation (Morristown, NJ), and in 1981 he moved to Air Products and Chemicals Corporation (Allentown, PA), where he held the position of chief scientist in inorganic chemistry. He has authored or coauthored 78 scientific publications; he is named as an inventor on 64 U.S. patents and was the recipient in 1994 of the ACS Award in Inorganic Chemistry. Guido has always had an interest in hydrogen, from hydrogenation catalysis using soluble metal complexes, to H_2 production via methane-steam reforming, as well as in hydrogen storage, initially employing carbon materials but then mostly using organic liquid carriers. In 2009, he retired from Air Products and has since been challenged with teaching organometallic and inorganic chemistry at Lehigh University and more recently at Barnard College of Columbia University (New York City). Guido enjoys skiing and sailing, and he and his wife, Terri, are also well occupied by their nine grandchildren.

Justin Purewal
Energy Technologies Department
HRL Laboratories, LLC
Malibu, California
jjpurewal@hrl.com

Dr. Justin J. Purewal was born in Orange, California. Justin received a BS degree in 2005 in applied physics from the California Institute of Technology (Pasadena). He received a PhD degree in materials science in 2010 from California Institute of Technology, under the direction of professor Brent Fultz. Between 2010 and 2012, he has worked as a postdoctoral researcher at the University of Michigan, studying heat transfer and gas storage properties of metal-organic framework materials and composites. His current research focuses on novel materials for adsorption-based hydrogen storage. Other research interests include studying the dynamics of adsorbed layers by quasielastic and inelastic neutron scattering. In his spare time, Justin enjoys playing inline roller hockey and classical piano.

Susan Schoenung
Longitude 122 West, Inc.
Menlo Park, California
susan.schoenung@sbcglobal.net

Dr. Susan Schoenung was born in Milwaukee, Wisconsin. She grew up in the Chicago suburbs. She has the following degrees: BS in physics, Iowa State University (Ames); and MS and PhD, mechanical engineering, Stanford University (Stanford, CA).

Susan is president of Longitude 122 West, Inc., a consulting firm in Menlo Park, California, specializing in energy systems analysis and environmental remote sensing. She is co-operating agent for the International Energy Agency Hydrogen Implementing Agreement Task 30—Global Hydrogen Systems Analysis. Previously, she was operating agent for Task 18—Evaluation of Integrated Hydrogen Demonstration Systems. She has performed energy systems analysis for the U.S. Department of Energy, the Electric Power Research Institute, and numerous industry clients. Previously, she worked for Chevron Research Laboratories, Bechtel Engineering, and Schafer Associates. Longitude 122 West is a member of the Energy Storage Association and the California Hydrogen Business Council.

Susan is currently training for her fifth Susan G. Komen 3-Day Walk-for-the Cure, an annual 60-mile walk raising funds for breast cancer research and community mammograms.

Paul Scott
Consultant to Ballard Power, Transportation Power, Inc.
San Diego, California
pbscott2@gmail.com

Dr. Paul B. Scott, Sc.D., vice president for Advanced Technologies, TransPower USA, left his home state, Michigan, to study at the Massachusetts Institute of Technology (MIT, Cambridge, MA) and then joined the professorial staff at MIT and later at the University of Southern California Department of Aerospace Engineering (Los Angeles). Consulting activities finally lured him from the university life; in time, he left work on spacecraft and high-altitude fluid mechanics in favor of medical x-ray imaging technology development, solar photovoltaic applications, and zero-emission vehicle development. His past two decades have been focused on hydrogen production, vehicles and fueling, and development of hybrid electric hydrogen-fueled buses and electric buses.

With wife, Lauren, he supports the San Diego City Ballet and serves on the boards of Energy Independence Now and the California Hydrogen Business Council (of which he was immediate past president).

Christine Sloane
General Motors (retired)
chrissloane1@gmail.com

Dr. Christine S. Sloane was born in Washington, D.C. She received her PhD from MIT in chemical physics. She served as the GM global lead for Hydrogen and Fuel Cell Safety Codes and Standards. She directed the global General Motors hydrogen and fuel cell safety codes and standards team. This included internal, industry/professional, and regulatory elements of standards for fuel cell vehicles and their interface with the environment, roadways, and hydrogen fuel infrastructure. She was responsible for technical strategy and R&D guidance to identify performance-based vehicle and component requirements for assurance of public safety, commercial feasibility, and infrastructure compatibility. Chris guided those elements of GM's broad portfolio of cooperative R&D and vehicle demonstration activities that were aimed at gaining real-world experience as a base for developing transportation codes and standards. Earlier, she served as the General Motors director for the FreedomCAR and Fuels Partnership, an R&D program through which domestic U.S. automakers and energy companies engaged in fuel cell and hydrogen infrastructure development in addition to advanced power train concepts.

Prior to 2002, Chris served as director of technology strategy for advanced technology vehicles focused on hybrid drive systems and as chief technologist for the development and demonstration team for Precept, GM's 80-mile-per-gallon five-passenger hybrid electric vehicle concept. She previously served as director of environmental policy for global climate issues and for mobile emission issues involving advanced technology vehicles.

Her early research interests included air quality and manufacturing and vehicle emissions. Chris has authored over 80 technical papers and coedited one book. She has served on several boards of professional organizations and numerous National Academy of Science panels and study groups. She currently provides consultation services to the U.S. Department of Energy through SloaneSolutions LLC. In her spare time, Chris enjoys skiing and spending time with her family.

Vitalie Stavila
Hydrogen and Combustion Technology Department
Sandia National Laboratories
Livermore, California
vnstavi@sandia.gov

Dr. Vitalie Stavila performed undergraduate and graduate studies at the State University of Moldova and received his BS and PhD degrees in 1996 and 2002, respectively. His PhD research involved synthesis of new heterometallic complexes as precursors for mixed oxides. He worked as a lecturer at the same institution between 2002 and 2004 teaching inorganic and coordination chemistry and conducting research on coordination and supramolecular compounds. He then did postdoctoral studies in France at the Ecole Normale Superieure de Lyon developing contrast agents for magnetic resonance imaging and in the U.S. at Rice University in Houston studying molecular precursors for nanostructured metal oxide and chalcogenide materials. Since 2008, Vitalie has been at Sandia National Laboratories in Livermore, California, working in the area of novel materials for energy storage. His hydrogen-related research efforts involve synthesis and characterization of novel complex metal hydrides for reversible hydrogen storage. During his free time, Vitalie enjoys traveling, sightseeing, reading, and playing soccer.

Ned Stetson
U.S. Department of Energy
Fuel Cell Technologies Program
Washington, DC
ned.stetson@ee.doe.gov

Dr. Ned T. Stetson is a native Vermonter with degrees in chemistry from the University of Vermont and the University of California, Davis. His career in hydrogen storage started when he accepted a postdoctoral position with Prof. Klaus Yvon at the University of Geneva, Switzerland, in 1991. In his postdoctoral position, he investigated the synthesis and structure of novel complex metal hydrides, including $BaReH_9$, one of the only known compounds with a hydrogen-to-nonhydrogen atom ratio exceeding methane. After his time in Switzerland, Ned returned to the United States to work at a small Michigan-based company, Energy Conversion Devices, Inc. (ECD). At ECD, he was involved with development of magnesium-based and intermetallic hydrogen storage materials, system design and development, and overcoming the barriers related to codes, standards, and regulations for hydrogen storage systems. This work led to the commercial introduction of the Ovonic solid hydrogen storage canisters. Ned was also the convener of the International Organization for Standardization working group that developed the first international standard for portable rechargeable metal hydride storage systems, ISO-16111:2008, "Transportable Gas Storage Devices—Hydrogen Absorbed in Reversible Metal Hydride." In 2007, he joined the U.S. Department of Energy (DOE) as part of the Hydrogen and Fuel Cells Technologies Program; he is the team lead for all of the hydrogen storage activities. The U.S. DOE hydrogen storage subprogram maintains a comprehensive portfolio of projects covering development of low-cost, high-pressure tanks; cryogenic hydrogen storage; advanced hydrogen storage materials development, including reversible metal hydrides, chemical hydrogen storage materials, and hydrogen sorbents; and system engineering. When he is not working on solving hydrogen storage problems, Ned enjoys a number of hobbies, including cooking, home brewing, wine tasting, and spending time on the lake in his boat.

John Vajo
Energy Technologies Department
HRL Laboratories, LLC
Malibu, California
jjvajo@hrl.com

Dr. John J. Vajo was born in the Bronx, New York, in 1958. He received a BS in chemistry from Rensselaer Polytechnic Institute (Troy, NY) in 1980 and a PhD in physical chemistry from the California Institute of Technology (Pasadena) in 1986, where he studied the surface chemistry of heterogeneous catalysis using ultra-high-vacuum surface science techniques. John conducted postdoctoral research in surface chemistry at SRI International (Menlo Park, CA) from 1986 to 1989. In 1989, he joined HRL Laboratories (Malibu, CA) as a member of the technical staff and is currently a senior scientist. He has studied solid-state materials for hydrogen storage since 1998, focusing on complex hydrides and methods to alter their thermodynamics and kinetics. John enjoys hiking and mountaineering using ice ax and crampons.

David Wolstenholme
Department of Chemistry
University of New Brunswick
Fredericton, New Brunswick, Canada
dwolsten@unb.ca

Dr. David J. Wolstenholme was born in Kitchener, Ontario, Canada, in 1979. He obtained his undergraduate and doctoral degrees at Dalhousie University (Halifax, NS, Canada; 2000–2008), the latter under the supervision of Prof. Stan Cameron. His PhD research involved an in-depth investigation of weak homopolar C-H⋯H-C interactions through combined experimental and theoretical charge density studies. He was subsequently awarded an Alexander von Humboldt Fellowship to work as a postdoctoral researcher at the Universität Augsburg, Germany (2008–2010), where he explored the chemistry of β-agostic transition metal complexes. He returned to Canada in 2010 as a senior researcher in Dr. McGrady's group at the University of New Brunswick (Fredericton), studying the structure-bonding-reactivity relationship of molecular and complex metal hydrides with the aim of developing more efficient and cost-effective hydrogen storage materials. David has published 18 peer-reviewed articles in high-profile journals, 13 as corresponding author, along with two book contributions on alternative energy storage.

Section I

The Need for Hydrogen in the 21st Century and Devices for Converting It to Power

1

The Need for Hydrogen-Based Energy Technologies in the 21st Century

Jay Keller, Lennie Klebanoff, Susan Schoenung, and Mary Gillie

CONTENTS

Introduction 3
Reasonable Standard of Living for All...4
Fuel Resource Insecurity 5
Political Energy Insecurity...9
Global Climate Change 10
Efficiency vs. Net Zero 15
Timescales for Change...17
Renewable Energy Resources..19
Options for Grid-Scale Energy Storage..20
Making Do with Nonrenewable Energy..22
Energy Carriers...23
Summary 27
References...27

Introduction

In this chapter, we motivate why we need to aggressively pursue diversified sustainable carbon-free energy technologies for our energy sector. Several of these technologies are founded on hydrogen, the storage of which is the central topic of this book. To broadly understand the need for hydrogen to address energy concerns, it is important to understand our energy circumstances in terms of available energy resources, their impact on the environment, the political ramifications, and our time constraints.

We root the discussion on projected global energy demand to 2050 and beyond. We also establish that any future energy system must be sustainable energetically, politically, and environmentally. By energy sustainability, we mean an energy system based on a source (or sources) that is effectively infinite—at least on timescales (thousands of years) that matter to the current animal and plant population on Earth. By environmental sustainability, we mean an energy system that operates autonomously from the environment with no release of criterion pollutants, no greenhouse gas (GHG) emissions, no release of ozone-depleting gases, or other significant environmental impacts. By politically sustainable, we mean available to all people.

The discussion and arguments we make here really apply for any carbon-free energy technology (e.g., solar, wind). However, the reader will come away convinced that hydrogen technology is in fact required if our goals for energy and environmental sustainability are to be met. We use the year 2050 as the target year to complete the evolution of our sustainable energy infrastructure. It will become clearer as the discussion evolves why we need to embark *now* to restructure our energy infrastructure and why the year 2050 is so critical.

The Editor's Introduction described how there are three primary reasons for considering a hydrogen-based energy system. These were fuel resource insecurity, political energy insecurity, and finally global climate change. Here, we take up these three issues in more depth, and describe the global energy-political-environmental circumstances faced by any new emerging carbon-free technology.

Although the resource, environmental, and political issues are listed separately, they are actually linked. For example, energy and environmental sustainability must exist for all people; otherwise, political insecurity will inevitably undercut any progress that is made. We start here with the issue of creating security around our fuel resources and the challenge of creating a reasonable standard of living for all. By standard of living, we mean access to energy and energy-based technology. This is an admittedly Western view of a living standard. Perhaps it would be better to call it a "technical standard of living," which correlates with long life expectancy.

Reasonable Standard of Living for All

Envision a world where every inhabitant enjoys the same standard of living that currently exists in the United States, Europe, Japan, and other technical societies. How much energy consumed per year will it take to accomplish this? As of 2010, the average energy consumption in the United States was 313 gigajoules/person/year [1, 2]. Let us assume that the 2010 U.S. energy consumption will suffice for all the inhabitants of the world in 2050. How much energy do we need to provide then? We are currently at 7.0 billion people. Some estimate that the maximum carrying capacity of the Earth could be as high as 16 billion people [3], with other estimates as high as 10.5 billion by 2050 [3, 4]. For the sake of this discussion, let us assume 11 billion human inhabitants for 2050 and beyond. With this population, we will need 2.19 zetajoule (ZJ) per year [5] to raise the entire world population in 2050 to the average per capita energy use of the United States in 2010. This is an enormous amount of energy required to sustain our way of life on this planet in a way that is equitable to all. Professor Nathan Lewis from Caltech captured this energy challenge well in the keynote speech delivered to the first California Clean Innovation Conference in 2007 [6]:

> Energy is the single most important technological challenge facing humanity today. Nothing else in science or technology comes close in comparison. ... With energy, we are in the middle of doing the biggest experiment that humans will have ever done, and we get to do that experiment exactly once. And there is no tomorrow, because in 20 years that experiment will be cast in stone. If we don't get this right, we can say as students of physics and chemistry that we know the world will, on a timescale comparable to modern human history, never be the same.

In the following discussion, we describe how our conventional fossil-based energy resources are limited, and it will be impossible to supply our energy needs through these conventional fuels.

Fuel Resource Insecurity

We base our discussion of the Earth's fossil fuel resources using the BP "2011 Statistical Review of World Energy" [1]. While one might be suspicious that a review of the topic from an oil company might be biased, this review was based on a number of other independent sources (documented in the review); in fact, it is in the interest of oil companies to estimate the Earth's reserves as accurately as possible. There is scatter in these estimates, but it is safe to say that for the most part there is agreement to within about 10–20% from the various sources. Moreover, the predictions of when we might run out of a particular resource are a function of how the rate of depletion (consumption) will vary in the future and how the discovery of the resources in the future unfolds. Both numbers are subject to assumption and uncertainty. We start here initially with a discussion that does not take into account estimates of "unproven" reserves since they are significantly less well known. We take up the question of future reserves a bit later in the discussion.

Estimated proven world reserves (at the end of 2010) can be found in the BP report [1] and are presented in Table 1.1. Proven reserves are [1] "those quantities that geological and engineering information indicates with reasonable certainty can be recovered in the future from known reservoirs under existing economic and operating conditions."

The ratio reserves/production (R/P) numbers gives the years it will take to completely deplete the proven reserves at the current consumption (production) rate [1]. Here, we equate consumption with production, as production is carefully controlled to be just slightly more than demand. We see that the time to total depletion of oil and natural gas (NG) is on the order of 50 years, with the time to total depletion for coal being about 120 years. The R/P figures for oil and NG have held rather steady at nearly 50 years for the past 25 years [1]. However, the R/P ratio for coal has been rapidly decreasing, dropping from 200 years in 1990 to 119 years in 2010 [1]. These estimated times for resource depletion, given by R/P, assume consumption holds at the 2010 levels and there is no further resource discovery.

In Table 1.1, we converted these fossil fuel reserves to energy content (HHV, higher heating value) and totaled them in the table. It is interesting to estimate when these total energy resources might be depleted, assuming that there is no distinction between how fossil fuel is used (i.e., demand for energy will be satisfied by any available fossil fuel resource). From Table 1.1, that time to total depletion is currently 78 years at the 2010 energy consumption rate.

TABLE 1.1

2010 Fossil Fuel Proven Reserves

Resource	Proven Reserves	Proven Reserves (HHV, ZJ)	2010 Production	2010 Production (HHV, ZJ)	Reserves/ Production (Years)
Coal	861 billion tonnes	21.5	7.3 billion tonnes	0.18	119
Natural gas	187 trillion m^3	7.8	3.2 trillion m^3	0.13	58
Oil (including oil sands)	1526 billion barrels	9.7	30 billion barrels	0.19	51
Total energy		39		0.5	78

Source: From BP. 2011 statistical review of world energy. http://www.bp.com/sectionbodycopy.do? categoryId=7500&contentId=7068481.

Note: 1 ZJ = 1 zetajoule = 1×10^{21} J.

This analysis does not consider new discoveries of fossil fuel reserves. A better estimate of the dynamics of this situation was developed by Dr. M. King Hubbert [7]. While the notion of when we will run out of our energy resources may seem the more obvious concern, the more important point in the resource depletion is when we have reached a peak in the production of that resource. After the peak in production, production slows even in the face of rising demand, and the resource will become much more expensive. Hubbert was the first to investigate the phenomenon in a rigorous way.

Hubbert gave a thorough discussion of his methodology in his 1962 *Report to the Committee on Natural Resources of the National Academy of Sciences* [7]. Hubbert's analysis was based on existing cumulative discovery data Q_D as a function of time and existing cumulative production data Q_P as a function of time, both of which were well known at the time and continue to be amongst the best-known data in the fossil fuel industry. For a finite resource, both of the curves $Q_D(t)$ and $Q_P(t)$ would be expected generally to start slowly, gradually accelerate, and finally level off to a maximum as all the fossil fuel resource that was there to be discovered would, over very long times, be found. Such behavior is said to follow a logistic growth curve, which can be described by the empirical expression for $Q_D(t)$ [7]:

$$Q_D(t) = Q_\infty/(1 + ae^{-bt}) \tag{1.1}$$

where Q_∞ is the asymptotic value of $Q_D(t)$ given long times t. In other words, Q_∞ is the total cumulative discoveries that could be made. Curves for both $Q_D(t)$ and more prominently the cumulative production curve $Q_P(t)$ could be determined from fitting the logistic growth curve to the known historic $Q_D(t)$ and $Q_P(t)$ data from the industry [7]. Rates of production $dQ_P(t)/dt$ could then be calculated and plotted, with the integrated area under the rate curve (derived from the fit) equaling the total resource Q_∞ that could be exploited [7]. This quantity Q_∞ is also called the ultimate recoverable resource (URR).

Although other methods have been developed to calculate the production rate $dQ_P(t)/dt$, the peaks in resource production rate predicted over time have come to be known as "Hubbert's Peaks," and the plots of resource production rate vs. time have come to be known as "Hubbert's Curves." The importance of the "peak" in these curves is that, at this point, the deleterious effects of the shortage (e.g., increases in price) thereafter become keenly felt, far before the last drop of oil or chunk of coal is consumed. There is a rich discussion in the literature on Hubbert's curve for every finite resource; we present here a discussion of only the fossil fuel resources petroleum, NG, and coal. We also briefly discuss fissionable resources.

It is instructive to examine the predicted production rate $dQ_P(t)/dt$ curves for our fossil fuel resources. The predictions for total world oil and total world NG are shown in Figure 1.1 (adapted from Bentley [8]). Figure 1.1 demonstrates that the Hubbert peak for conventional oil is already on us, and we can expect diminishing production of conventional oil in the future. The peak for NG would also seem to be about now, with a future of further-diminishing resources looking forward.

Figure 1.2 shows predictions by Mohr and Evans [9] for production rates of both conventional and total oil (conventional + unconventional) using a method different from Hubbert's. Unconventional oil is derived from extra heavy oil, natural bitumen (oil sands, tar sands), and oil shale. Table 1.1 shows that the world's *proven* reserves of total oil are currently estimated to be 1526 billion barrels. However, in the method of Reference 9, forecasting a peak in production involves making estimates of the URR. The URR for unconventional oil is highly uncertain, leading the authors to make pessimistic, optimistic, and

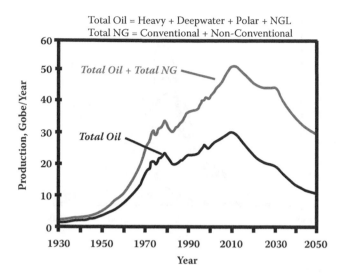

FIGURE 1.1 (See color insert.)
Production predictions for total world oil and total world oil + total world NG. (Reproduced with permission from Reference [8].)

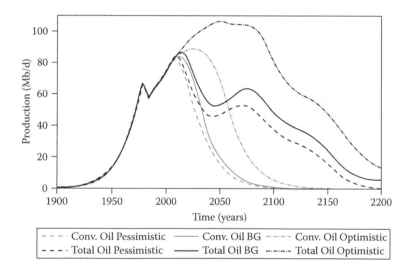

FIGURE 1.2
Predictions for world production of conventional oil and combined conventional plus unconventional oil (i.e., total oil) using pessimistic, best guess (BG), and optimistic assumptions. (Reproduced with permission from S.H. Mohr and G.M. Evans, *Energy Policy* **38**, 265 (2010).)

"best guess" assumptions about this key aspect of the analysis. For the best guess estimate, they assumed a URR of 2500 billion barrels for unconventional oil [9].

For both the pessimistic (URR = 2000 billion barrels) and best guess scenarios (URR = 2500 billion barrels) for unconventional oil, Figure 1.2 shows that total oil production rate peaks at 2010 and 2014, respectively. In the optimistic unconventional oil scenario (URR = 3750 billion barrels), total oil production peaks around 2050 but extends out in time with

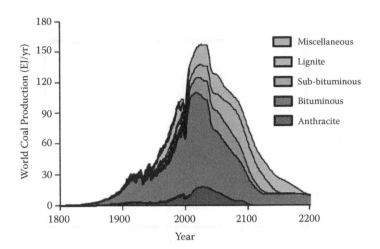

FIGURE 1.3 (See color insert.)
World coal production prediction for different coal types in EJ/y. (Reproduced with permission from Reference [10].)

decent production rates for another 100 years. These authors stated that all the results of Figure 1.2 are economically conservative in the sense that the model did not involve any economic restraints, and exploitation of these unconventional resources is expensive and energy intensive [9]. A further restraint on the unconventional resource is the extensive surface environmental damage caused by its extraction. Mohr and Evans concluded [9] that "the analysis of unconventional oil indicates that at the absolute best it can only delay the peaking of world oil production by about 25 years."

Figure 1.3 predicts annual production for the world's resource of coal as reported by Mohr and Evans [10]. These authors predict, for a best guess at the coal URR, a peak in coal production about 2030. Patzek and Croft [11] predicted a Hubbert peak for coal in 2011. Overall, there is good agreement for these most recent analyses of coal [10–12] that the peak in worldwide coal production will occur in the 2011–2030 time frame. This led Mohr and Evans to conclude [10] that "the notion that coal is widely abundant therefore appears to be unjustified."

Summarizing, Table 1.1 and Figures 1.1–1.3 show that while the amount of fossil fuel resources extends beyond 2050, the "peak" of the production of these resources will occur near or before the 2050 timeframe of our discussion. The implications of the peak are that beyond it, production will not be meeting demand, and prices will rise. The deleterious effects of the shortage and the stressing of societies will occur long before the resources are completely exhausted.

The transportation sector is a particularly weak and vulnerable link in our energy infrastructure. Globally, about 38% of petroleum use stems from the transportation sector. It is projected to decline to about 33% by 2030, but transportation will remain the largest single consumer of petroleum [13]. To make matters worse, the transportation sector is 97% dependent on petroleum [13]. This sensitivity of the transportation sector is one reason why new alternative energy technologies have been developed with transportation (vehicles) in mind. This includes new hydrogen-based technologies and hydrogen storage in particular.

Like our fossil-fuel-based power technologies, nuclear power technology is based on a finite uranium resource. One can find in the literature predictions of proven reserves, with Hubbert analyses yielding predictions of "peak uranium." However, the nuclear industry will argue that the amount of recoverable uranium is a matter of cost. Uranium is abundant in the world's oceans and crust in levels of 2–3 parts per million. The matter of supply depends on how much one is willing to pay. Unlike the fossil energy industry, the cost of the fuel for the nuclear industry is a very small fraction of the cost of the electricity produced. This industry can afford a significant cost increase of a factor of 3 in the cost of fuel before it affects the electricity price. The August 2012 price of uranium was $109/kg. At the current consumption rate, we have about 85 years of supply remaining. If we were to use advanced nuclear technologies to breed fuel, then the predictions are that we have thousands of years of resource available with this technology.

Political Energy Insecurity

The world has long experienced difficulties when one country's energy needs are dependent on another country. The petroleum shortage of the 1970s, which resulted in long gas lines, rationing, and higher prices, is commonly recognized as caused by geopolitical forces rather than resource limitation [14]. Geopolitical forces can (and frequently do) cause significant disruption to otherwise stable economies. The recent wars in the Middle East (Iraq) can also be attributed in part to the Western world's thirst for oil. Any new and sustainable energy infrastructure that is energetically (resource) and environmentally sustainable must also be politically secure. Political energy security requires availability to all peoples of the Earth. The cell phone is an example of a valuable resource that has no political insecurity about it. Nobody is going to war over their cell coverage. The technology is available virtually to all people.

In the last few years, there has been evidence of how far the effect of political instability can spread. In 2006 and 2009, disputes over the price Ukraine paid for NG led to Russia cutting off supplies to the Ukraine. Pipelines feeding a fifth of the gas supply to the rest of Europe traverse the Ukraine. Despite Ukraine and Russia claiming that supplies to the rest of Europe would remain stable, many countries were concerned that there would be a loss of pressure as the amount of NG was reduced. The dispute had a political dimension as Russia attempted to maintain influence over Ukraine. It highlighted the vulnerability of countries that were reliant on Russian NG. Furthermore, it demonstrated how oil and gas pipelines that traverse a number of different countries can bring political energy insecurity to all of them. The Trans-Siberian pipeline passes through Russia and Ukraine. Other pipelines pass through more countries. The Yamal-Europe pipeline crosses Russia, Belarus, and Poland. The Central Asia–Center gas pipeline system runs from Turkmenistan, though Uzbekistan and Kazakhstan, to Russia.

Political insecurity can be both "international" and "intranational" in nature. Recently, political unrest in the Middle East reduced the output of oil, and the price of gas futures increased by 14%. Libyan production, which supplies 2% of the world's oil, was reduced to almost zero. Staff shortages resulting from the unrest in Bahrain caused a cutback in production by 250,000 barrels per day. There was concern that if the unrest spread to Saudi Arabia, which supplies 12% of the world's oil, output would be reduced still further. Brent crude futures went above $113 per barrel as the market reacted to uncertainties as well as a shortage of supply against demand. Other potential hot spots are Iran, which supplies 5.3% of world production, and Angola, which supplies 2.3%. These examples of political insecurity highlight the advantage of being self-sufficient in energy from renewables that are more immune to the vagaries of our poor ability to manage our political affairs, both within states and between states.

Breeder technology was mentioned above in connection with nuclear power. An issue with "breeder technology" is that it produces "weapon-grade" fissionable material, thus introducing yet another kind of political insecurity, the insecurity of nuclear weapons proliferation.

Global Climate Change

While Nathan S. Lewis [6] felt that energy is the single most important technological challenge facing humanity today, we maintain that

> Climate change is the most profound constraint on energy use facing humanity today.

The Intergovernmental Panel on Climate Change (IPCC) [15], the foremost authoritative body on climate change, has establish GHG reduction targets of 85% below 2000 levels to cap CO_2-equivalent concentrations by 2050 and beyond to 445–490 ppm. This cap would limit the predicted global temperature rise to 2–2.4°C. Others have suggested that a level of 350 ppm is required to stabilize the climate [16], requiring GHG emission to go to zero, and the sooner the better. The U.S. National Academy of Sciences (NAS) National Research Council (NRC) conducted an independent study, *America's Climate Choices* [17]. In this work, the NRC addressed many critical questions, including the following: (1) Is climate changing? (2) Are humans causing it? and (3) How bad is it? The answers to these questions from this independent NAS study are as follows: (1) "Climate change is occurring"; (2) "it is caused largely by human activities"; and (3) "it poses significant risks for, and in many cases is already, affecting a broad range of human and natural systems" [18].

Taken together, these independent analyses performed by the IPCC and the NAS/NRC provide the robust technical foundation needed to establish that climate is changing, and humans are causing it without question. All scientific evidence and analyses presented by these works indicate that to avoid significant global warming we need to stabilize GHG concentrations [19] by 2050 and beyond. The level of acceptable GHG concentration in the atmosphere is still under study, with the value in the range 350 ppm [16] to 490 ppm [15]. The current (August 2012) level of CO_2 in the atmosphere is 395 ppm.

At this point, it is instructive to point out that concern over rising atmospheric CO_2 and the potential effects this might have on our climate has been discussed for over 100 years. A brief review of this history is provided here. To properly understand the severity of our carbon dioxide emissions problem, it is interesting and revealing to review some of the early history of the scientific developments that led to our understanding of the role of CO_2 and climate change. We focus on the history up to the early 1960s. It turns out that much about this problem was known by 1962, and that fact alone indicates that CO_2 emission (and more generally emission of GHGs) is an old problem.

The first published thinking about the atmosphere playing the same role as the glass in a greenhouse can be attributed to Fourier [20], with elaboration by Pouillet [21]. However, the brilliant, but perhaps underappreciated, English scientist John Tyndall was the first to attempt systematic experiments that examined the role of atmospheric gases in the absorption and emission of infrared radiation. In 1861, while the United States was embroiled in a barbarous civil war, in England Tyndall was making initial observations of radiant heat transfer in gases [22]. Although Tyndall admittedly did not understand the molecular basis for the behavior he saw (invoking as he did "the ether" in his explanations), he was able to gather some correct overall conclusions concerning radiation heat transfer in a number of gases, including water vapor (referred to then as "aqueous vapor") and CO_2 (referred to then as "carbonic acid").

In reviewing his work in his memoirs of 1872 [23], Tyndall remarked:

> This aqueous vapor is a blanket more necessary to the vegetable life of England than clothing is to man. Remove for a single summer night the aqueous vapour from the air which overspreads this country, and you would assuredly destroy every plant capable of being destroyed by a freezing temperature. The warmth of our fields and gardens would pour itself unrequited into space, and the sun would rise up on an island held fast in the iron grip of frost. The aqueous vapour constitutes a local dam, by which the temperature at the earth's surface is deepened: the dam, however, finally overflows, and we give to space all that we receive from the sun.

The particular influence of carbon dioxide was advanced considerably by the work of the Swedish scientist Svante Arrhenius [24]. Arrhenius was the first to calculate the effect of increasing the carbon dioxide concentration in the atmosphere on the surface temperature of Earth. His work, published in 1896, involved tens of thousands of calculations by hand [25]. Arrhenius stated [24], "I should certainly not have undertaken these tedious calculations if an extraordinary interest had not been connected with them." The extraordinary interest was a better understanding of the prior Ice Age.

Arrhenius adopted the value 300 ppm for the concentration of carbon dioxide in the atmosphere. Arrhenius calculated that doubling the concentration of CO_2 to 600 ppm would increase the surface temperature of Earth by 5–6°C, depending on latitude. Contrary to some claims, Arrhenius was actually silent on the contribution of human fossil fuel combustion on atmospheric levels of CO_2. At that time, there had been no observation of an increase in atmospheric CO_2 levels. However, Arrhenius did quote the memoir of his colleague Professor Högbom [26], which examined the quantity of organic matter being "transformed" (i.e., burned) into CO_2 by the technology of man. Högbom wrote [26]: "The world's present production of coal reaches in round numbers 500 million tons per annum, or 1 ton per square kilometer of the earth's surface. Transformed into carbonic acid, this quantity would correspond to a thousandth part of the carbonic acid in the atmosphere." This represents probably the first written speculation comparing CO_2 levels in the atmosphere and human industrial activity.

In 1938, the steam technologist G. S. Callendar published a remarkable paper [27], "The Artificial Production of Carbon Dioxide and Its Influence on Temperature." In this work, Callendar confirmed the work of Arrhenius in predicting that an increase in atmospheric carbon dioxide would increase the surface temperature of Earth. Callendar predicted a 2°C surface temperature increase for a CO_2 level of 600 ppm level in the atmosphere, smaller than the prediction of Arrhenius. Callendar was the first to examine historical temperature readings over time and to explain them with changes in atmospheric CO_2. His report of temperature readings at the time suggested that departures of the average surface temperature from levels observed between 1820 and 1900 started to occur about 1910 [27]. The observed rate of temperature increase (observed in 1938) due to the artificial production of CO_2 was calculated by his theory to be 0.003°C per year, while the observed rate of temperature increase was 0.005°C per year.

Interestingly, Callendar estimated the CO_2 levels in the year 2000 to be 335 ppm, assuming that the rate of CO_2 emissions into the atmosphere held constant to 1938 values. At the time, Callendar wrote: "It may be supposed that the artificial production of this gas (CO_2) will increase considerably during such period, but against this is the ever increasing efficiency of fuel utilization, which has tended to stabilize carbon dioxide production at around 4,000 million tons during the last 20 years, in spite of the greatly increased number of heat units turned to useful purpose." This early hope of energy efficiency holding back the artificial production of CO_2 emission has clearly not been realized.

One might think that this early work of Callendar was a clarion call to the dangers of CO_2 release by human technology. Actually, quite the opposite is evident. In his conclusion, Callendar wrote:

> It may be said that the combustion of fossil fuel, whether it be peat from the surface or oil from 10,000 feet below, is likely to prove beneficial to mankind in several ways, besides the provision of heat and power. For instance the above mentioned small increases in mean temperature would be important at the northern margin of cultivation, and the growth of favourably situated plants is directly proportional to the carbon dioxide pressure. In any case the return of the deadly glaciers should be delayed indefinitely.

In retrospect, his speculation that CO_2 would prevent future ice ages may have been Callendar's way of finding the bright side.

In 1956, Gilbert Plass published a paper [28], "The Carbon Dioxide Theory of Climatic Change." In this paper, Plass confirmed earlier work connecting increases in CO_2 concentration in the atmosphere to increases in mean surface temperature of Earth. Plass stated, "The radiation calculations predict a definite temperature change for every variation in CO_2 amount in the atmosphere. These temperature changes are sufficiently large to have an appreciable influence on the climate. A relatively small change in the average temperature can have a large effect on the climate." In his work, he found that doubling the CO_2 concentration in the atmosphere beyond preindustrial levels increased the mean surface temperature 3.6°C. Furthermore, Plass considered the role of the oceans in mitigating such atmospheric increases. Plass argued that the deep oceans require on the order of 10,000 years to reach true equilibrium with a rapidly established atmospheric concentration of CO_2. Summarizing, Plass stated:

> Some of the extra CO_2 will be absorbed by the oceans. Because of the slow circulation of the oceans it would probably take at least 10,000 years for the atmosphere-ocean system to come to equilibrium after a change in the atmospheric CO_2 amount.
>
> The surface layers of the ocean start absorbing some of the extra CO_2 from the atmosphere soon as the P_{CO2} is greater than the equilibrium amount. The rate at which this absorption takes place is not known accurately, but it is probably true that the surface layers can absorb only a small fraction of the extra CO_2 in a period of several hundred years. Thus, it appears that most of the additional CO_2 that is released into the atmosphere will stay there for at least several centuries. Even if the oceans absorb CO_2 much more rapidly than has been assumed here, the accumulation of CO_2 in the atmosphere will become an increasingly important problem through the centuries.

In 1958, Callendar published more work [29], further demonstrating increasing levels of CO_2 in the years since his original work 20 years previously, and concomitant increases in the observed mean surface temperature of the Earth. The increases in CO_2 clearly demonstrate that the oceans are not completely absorbing the excess CO_2, and extended analysis of temperature records affirmed his earlier report that the measurable increases in CO_2 started to appear around 1910 [29]. Bray published work [30] in 1958 providing an analysis of the statistical significance to temperature increases and found in fact the mean temperature from 1907 to 1956 was, in a statistically significant way, higher than the mean surface temperature from 1857 to 1881.

In 1959, Bolin published work [31] in which the role of ocean CO_2 absorption on the atmospheric levels of CO_2 was considered explicitly. In support of the arguments by Plass, Bolin considered rates of exchange between atmospheric CO_2 and the surface of the ocean, and furthermore the exchange between the ocean surface and the deeper ocean reservoir. Figure 4 in his paper showed [31] that the rate of CO_2 exchange within the ocean layers had the strongest influence of the levels of CO_2 maintained in the atmosphere, but for exchange times beyond 1000 years, the effect on atmospheric concentrations was minimal. Bolin also found that the level of CO_2 in the atmosphere was not affected significantly by the rate of exchange between the atmosphere and the surface of the sea.

Concluding, Bolin wrote: "In view of the fairly slow circulation of the ocean it is then easily understandable that a major portion of the CO_2 released by combustion may still stay in the atmosphere in an approximate balance with the sea in spite of a rapid exchange between the atmosphere and the sea."

One might ask, given this scientific exploration of the role of CO_2 and warming of Earth's surface since 1896, when the concept of CO_2-induced temperature increases started to circulate among the upper echelons of the scientific leadership in the United States. In the United States, the National Academy of Sciences (NAS) is the most prestigious technical body, with a special role of advising the U.S. government on scientific matters. We ask: When did concern about CO_2 emissions start to become apparent in publications from the NAS?

In 1957, the NAS National Research Council (NRC) Division of Earth Sciences published a report [32] from the Committee on Climatology. In its "First General Report on Climatology to the Chief of the Weather Bureau," this committee reported:

> The climatic bench-mark program is timely and is, of course, a very long-range program. It is essential that such records be available fifty or a hundred years from now if there is to be rational approach to the question of long-term climatic trends. It has been remarked that, in consuming our fossil fuels at a prodigious rate, our civilization is conducting a grandiose scientific experiment which will never be repeated. The question of the influence of an increase of carbon dioxide in the atmosphere on our climate can perhaps best be answered by recourse to such records as will be available from the bench-mark stations.

This is probably the first written comment at the level of the NAS that fossil fuel combustion may be increasing atmospheric CO_2 levels in a way important for climate. As can be seen from the passage, the recommendation is for more monitoring (the "benchmark stations").

The "remark" concerning the "grandiose scientific experiment" referenced above originated from a 1957 paper published by Revelle and Suess [33], in which C^{14}/C^{12} and C^{13}/C^{12} ratios were examined to probe CO_2 exchange between the atmosphere and the ocean. In this paper, in an often-quoted passage, the authors stated:

> Thus, human beings are now carrying out a large-scale geophysical experiment of a kind that could not have happened in the past nor be reproduced in the future. Within a few centuries we are returning to the atmosphere and oceans the concentrated organic carbon stored in sedimentary rocks over hundreds of millions of years.

Authors Revelle and Suess had a rather dispassionate view of the subject. In the next sentence, the authors stated, "This experiment, if adequately documented, may yield a far-reaching insight into the processes determining weather and climate." In their conclusions, the authors stated: "In contemplating the probably large increase in CO_2 production by fossil fuel combustion in coming decades we conclude that a total increase of 20 to 40% in atmospheric CO_2 can be anticipated. This should certainly be adequate to allow a determination of the effects, if any, of changes in atmospheric carbon dioxide on weather and climate throughout the earth." In retrospect, this statement considered man-made CO_2 emissions to be an experiment worthy of tracking and documenting, rather than preventing.

As indicated [32], in 1957 we had the earliest written connections between industrial combustion of fossil fuel and CO_2 emissions at the level of the NAS. By 1962, the problem was clearly understood and finding its way into the most important NAS publications. In the document *Energy Resources—A Report to the Committee on Natural Resources* [34], we find the first written warning in the upper echelons of the U.S. scientific establishment that CO_2 emissions were a serious problem:

> There is evidence that the greatly increasing use of the fossil fuels, whose material contents after combustion are principally H_2O and CO_2, is seriously contaminating the earth's atmosphere with CO_2. Analyses indicate that the CO_2 content of the atmosphere since 1900 has increased 10 percent. Since CO_2 absorbs long-wavelength radiation, it is possible that this is already producing a secular climatic change in the direction of higher average temperatures. This could have profound effects both on the weather and on the ecological balances.

This passage is remarkable for the following reasons: First, it appeared in a report of the U.S. NAS, the most august technical body in the United States. Second, it was authored by M. King Hubbert, the same King Hubbert responsible for Hubbert's curve who successfully predicted the peak of U.S. oil production to be in 1972 [35]. By 1962, Hubbert was one of the most prominent Earth scientists in the world. Third, the passage stated in remarkably clear and succinct fashion the observations, mechanistic cause, and likely impact of man's combustion of fossil fuels. It could not have been said any better if it were written today. Finally, the passage was written in 1962, 50 years ago.

It is clear that the warming of the Earth's surface by increased atmospheric CO_2 and other GHGs *is an old problem,* and it has been known to be a problem at the highest technical levels (at least in the United States) since 1962. In a sense, we have the Ice Age to thank for this early awareness of the problem of CO_2 levels in the atmosphere. As described previously, the earliest studies of CO_2 levels were really motivated by trying to explain past deviations of climate that were known to exist, deviations in the direction of lower temperatures worldwide. Without the Ice Age, perhaps our awareness of CO_2 concentrations would have been delayed several decades.

This review, ending as it does in 1962, is not meant to be a complete review of the history of global warming. Excellent historical accounts are available [36–38] that in particular document the extensive work done in this field since the early 1960s. Rather, our purpose is to track when the key early findings were reported and what was known when within the U.S. scientific leadership. Surface warming by CO_2 is an old problem that has been known to be an issue for a long time. The fact that it is such an old problem, known to be a problem among the national scientific leadership 50 years ago, and originating from a time 100 years ago when the population was vastly smaller, with far smaller uses of fossil fuel, drives home more than any other fact how worrisome this problem really is and how massive the challenge will be to solve it.

The work of Callendar [27] reported that the earliest observable increases in the level of CO_2 in the atmosphere occurred around 1910. What was the world like back then? The world population was 1.75 billion [39], compared to today's (August 2012) population of 7.03 billion [40]. World emissions of CO_2 in 1910 were approximately 4.0 Gt/year (1 Gt = 1 billion tonnes) [41]. World emissions of "CO_2 equivalent" (taking into account all GHG emissions) in 2010 were reported to be 30.6 Gt [42]. It is interesting to compare the CO_2 emissions/population ratio for the two eras. In 1910, that ratio was 2.3 Gt/billion people. In 2010, that ratio is 4.4 Gt/billion people. Evidently, efforts to increase the efficiency with which we are extracting energy from fossil fuel *is not* keeping pace with the intensity with which humans use energy. This is a remarkable result. It points to the ultimate futility of trying to make further energy-efficiency increases in those energy technologies based on fossil fuel as a strategy for seriously addressing the CO_2 emissions problem.

What level of CO_2 reduction is needed to rectify this problem? An initial response would be that we need to reduce emissions to the point at which we cannot reasonably detect CO_2 increases in the atmosphere. Already, levels were increasing in 1910 and were observable even given the relatively poor measurement technology available 100 years ago. For the sake of this discussion, let us assume we need to get to the 1910 levels of emissions. This would require that, as a world civilization, we return to 4.0 Gt/year world CO_2 emissions. This constitutes a reduction from current emissions of 4.0/30.6 = 0.13, or an 87% reduction in emissions. This is in good agreement with the IPCC recommended cut described previously. Because there was already an increase in atmospheric CO_2 in 1910, even further reductions are likely needed—90% or more.

This reasoning indicates the magnitude of the problem in what we hope to be a straightforward way. Massive reductions in CO_2 and other GHG emissions are required, and indeed a net-zero GHG energy system is really what is needed to make progress robust against growth. This early history, particularly the revelation of how old this problem is, should strike fear in the heart of any attentive reader with a concern for the world and strongly motivate those doing technical work in this area.

We now make the case that it is essential to get to net-zero carbon emission technologies; otherwise, growths in human population and increases in the average level of energy use will eventually overwhelm any progress made.

Efficiency vs. Net Zero

We have just shown that massive reductions (87%) are required to get to 1910 levels of CO_2 emissions. The sad fact is that, in retrospect, increases in CO_2 were being observed even in 1910 [27], and it is really necessary to drive zero or "net-zero" levels of emissions. Here, we talk about this in a more quantitative way.

Increasing the efficiency with which we use our raw fossil fuel energy feedstocks may seem like a no-regrets action. Future energy systems are likely to be more costly than those in place today, requiring us to change how we use fossil fuels in the transition period. However, the danger in investing too heavily in increasing the efficiency of fossil fuel technology is that it can misguide us into thinking that we are properly addressing the problem, when resources thus expended could be directed toward developing technologies that are net zero that have the needed impact. In this sense, doing exclusively something that partially addresses the problem (e.g., cutting CO_2 emissions by 20%) can be worse than doing nothing, as it can tie up precious financial resources that need to be directed to net-zero energy technologies.

When considering the effects on climate, we need to think beyond improving efficiency. The second law of thermodynamics dictates that we will never reach 100% in any energy conversion we attempt. The laws of thermodynamics apply to all systems, so we are stuck. What this implies is that in any system that experiences growth (like the annual growth of CO_2 emissions due to increase in energy use), as soon as we have achieved the maximum gains due to improvements in efficiency, the system will then revert to the original growth curve. Two examples of growth are population growth and the intensity with which each person uses energy.

Consider Figure 1.4, which shows U.S. historical and projected CO_2 emissions (simple linear extrapolation) and the CO_2 emissions for the light-duty vehicle fleet as we implement several "efficiency" or other system "efficiency" improvements [43]. The flat line at the bottom shows a goal of 80% reduction of CO_2 emissions from 2010 levels. Note that this reduction target is less than that recommended by the IPCC. The first efficiency improvement is implementing the legislated improvement to 35 mpg in corporate average fuel economy (CAFE) standards [44]. These were to be implemented by 2016, resulting in the corporate sales fleet complying with the improved fuel economy standards. However, note that as soon as those vehicles penetrate the installed fleet and the vehicle population becomes "saturated," the CO_2 emissions cease to decrease and trend upward, paralleling the original CO_2 emission growth curve. If there is any CO_2 in the exhaust, as soon as the technology becomes saturated, the emissions slope will revert to the original growth curve. The CO_2 reduction is not stable against growth and is therefore not environmentally sustainable.

Figure 1.4 shows that hydrogen derived from steam reforming of fossil-based methane cannot achieve the goals by 2050 due to CO_2 release originating from the CH_4 feedstock. If, on the other hand, we eliminate C (or net zero) from the system, such as H_2 production

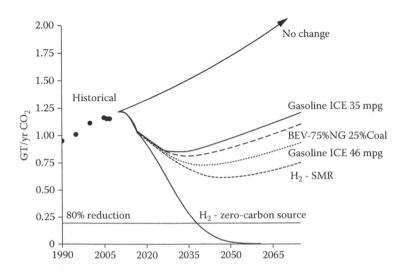

FIGURE 1.4
Variation of CO_2 emissions with changes in vehicle efficiency and the introduction of new technology. (From Reference [43].)

from a zero-carbon source, we can achieve the desired CO_2 emissions reduction by 2050 and for all time. We note here that biofuels (not shown) cannot achieve the goals by 2050 as they still involve some emissions (0.9 kg C/gallon) [45]. Biofuel deployment will be limited as U.S. biofuel production capacity could *at most* produce about 30% of the transportation oil predicted to be used in 2030 [45].

Summarizing the results of Figure 1.4, efficiency improvements to vehicles with 35 mpg cannot get us where we want to go; battery electric vehicles (BEV) cannot get us there with the current emissions associated with the U.S. electric grid; improving vehicle efficiency to 46 mpg cannot get us there; using biofuels cannot get us there; and even using hydrogen fuel cell vehicles with the H_2 derived from steam methane reforming (SMR) of fossil-based NG cannot get us there. None of these technologies is a net-zero technology. However, a net-zero technology such as a hydrogen-based vehicle with the H_2 derived from a carbon-free source (nuclear, hydrocarbon fuels with carbon capture and storage, or renewables) is the only route that achieves our goals.

Timescales for Change

The previous discussion describes "how much time do we have?" from an energy resource and carbon emissions perspective and describes the circumstances facing a new energy technology. Another question is: "How much time does it take to implement new technological advances?"

Time is a poorly appreciated physical phenomenon. We all try to ignore it as we add years to our age, but time marches on, and adding another year is inevitable. Time marches so fast that it readily passes us by. Take note of Nathan Lewis's previous quotation [6]. In it, he referred to action in 20 years. Without paying attention, the reader would quickly assume that we have 20 years to respond—be careful. As of this writing, we have only 15 years left according to his quotation. The passage of time is particularly pertinent when conditions are changing at a rate that is predetermined by actions of the past. This is precisely the situation we are in with the energy problem as well as the climate problem.

The notion of timescales for events to take place is very powerful. It allows us to put the notion of time into our thinking and into our perspective. At the beginning of this chapter, we stipulated the period of time of interest for this discussion was to the year 2050 and beyond. Is this a long time into the future? In 2050, our children will be alive; our grandchildren will be in a midlife crisis. If we do nothing and maintain "business as usual," we will be well along the downhill slope of all the fossil fuel Hubbert curves. We will see a doubling of CO_2 concentration in the atmosphere and witness dramatic changes in the Earth's climate and coastal sea levels [18]. With no other choices available, we will be deep into adaptation as a result of the effects of climate change, and none of these adaptations will be desirable.

The year 2050 is not far into the future. It has an impact on what we care about today and touches our personal lives. Very successful petroleum companies have a time horizon of typically 30 years. One can find many predictions in the oil and gas industry for which the graph simply terminates 30 years out, with no vision beyond. Well, 30 years from this writing is 2042, within our timescale. So, how long does it take to change things?

Here are some historical examples of the timescale associated with changes in technology:

- It took 86 years to commercialize the zipper [46].
- It took 30 years to remove lead from gasoline [47].
- Fuel cells were discovered in 1839 and are still being commercialized (173 years and counting) [48].
- There are 50-year lifetimes for many critical infrastructure components.
- The median lifetime for light-duty vehicles is about 15 years; for aircraft, it is about 20 years.
- The first ARPANET (Advanced Research Projects Agency Network) link was successfully established November 21, 1969; however it took 43 years to develop the Internet as we know it today.
- The Super Car Clean Car program (a U.S. program) was initiated in 1992 [49], the first hybrid car (Toyota Prius) was introduced in the U.S. commercial market in 2001 [50], and about 10 years later in 2010 the market penetration into the fleet for all hybrids was only 2%. The first hybrid car was actually built in 1898 (114 years ago), and it still does not have significant market share in 2012 [51].

There are about 235 million vehicles on the road in the United States today. So, how long will it take to change the installed light-duty vehicle fleet in the United States using a zero-emission technology such as a hydrogen fuel cell vehicle? The answer is shown in Figure 1.5 [52].

The time it takes to retire the installed fleet is about 40 years. This is assuming the current scrap rate of vehicles and a sales penetration rate by which new vehicles are assumed to make up 50% of the sales fleet by 2030. Remember that the Prius was introduced into the United States in 2001, and in 2010 *all* hybrids had only 2% of the market share—about 10 years later. We have to go dramatically faster than we have done in the past.

We need to put into place starting *now* the construction of our future sustainable global energy system. The system must be energetically sustainable (based on "infinite resource"),

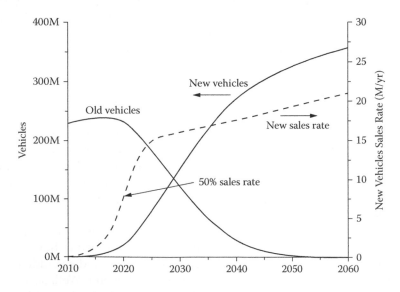

FIGURE 1.5
Time to roll over an existing vehicle fleet with new vehicles.

environmentally sustainable (net-zero emissions), and politically sustainable (available to all). It will take us about 50 years to effect this change; meanwhile, we are racing toward the peak in Hubbert's curve for all resources in only about 40 years. A particularly clear discussion of the Earth's energy resources can be found in Hubbert's 1962 report to the NAS [7]. Sadly, remarkably little has changed in our energy picture in the 50 years since Hubbert's report [7]. It will become clear as the discussion proceeds the role that hydrogen technologies must play, and hydrogen storage in particular must play, to meet the challenges in energy and climate that we face.

Renewable Energy Resources

As discussed, there is plenty of useful renewable solar energy incident on the Earth to satisfy all of our current and projected global energy needs. There is about 660 ZJ [53] of useful solar energy incident on the global landmass per year. At a collection efficiency of 15%, we can collect 99 ZJ/year. Thus, to provide every person on Earth in 2050 with a standard of living equivalent to that enjoyed by the 2010 U.S. population (a global total energy need of 2.19 ZJ/year), we would need to cover only 2.2% of the landmass with solar collectors at today's collection efficiency. To put this coverage in context, human settlements today occupy about 3%, and crops occupy about 11% of the landmass. We have orders of magnitude more accessible solar energy than is needed to fuel the energy demands of the human race.

These figures assume a 15% solar-to-electric conversion rate. Biomass is also a form of solar energy, but its overall solar-to-biomass energy conversion efficiency is low, on the order of 1%. As such, we need to be targeted on the application of biomass energy feedstocks to optimize our land use. Further in this chapter, we discuss the application space for transportation.

Solar energy comes in many derivative forms of "renewable energy," namely, wind, tidal, ocean currents, and direct solar insulation. As such, these renewable energies are abundant. However, these forms are also intermittent (being derived from solar radiation) and often come in a geographical location and at a time when the demand is low and manifests itself in a form that requires conversion to a more useful form. Also, the demand side and renewable energy supply side are frequently not in phase or at the same geographical location. This situation is shown in Figure 1.6 [54], in which variations in the load of the eastern Germany high-voltage grid in 2007 are seen to be poorly correlated with available wind power in the same region [54].

Renewable energy requires energy storage, transportation, and conversion into the energy form that is useful for the demand. For example, incident radiant solar energy can be harvested as heat (from the infrared part of the spectrum), as electrons (photovoltaic from the near ultraviolet to the visible part of the spectrum), or as photosynthesis (biological systems). We need the ability to capture the primary energy, store it, transport it, and convert it back to a useful form where and when the demand requires. This is one of the uses of hydrogen in a renewable energy infrastructure, as an energy storage medium. Inexpensive and high-density (volumetric, gravimetric) hydrogen storage on a grid scale would enable the use of intermittent renewable energy sources. An examination of energy storage needs and opportunities is instructive at this point.

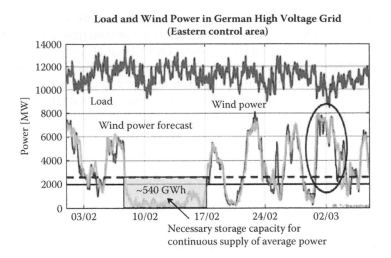

FIGURE 1.6 (See color insert.)
Load and wind power in the German high voltage grid indicating the energy storage required to make wind a dispatchable electric resource. The horizontal axis shows 2007 dates in Febuary (02) and March (03). (From Reference [54].)

If large-capacity renewables, especially wind, are built to meet electric power load, it is often the case that the output is not a good match for the demand, as shown in Figure 1.6. Germany is currently experiencing this problem as a direct result of their significant introduction of renewable primary energy into their power system [55]. Wind often blows at night, not during peak load hours of the day. Wind farms and solar panels or concentrator plants are often built in locations far from city centers or other load centers. Wind especially is unpredictable, such that forecasts can be wildly wrong, leaving demand unmet. This poor reliability and predictability requires the installation of backup generators. All of these situations can and are being addressed with energy storage of various types to accept excess energy when available and deliver it back to the grid when required.

Options for Grid-Scale Energy Storage

The simplest type of energy storage is a reversible battery, charged directly from a power generator, such as a wind turbine, or from the electric grid. When power is needed, the battery discharges—simple enough. However, it turns out that few battery types are cost effective in sizes large enough to provide bulk storage capacity for a large intermittent system [56].

Currently, the largest-scale energy storage technology is pumped hydro storage. When one has excess energy that needs to be stored, one uses the excess power to pump water to a higher elevation. When the energy is needed, the water is released, turning turbines as it falls, generating power. This type of energy storage is a mature technology and can be found in many parts of the world, often in conjunction with other hydropower resources [57]. It is difficult to site new pumped-hydro storage facilities for environmental reasons.

Another energy storage technology that can store fairly large amounts of energy at reasonable cost is compressed air energy storage, known as CAES. In this technology, available

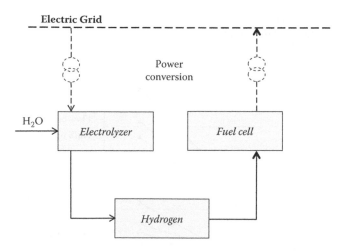

FIGURE 1.7
Renewable grid energy storage configuration using hydrogen and a fuel cell.

excess power is used to compress air into an underground storage reservoir, such as an aquifer or salt cavern [58]. When the stored energy is needed, the pressurized air is released to run a conventional turbine generator. A large CAES plant currently operates in Germany, one in the United States, and several additional plants are under construction. The reservoir or caverns have low marginal cost for bulk storage, and many possible sites exist. For lowest-cost power, the turbine is run on NG, so the technology is not carbon neutral.

A similar approach to bulk energy storage is to use excess renewable power to produce hydrogen using an electrolyzer [59]. When excess energy from the grid needs to be stored, the power is used to run an electrolyzer to split water into hydrogen and oxygen, with the hydrogen being stored. When needed, the stored energy is released by using the hydrogen to power a fuel cell or hydrogen engine generator to produce power or possibly useful heat. Alternatively, or jointly, the hydrogen can be used to fuel hydrogen-powered vehicles. The hydrogen can also be stored underground in caverns or compressed into storage tanks. The production and storage of hydrogen to stabilize a renewable power grid is depicted in Figure 1.7.

All of the energy storage technologies described have been studied in connection with wind energy situations [60]. The use of very large-scale underground geologic storage of hydrogen is being pursued in Europe, with the intent to serve electric load from hydrogen storage when wind power does not meet demand. In addition, Germany is moving ahead, having integrated into the electric system in the city of Prenzlau an electrolyzer plant that produces hydrogen from excess wind and uses the hydrogen as both motor fuel and combustible fuel in a combined heat and power (CHP) electric plant [61].

While the United States is also looking at geologic hydrogen storage [62], smaller-scale tank storage may also prove cost-effective when combined with wind systems that generate considerable excess power at the wrong time of day. An analysis based on wind generation in Texas suggested a viable business case for hydrogen energy storage [63].

Figure 1.8 shows a comparison of annual costs ($/kW) for a variety of energy storage technologies for the case of excess wind up to 6 h/night in a U.S. location [64]. The least-expensive options for renewable energy storage (in this case wind energy) are hydrogen-based technologies along with CAES and pumped hydro storage of energy.

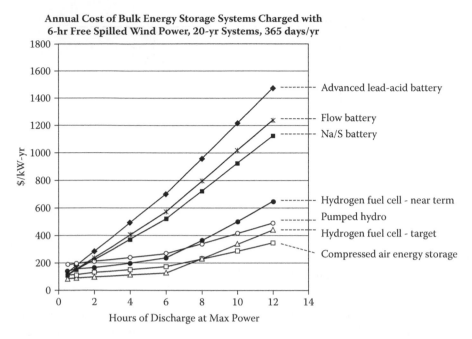

FIGURE 1.8
Annual costs ($/kW) for large-scale energy storage charged with excess (spilled) wind for up to 6 h per 24 h. Size is shown in hours of storage at maximum discharge power. (Reproduced with permission from S. Schoenung, *Economic Analysis of Large-Scale Hydrogen Storage for Renewable Utility Applications*, Sandia Report SAND11-4845, June 2011.)

Additional studies of the feasibility of hydrogen storage within the utility sector are under way at the National Renewable Energy Laboratory (NREL) and sponsored by the Department of Energy (DOE) Hydrogen Technology Advisory Committee (HTAC) [65].

Making Do with Nonrenewable Energy

It has been shown in this chapter that our current use of fossil fuels cannot continue indefinitely as these fuels are insecure from a resource point of view, environmentally unsustainable (global climate change), and politically unstable. However, it needs to be recognized that renewable energy currently makes up only a small fraction of the global energy supply [1]. Renewable energy (in oil equivalent) was only 4% of oil consumption in 2010 [1]. This is because renewable energy is more expensive. Unless we lower the costs for renewable energy, human nature (and government policy) is such that fossil energy feedstocks will be used until exhaustion or until the point at which the pain (economic, environmental, or political) caused by using them becomes unbearable.

This begs the observation that we need to decarbonize our conventional energy system, enabling a net-zero or near-net-zero GHG emission system as a transition to our sustainable system for the future. The technology to decarbonize the power sector is in the demonstration phase, funded as the FutureGen program from the U.S. DOE Fossil Energy Program [66]. There is a demonstration plant being constructed at Meredosia, Illinois, that will demonstrate gasification of hydrocarbon feedstock to hydrogen and CO_2, followed by sequestration of the CO_2 in a spent oil field. The H_2 thus produced would be used to power a turbine to create electricity. This plant will make hydrogen, electricity, and heat. Surplus hydrogen could also be used to power the transportation sector through the use of hydrogen-powered vehicles.

Of course, it must be pointed out that this approach relieves only the environmental unsustainability of fossil fuel use. The fuel insecurity and political insecurity issues remain, and as such, this approach is not broadly sustainable in the manner we have been describing.

Nuclear power could be a solution to the energy problem for the foreseeable future if we were to deploy breeder technologies on a worldwide scale. Nuclear power can also be used to generate hydrogen through thermochemical cycles or electrolysis, electricity and heat, enabling a net-zero transportation system. A comprehensive discussion of nuclear power is outside the scope of the present discussion.

Energy Carriers

Although only renewable (intermittent) energy sources require energy storage for their base operation, all primary energy resources need to be coupled to suitable energy carriers so the energy can be (1) stored for use later, (2) transported to meet the time and geographical demand, and (3) used in the transportation sector. There are only four energy carriers available to us: hydrogen, electricity, biomass (a form of solar energy), and manufactured synthetic hydrocarbon fuels.

Manufactured synthetic hydrocarbon fuels need a little explanation. There are a number of research and development projects under way [67] to develop the technologies necessary to take CO_2 and dissociate it to carbon and oxygen, followed by processing of the carbon with hydrogen to make a liquid hydrocarbon fuel. Some of this technology readily exists. The Fischer-Tropsch process was invented by Franz Fischer and Hans Tropsch at the Kaiser Wilhelm Institute in the 1920s. Since then, it has been widely used to produce long-chain fuels for automobile fuels. Germany used this process to supply about 9% of their fuel needs during World War II. The challenge today is to create "drop-in" hydrocarbon fuels using sustainable (renewable) energy with CO_2 extracted from the atmosphere to make it a net-zero CO_2 emissions technology. Getting CO_2 from any other source (like coal-fired exhaust) will reduce the CO_2 emission by the amount of the fuel the synfuel replaces. If the synfuel is used in transportation, then it displaces some petroleum, and the amount of CO_2 emitted is reduced by a commensurate amount. This is *not* a net-zero technology, and as pointed out previously, to solve our climate problems we need net-zero technology solutions so that growth does not overwhelm any gains made. It is interesting to note that hydrogen made from net-zero carbon sources would be needed to make these synthetic fuels in any event.

As indicated, the transportation sector is particularly dependent on petroleum and is a major consumer of that precious resource. Arguably, while we have a serious energy problem in all sectors, without a doubt the problem in the transportation sector is most acute. A closer look at how various energy carriers might have an impact on the transportation sector in terms of operational cost, infrastructure, and vehicle rollout is instructive at this point. We exclude biomass from the remainder of this discussion because even with cellulosic ethanol, there simply is not enough land area to grow the crops necessary to

TABLE 1.2

Comparison of Relative Costs for Gasoline, Synthetic Hydrocarbon Fuel, and Hydrogen

Fuel	Fuel Cost/ Relative to Today	Advanced Technology Range Improvement per GGE (Efficiency)	Cost per Mile (Relative to Gasoline)	Roll Over the Fleet? (30–40 yr)
Gasoline	1 ($3.00/gal)	1	1	No
Synfuel (C8)	8.2 ($24.62/GGE)	1	8.2	No
Synfuel (C10)	7.66 ($22.98/GGE)	2	3.83	Yes
Hydrogen	1.67 ($5.00/GGE)	2	0.84	Yes
Electricity from the grid	1.75 ($5.27/GGE)	2.56	0.69	Yes

produce enough fuel to satisfy transportation needs. A 2009 study [45] by Sandia National Laboratories showed that even in the United States, with large expanses of empty land, at most only 30% of transportation fuel demand in 2030 could be supplied as biofuel. Biofuel resources need to be focused on those applications for which the high energy density that liquid fuels offer is truly required. Examples of such applications include long-haul class 8 trucks (the short-haul class 8 trucks do not really require energy densities commensurate with liquid fuels since their average daily range is only ~100 miles), short-mission aviation, rail, and some marine applications. Understanding this, the remainder of the discussion focuses on the light-duty ground transportation sector.

Table 1.2 presents some costs relative to gasoline [68] on a per mile basis comparing synthetic hydrocarbon fuels to a hydrogen-fueled vehicle on a per mile basis. Shown in parentheses is the estimated cost of the particular fuel on a gallon of gasoline energy equivalent (GGE) basis. Note that the fuel cost (relative to gasoline) is greater than one for all fuel types. However, when considering the performance of the fuel combined with the advanced power train options, the comparative cost on a per mile basis tells quite a different story. The column, "Advanced Technology Range Improvement" compares just the improvement the advanced technology enables. For the synthetic C8 fuels (gasoline), no improvement over the contemporary gasoline technology was assumed. However, for the C10 fuel an advanced high-efficient compression ignition direct injection (CIDI) engine was assumed; for the hydrogen vehicle a hydrogen fuel cell electric vehicle was assumed; and for the electricity from the grid a battery electric vehicle was assumed. Note that on a well-to-wheels comparison the cost per mile for the two synthetic fuel pathways results in greater cost on a per mile basis than today's gasoline vehicle, whereas for the hydrogen and the electricity from the grid options both are less expensive on a per mile basis than the conventional gasoline option. Note also that both the hydrogen and grid options require a fleet rollover and an infrastructure development.

In the governmental and automotive research communities around the world, there is a major investment thrust to develop the "advanced" battery, which many believe will enable electrification of the fleet. The electrification of the fleet (short of long-haul class 8 trucks) is a feature that the global automobile manufacturers are aggressively moving toward. Bill Ford reportedly said that while his company continues to invest in biofuels, hydrogen, and more efficient internal combustion engines (ICEs), the electrification of the U.S. fleet is inevitable. It should be noted that electrification does not imply the use of batteries but rather electrification of the power train. Batteries are simply an energy storage device.

The refueling times for different energy carriers vary markedly. Since there is great debate over the performance expectations of future advanced batteries, let us assume for this discussion that we indeed have developed a battery that has completely satisfactory

TABLE 1.3

Refueling Times for a 300-Mile Range Toyota Highlander for Various Energy Carriers and Infrastructure Compatibility

Energy Carrier	Refueling System	Refueling Time	Infrastructure Change Needed
Gasoline (ICE at 17% combined highway and city)	13.5-gal liquid tank	~5 min	No
Hydrogen (FCEV at ~50% combined highway and city) [69]	4.6 kg at 70 MPa	~8 min	Yes At an estimated cost of ~$1000 per vehicle at today's refueling station volume (100,000 miles × $0.01/mile).
Battery (BEV at ~65% combined highway and city) [70]	175 kWH, type I charger (120 V, 15 amp, single phase)	97.2 h	No This is a standard U.S. electrical outlet.
Battery (BEV at ~65% combined highway and city)	175 kWH, type II charger (120 V, 20 amp, single phase)	72.9 h	Possibly This type of circuit exists only in locations where slightly higher power (over a 15-amp circuit) is anticipated. If a new circuit needs to be installed, it will cost the homeowner ~$2000.00 on average [71].
Battery (BEV at ~65% combined highway and city)	175 kWH, typical residential power in Europe (220V, 13 amp, single phase)	61.2 h	No This is a power rating for a typical wall circuit in Europe at 2.86 kW vs. 1.8 kW in the United States.
Battery (BEV at ~65% combined highway and city)	175 kWH, type III charger (240 V, 40 amp, single phase)	18.2 h	Most likely This type of circuit exists in most homes but only for the clothes dryer, oven, etc.; a new dedicated circuit needs to be installed for the charger. It will cost the homeowner ~$2000.00 on average [72].
Battery (BEV at ~65% combined highway and city)	175 kWH, type IV charger (480 V, 500 amp, three phase)	43.7 min	New infrastructure This kind of power is industrial and is not available for private residence; type IV chargers will be restricted to "industrial"-type applications, parking lots, etc.

Note: BEV, battery electric vehicle; FCEV, fuel cell electric vehicle.

power and specific energy density. Doing this will allow us to examine the isolated consideration of refueling times and remove speculative improvement in battery technology from this discussion. The base vehicle for this analysis was chosen to be the Toyota Highlander, which comes in a gasoline and hydrogen fuel cell version. This was chosen since it represents a "full-size," "fully functional" vehicle, and we have real performance data on the power trains needed to make a real comparison. On-the-road performance data from both the gasoline and fuel cell electric vehicle were used to determine the energy requirements for a 300-mile vehicle driving range. Refueling times for the various energy carriers and that available in our infrastructure are given in Table 1.3.

It is clear from Table 1.3 that battery-based systems for vehicles that perform at the same levels we expect from our current vehicles simply take too long to recharge. However, hydrogen-powered fuel cell vehicles can in principle be recharged in 8 min. As a result, from an infrastructure perspective, it is clear from Table 1.3 that electric vehicles that rely on the grid for their energy are restricted in application space where the energy requirements per charge are low. For light-duty transportation needs, this means that grid-fed battery electric vehicles will be restricted to small-vehicle, short-range applications, not because of the storage capacity, but due to the kinetics of recharging. Vehicles and duty cycles that require increased power and range will require a different energy storage technology if they are to be refueled in reasonable times. The one the auto industry is aggressively pursuing is a hydrogen-fueled proton exchange membrane (PEM) fuel cell [73]. This is an electric vehicle having all the attributes of a battery electric vehicle without the range or recharge time limitations. Note that the recharge time limitation is both a battery limitation and an infrastructure limitation. These difficulties will not be resolved unless the consumer is willing to install a dedicated 240-V, 40-A circuit for a type III charger at a cost of several thousand dollars and tolerate 18-h recharge times, as in the Toyota Highlander example in Table 1.3.

It is important to note that Table 1.3 calculations have taken advanced battery development out of the picture; we assumed a 100% efficient battery with satisfactory power density and specific power. This is the best we can do with the power available with our current utility grid.

Figure 1.9 shows the technology and energy carrier operating domain that GM is pursuing [74]. Other automobile manufacturers have similar road maps. Shown in this figure are the technology power train applications as a function of range and power requirements.

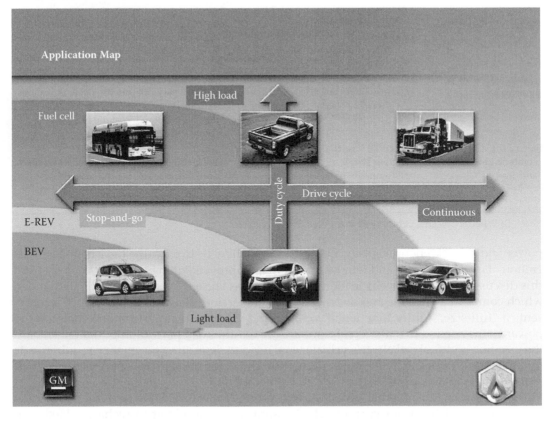

FIGURE 1.9 (See color insert.)
Road map for vehicles for GM as of 2010. (From R.C. Kauling, Electrification of the Vehicle: Critical Role of On-Board Hydrogen Storage Continues, Hydrogen Storage Materials Workshop, Ottawa, Canada, April 28, 2010. Reproduced with permission from General Motors.)

Figure 1.9 points out the large space of range and power that hydrogen fuel cell vehicles provide. The long-haul class 8 trucks (>1000-mile range) are not well suited for fuel cell applications. The range requirement simply cannot be met with hydrogen fuel cell technologies. However, the short-haul class 8 truck applications (<100-mile range) most certainly can be met with hydrogen fuel cell technologies.

The next chapter describes in more detail these hydrogen energy conversion devices, such as fuel cells and hydrogen-optimized ICEs, and explores the hydrogen fuel cell vehicle and H_2 ICE as the basis for a net-zero-carbon transportation system.

Summary

As a global society, we need to aggressively develop sustainable renewable integrated energy systems. The technical community agrees that our climate system cannot absorb the carbon (in the form of CO_2) that would result in the use of our remaining hydrocarbon energy reserves without catastrophic consequences. Our next-generation energy systems must be sustainable from fuel resource, environmental, and political perspectives. For the next energy grid to be truly sustainable, it must be based on essentially an infinite resource (wind, solar, nuclear), be truly zero emissions in CO_2 and other GHGs, and be available to all. We must be able to store energy from the intermittent renewable resources that produce excess energy exceeding demand at certain times of the day. We must also have in place a renewable transportation energy system that is similarly sustainable in the resource, environmental, and political spaces. Hydrogen enables renewable grid-scale power production by providing a technology to buffer the grid and to transport the renewable energy to where the demand exists. In addition, hydrogen can be used as an energy carrier, used in conjunction with fuel cells to enable the electrification of the transportation sector where the range, recharge times, and power requirements are greater than grid-fed battery technologies can satisfy. Hydrogen-powered vehicles are the technology of choice for long-range vehicle applications with zero emissions in the future.

Hydrogen can also be used effectively as a long-mission aviation fuel [75], and hydrogen-fed fuel cells enable net-zero carbon emission distributed CHP applications for stationary domestic power and heating requirements. These applications further support the argument that hydrogen will play a critical role in the truly sustainable energy infrastructure of tomorrow. Indeed, hydrogen technologies will play a dominant role alongside other energy carrier technologies, motivating the advanced methods of hydrogen storage, which are described shortly.

References

1. BP. 2011 statistical review of world energy. http://www.bp.com/sectionbodycopy.do?categoryId=7500&contentId=7068481.
2. Figure calculated from the resource data in Reference 1.
3. UN world population report 2001 (PDF), p. 31; retrieved December 16, 2008.
4. Population Division of the Department of Economic and Social Affairs of the United Nations Secretariat, World population prospects: the 2008 revision. June 2009.

5. A zetajoule (ZJ) is 10^{21} joules.
6. Nathan S. Lewis, Powering the planet, keynote speech at the first annual California Clean Innovation Conference, California Institute of Technology, Pasadena, May 11, 2007.
7. M. King Hubbert, chairman, *Energy Resources, a Report to the Committee on Natural Resources of the National Academy of Sciences–National Research Council*, Publication 1000-D, 1962, p. 50 and references therein.
8. R.W. Bentley, *Energy Policy* **30**, 189 (2002).
9. S.H. Mohr and G.M. Evans, *Energy Policy* **38**, 265 (2010).
10. S.H. Mohr and G.M. Evans, *Fuel* **88**, 2059 (2009).
11. T.W. Patzek and G.D. Croft, *Energy* **35**, 3109 (2010).
12. W. Zittel and J. Schindler, Coal: resources and future production, Energy Watch Group Paper No. 1/07, 2007. http://www.energywatchgroup.org/fileadmin/global/pdf/EWG_Report_Coal_10–07–2007ms.pdf.
13. EIA 2003-AEO, Table A2, p. 120.
14. http://news.bbc.co.uk/1/hi/world/europe/7806870.stm.
15. Intergovernmental Panel on Climate Change (IPCC), *Intergovernmental Panel on Climate Change Report: Climate Change 2007: The Physical Science Basis*, Contribution of Working Group I to the Fourth Assessment Report of the IPCC, S. Solomon, D. Qin, M. Manning, Z. Chen, M. Marquis, K.B. Averyt, M. Tignor, and H.L. Miller, eds., Cambridge University Press, Cambridge, UK, 996 pp.
16. J. Hansen, M. Sato, P. Kharecha, et al., *Open Atmos. Sci. J.* **2**, 217 (2008).
17. *America's Climate Choices*, Panel on Advancing the Science of Climate Change, National Research Council, ISBN 0-309-14589-9 (2010).
18. The compelling case for these conclusions is provided in "Advancing the Science of Climate Change," part of the congressionally requested suite of studies known as *America's Climate Choices* (see Reference 17).
19. The physical quantity driving climate change is the atmospheric concentration of GHGs, measured in parts per million. The acceptable GHG emission rate is a more complicated question as it requires a discussion of emission rates vs. sink rates, the sum of which determines the rate of change in GHG atmospheric concentration.
20. J. Fourier, *Mem. de l'Ac. R. d. Sci. de l'Inst. de France*, t, vii (1827).
21. C. Pouillet, *Comptes Rendus* t., vii, p. 41 (1838).
22. John Tyndall's paper, "On the Absorption and Radiation of Heat by Gases and Vapours, and on the Physical Connexion of Radiation, Absorption, and Conduction" was delivered on February 7, 1861, as the Bakerian Lecture to the Philosophical Society and appears in *Philosophical Transactions*.
23. J. Tyndall, "Contributions to the Molecular Physics in the Domain of Radiant Heat," a series of memoirs published in *Philosophical Transactions* and *Philosophical Magazine*, Longman's Green, London, 1872.
24. S. Arrhenius, *Philos. Mag. J. Sci.* Series 5, **41**, 237 (1986).
25. J. Uppenbrink, *Science* **272**, 1122 (1996).
26. P. Högbom, *Svensk kemisk Tidskrift*, Bd. vi. p. 169 (1894).
27. G.S. Callendar, *Q. J. R. Meteorol. Soc.* **64**, 223 (1938).
28. G.N. Plass, *Tellus* **8**, 140 (1956).
29. G.S. Callendar, *Tellus* **10**, 243 (1958).
30. J.R. Bray, *Tellus* **11**, 220 (1958).
31. B. Bolin, *Proc. Natl. Acad. Sci. U.S.A.* **45**, 1663 (1959).
32. T.F. Malone, chairman, *First General Report on Climatology to the Chief of the Weather Bureau, Committee on Climatology, Advisory to the United States Weather Bureau, National Academy of Sciences–National Research Council*, 1957.
33. R. Revelle and H.E. Suess, *Tellus* **9**, 18 (1956).
34. M. King Hubbert, chairman, *Energy Resources—A Report to the Committee on Natural Resources of the National Academy of Sciences National Research Council*, Publication 1000-D, 1962, p. 96.

35. Ibid., p. 63.
36. Spencer R. Weart, in *The Discovery of Global Warming*, Harvard University Press, Cambridge, MA, 2003.
37. W.W. Kellogg, *Climatic Change* **10**, 113 (1987).
38. H. Le Treut, R. Somerville, U. Cubasch, Y. Ding, C. Mauritzen, A. Mokssit, T. Peterson, and M. Prather, Historical overview of climate change. In *Climate Change 2007: The Physical Science Basis*, Contribution of Working Group I to the Fourth Assessment Report of the Intergovernmental Panel on Climate Change (S. Solomon, D. Qin, M. Manning, Z. Chen, M. Marquis, K.B. Averyt, M. Tignor, and H.L. Miller, eds.), Cambridge University Press, Cambridge, UK.
39. From the report, The world at six billion. http://www.un.org/esa/population/publications/sixbillion/sixbilpart1.pdf.
40. Current estimates of world population are available from the U.S. Census Bureau. http://www.census.gov/main/www/popclock.html.
41. From Figure 2.2 of the *America's Climate Choices, Panel on Advancing the Science of Climate Change*, National Research Council, ISBN 0-309-14589-9 (2010).
42. From the International Energy Agency (IEA). http://www.iea.org/index_info.asp?id=1959.
43. D.S. Reichmuth, A.E. Lutz, D.K. Manley, J.O. Keller, Sandia Report SAND2012-1092J, February 2012, Sandia National Laboratories, Albuquerque, New Mexico.
44. A description of CAFE standards can be found at the Web site of the U.S. National Highway Traffic Safety Administration (NHTSA). http://www.nhtsa.gov/fuel-economy.
45. T. West, K. Dunphy-Guzman, A. Sun, L. Malczynski, D. Reichmuth, R. Larson, J. Ellison, R. Taylor, V. Tidwell, L.E. Klebanoff, P. Hough, A. Lutz, C. Shaddix, N. Brinkman, C. Wheeler, and D. O'Toole, Sandia Technical Report SAND 2009-3076J (2009).
46. P.J. Federico, *J. Patent Office Society* **28**, 855 (1946).
47. H.L. Needleman, *Environ. Res.* **84**, 20 (2000).
48. For a good review of fuel cell systems, see James Larminie and Andrew Dicks, *Fuel Cell Systems Explained*, Wiley, Chichester, UK, 2000, as well as Chapter 2 of this book.
49. This program was renamed the "Partnership for a New Generation of Vehicles" (PNGV).
50. Interestingly, Toyota was excluded from PNGV.
51. The Austrian Dr. Ferdinand Porsche, at age 23, built his first car, the Lohner electric chaise. It was the world's first front-wheel-drive car. Porsche's second car was a hybrid, using an internal combustion engine to spin a generator that provided power to electric motors located in the wheel hubs. On battery alone, the car could travel nearly 40 miles.
52. Unpublished results from D. Reichmuth (Sandia National Laboratories) communicated to J.O. Keller. June 1, 2011.
53. Hubbert, *Energy Resources*, p. 4.
54. H. Meiwes and D. Sauer, "Technical and Economic Assessment of Storage Technologies for Power-Supply Grids," presented at Electrical Energy Storage Applications and Technologies (EESAT) Conference 2009, Seattle, Washington, October 2009.
55. K. Lundgren and L. Paulsson, Utilities giving away power as wind, sun flood European grid, *Bloomberg News*, September 30, 2011.
56. H. Landinger, Energiespeicher in Stromversorgungssystemen mit hohem Anteil erneuerbarer Energieträger, presented at VDE Energy technology working group, Munich, April 2010.
57. For a description of pumped hydro energy storage, please see http://www.electricitystorage.org/technology/storage_technologies/pumped_hydro/.
58. For a description of CAES, please see the Electricity Storage Association Web site. http://www.electricitystorage.org/technology/storage_technologies/caes/.
59. For a description of renewable electrolysis, please see the Web site of the National Renewable Energy Laboratory (NREL). http://www.nrel.gov/hydrogen/renew_electrolysis.html.
60. F. Crotogino et al., Large-scale hydrogen underground storage for securing future energy supplies, Proceedings of the World Hydrogen Energy Conference, Essen, Germany, 2010.
61. P. Hoffman, World's first renewables hydrogen hybrid power plant starts production, *Hydrogen Fuel Cell Lett.*, November 2011.

62. A.S. Lord, P.H. Kobos, and D.J. Borns, Underground storage of hydrogen: assessing geostorage options with a life cycle based systems approach, 28th USAEE/IAEE North American Conference, New Orleans, LA, SAND2009-7739C.

63. M. Kapner, Electric energy storage and wind, presented at Electricity Storage: Business and Policy Drivers, EUCI Conference, Houston, TX, January 24–25, 2011.

64. S. Schoenung, *Economic Analysis of Large-Scale Hydrogen Storage for Renewable Utility Applications*, Sandia Report SAND11-4845, June 2011, Sandia National Laboratories, Albuquerque, New Mexico.

65. F. Novachek, Hydrogen Enabling Renewables Working Group, Proceedings of the DOE HTAC meeting, November 3–4, 2011. http://www.hydrogen.energy.gov/htac_meeting_nov11.html.

66. For a description of the FutureGen program, see http://www.futuregenalliance.org/the-alliance/.

67. M. Schaefer, M. Behrendt, and T. Hammer, *Front. Chem. Eng. China* **4**, 172 (2010).

68. A.E. Lutz, University of Pacific, private communication with J.O. Keller, June 10, 2010.

69. Efficiency was calculated by assuming 60% for the fuel cell, 90% for both the power electronics and the motor, making the overall efficiency 0.6*0.9*0.9 = 0.49.

70. Efficiency was calculated by assuming 81% round-trip efficiency for the batteries (0.9*0.9), 90% for both the power electronics and the motor, making the overall efficiency = 0.81*0.9*0.9 = 0.66.

71. K. Morrow, D. Karner, and J. Francfort, *Plug-in Electric Hybrid Charging Infrastructure Review*, final report for U.S. DOE Vehicle Technologies Program, Advance Vehicle Testing Activity, 2008.

72. This type of circuit is common in Europe.

73. This is described in more detail in Chapter 2.

74. R.C. Kauling, Electrification of the vehicle: critical role of on-board hydrogen storage continues, Hydrogen Storage Materials Workshop, Ottawa, Canada, April 28, 2010.

75. For an excellent technical discussion of the use of hydrogen as a primary fuel for aviation, see G. Daniel Brewer, *Hydrogen Aircraft Technology*, CRC Press, London, 1991, and references therein.

2

Hydrogen Conversion Technologies and Automotive Applications

Lennie Klebanoff, Jay Keller, Matt Fronk, and Paul Scott

CONTENTS

Introduction 31
Hydrogen Internal Combustion Engines: Spark Ignition Engines 32
Hydrogen Internal Combustion Engines: Gas Turbines 40
Hydrogen Fuel Cells 44
H$_2$ PEM Fuel Cell 47
Operational Effects on PEM Fuel Cells 50
Membrane Degradation 53
Stability of the Catalyst 54
Gas Diffusion Layer 56
Automotive OEM View of Fuel Cell Attractiveness 58
Future Generations 60
Acknowledgments 61
References 62

Introduction

Chapter 1 described the motivations for developing a hydrogen-based energy infrastructure. Here, we discuss the hydrogen-based power technologies that convert hydrogen into electricity and useful heat or shaft power and provide the context for the need for improved hydrogen storage materials.

Ultimately, hydrogen power technology seeks to take advantage of the following chemical reaction:

$$2H_2 + O_2 \rightarrow 2H_2O + Q \qquad (2.1)$$

That's it. All of hydrogen power technology we are about to describe aims to implement that chemical reaction in the lowest-cost and most efficient manner possible. It turns out, however, that if you put hydrogen and oxygen in a balloon and let it sit at room temperature, the mixture will remain unreacted essentially indefinitely. The reaction as written in Reaction 2.1 does not proceed because the activation energy for the reaction is too high at room temperature for the reaction to take place—nothing happens.

However, if you do something that either changes the activation barrier of the reaction (add a catalyst) or provides an external source of energy (like an electric spark) to enable the reaction to surmount the activation barrier, the reaction proceeds to the right, forming water and releasing heat (denoted as Q). If you fill a balloon with a stoichiometric mix of hydrogen and oxygen (as in Equation 2.1) and hold a match under it, the reactants will be excited above the activation energy, and the reaction then proceeds to the right, producing water and heat. Overcoming the activation energy barrier of Reaction 2.1 is the basis for extracting heat (Q) to perform useful work. This heat can be transformed into shaft power using a heat engine such as a reciprocating internal combustion engine, a turbine, or a sterling engine, for example.

Alternatively, if one places a catalyst in the balloon, like a piece of Pt metal, one changes the reaction pathway such that the activation energy is lowered, and the reaction proceeds rapidly at room temperature. This catalytic acceleration is the basis for hydrogen catalytic heaters and hydrogen fuel cells. Hydrogen internal combustion engines (ICEs) (turbines and reciprocating engines) [1] and hydrogen fuel cells [2] are two of the primary means of converting the reaction energy of Equation 2.1 into useful shaft power and electrical energy, respectively. These energy conversion devices are described in more detail in this chapter.

Hydrogen is an attractive fuel for ICEs [1]. Hydrogen has physical and chemical properties that enable conversion devices such as ICEs that are more efficient than their hydrocarbon counterparts. Both H_2 ICEs and hydrogen fuel cell systems, when designed properly, produce near-zero emissions, with the only substantial emission being water vapor.

Hydrogen Internal Combustion Engines: Spark Ignition Engines

The earliest attempt at developing a hydrogen-fueled ICE was by Reverend W. Cecil in 1820. N. A. Otto in the 1860s and 1870s used a synthetic gas with a high hydrogen content in some of his work developing the reciprocating ICE. Otto is most famous for the development of the Otto cycle, which is a common cycle used in reciprocating internal engines of today; the other common thermodynamic cycle is the diesel cycle. Figure 2.1 shows a concept from an early patent by Erren [3] from 1939, showing an "internal combustion engine using hydrogen as a fuel." In this patent, hydrogen is mixed with oxygen (from air or another source) inside the cylinder of an engine. Hydrogen mixes rapidly with the oxygen, a spark ignites the mixture, the gas expands, performing work on the piston, and the waste products water vapor, any NO_x formed (if air is used as the oxidant), or any oxides of carbon (CO_2, CO) originating from burned lubricating oil are ejected out the exhaust valve.

The work on the H_2 ICE for nearly 100 years has been directed toward maximizing the power available from the hydrogen combustion, minimizing the emission of NO_x or any oxides of carbon, and improving reliability. Excellent reviews of the development of hydrogen ICEs can be found in the literature. A historical review has been written by Das [4], and an exceptionally clear and accessible review of the developments since that of Das has been written by White and coworkers [1]. Our description leans heavily on the review of White et al. [1].

The chemical and combustion properties of hydrogen allow hydrogen-fueled ICEs to operate cleanly. These properties also, unfortunately, lead to hydrogen engines with less power density than their hydrocarbon-based cousins. The lack of carbon in the primary fuel, hydrogen, naturally leads to a complete elimination of CO_2 from the burning of the primary fuel. Any remaining CO_2 produced would come from the combustion of lubricating oil in the engine, which in practice is produced at near-zero levels [1]. However, if air is

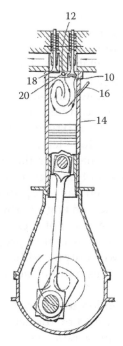

12

10. Air inlet valve
12. Cylinder head
14. Cylinder wall
16. H_2 injection pipe
18. Exhaust valve
20. Spark plug

FIGURE 2.1
Design of a hydrogen internal combustion engine, circa 1939. (Figure reproduced with permission from R.A. Erren, Internal combustion engine using hydrogen as a fuel, U.S. Patent No. 2,183,674, December 19, 1939.)

used as the oxidant, there is still the concern of oxidation of nitrogen within the combustion engine cylinder, leading to undesired NO_x emission. Recent research on H_2 ICEs has focused on improving the power density while minimizing NO_x release.

Table 2.1 gives the combustion properties of hydrogen and compares it with methane and gasoline [4]. Table 2.1 shows that the autoignition temperature of hydrogen, the minimum temperature at which a hydrogen air mixture will self-combust without aid of a

TABLE 2.1

Combustion Properties of Hydrogen, Methane, and Gasoline

Property	Hydrogen	Methane	Gasoline
Limits of flammability in air, vol%	4.0–75.0	5.3–15.0	1.0–7.6
Stoichiometric composition in air, vol%	29.53	9.48	1.76
Minimum energy for ignition in air, MJ	0.02	0.29	0.24
Autoignition temperature, K	858	813	501–744
Flame temperature in air, K	2318	2148	2470
Burning velocity in NTP air, cm s^{-1}	265–325	37–45	37–43
Quenching gap in NTP air, cm	0.064	0.203	0.2
Percentage of thermal energy radiated from flame to surrounding, %	17–25	23–32	30–42
Diffusivity in air, cm^2 s^{-1}	0.63	0.2	0.08
Normalized flame emissivity, 2000 K, 1 atm	1.00	1.7	1.7
Limits of flammability (equivalence ratio)	0.1–7.1	0.53–1.7	0.7–3.8

Source: Reproduced with permission from L.M. Das, *Int. J. Hydrogen Energy* **15**, 425 (1990).
Note: Normal Temperature and Pressure (NTP): T = 293.15 K, P = 1.013 bar.

spark or catalyst, is 858 K, or 585°C. Hydrogen's high autoignition temperature explains why a balloon filled with hydrogen and oxygen at room temperature does not spontaneously combust. The high autoignition temperature for hydrogen makes it less suitable for compression ignition (CI) engines and more suitable for spark ignition (SI) engines. The remainder of this discussion focuses on hydrogen SI engines.

One particularly important characteristic shown in Table 2.1 is the limits of flammability of hydrogen. The limits of flammability can be expressed as "volume percent," where 4.0% means 4.0% of the *total* volume of a mixture is hydrogen. Below the lower flammability limit (LFL), the fuel-to-air mix is too "lean," and there just is not enough hydrogen present to sustain a flame. Above the upper flammability limit (UFL), the fuel-to-air mix is too rich, and there is just not enough oxygen present to sustain a flame. One can see from Table 2.1 that hydrogen has exceptionally wide limits of flammability in air, from 4.0 vol% to 75 vol%. Inside an engine, the wide flammability limit (FL) allows the H_2 ICE to operate stably over a range of fuel-to-air mixtures, which can improve engine control and operation. Indeed, it is this property of H_2/air combustion that enables control over H_2 ICE's power output without the need for an intake manifold throttle. Controlling the engine in this manner improves the efficiency at low power levels.

In the combustion community, it is convention to describe the fuel-to-air mixture in terms of the equivalence ratio ϕ, where ϕ is defined by the actual fuel/air mass ratio for a given situation to the stoichiometric fuel/air mass ratio associated with Reaction 2.1. The equivalence ratio is given by

$$\phi = [Fuel/Air]/[Fuel/Air]_{stoichiometric} \qquad (2.2)$$

For Reaction 2.1, consider the combustion of 2 moles of hydrogen (with mass 4.0 g). The amount of oxygen needed to fully combust this amount of hydrogen is 1 mole of oxygen with mass 32 g. However, to get 32 g of oxygen in air, one needs a total mass of air of 137.9 g. Thus, the stoichiometric fuel/air mass ratio for hydrogen is (4.0 g)/(137.9 g) = 0.029 = 2.9%. This fuel/air mass ratio defines $\phi = 1.0$. Leaner fuel mixtures with less hydrogen have $\phi < 1$; fuel mixtures richer in hydrogen than stoichiometric have $\phi > 1$. The limits of flammability for hydrogen expressed in terms of ϕ are 0.1 to 7.1, as shown in Table 2.1.

Although hydrogen has a high autoignition temperature, it turns out that the minimum energy required to initiate a hydrogen flame in air is significantly lower than for hydrocarbon-air mixtures. This is shown in Figure 2.2 [1], where the minimum ignition energies for hydrogen-air mixtures (at atmospheric pressure) are compared to methane-air and heptane-air mixtures, all as a function of the equivalence ratio ϕ.

Note the logarithmic scale on the ordinate of Figure 2.2. Hydrogen-air mixtures require an order of magnitude less input energy to initiate combustion than hydrocarbons, as was also reported in Table 2.1 for the minimum ignition energy values. This is a remarkable feature of hydrogen.

Since hydrogen-air mixtures have a very low ignition energy, such mixtures can be inadvertently ignited by a number of uncontrolled sources in the combustion chamber. The problem has been described by White et al. [1]. The low ignition energies of hydrogen-air mixtures mean that H_2 ICEs are vulnerable to "preignition," with preignition defined as combustion prior to spark discharge and is caused by ignition from spurious hot surfaces. Such spurious surfaces include hot surfaces on the spark electrodes, the valves, or heated contamination deposits within the cylinder. Preignition events are not correlated with the spark plug firing control system and produce combustion events that increase the

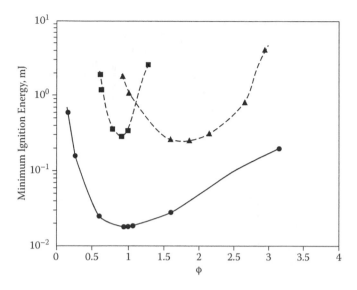

FIGURE 2.2
Minimum ignition energies of (●) hydrogen-air, (■) methane-air, and (▲) heptane-air mixtures in relation to ϕ at atmospheric pressure. (From C.M. White, R.R. Steeper, and A.E. Lutz, *Int. J. Hydrogen Energy* **31**, 1292 (2006). With permission.)

chemical heat release rate, thereby increasing the pressure in the cylinder. A lot of preignition can increase the inner cylinder wall temperature to near the autoignition temperature, which complicates the combustion even further. All of these processes (and more) associated with preignition can lead eventually to engine failure, making preignition a thing to be avoided [1, 4] in hydrogen ICE engines and in all engines for that matter.

Prior work has shown that a properly functioning engine must limit preignition to fewer than 1% of the combustion cycles. Preignition can be heard as an audible pinging in the engine but is conceptually different from another combustion problem called "knock." Studies have shown that the maximum hydrogen/air mix for which an H_2 ICE engine can be operated free of preignition is about $\phi = 0.6$. It turns out that the limiting ϕ is determined by the engine-out NO_x values and not by the preignition issue, and is discussed in the following.

Knock is the autoignition caused by the compression of the fuel/air mixture ahead of the flame front that originated from the spark. The parameter used to measure a fuel's resistance to autoignition due to compression heating is the octane number. High compression racing engines require high-octane-rated fuels. Conventional engines operate with compression ratios from about 8:1 to maybe 10:1. Conventional gasoline fuels available at gas stations range from an octane rating of 85–91, with some variation from supplier to supplier. This combination of compression ratios and octane rating controls engine knock problems in conventional engines to first order. Hydrogen has an octane rating of about 120, which makes it even less susceptible to engine knock as long as the engine compression ratio is not too elevated.

It was mentioned that while hydrogen does not contain carbon, and therefore CO_2 emissions out of the engine are essentially zero, it is still possible to produce oxides of nitrogen (NO_x) as hydrogen is burning in air in the cylinder. A great deal of research has been devoted to reducing NO_x emissions from H_2 ICE engines. Figure 2.3 shows a summary of NO_x emissions for various studies of hydrogen ICE engines, as compiled by White [1].

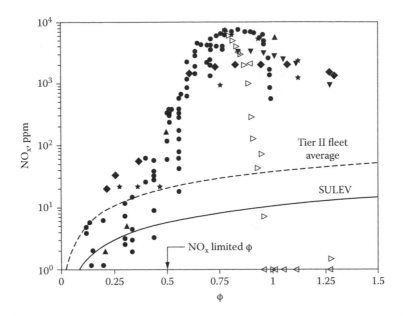

FIGURE 2.3

NO_x emissions as a function of equivalence ratio for engine-out (closed symbols) and tailpipe with exhaust gas after treatment (open symbols) from various studies with varying compression ratios. The dashed and solid lines represent the U.S. Federal Tier II manufacturer fleet average NO_x standard and CARB LEV II SULEV NO_x standard, respectively, for a fuel efficiency of 35 mpg. (Figure reproduced from C.M. White, R.R. Steeper, and A.E. Lutz, *Int. J. Hydrogen Energy* **31**, 1292 (2006), with data sources identified therein. With permission.)

Note the logarithmic scale on the ordinate of Figure 2.3. The NO_x production for hydrogen/air flames is determined by the thermal NO_x production mechanism. This has been well studied and can be found in a number of fundamental textbooks on combustion. The thermal NO_x mechanism is an exponential function of temperature and is linear in residence time. The combustion temperature varies with the equivalence ratio. The combustion temperature is low for fuel-lean conditions ($\phi < 1$), rises to a peak near $\phi = 1$, and then drops again for fuel-rich conditions ($\phi > 1$). The NO_x production tracks this behavior. Figure 2.3 shows that for low values of ϕ, where the combustion temperature remains sufficiently low, NO_x production is minimized. However, as the temperature rises, NO_x production, as an exponential function of temperature, increases dramatically. NO_x production then drops off on the rich side of a stoichiometric burn as the temperature drops. The open symbol triangles in the figure show dramatic NO_x reduction (values near zero for tail pipe emissions). They represent a stoichiometric (or slightly rich) combustion event with lots of exhaust gas recirculation (EGR), which is used as a thermal diluent (similar to operating lean). However, now there is no oxygen in the exhaust, enabling the use of a conventional three-way catalyst. The resulting NO_x emissions from the tail pipe were measured to be on the order of 50 parts per billion (ppb), a near-zero value.

It should be noted that as the equivalence ratio drops from $\phi = 1$, NO_x also reduces by orders of magnitude until $\phi \sim 0.4$, at which point the NO_x engine-out emission without any treatment of the exhaust gas is ~5–40 ppm (Figure 2.3). This NO_x level compares well with the U.S. Federal Tier II limits and the California Air Resources Board (CARB) vehicle emissions standards for the super ultra low emissions vehicles (SULEV) in effect in 2006. These regulated limits are shown by the lines in Figure 2.3. With no treatment of the exhaust

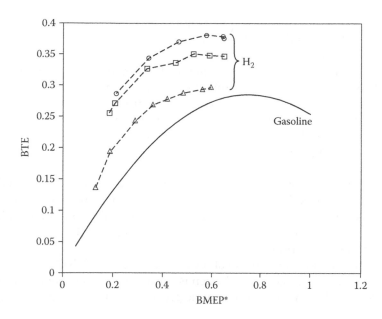

FIGURE 2.4
Brake thermal efficiency (BTE) as a function of normalized brake mean effective pressure (BMEP), where BMEP* = BMEP/BMEPmax gasoline. (Figure reproduced from C.M. White, R.R. Steeper, and A.E. Lutz, *Int. J. Hydrogen Energy* **31**, 1292 (2006).)

gas to remove NO_x, we need to have $\phi < 0.5$. Thus, the operational ϕ is limited by NO_x production and not by preignition ($\phi < 0.6$).

A somewhat separate topic from overall engine power is the thermal efficiency of the engine and is characterized by the brake thermal efficiency (BTE). BTE is defined as the ratio of useful crankshaft work available divided by the fuel combustion energy required to perform the work. BTE captures how well the energy available from the fuel can be captured by the piston/cylinder assembly to deliver useful work. The BTE [1] data for a hydrogen engine are shown in Figure 2.4.

The different curves for hydrogen are taken at varying compression ratios [1]. However, one can directly compare the triangle (Δ) hydrogen data with the gasoline data as they are both for a compression ratio of 9:1. In Figure 2.4, the thermal efficiencies of the hydrogen ICE engine and a gasoline ICE are compared via the brake mean effective pressure (BMEP). BMEP is an effective concept for comparing the performance of one engine to another. The definition of BMEP is the average (mean) pressure that, if imposed on the pistons uniformly from the top to the bottom of each power stroke, would produce the measured (brake) power output. In Figure 2.4, the BTE is plotted against an abscissa normalized to the maximum BMEP for gasoline.

There are a few things to note about Figure 2.4. First, the BTE of the ideal thermodynamic cycle is a function primarily of compression ratio or, more precisely, expansion ratio. This can be seen in Figure 2.4 where the BTE of the H_2 ICE increases markedly with compression (BMEP). Since the effective octane rating of H_2 and air combustion is significantly higher than that of conventional gasoline, a H_2 fuel ICE can be designed with a higher compression ratio, hence improving the thermal efficiency of the engine. Second, comparing the data for the H_2 ICE with the gasoline engine at the same compression ratio (9:1), the hydrogen ICE is about 30% more thermally efficient than the gasoline engine.

Third, the thermal efficiency drops off as the BMEP increases, most likely due to increased heat conduction to the cylinder walls for the hydrogen engine and increased mechanical losses due to the higher cylinder pressures.

So, summarizing the discussion, we see that if a hydrogen ICE engine is operated below the engine-out NO_x limit ($\phi \sim 0.5$), preignition problems are a nonissue, and the engine-out NO_x, CO, and CO_2 are near zero. Furthermore, the intrinsic efficiency (BTE) is quite high, about 30% higher than the comparable gasoline engine, with BTEs approaching 40% at higher compression ratios [1]. Measured BTE from research engines has reached 47%, with expectations that with aggressive friction control, efficiencies might exceed 50%. These are all good things.

However, to achieve this operation, we have to operate at $\phi \sim 0.5$, causing a drop in total engine power density. In other words, the engine is efficient with its fuel resource and is near-zero emission, but the total power density obtainable from a naturally aspirated engine may be comparatively low. Why is the power density low, how low is it, and is there anything that can be done about it?

In addition to the requirement to run at $\phi \leq 0.5$ (the engine-out NO_x limit), another difficulty with hydrogen engines is that hydrogen is a low-density gas. Let us suppose one has a 200-cc cylinder filled with air at atmospheric pressure and room temperature, as shown in Figure 2.5a.

The mass of air in that cylinder is 0.258 g. Now, if one introduces a stoichiometric amount of hydrogen, if one maintains standard conditions of temperature and pressure, the hydrogen occupies 30% of the total cylinder volume, thereby displacing air that was already in there. The situation is now as shown in Figure 2.5b. There is significantly less oxygen in the cylinder that can be burned with H_2, with the mass of air in the remaining 70% being 0.181 g. However, if a stoichiometric amount of fully vaporized gasoline is drawn into the cylinder full of air, the gasoline vapor need occupy only 2% of the total cylinder volume, as shown in Figure 2.5c. There remains 0.253 g (or 98%) of air in the cylinder that can be combusted with gasoline. The stoichiometric heat of combustion per standard kilogram of air is 3.37 MJ and 2.83 MJ for hydrogen and gasoline, respectively [1]. The combustion energy available from the hydrogen situation of Figure 2.5b is 610 J, whereas the combustion energy available from the gasoline situation of Figure 2.5c is 716 J. Thus, the

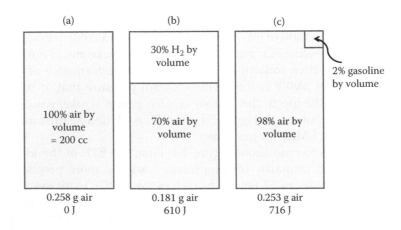

FIGURE 2.5
Comparison of combustion energy contained in a 200-cc cylinder for a stoichiometric mix of air and (a) nothing, (b) hydrogen, and (c) gasoline.

stoichiometric hydrogen mix has only 85% of the energy of its gasoline counterpart at $\phi = 1$. We have already discussed that in order to limit engine-out NO_x production, we must operate at $\phi < 0.5$. This reduces the power density in the cylinder even further, to 305 J or 43% that of the gasoline engine. This is a drop in power density that can be considered significant when compared to a stoichiometric burn gasoline engine. On the other hand, the power density of the NO_x-limited H_2 ICE is about the same as conventional diesel engines and higher than fuel cell systems.

The primary approach to increasing H_2 ICE power density is to force more hydrogen/ air mixture into the cylinder by turbocharging or supercharging. Boosting the pressure of the intake fuel/air mix is an established strategy for increasing peak engine power in standard petroleum-fueled ICEs. The extension to hydrogen is straightforward and has been reviewed by White [1]. Notably, substantial development of supercharged and direct-injected hydrogen SI engines has been performed by BMW [5] and Ford [6, 7]. Because of the increased pressures and temperatures, the issues of preignition and NO_x formation are exacerbated somewhat. Nonetheless, these issues can be brought under control, and road-worthy hydrogen engines have been built and driven that have nearly optimized combustion characteristics. A photograph of a modern hydrogen ICE engine power train from Ford [7] is shown in Figure 2.6.

We have come a long way since the design in Figure 2.1. The Ford- and BMW-developed hydrogen-fueled engines employ a traditional SI reciprocating engine configuration, whereas Mazda designed an H_2 engine based on their SI rotary engine designs. These engines are capable of measured 45% peak BTE, power densities equal or exceeding naturally aspirated gasoline engines, NO_x emissions that meet or exceed the most stringent NO_x emission level (California SULEV or 0.004 g/mile), and 99+% reduced CO_2. Ford, BMW, and Mazda all have "commercial-ready" engines. Indeed, BMW and Mazda have made small production runs of their vehicles.

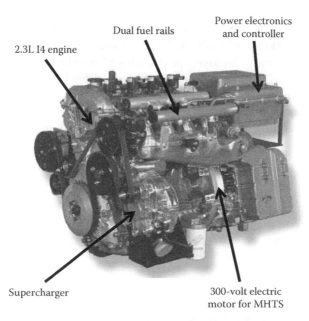

FIGURE 2.6
Photograph of Ford hydrogen ICE engine. (Figure reproduced with permission from A.K. Jaura, W. Ortmann, R. Stuntz, R.J. Natkin, and T. Grabowski, SAE Paper 2004: 2004-01-0058.)

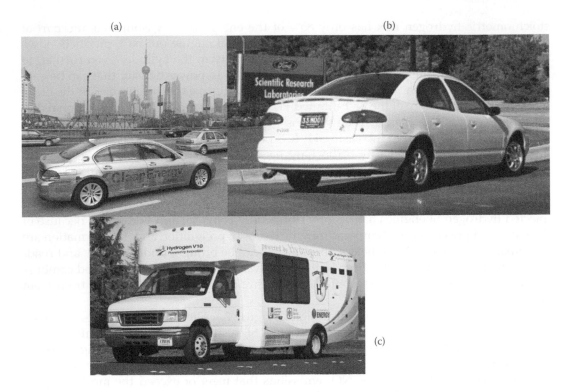

FIGURE 2.7
(a) The BMW Hydrogen 7 ICE vehicle; (b) the Ford P2000 ICE vehicle; and (c) the Ford Hydrogen V-10 E450 ICE Shuttle Bus. (The BMW Hydrogen 7 photo courtesy of BMW; the Ford P2000 picture courtesy of the Ford Motor Company.)

Some of these vehicles are shown in Figure 2.7, including the BMW Hydrogen 7 vehicle [8] and the Ford P2000 H_2 ICE vehicle [9]. Also shown is a Ford Hydrogen V-10 E450 Shuttle Bus that was recently deployed at the sites of Sandia National Laboratories and the Lawrence Livermore National Laboratory as part of Fuel Cell Market Transformation project funded by the Department of Energy (DOE).

Summarizing this description of hydrogen ICE engines, it is fair to say that these engines are high performance, high efficiency, and near zero emission and can be economical and viable in the near term. Although further work is needed to fully optimize the exploitation of hydrogen's unique combustion characteristics, there is little doubt that these hydrogen SI ICE engines have played and may continue to play an important role in the early roll-out of the hydrogen-based transportation sector, along with hydrogen fuel cell vehicles. Many argue that it makes sense to leverage the massive investment the automobile industry has made in manufacturing very cost effective (\$45/kW) ICEs and deploy H_2 ICE vehicles now to help accelerate the development of a hydrogen fueling infrastructure.

Hydrogen Internal Combustion Engines: Gas Turbines

Hydrogen combustion in gas turbines is finding increasing interest as a step in the cleaning up of coal-based electricity generation. We describe aspects here for completeness.

However, hydrogen combustion in stationary power gas turbines is not a driver for the development of hydrogen storage materials, as "storage" is not a requirement for this combustion technology since the hydrogen can be produced and utilized "on the fly."

Hydrogen-burning turbines were originally developed by the jet aircraft industry. A fascinating review of hydrogen aircraft technology can be found in the book of the same title by G. Daniel Brewer [10]. The interest in using hydrogen as a primary fuel in aviation derives from hydrogen's remarkable specific combustion energy (120 kJ/g), which is 2.8 times higher than for conventional jet fuel. Thus, hydrogen is an exceptionally light aviation fuel, and its thermal conduction properties also enable high-speed hypersonic flight [10]. We give a brief review of that history here since airplanes are cool. It should be noted the major aircraft manufacturers Boeing [11] and Airbus [12] have both examined the use of hydrogen as a primary fuel for commercial jet aircraft.

There was early understanding of the advantages of the use of hydrogen for in-atmosphere flight as well as space flight engines due to the high specific impulse (as a consequence of the unique light weight, hence high sound velocity of the lightest element). In 1945, with the success of the earliest aircraft jet engines, studies were initiated of hydrogen-fueled combustion in gas turbines.

About 1954, the NACA (National Advisory Committee for Aeronautics, later renamed NASA) initiated early work at the NACA Lewis Flight Propulsion Laboratory (in Cleveland, OH) using gaseous hydrogen in a turbojet combustor [10]. For almost all aircraft applications, the hydrogen is stored as liquid hydrogen (LH_2) to minimize weight. This work culminated in a flight demonstration in which one engine of a B-57 medium bomber was fueled with hydrogen for a short portion of a test flight [10].

Flight testing with liquid hydrogen on board (in a wingtip tank) began in late 1956, with successful engine operation using hydrogen beginning in February 1957. The procedure was to climb to an altitude of 50,000 ft using conventional JP4 fuel, then add hydrogen to operate the port engine on a hydrogen-JP4 mixture for 2 min, and finally replace the JP4 flow and operate the port engine only on hydrogen. The quantity of hydrogen stored onboard was sufficient to power the aircraft for 20 min at a flow of 4 kg/min [13].

The Lewis group found the hydrogen combustor could be 1/3 shorter than one burning JP4. In high-altitude wind tunnel tests, they found the hydrogen engine ran with stable combustion to the 27,000-m limit of the facility, while the engine on JP4 would become unstable at a 20,000-m (66,000-ft) altitude. Further, the thrust with the hydrogen fuel was 2–4% higher, and the specific fuel consumption was 60–70% lower [13]. These are compelling aviation *performance* reasons for the use of a hydrogen-burning gas turbine. Note that none of these investigations was motivated by fuel resource insecurities or environmental concerns.

The U.S Air Force was supportive of the work at Lewis Laboratory and was promoting high-altitude hydrogen-fueled engine work with industry as well, as described in the account of the early engine development work at Pratt and Whitney [14], which was supporting the Lockheed Skunk Works code name "Suntan" program, also referred to as the CL-400 program. This hydrogen expander engine was described well by Mulready et al. and represents one of the more impressive achievements of the engine development community [14]. There was continuing work on hydrogen-fueled gas turbine aircraft into the 1990s, with a Russian consortium investigating a transport application [15]. Most all work currently being performed on hydrogen combustion in turbines is motivated by the ground power application.

There are two motivating reasons for using hydrogen in turbine applications for ground power applications. The first and probably most significant reason is that hydrogen use in turbines enables a zero or net-zero carbon system in a well-established combustion system such as the gas turbine. Also, just like in the H_2 ICE, because of broad flammability limits of hydrogen-air mixtures, the turbine can be operated with a lean premixed combustion strategy to mitigate engine-out NO_x production. Limiting NO_x in this way eliminates the need for postcombustion NO_x cleanup technologies, like selective catalytic converters. This is discussed further in this chapter.

The FutureGen Industrial Alliance was formed [16] to partner with the U.S. DOE, whose goal is the demonstration of near-zero emission coal-based power generation technology. The basic idea is to employ coal gasification to allow electricity production in a way compatible with geologic sequestration of CO_2. Hydrogen is a key by-product of this process, and its combustion in gas turbines ultimately allows electricity generation in a much cleaner way. The U.S. DOE Fossil Energy Web site has some good information about the role of hydrogen and gas turbines in enabling CO_2 capture and sequestration from coal combustion for electricity production [17, 18].

Rather than burning coal directly, the coal is subjected to "gasification" with steam and oxygen at high temperatures and pressures to produce a mixture of carbon monoxide, hydrogen (syngas), and other gases. Gasification, in fact, may be one of the future technologies for general hydrogen production [17]. The syngas thus produced can then be burned in a gas turbine to produce electricity. A diagram of a typical gas turbine is shown in Figure 2.8 [18]. A picture showing the installation of a real operational turbine for ground power generation is shown in Figure 2.9 [19].

All gas turbines, including those burning hydrogen, operate in basically the same way. Referring to Figure 2.8, a compressor draws air into the engine, pressurizes it, and feeds it to the combustion chamber. Typical compression ratios are from 20 to 30. This pressurized air is then mixed with fuel in the combustion system through the use of a ring of fuel injectors and is then burned at temperatures exceeding 2000°F. The combustion produces a high-temperature, high-pressure gas stream that enters and expands through the turbine section. The turbine is an array of alternate stationary and rotating propeller-shaped blades. As hot combustion gas expands through the turbine, it spins the rotating blades.

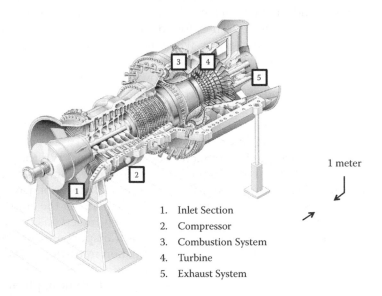

1 meter

1. Inlet Section
2. Compressor
3. Combustion System
4. Turbine
5. Exhaust System

FIGURE 2.8 (See color insert.)
Diagram of a gas turbine for power generation. (Cut-away turbine image courtesy of Siemens.)

FIGURE 2.9
Installation of a real Siemens SGTS-8000H power generating turbine in Dusseldorf Germany. (Photo reproduced with permission from Siemens.)

The rotating blades drive the compressor (to draw more air into the combustion section) and turn a generator to produce electricity [18].

A simple cycle gas turbine can achieve energy conversion efficiencies ranging between 20% and 35% [18]. Higher combustion temperatures, achieved through the use of advanced materials and cooling strategies, can in principle allow thermal efficiencies up to about 40% or more for a hydrogen- and syngas-fired gas turbine [18]. When waste heat is captured from these turbine systems for heating or industrial purposes, the overall energy cycle efficiency can approach 80%, in a manner similar to combined heat and power (CHP) operation for stationary fuel cells.

Just like it was for hydrogen combustion in SI engines, NO_x generation is a concern for hydrogen combustion in turbines using air as the oxidizer. Chiesa et al. [20] offered an accessible discussion of the NO_x control issues that arise in using hydrogen in gas turbines designed for operation on natural gas, as well as other aspects of the engine design. The NO_x generation can be avoided by use of a very lean mixture (low ϕ), just as in the H_2 ICE SI engine. H_2/air mixtures as lean as 4% by volume can be ignited, with the dilute hydrogen mixture resulting in a cool burn with low NO_x [20]. High temperatures are needed to produce an efficient engine, but massive amounts of NO_x can result as the hydrogen-air mixture approaches stoichiometric, as was observed for the SI engines discussed previously [1]. Chiesa et al. also examined the use of fuel dilution by steam or nitrogen as a means of lowering the combustion temperature. Either type of dilution provides the necessary temperature reduction, producing NO_x reductions with acceptably small efficiency penalty. Some data from these studies are shown in Figure 2.10 [20].

As in H_2 reciprocating ICEs, NO_x production for H_2 turbines is governed by the exponential thermal NO_x mechanism. Figure 2.10 shows the exponential dependence of NO_x production on flame temperature for various fuels. Since turbines operate with varying degrees of dilution (excess air), these NO_x emission numbers have all been corrected to an exhaust stream with 15% oxygen concentration, so the NO_x values can be compared

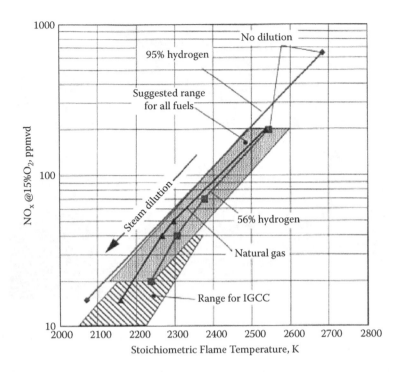

FIGURE 2.10

NO_x turbine emission results. (Figure reproduced with permission from P. Chiesa, G. Lozza, and L. Mazzocchi, *J. Eng. Gas Turbines Power* **127**, 73 (2005).)

without the effects of varying dilution factors. While NO_x production by the thermal mechanism is an exponential function of temperature, it is also linear in residence time. So, an alternative approach to NO_x reduction is fast combustion, including supersonic combustion—a feature of the "scramjet" [21]. This is likely of interest for reducing NO_x only in flight applications.

The limitations of materials can inhibit some approaches to eliminating nitrogen from the combustion chamber, thereby eliminating NO_x. For example, one possible approach to NO_x control is to react the hydrogen with pure oxygen. At stoichiometric levels, this would result in extremely high temperatures—reportedly 3200°C—well above the working temperatures of conventional combustors and turbine blades.

Work continues on other means of NO_x control in hydrogen gas turbine engines. The reaction of nitrogen with oxygen to make NO_x does take time as well as high temperatures, and hence one approach has been to make a very small flame with enhanced mixing. This appears to be the focus of recent research by a Siemens group [22], which is reportedly using swirl to promote fast combustion as well as enhanced mixing to reduce the size of the flame.

Hydrogen Fuel Cells

An excellent presentation of the science and engineering of fuel cells can be found in the book *Fuel Cell Systems Explained* by Larminie and Dicks [2]. A hydrogen fuel cell is an electrochemical device that executes the hydrogen/oxygen Reaction 2.1 without direct combustion. Rather, Reaction 2.1 is made to happen in two spatially separated half-reactions:

$$2H_2 \text{ (g)} \rightarrow 4H^+ + 4 \text{ e}^- \text{ (occurring at the anode)} \qquad (2.3)$$

$$O_2 \text{ (g)} + 4 \text{ e}^- \rightarrow 2O^{2-} \text{ (occurring at the cathode)} \qquad (2.4)$$

Initially, the produced H^+ and O^{2-} are created at separate electrode sites, but subsequently H^+ diffuses from the anode to the cathode through an H^+-conducting membrane. The last step of the reaction occurs spontaneously at the cathode, with

$$4H^+ + 2O^{2-} \rightarrow 2H_2O \qquad (2.5)$$

These reaction steps of oxidation of H_2 (Reaction 2.3) and reduction of O_2 (Reaction 2.4) happen not in the gas phase, but at physically separate electrode surfaces. The physical separation is the key point, and this separation allows the thermodynamic tendency for each half-reaction to express itself as a half-cell voltage. The reactive sequence is actually similar to a catalytic reactor, where H_2 becomes dissociated at a catalyst surface, and the hydrogen atoms subsequently react with oxygen on the catalyst surface to form H_2O. There are catalysts on the anode and cathode of the fuel cell to accelerate the kinetics of the half-reactions. However, unlike the catalytic reactor, in the fuel cell the two steps are physically separated, allowing a voltage and external current to be developed.

In 1842, Sir William Grove developed the first fuel cell, which combined hydrogen and oxygen to produce electricity. In its simplest form, these first experiments set the stage for the work that has been done over the last 170 years. Some of the basic fundamentals of electrochemistry were elucidated in these experiments. The efficiency of the electrochemical process can be significantly higher than traditional combustion. However, given no constraints on the temperature of combustion, both ICE and fuel cells have very high and in fact equivalent thermal efficiencies, as discussed by Lutz and Keller [23] and others [2, 24]. Whereas traditional gasoline combustion has a thermal efficiency of about 35%, limited primarily by the temperatures achievable in traditional combustion systems, the thermal efficiency of the electrochemical process can be about 50%. Thus, 50% of the reaction energy can be converted to electricity, with the remaining 50% constituting "waste heat," which is removed from the system by cooling air or liquid.

Over the years, there have emerged five general classes of fuel cell systems, which are viable and commercially available. These fuel cells differ from each other primarily by the electrolyte they use to perform ion conduction within the fuel cell and the corresponding temperature ranges of operation. Each type of fuel cell has an application that is most suited for it, from the operating conditions required; to the size, weight, and efficiency of the units; to the high-volume cost potential as well. Fuel cells and batteries share some of the same principles regarding the generation of electricity, with a basic difference being that the fuel cell consumes fuel (i.e., hydrogen and oxygen from air) from external sources, whereas a battery discharges itself and must be "recharged" to perform again. Table 2.2 briefly describes these five types of fuel cells [2].

The fuel cell types can be divided into two regimes of operating temperature: low-temperature fuel cells that operate in the range 50°C–220°C (proton exchange membrane [PEM], alkaline, and phosphoric acid fuel cells) and high-temperature fuel cells that operate above 650°C (molten carbonate and solid oxide fuel cells).

TABLE 2.2

Types of Fuel Cells

Fuel Cell Type	Mobile Ion	Operating Temperature
Proton-exchange membrane (PEM)	H^+	50–100°C
Alkaline (AFC)	OH^-	50–200°C
Phosphoric acid (PAFC)	H^+	~220°C
Molten carbonate (MCFC)	CO_3^{2-}	~650°C
Solid oxide (SOFC)	O^{2-}	500–1000°C

Proton Exchange Membrane. The PEM fuel cell is perhaps the simplest of the fuel cells. It uses pure hydrogen fuel at the anode and can operate at quite low temperatures (50–100°C), using a catalyst (typically Pt) to increase the reaction kinetics. The PEM employs a solid polymer that conducts hydrogen ions from the anode to the cathode. The PEM fuel cell offers high power density, high efficiency, good cold and transient performance and shows the best potential to meet future automotive high-volume cost requirements.

Alkaline. Alkaline fuel cells (AFCs) are among the highest-efficiency fuel cells and have been used by NASA in space missions since the 1960s in both the Apollo and Space Shuttle applications using pure hydrogen and oxygen. The electrolyte is an alkaline solution, typically potassium hydroxide. AFCs operate about 50–200°C and use a different sequence of reactions (involving hydroxide anions OH^-) to affect the overall reaction of hydrogen with oxygen. These fuel cells are not as convenient to use as PEM fuel cells due to the need for highly corrosive alkaline solutions and the higher-temperature operation. As such, they are not currently a popular fuel cell technology option.

Phosphoric Acid. The phosphoric acid fuel cell (PAFC) functions in a similar chemical way as the PEM fuel cell, accepting pure hydrogen and using a proton-conducting electrolyte to complete the circuit. In the case of the PAFC, the proton-conducting medium is phosphoric acid. Typical operating temperatures are in the 200°C range, and as such, PAFCs require precious metal catalysts like the PEM to accelerate reaction kinetics. PAFCs are typically deployed in the 200- to 400-kW range for power generation in a variety of applications. They can operate efficiently when CHP is employed and when the fuel cell is run in a continuous fashion. These large PAFC units typically take natural gas as the feedstock and perform an "on-board" steam methane-reforming step to generate hydrogen needed for the PAFC fuel cell operation.

Solid Oxide. The solid oxide fuel cell (SOFC) operates on similar chemical principles as the PEM fuel cell (in terms of the half-cell reactions) but operates at much higher temperatures and uses a thermally robust ceramic O^{2-}-conducting medium as the electrolyte. Operating temperatures are in the 850°C–1000°C range. Due to the high temperatures, the reaction rates are sufficiently high that expensive precious metal catalysts are not needed. There are several variants in SOFC cell design and architecture. SOFCs are typically used in high-power applications such as power-generating substations or facility powering, but there has been interest in the last few years to apply this technology to mobile auxiliary power units (APUs) for a variety of road trucks, including military vehicles. The long warm-up time and high operating temperature would make it difficult to place an SOFC in automobiles. Typically, the feedstock of choice is natural gas, with the reforming of the natural gas to hydrogen occurring internally on the hot anode surface.

Molten Carbonate. The electrolyte of a molten carbonate fuel cell (MCFC) is, as the name implies, a molten mixture of alkali metal carbonates. Typically, it is a mixture of lithium and potassium carbonates retained in a ceramic matrix of $LiAlO_2$. At the high temperatures utilized, typically 600–700°C, the alkali carbonates form a highly conductive molten salt, with the CO_3^{2-} ion providing charge transfer from the anode to cathode. Unlike other fuel cells, in the MCFC, carbon dioxide must be supplied to the cathode along with oxygen. The high operating efficiencies and operating temperatures (on the order of 650°C) make this type ideally suited for large-scale power generation where one can take advantage of CHP to increase the overall system energy efficiency.

The use of fossil-fuel-derived natural gas as the hydrogen feedstock in the larger PAFCs, MCFCs, and SOFCs limits their capacity to reduce CO_2 emissions, with the reduction being only about 30% when compared to analogous gasoline or diesel sources of electric power. Of course, if a bio-derived source of methane is used, the carbon reduction can be much larger. Another carbon-based feedstock for fuel cells is methanol. The "direct methanol" fuel cell is a variation on the PEM theme, albeit with a lower efficiency. Direct methanol fuel cells have found utility in very-low-power applications for battery replacements in portable electronics. The fuel is currently more convenient than hydrogen, but the low power density of the current technology makes it difficult to scale up for larger-scale applications. Also, the use of a carbon-based fuel increases the CO_2 output of the technology significantly.

We eliminate large, high-temperature fuel cells from further discussion as they are typically fueled with natural gas, and this is a book about hydrogen storage. For automotive applications, which have driven a lot of the hydrogen storage research work recently, it would take too long to bring a high-temperature fuel cell stack up to operating temperature. Although the phosphoric acid system can be considered a relatively low-temperature fuel cell, the 220°C temperature required for its operation eliminates it as a viable candidate fuel cell for vehicles due to prolonged startup times. This leaves the PEM fuel cell and the AFC for rapid-start applications. As mentioned, the AFC is not nearly so convenient to use as the PEM due to the presence of caustic alkaline solutions and the management of those liquids. This leaves the PEM fuel cell, which is the basis for automotive fuel cell vehicle development.

H$_2$ PEM Fuel Cell

Figure 2.11 shows the relevant reactions in a H_2 PEM fuel cell. At the PEM anode (site of oxidation), hydrogen gas ionizes (oxidizes), releasing protons and electrons for the external circuit. At the cathode (site of reduction), oxygen molecules are reduced in an acidic environment by electrons from the circuit, forming water molecules. Protons pass through the PEM, from anode to cathode, completing the circuit.

PEM fuel cells deliver high power density and offer lighter weight and smaller volume than other fuel cell systems. Traditional PEM fuel cells use a solid proton-conducting

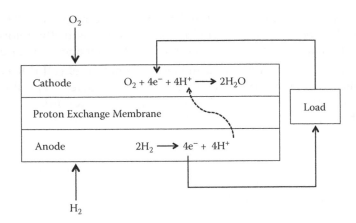

FIGURE 2.11
Schematic diagram of a proton exchange membrane (PEM) fuel cell. (Derived from J. Larminie and A. Dicks, in *Fuel Cell Systems Explained*, Wiley, New York, 2000.)

polymer membrane called Nafion®, a type of polyfluorinated sulfonic acid (PFSA) material, which allows proton transfer between the anode and cathode without the need for corrosive electrolyte fluids such as that needed by the AFC systems [2].

Nafion-based fuel cells operate at low temperatures, around 80°C. The low-temperature operation provides for rapid startup, which is essential for most low-power or mobile applications. However, for temperatures at or below about 80°C, the reaction product is liquid water, making management of liquid water an important issue.

A more detailed diagram of a PEM fuel cell is shown in Figure 2.12 [25].

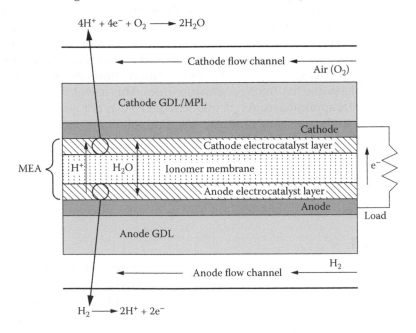

FIGURE 2.12
Detailed diagram of a PEM fuel cell. (Figure reproduced with permission from R. Borup et al., *Chem. Rev.* **107**, 3904 (2007).)

A single cell of a PEM fuel cell consists of a porous cathode and anode (typically composed of carbon cloth), each connected to a leg of the load in Figure 2.12. Both the anode and cathode have an impregnated catalyst (electrocatalyst) layer and a gas diffusion layer (GDL). The two electrocatalyst layers/electrodes are separated by a polymer ionomer membrane. The electrochemical half-reactions of hydrogen oxidation and oxygen reduction take place within the cell's inner three layers of ionomer membrane, electrocatalyst layer/anode, and electrocatalyst layer/cathode. The three-layer assembly is commonly known as a *membrane electrode assembly* or MEA.

Adjacent to the cathode and anode are GDLs. The GDLs allow gas to approach the catalyst layers. They are also electrically conducting and provide a path for the liquid water to move from the MEA to the flow channel in the bipolar plates. The GDL consists of a macroporous substrate layer and a microporous layer (MPL). The macroporous substrate layer is typically a carbon fiber matrix, and the MPL is a hydrophobic film consisting of carbon black mixed with fluoropolymer.

Each GDL is situated against a bipolar plate, which provides a H_2 flow channel for the anode and an O_2 (air) flow channel for the cathode. The bipolar plates have much more structural detail than shown in Figure 2.12. These plates not only provide gas flow to the anode and cathode but also serve to connect the individual fuel cells in series, thereby providing a conduction path for electrons between individual cells. The bipolar plates also provide cooling channels for the removal of waste heat and give the overall fuel cell stack structural strength.

The automotive original equipment manufacturers (OEMs) have been heavily involved in developing bipolar plate materials and manufacturing approaches, in particular the stamping of metals and the development of sealing approaches. These designs have to employ effective sealing methods and assembly techniques for 3–400 cell stacks. These are developments that have become critical to the success of the technology in the automotive market. These stamping processes are clearly high volume and have an impact on both the design and the performance.

In modern MEAs, the electrocatalysts are Pt or Pt-alloy nanoparticles deposited on high-surface-area (nanoparticle) carbon supports. The nanoparticles increase the active catalytic surface area per unit mass of platinum, increasing the catalytic activity. The ionomer is a polymeric material with sulfonic acid side chains and a fluorocarbon or hydrocarbon backbone. The most well known and best studied of these PFSA-based materials is Nafion and remains the industry standard for the PEM fuel cell ionomer [2].

The research-and-development (R&D) situation with H_2 PEM fuel cells has been different than for the H_2 ICE technology development. For H_2 ICE research, the hardware and materials associated with ICEs are quite developed, but the chemical processes (preignition, NO_x generation, thermal efficiency) have needed to be understood better. For the H_2 PEM fuel cell, the electrochemical process is basically understood, but it is the hardware and materials that have needed further improvement, particularly with regard to durability and cost.

As can be seen from Figure 2.12, there are a variety of different types of materials within a PEM fuel cell, and they are all sources of potential durability problems. We review here recent work characterizing the mechanisms of degradation operative in PEM fuel cells and steps that have been taken to mitigate them. We believe that a review of these processes leads to a deeper understanding of how a PEM fuel cell works and acknowledges the intense R&D effort to bring the degradation mechanisms under control. Our review leans heavily on a superb review of the scientific aspects of PEM fuel cell durability and degradation by Borup et al. [25].

The materials in a PEM fuel cell are exposed to strongly acidic aqueous solutions, strong oxidizing conditions at the anode, strong reducing conditions at the cathode, considerable electric currents, and potential gradients. Chemical mechanisms derived from these conditions degrade fuel cell performance. In an analogous program within the U.S. DOE to develop hydrogen storage, the U.S. DOE Energy Efficiency and Renewable Energy (EERE) developed targets for fuel cell operation [26].

There are a variety of influences on PEM fuel cell performance. These include

1. *Operational Effects*—the manner in which the fuel cell is used; includes effects of impurities in the hydrogen feed, impurities in the air feed, cycling, fuel starvation, and exposure to subfreezing environments;
2. *Chemical and Physical Degradation of the PEM*—the manner in which chemical species attack the chemical integrity of the membrane itself (chemical degradation) at the atomic level or how membrane creep (thinning) and development of cracks affect the physical integrity of the membrane;
3. *Degradation of the Electrocatalyst*—chemical properties affecting the all-important fuel cell catalyst, including dissolving of the catalyst in the aqueous environment and agglomeration and movement of catalyst nanoparticles with repeated fuel cell cycling;
4. *Corrosion of the Catalyst Support*—the manner in which chemical species generated during fuel cell operation can, under loss of electrochemical control, attack and irreversibly damage the catalyst support; and
5. *Degradation of the GDL*—chemical changes in the GDL that affect its ability to transport the reaction product water.

Although this list may seem long and dreary, it turns out that a lot has been learned about these destructive processes, with mitigation strategies developed. The result has been the commercial availability of PEM fuel cells with increasingly robust character. A full discussion of these effects can be found in Reference 25, but we provide a brief review here to give a sense of the recent R&D that has been conducted on PEM fuel cells and to provide a deeper understanding of how PEM fuel cells work in practice.

Operational Effects on PEM Fuel Cells

One of the most important influences on PEM fuel cell operation is simply the purity of hydrogen and oxygen reactants. Impurities adsorb onto the catalyst surfaces, making them unavailable to catalyze the hydrogen-oxygen reaction to form water. This is called catalyst poisoning and is an important issue for PEM fuel cells. Carbon monoxide (CO) is one of the best-studied catalyst poisons due to its production in the intermediate step of the water gas shift reaction used to make commercial hydrogen. CO forms a strong bond with catalysts such as Pt, blocking reactive sites important for the adsorption of hydrogen and oxygen.

In addition to CO, ammonia (NH_3) is an important poison for fuel cell operation. Traces of NH_3 are also formed in hydrogen production, and its presence even at the parts-per-million

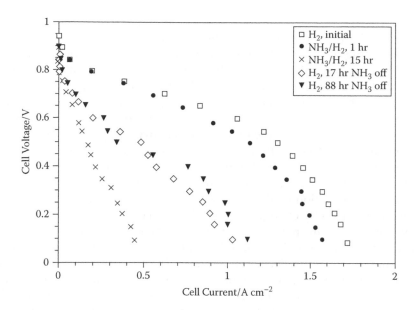

FIGURE 2.13

Polarization curves of a PEM fuel cell showing the effect of 30 ppm NH_3 injected into the hydrogen fuel stream for a total of 15 h. After this period, the cell continued operation with pure H_2 for an additional 88 h, showing the cell degradation by ammonia was irreversible. (Figure reproduced with permission from R. Borup et al., *Chem. Rev.* **107**, 3904 (2007).)

level can seriously degrade PEM fuel cell performance, as shown in Figure 2.13. Research has suggested that the primary degradation caused by ammonia is related to a loss of ionic conduction in the membrane and not to a poisoning of the fuel cell catalyst. The mechanism is still under study.

The sensitivity of PEM fuel cell operation to NH_3 poses a vitally important issue for solid-state hydrogen storage. As we shall see, a number of the more prominent high-capacity hydrogen storage materials are based on nitrogen, primarily because nitrogen is a relatively light element with very flexible chemical properties. The formation of NH_3 concomitant with the hydrogen release from these materials poses a substantial risk to fuel cell operation, in addition to representing a material loss problem for the hydrogen storage material itself.

Hydrocarbon impurities are common in reforming reactions. Methane has no catalyst poisoning effect. Other hydrocarbons, especially liquid hydrocarbons such as benzene, can physically smother the H_2 side (anode) of the fuel cell if present in large amounts. Overall, hydrocarbon contamination is much less a concern than for NH_3 contamination.

Whereas the contaminants CO and NH_3 can be present in the merchant hydrogen fueling the fuel cell, other impurities can be introduced from the air side of the system. These impurities include common air contaminants such as SO_2 and NO_x. Both NO_2 and SO_2 can degrade catalysis at the cathode. NO_2 has been shown to dramatically affect fuel cell operation, although the NO_2-based effects are largely reversible. The NO_2-induced degradation seems not to be due to catalyst poisoning; the mechanism is still not understood [25].

As water is the only reaction product during fuel cell operation, freezing of water is a concern for PEM fuel cell durability. It is not so much a problem during fuel cell operation as the 50% waste heat from the fuel cell keeps the unit above 0°C. However, if the PEM fuel cell is kept unpowered in a subfreezing environment and then an attempt is made to start it, there can potentially be problems if ice is blocking the gas flow channels. Also, repeated freezing and thawing of water in the fuel cell can lead to delamination of the electrocatalyst layer and breaking of carbon fibers in the GDL. The problems with freezing and thawing are mitigated by purging the fuel cell with hydrogen or nitrogen before shutdown or by vacuum drying the fuel cell. Alternatively, during startup, heat can be provided to the fuel cell from either a battery or catalytically combusting some of the hydrogen. Such heat can be used to heat the interior of the PEM fuel cell above 0°C prior to fuel cell startup.

Another "operational effect" on PEM fuel cells is voltage cycling. Voltage cycling changes the electrochemical environment surrounding all the materials comprising the fuel cell. Cycling-induced voltage effects can include dissolution of Pt in the acidic aqueous media of the fuel cell and oxidation of Pt. Any commercially viable fuel cell catalyst must be robust against the constantly changing electrochemical environment.

In some instances, certain regions of a fuel cell may not be receiving a proper supply of hydrogen. Perhaps a channel is blocked, or there is some other transport problem for hydrogen getting to the anode electrocatalyst layer. In this case, where there is gross local fuel starvation, there is insufficient current at the anode side from hydrogen oxidation, and the cell potential rises until some other oxidation occurs, such as oxidation of the carbon support in the catalyst layer to CO_2. This represents irreversible damage to the anode catalyst layer and needs to be avoided. In the event of hydrogen starvation, the fuel cell voltage can actually change sign, as shown in Figure 2.14.

Clearly, the design of metal electrode assemblies needs to ensure that flow channels have minimal chance of being obstructed, and that the delivery of reactant gases H_2 and O_2 to the fuel cell is as uniform as possible.

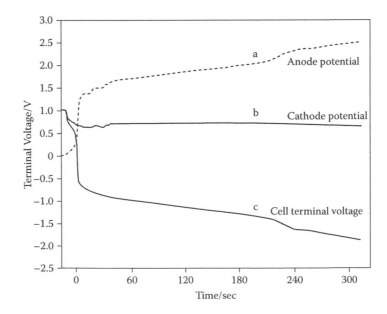

FIGURE 2.14
Electrode potentials of a PEM fuel cell driven to pass current after hydrogen flow is interrupted. (Figure reproduced with permission from R. Borup et al., *Chem. Rev.* **107**, 3904 (2007).)

Membrane Degradation

The membrane in a PEM fuel cell is typically a proton-conducting polymer film. The most successful of these has been Nafion, an example of a PFSA material developed by Dupont in the mid-1960s. Many PFSA membranes are commercially available from several membrane manufacturers. Figure 2.15 shows the chemical structure of the different PFSA materials Nafion, Flemion®, and Aciplex®.

These membranes suffer both chemical and physical degradation with fuel cell use. Chemical degradation arises from chemical attack by radicals generated by fuel cell operation. Physical degradation involves membrane creep, morphological changes, and microcrack and pinhole formation.

One of the principle routes of chemical degradation is caused by hydroperoxy radicals. It is believed the process starts with the permeation of O_2 through the membrane to the anode catalyst, forming the hydroperoxy radical: $Pt-H + O_2 \rightarrow {}^\bullet OOH$. The radical thus formed can react further with adsorbed hydrogen atoms at the Pt catalyst to form hydrogen peroxide $HOOH = H_2O_2$. Although PFSA membranes are generally stable against H_2O_2 attack, in the presence of trace metal ions such as Fe^{2+} and Cu^{2+} derived from the metal end plates, both ${}^\bullet OH$ and ${}^\bullet OOH$ radicals are formed, which dramatically degrades the PFSA membrane via formation of carbon radicals and subsequent disruption of the membrane structure. This radical attack has been the primary chemical stability concern for PEM fuel cell membranes [25].

Membrane manufacturers have implemented ingenious methods to protect against hydroperoxy attack. For Nafion, changing susceptible C-H bonds at end groups with more stable C-F moieties improves the chemical stability against radicals. Chemical degradation can also be improved by preventing the O_2 permeation in the first place. Several researchers have examined using cross-linking strategies within the membrane or increasing the membrane thickness to reduce O_2 crossover from the cathode side to the anode side. These approaches typically involve a trade-off between radical protection and an increase in the membrane electrical resistance, which lowers fuel cell performance [25].

The physical integrity of the PEM membrane needs to be sufficient to ensure stable long-term operation. However, in the normal operation of a fuel cell, the MEA is put under compressive stress between the bipolar places. Under this stress, the polymer membrane can suffer time-dependent deformation (creep), leading to membrane thinning and pinhole

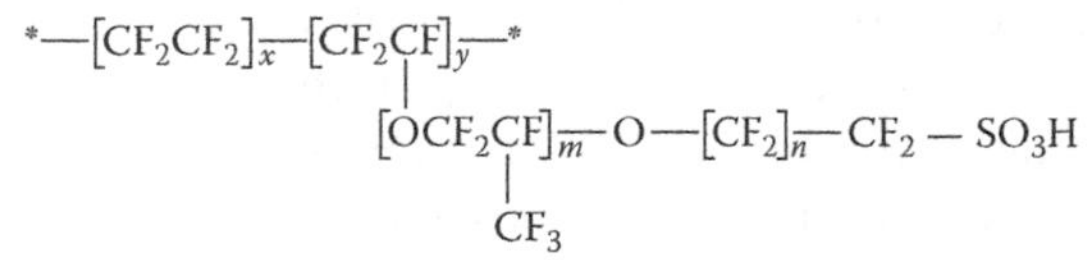

$$*-[CF_2CF_2]_{\overline{x}}[CF_2CF]_{\overline{y}}-*$$
$$[OCF_2CF]_{\overline{m}}-O-[CF_2]_{\overline{n}}-CF_2-SO_3H$$
$$| \quad CF_3$$

Nafion® ($m \geq 1$, $n = 2$, $x = 5\text{--}13.5$)
Flemion® ($m = 0, 1$, $n = 1\text{--}5$)
Aciplex® ($m = 0, 3$, $n = 2\text{--}5$, $x = 1.5\text{--}14$)

PFSA membranes

FIGURE 2.15
Structure of some PFSA membranes. (Figure reproduced with permission from R. Borup et al., *Chem. Rev.* **107**, 3904 (2007).)

formation. Pinhole formation represents catastrophic loss of fuel cell membrane integrity. Nafion creeps at a very slow rate, so thinning by this mechanism can take thousands of hours of operation.

In regions of the fuel cell where local stresses have built up, microcracks can form. The mechanical properties of the membranes can be improved with "reinforcing materials." W. L. Gore has used porous polytetrafluoroethylene (PTFE) fibrils to increase the durability of Nafion under mechanical stresses. Carbon nanotubes have also been investigated to reinforce the mechanical properties of Nafion [25].

Stability of the Catalyst

The degradation mechanism that has probably received the most R&D attention is preserving the physical and chemical integrity of the fuel cell catalyst, typically nanosize Pt particles. Pt is an expensive metal. As of August 13, 2012, the price of Pt was $1400 per ounce. In May 1999, the cost was $360 per ounce. Clearly, there is a commercial motivation to minimize the use of Pt in PEM fuel cells. The worldwide R&D effort on fuel cells has produced a remarkable optimization/reduction of the amount of Pt required for stable fuel cell operation. However, that quantity of Pt must also be stable against the chemical and physical changes occurring in the catalyst environment during fuel cell operation. Electrocatalyst stability is a prime determining factor in PEM fuel cell lifetime.

One catalyst degradation mechanism is dissolution of Pt. When a fuel cell is operated with little load, the output voltage is maximum. When the fuel cell is operated with significant load, the voltage will drop. Thus, a changing voltage occurs at both electrodes during practical fuel cell use. Research has shown that when the fuel cell voltage is cycling, that platinum particles can dissolve to some extent in the acidic aqueous media and deposit on other Pt particles (forming larger particles) or deposit on electrochemically inactive interior surfaces of the fuel cell. The exact mechanism of the dissolution depends on the extent to which Pt oxidizes in the fuel cell. Prior work has shown that Pt is in fact oxidized at the relevant potentials for fuel cell operation, and that the nascent Pt oxide can dissolve [25].

Another prominent change in the catalyst can be agglomeration/growth of the nanoparticles. The catalyst is typically prepared as nanosize particles to increase the effective surface area, thereby reducing Pt mass while maximizing catalytic activity. A great deal of work has been devoted to understanding the causes of catalyst particle growth during fuel cell use. Possible mechanisms for nanoparticle growth are coalescence of agglomerated particles and redeposition of dissolved Pt. The transmission electron microscope (TEM) has been a valuable tool in the unraveling of catalyst particle growth during fuel cell cycling. Figure 2.16 shows an example of Pt particle agglomeration observed after repeated high-voltage cycling to 1.2 V. As can be seen, catalyst particles that were previously small and distributed about carbon support nanoparticles (Figure 2.16a) are, after cycling, much larger and have moved into the ionomer region.

TEM also has allowed an assessment of the change in the catalyst particle size distribution with fuel cell operation. Such a study is shown in Figure 2.17, which compares Pt particle size distribution for a fresh PEM cathode with that of the same cathode having been cycled 1500 times from 0.1 to 1.2 V. The Pt size distribution becomes wider with repeated voltage cycles.

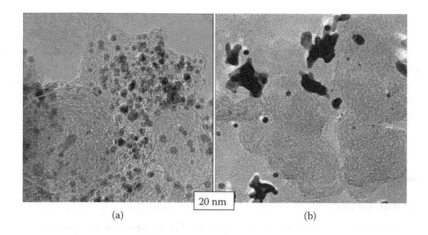

(a)   (b)

FIGURE 2.16
(a) TEM of Pt nanoparticle in a PEM fuel cell before (a) and after (b) potential cycling at 80°C to 1.2 V for 1500 cycles. (Figure reproduced with permission from R. Borup et al., *Chem. Rev.* **107**, 3904 (2007).)

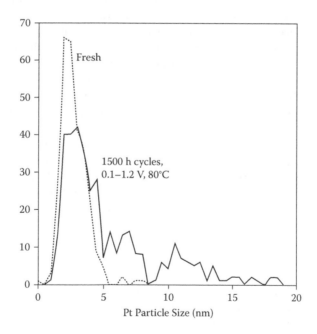

FIGURE 2.17
Pt particle size distributions comparing the Pt particles in a fresh PEM fuel cell cathode with Pt in the cathodes subjected to 0.1–1.2 V potential cycling for 1500 h, 80°C, and 50% relative humidity. (Figure reproduced with permission from R. Borup et al., *Chem. Rev.* **107**, 3904 (2007).)

It is important to note that in Figure 2.12 the anode and cathode electrode catalyst layer is comprised of catalyst nanoparticles supported by a carbon support, which anchors the Pt nanoparticle in place and provides electrical connection to the fuel cell electrodes themselves. This carbon support layer also must be robust against chemical and physical attack. Fundamental research has been conducted examining the chemical modifications of carbon in acidic aqueous media, as exists in the PEM fuel cell. The principle concern is oxidation of carbon to form CO_2, thereby leading to loss of carbon. The phenomenon was referred to here during conditions of H_2 fuel starvation. As carbon corrodes, it changes morphology, which as a practical matter changes the morphology of the catalyst being supported. A great deal of work has been conducted examining different types of catalyst support materials, including carbon nanotubes, inorganic oxides, silicon, conducting polymers of different types, diamond, and novel morphologies such as whiskers.

Gas Diffusion Layer

The GDL consists of two layers bonded together. These are a macroporous layer made of conductive carbon fibers and an MPL made of carbon particles and a PTFE binder. The MPL provides electrical connectivity and allows water to be removed from the catalyst sites. A properly functioning GDL is hydrophobic, repelling water so it can be transported away.

It is observed that the ability of the GDL to transport water out of the system slowly degrades with fuel cell use. It seems that the operational environment of the fuel cell changes the GDL material from the desired hydrophobic (water repelling) to hydrophilic (water attracting). There are also changes in the gas conduction within the GDL with fuel cell use. The mechanism for the change is that after prolonged exposure to oxygen at the cathode side of the fuel cell, carbon atoms in the GDL oxidize to form carboxyl groups or phenols, both of which are hydrophilic moieties, attracting water via hydrogen bonding. These chemical changes cause a gradual increase in cathode water uptake as the fuel cell is used [25].

Summarizing this work on understanding the degradation mechanisms, a great deal has been done to elucidate the fundamental operating chemical processes happening in a PEM fuel cell. In many cases, proprietary mitigation strategies have been developed that allow these fuel cells to operate for thousands of hours. Still, further improvements are needed in all aspects of fuel cell durability for it to match the durability standard set by the ICE, which has had 100 years of development.

One of the drivers for fuel cell R&D has been to reduce system cost, which often has focused on reducing the quantity of precious Pt catalyst in the system. There has been a remarkable reduction in the amount of Pt catalyst in PEM fuel cells, from about 2 mg/cm^2 to about 0.2 mg/cm^2, without significant sacrifice of fuel cell performance. Unfortunately, these gains have been largely offset by the increase in Pt cost the past 10 years. Other metals have been examined for use in fuel cells, particularly Pd- and Ru-based catalysts. The price of these metals on August 13, 2012 was about $580 per ounce for Pd and about $100 per ounce for Ru. The use of nonprecious metal catalysts for the oxygen reduction at the fuel cell anode has been examined. Unfortunately, many of these alternative catalysts showed poor stability in the acidic environment of the PEM fuel cell cathode.

Beyond the electrical power that an application needs, there are other electrical energy demands in a real fuel cell system that the hydrogen must provide. For example, in any real fuel cell system, there are power requirements for electronics to control the system, blowers for moving gases around, fans to blow cooling air through the unit (if an air-cooled fuel cell is deployed), power conditioning units, and other devices. The extent of the additional electrical demands is highly design dependent. These "balance-of-plant" (BOP)

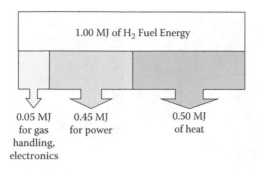

FIGURE 2.18
An example of a "Sankey" diagram for a PEM fuel cell. The diagram indicates the various energy flows in the system, as described in Reference [2].

energy demands add an additional approximately 5% energy requirement to the system on top of the load that the application actually demands. The energy balance situation in the fuel cell is shown in Figure 2.18, a Sankey diagram [2].

After reviewing all of the material degradation mechanisms, one might wonder if a PEM fuel cell could be made to work at all! Actually, all of these degradation mechanisms slowly degrade PEM fuel cell performance. Most of them can be brought under sufficient control so that commercially viable and robust fuel cells are now available. Figure 2.19 shows a picture of a commercial 5-kW PEM fuel cell from Altergy Systems.

This PEM fuel cell generates electricity with a thermal efficiency (electrical work out/ fuel energy in) of 47%. This is a large improvement over the thermal efficiency of diesel generators for equipment (typically 25%) and is somewhat larger than that attained by hydrogen ICE engine technology. Typically, 53% of the waste heat is discharged to the atmosphere using a cooling fan. This fuel cell receives hydrogen with an inlet pressure of about 2.5 bar.

PEM fuel cells, operating with waste-heat-cooling and electronic-cooling fans, are dramatically quieter than internal combustion technology. In a recent Fuel Cell Mobile Light project, sponsored by the DOE Fuel Cell Technologies (FCT) program Market Transformation activity, a 5-kW Fuel Cell Mobile Light system had a noise specification of

FIGURE 2.19
A 5-kW PEM fuel cell made by Altergy Systems. (Photo courtesy of Chris Radley and Mickey Oros, Altergy Systems.)

FIGURE 2.20 (See color insert.)
(Left) Storage of the Fuel Cell Mobile Light in winter conditions. (Right) Use of the Fuel Cell Mobile Light by Caltrans road crews in Kingvale, California, 10 miles west of Lake Tahoe, California, in the winter weather of March 2012. The unit is illuminating a Caltrans chain control checkpoint. (Photos courtesy of Wil White, UC Davis, and Tom Damberger, Golden State Energy.)

43 dB at 23 ft, compared to 73 dB for a comparable diesel-based system. This is a dramatic reduction in noise. Furthermore, since there is no combustion occurring, there is zero NO_x emission, zero SO_x, and zero particulate emission. The fuel cell in Figure 2.19 is certified as a zero-emissions power system by the CARB.

Fuel cell development has also allowed them to operate in subfreezing weather. Figure 2.20 shows the storage and operation of the Fuel Cell Mobile Light system by the California Department of Transportation (Caltrans) in the Sierra Mountains near Lake Tahoe, California. This particular Fuel Cell Mobile Light is outfitted with a proprietary "cold weather package" that allows starting the fuel cell down to –20°C. The PEM fuel cell technology has become quite robust. The Altergy PEM unit that lies at the heart of the Fuel Cell Mobile Light has performed flawlessly despite being towed (with very little suspension) on freeways for over 1100 miles and being used in subfreezing conditions. The Fuel Cell Mobile Light technology will be commercially available on the H_2LT light tower by Multiquip Inc. at the end of 2012.

Automotive OEM View of Fuel Cell Attractiveness

Since a great deal of the hydrogen storage work presented in this book originated from the problem of storing hydrogen on light-duty vehicles, we describe here the role that the automotive OEMs have played and their general perspective on PEM fuel cell technology.

The major world automotive companies have been involved with the development of PEM fuel cells since the early 1990s, just as they were involved with the development of the H_2 ICE engine discussed previously. This early work in the United States was sponsored by the DOE, and three U.S.-based programs emerged. Ford worked closely with United Technologies Corporation (UTC) on a direct hydrogen version [27]. Chrysler worked with Honeywell on a direct hydrogen version as well [28], and GM worked with Los Alamos National Lab in the development of a methanol reformer-based system [29]. In parallel to this, Daimler in Germany [29] as well as Toyota [29] and Honda [29] in Japan began programs in their respective countries. Informative presentations of the automotive fuel cell programs of GM, Toyota, Honda, Hyundai, Daimler/Mercedez-Benz, and Mazda can be found at the DOE Web site [29], documenting the 2009 Hydrogen and Fuel Cell Technical Advisory Committee Meeting held in November 2009.

Each of the automotive companies started its programs for different reasons. Some OEMs looked at the long-range implications of transitioning away from oil/gasoline; some looked at ensuring dominance in their respective markets with replacement technology for an aging petroleum-based ICE, and still others understood the fundamental limitations (energy density, recharging kinetics) of battery-only vehicles. Most OEMs have thought through impacts on global climate change, as discussed in Chapter 1 and the Editor's Introduction, and the consequences we face if we continue on the current path. Most of the OEMs realize the automobile's increasing contribution to global CO_2 as emerging nations develop. History has shown that developing nations will naturally want more and more of what the developed countries already have had.

The reality was a sum of many of these reasons plus the fact that various parts of the United States and world were imposing different emission regulations for different time periods in the futurew. It could only be imagined what various power trains and calibrations would be required to meet all of these different objectives in all of the petroleum-based OEM vehicle offerings. Simply put: The hydrogen PEM fuel cell used in an automobile, once all of the durability and cost challenges were addressed, would emit zero tail pipe emissions (some would argue that water will become an emission needing to be regulated at some point) and allow its fuel source to be generated from a number of sources and with a variety of processes, including the use of renewables.

A hydrogen-powered PEM fuel cell automobile is no longer being environmentally debated. It will be an on-the-road, zero-emission vehicle, giving the customer the expected and accustomed range and cold start performance, as well as having a life corresponding to known technologies once it hits its commercial introduction. Almost all major OEMs have some type of fuel cell electric vehicle (FCEV) fleet in operation—some for many years and with millions of miles accumulated as well as many thousands of H_2 refills, which are providing feedback on operation to improve future designs. Four such vehicles are shown in Figure 2.21 for the GM Equinox, Honda FCX Clarity, Mercedes-Benz B-Class F-CELL and Toyota FCHV-adv FCEVs.

FCEV technology will become routine to the customer with the addition (and this will take a collective will from each of the countries) of a suitable hydrogen supply infrastructure. Germany and Japan have already taken global leadership roles in hydrogen infrastructure development, supporting deployment of FCEVs in 2015.

As described in Chapter 1, the automotive community sees the future vehicle landscape including several technologies spread across the various market sectors as well as global regions. OEM long-term visions are for full electrification of the automobile. Clearly, battery-electric vehicles (BEVs) can be very small vehicles with limited range requirements. They will become a niche vehicle or a second car in multicar families at best from a practical perspective, considering range and recharge times. Hybrid and plug-in/hybrid vehicles that continue to utilize ICEs will still contribute to the CO_2 footprint for automobiles (to a greater extent with biofuels) and are on the path to full vehicle electrification. There is currently interest in using fuel cells for extending the range of BEVs, and in California, such range-extended vehicles are called BEV-x. Fuel cell vehicles are considered the ultimate vehicle because of the ability to meet all of the requirements for tail pipe emissions, diversity of hydrogen production routes, and the ability to meet customer expectations from vehicle-operating and performance perspectives.

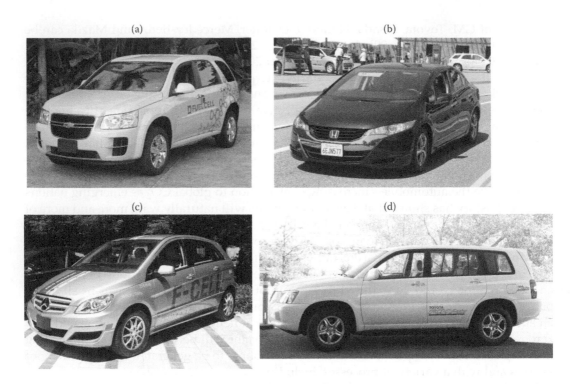

FIGURE 2.21
Near-commercial fuel cell electric vehicles: (a) GM Equinox; (b) Honda FCX Clarity; (c) Mercedes-Benz B-Class F-CELL; and (d) Toyota FCHV-adv. (Photo of the GM Equinox courtesy of GM. All other photos courtesy of the California Fuel Cell Partnership.)

The last decade has seen fundamental work focus on both durability improvements of the cell and developing simpler and more eloquent systems to surround the cell that can interface efficiently with electric propulsion systems. The simultaneous solving of these types of challenges involves materials, design, and systems/controls. All are required to meet stringent automotive requirements.

Future Generations

As PEM fuel cells begin initial penetration in the automotive market, the technology will begin its evolution in concert with electric propulsion systems and on-board hydrogen storage systems. One should look back at history in the automotive space and look at the advancements that have occurred over more than 100 years, going from hand cranks, downdraft carburetors, and manual transmissions to the sophisticated power trains of today. The engineers of those early mechanical horse replacement systems could not have imagined computer-controlled, fuel-injected, multivalve engines with eight-speed automatic transmissions that are on our roads today.

As the evolution begins, one can imagine improved membranes with better conductivity at higher temperatures promoting simpler operation during freezing conditions as well as in ambient environments. One could also imagine nonprecious metal catalyst systems that can lower cost and improve durability. Each of these types of improvements will help make the overall system simpler and more efficient. This in turn will drive costs lower and lower, which will further open markets for this technology that could not afford the high development costs of first-mover applications. Much like the use today of high-volume automotive engines for portable power generation (because of their low cost and high reliability) systems, one can envision the use of fuel cells in these types of markets world-wide—at very high volume.

Cell stack sealing designs that are compatible with −40°C and that can be applied in a high-volume fashion are clearly a challenge. This will be both a materials issue and one of design stackups over both life and temperature. The critical nature of sealing against hydrogen will drive innovative designs much like some of the multilayer automotive seals used in head gaskets and intake manifold sealing, for which there are material differences and a variety of fluids in contact over temperature ranges of −40°C to 120°C. Much work is ongoing to ensure the integrity of the stack as well as minimize leaching of any contaminants from the seal into the working cell. Some of this work goes back to the 1960s from a materials perspective, when die-cut seals were used in some designs. High-volume designs will require some type concept of forming or curing in place.

The MEAs shown in Figure 2.12 must be able to be mass manufactured if widespread deployment in automobiles is to be achieved. Roll-to-roll processing has driven designs to be compatible with this type of process. In turn, this type of process has enabled more repeatable performance and durability from the MEA as well as more consistent dimensional control of the electrodes. More work will continue in this area to improve yield and optimize performance, repeatability, reliability, as well as overall cell assembly ease.

Today, we are at the tipping point of the technology. Major automotive companies are investing hundreds of millions of dollars each, there are fleets of FCEVs around the globe, and a plan to support these vehicles with the required hydrogen infrastructure is in progress in the most forward-thinking places in the world. Commercial introduction is planned for 2015. Automotive history is already changing with the introductions of hybrid, plug-in hybrid, and-all electric vehicles. FCEVs will complete the transformation.

Summarizing this chapter, we have reviewed those energy conversion technologies that take the combustion energy inherent in Reaction 2.1 and converted that fuel energy to useful mechanical work or electricity. These technologies must in and of themselves support the three aims of fuel resource security, environmental sustainability, and political stability if our pressing energy and climate problems are to be resolved. The H_2 ICE, H_2 turbine, and H_2 PEM fuel cell technologies not only utilize carbon-free hydrogen, making them CO_2 free at the point of use, but also themselves offer dramatic advances in thermal efficiency and pollution control over their petroleum-fueled competitors. These energy conversion devices are the links between the hydrogen storage methods described in the remainder of this book and the energy goals we wish to achieve.

Acknowledgments

Lennie Klebanoff wishes to thank Nico Bouwkamp of the California Fuel Cell Partnership (CAFCP) for helpful conversations about FCEVs, and Mike Veenstra of Ford for communications about H_2 ICE vehicles.

References

1. C.M. White, R.R. Steeper, and A.E. Lutz, *Int. J. Hydrogen Energy* **31**, 1292 (2006).
2. J. Larminie and A. Dicks, in *Fuel Cell Systems Explained*, Wiley, New York, 2000.
3. R.A. Erren, Internal combustion engine using hydrogen as a fuel, U.S. Patent No. 2,183,674, December 19, 1939.
4. L.M. Das, *Int. J. Hydrogen Energy* **15**, 425 (1990).
5. M. Berckmuller, H. Rottengruber, A. Eder, N. Brehm, G. Elsasser, and G. Muller-Alander, SAE Paper 2003: 2003-01-3210.
6. R.J. Natkin, X. Tang, B. Boyer, B. Oltmans, A. Denlinger, and J.W. Heffel, SAE Paper 2003: 2003-01-0631.
7. A.K. Jaura, W. Ortmann, R. Stuntz, R.J. Natkin, and T. Grabowski, SAE Paper 2004: 2004-01-0058.
8. T. Wallner, H. Hohse-Busch, S. Gurski, M. Duoba, W. Thiel, D. Mailen, and T. Korn, *Int. J. Hydrogen Energy* **33**, 7607 (2008).
9. X. Tang, D.M. Kabat, R.J. Natkin, W.F Stockhausen, and J. Heffel, SAE Paper 2002: 2002-01-0242.
10. G.D. Brewer, in *Hydrogen Aircraft Technology*, CRC Press, Boca Raton, FL, 1992.
11. D. Daggett, O. Hadaller, R. Hendricks, and R. Wather, NASA Report NASA/TM-2006-214365.
12. S. Marquart, R. Sausen, M. Ponater, and V. Grewe, *Aerosp. Sci. Technol.* **5**, 73 (2001).
13. A. Silverstein and E.W. Hall, NACA RM E55C28a, NACA Lewis Flight Propulsion Lab, 1955.
14. The Pratt and Whitney work is described in the very readable account *Advanced Engine Development at Pratt and Whitney, the Inside Story of Eight Special Projects 1946–1971*, by Dick Mulready, SAE Press, Dedham, MA, 2001.
15. A brief summary of the Russian Tu-155 LH2 aircraft project can be found at http://www.tupolev.ru/english/Show.asp?SectionPubID=1782.
16. For a description of the FutureGen Program, see http://www.futuregenalliance.org/the-alliance/.
17. For a description of coal gasification, see http://www.fossil.energy.gov/programs/powersystems/gasification/index.html.
18. For a description of turbine operation, see http://www.fossil.energy.gov/programs/powersystems/turbines/turbines_howitworks.html.
19. See http://www.dlr.de/blogs/en/desktopdefault.aspx/tabid-6192/10184_read-255/.
20. P. Chiesa, G. Lozza, and L. Mazzocchi, *J. Eng. Gas Turbines Power* **127**, 73 (2005).
21. Current scramjet work is focused on the XP51a, which attempts to use hydrocarbon fuels at speeds to Mach 6. http://www.theregister.co.uk/2011/06/16/x51a_second_test_notsogood/.
22. E. Bancalari, P. Chan, and I.S. Diakonchak, *Advanced Hydrogen Turbine Development*, http://www.energy.siemens.com/co/pool/hq/energy-topics/pdfs/en/igcc/6_Avanced_Hydrogen.pdf.
23. A.E. Lutz, R.S. Larson, and J.O. Keller, *Int. J. Hydrogen Energy* **27**, 1103 (2002).
24. S.E. Wright, *Renewable Energy* **29**, 179 (2004).
25. R. Borup et al., *Chem. Rev.* **107**, 3904 (2007).
26. The DOE fuel cell technical targets can be found toward the bottom of the following Web page: http://www1.eere.energy.gov/hydrogenandfuelcells/fuelcells/fc_challenges.html.
27. For an account of Ford activities, see http://green.autoblog.com/2008/08/19/ford-expands-fuel-cell-test-fleet-tests-by-two-years/.
28. For an account, see http://inventors.about.com/library/inventors/blfuelcells3.htm.
29. http://www.hydrogen.energy.gov/htac_meeting_nov09.html.

Section II

Hydrogen Storage Materials and Technologies

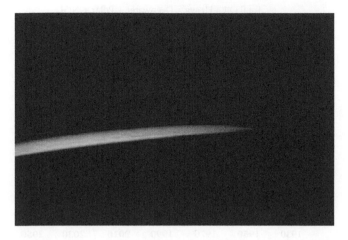

COLOR FIGURE I.1
The Earth's atmosphere photographed from space. (Photo courtesy of NASA.)

COLOR FIGURE I.2
The Earth photographed from the moon, *Apollo 8*, 1968. (Photo courtesy of NASA.)

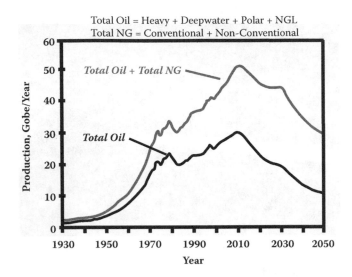

COLOR FIGURE 1.1
Production predictions for total world oil and total world oil + total world NG. (Reproduced with permission from Reference [8].)

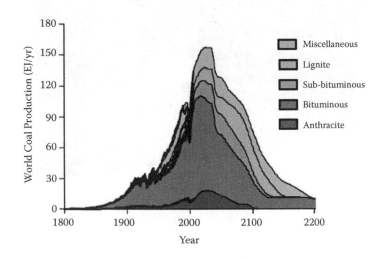

COLOR FIGURE 1.3
World coal production prediction for different coal types in EJ/y. (Reproduced with permission from Reference [10].)

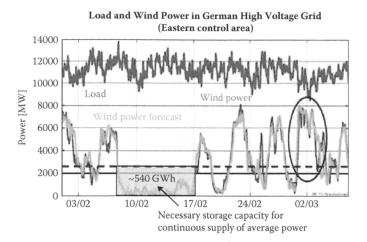

COLOR FIGURE 1.6

Load and wind power in the German high voltage grid indicating the energy storage required to make wind a dispatchable electric resource. The horizontal axis shows 2007 dates in Febuary (02) and March (03). (From Reference [54].)

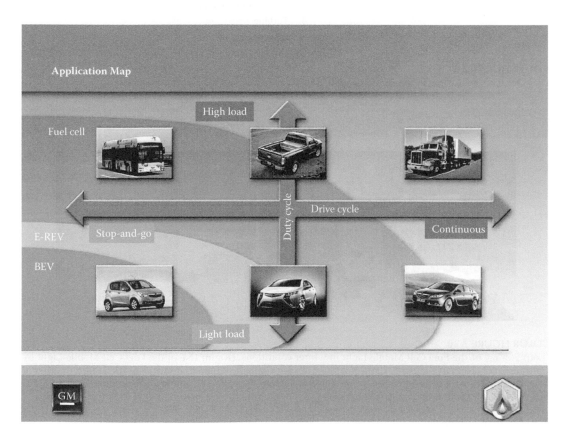

COLOR FIGURE 1.9

Road map for vehicles for GM as of 2010. (From R.C. Kauling, Electrification of the Vehicle: Critical Role of On-Board Hydrogen Storage Continues, Hydrogen Storage Materials Workshop, Ottawa, Canada, April 28, 2010. Reproduced with permission from General Motors.)

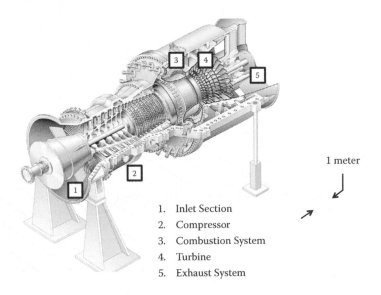

1 meter

1. Inlet Section
2. Compressor
3. Combustion System
4. Turbine
5. Exhaust System

COLOR FIGURE 2.8
Diagram of a gas turbine for power generation. (Cut-away turbine image courtesy of Siemens.)

COLOR FIGURE 2.20
(Left) Storage of the Fuel Cell Mobile Light in winter conditions. (Right) Use of the Fuel Cell Mobile Light by Caltrans road crews in Kingvale, California, 10 miles west of Lake Tahoe, California, in the winter weather of March 2012. The unit is illuminating a Caltrans chain control checkpoint. (Photos courtesy of Wil White, UC Davis, and Tom Damberger, Golden State Energy.)

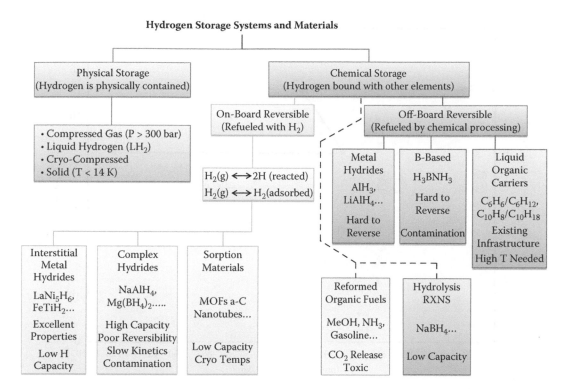

COLOR FIGURE 3.1
Overview of hydrogen storage systems and materials.

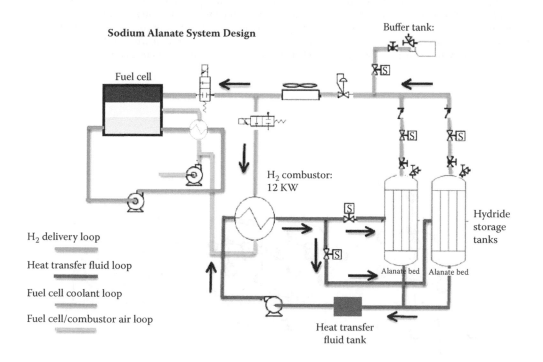

COLOR FIGURE 3.7
Schematic configuration for a reversible hydrogen storage system in fuel-cell-powered passenger cars that uses NaAlH$_4$ as the storage medium. (Figure from Pasini, J.M., B.A. van Hassel, D.A. Mosher, and M.J. Veenstra. 2012. System modeling methodology and analyses for materials-based hydrogen storage. *Int. J. Hydrogen Energy* **37**: 2874–2884 with permission of Elsevier.)

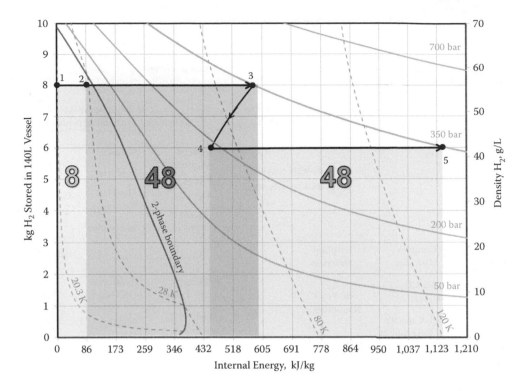

COLOR FIGURE 4.4
Phase diagram for H_2 showing density (right vertical axis) and internal energy (horizontal axis), with lines for constant pressure (solid) and temperature (dashed). Curves for entropy are not shown. A left axis shows the mass of H_2 contained in a vessel with 140-L internal volume, which would store 10 kg of LH_2 at 20 K and 1 bar. The figure also shows points and areas representing dormancy (in watt-days) of conventional LH_2 tanks (light green) and cryogenic pressure vessels (all shadings). Thermodynamic states that are "inside" (to the left of) the 2-phase boundary line consist of two phases; whereas states to the right of this line consist of one supercritical phase.

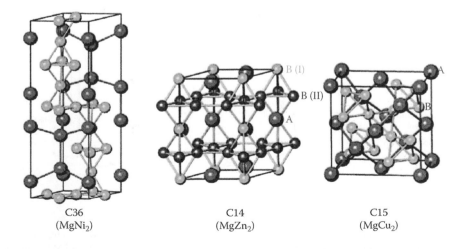

COLOR FIGURE 5.6
Schematics of three Laves structures in AB_2-type intermetallic compounds.

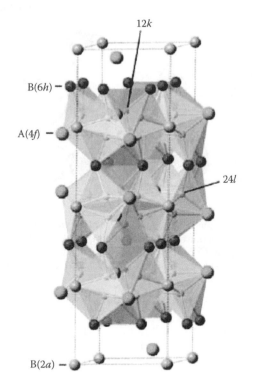

COLOR FIGURE 5.8
A schematic diagram of C14 structure indicating the two preferred [A_2B_2] sites for hydrogen. (Reproduced from I. Levin, V. Krayzman, C. Chiu, K.-W. Moon, and L.A. Bendersky, *Acta Mater.* 60, 645 (2012) with permission of Elsevier.)

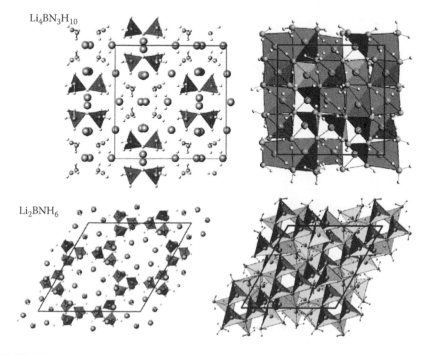

COLOR FIGURE 6.22
Crystal structure of $Li_4BN_3H_{10}$ and Li_2BNH_6. (Reprinted from H. Wu, W. Zhou, T.J. Udovic, J.J. Rush, T. Yildirim, *Chemistry of Materials*, 20 (2008) 1245–1247 with permission of American Chemical Society.)

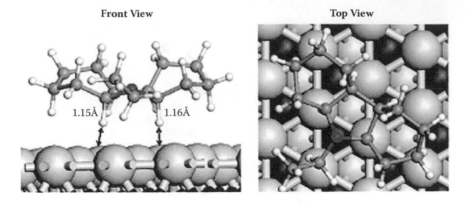

Front View   **Top View**

COLOR FIGURE 8.41
Calculated optimized adsorption geometries of a molecule of N-ethyldodecahydrocarbazole lying "flat" over a Pd(111) surface. Displayed is the interaction of the two H atoms of the C-H bonds in the central ring that are adjacent to nitrogen (blue). White, H; gray, C; light gray, first layer of Pd atoms; and black, second layer of Pd atoms. (Reproduced and adapted from Sotoodeh, F.; Smith, K.J.; *J. Catal.* 2011, *279*, 36, with permission.)

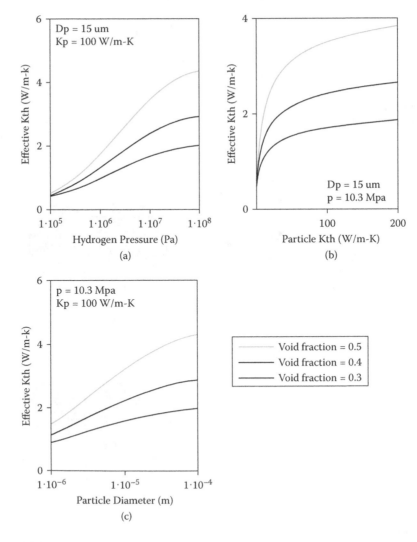

COLOR FIGURE 9.3
Calculated effective thermal conductivity of a packed particle bed as a function of void fraction, hydrogen pressure, and particle characteristics.

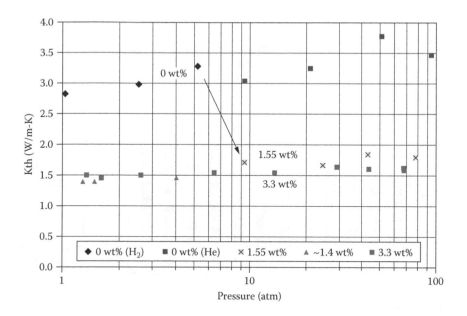

COLOR FIGURE 9.9

Effective thermal conductivity of a fully cycled NaAlH$_4$ bed, with 12% by mass added Al, as a function of gas pressure and H$_2$ capacity.

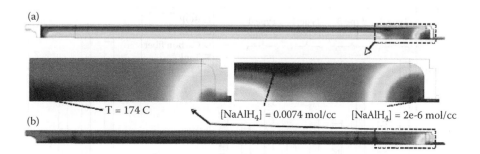

COLOR FIGURE 10.14

Two-dimensional axisymmetric model of GM/Sandia vessel containing sodium alanate: (a) phase distribution and (b) temperature distribution. The dotted regions in each distribution of (a) and (b) are blown up in the middle figure. The figure shows a snapshot in time during the second step of a hydrogen absorption simulation. Figure 10.14(a) shows the instantaneous concentration of NaAlH$_4$ and Figure 10.14(b) shows the temperature distribution. Note that the hydrogen gas cools the bed at the entrance on the right-hand side while the absorption reaction heats the bed producing peak temperatures along the axis. The temperature gradients produce concentration gradients as hydrogen is absorbed at different rates depending on local temperature.

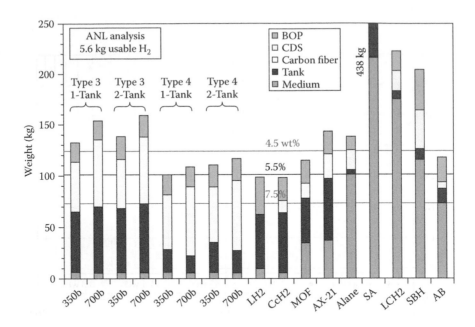

COLOR FIGURE 10.18
System weight for 5.6-kg H_2 storage for various storage options. Definitions: SA—sodium alanate, LCH2—liquid organic carrier, SBH—sodium borohydride, AB—ammonia borane. (Reproduced from R.K. Ahluwalia, T.Q. Hua, J.K. Peng, On-board and off-board performance of hydrogen storage options for light-duty vehicles, *Int. J. Hydrogen Energy* **37**, 2891 (2012) with permission of Elsevier.)

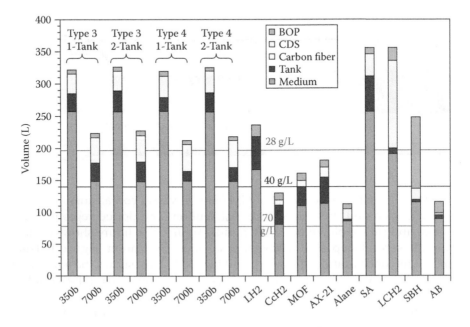

COLOR FIGURE 10.19
System volume for 5.6-kg H_2 storage for various storage options. (Reproduced from R.K. Ahluwalia, T.Q. Hua, J.K. Peng, On-board and off-board performance of hydrogen storage options for light-duty vehicles, *Int. J. Hydrogen Energy* **37**, 2891 (2012) with permission of Elsevier.)

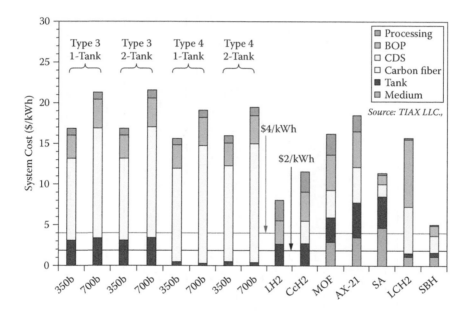

COLOR FIGURE 10.20
Projected cost of hydrogen storage systems at high-volume manufacturing. (Reproduced from R.K. Ahluwalia, T.Q. Hua, J.K. Peng, On-board and off-board performance of hydrogen storage options for light-duty vehicles, *Int. J. Hydrogen Energy* **37**, 2891 (2012) with permission of Elsevier.)

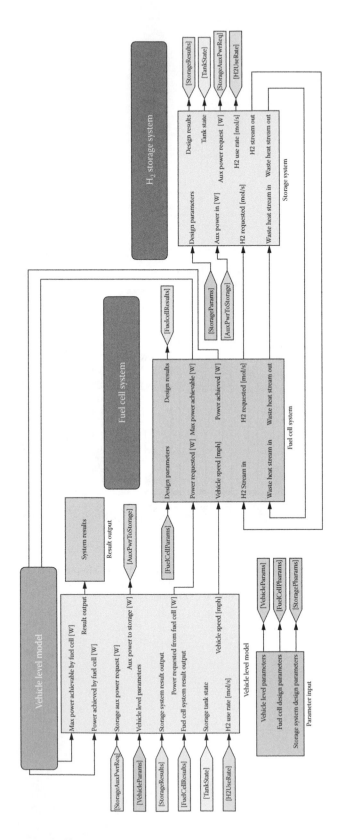

COLOR FIGURE 11.1

Integrated system model: hierarchy of modules that was developed and utilized by the Hydrogen Storage Engineering Center of Excellence [HSECoE] [Reference 20].

3

Historical Perspectives on Hydrogen, Its Storage, and Its Applications

Bob Bowman and Lennie Klebanoff

CONTENTS

Introduction 65
Compressed Gas Hydrogen Storage 67
Cryogenic Hydrogen Storage Systems 69
On-Board Reversible Hydrogen Storage Materials and Systems 74
    Interstitial Metal Hydrides... 74
    Complex Metal Hydrides.. 77
    Adsorption Materials.. 78
Off-Board Reversible Hydrogen Storage Materials and Systems 80
    Liquid Organic Hydrogen Carriers 81
    Hydrolysis Reactions 81
    Metal Hydrides and Ammonia Borane... 84
Concluding Remarks.. 84
Acknowledgments.. 85
References... 85

Introduction

Hydrogen (H_2) possesses a number of chemical and physical properties that have allowed it to be used widely in the synthesis and processing of materials in research laboratories as well as in industrial manufacturing. These uses are in addition to the emerging use of hydrogen in power generation technology, including transportation, as discussed in Chapter 2. Except for situations such as in the petrochemical industry, in which hydrogen is utilized more or less immediately after its production from a hydrocarbon feedstock, most applications require that hydrogen be stored after its production by some method prior to its eventual endpoint usage. Many factors influence the physical state (i.e., gas, liquid, solid, or contained in a chemical substance) in which hydrogen will be stored and how it will be supplied for a specific application.

In those situations where hydrogen has been a well-established commodity, the physical state is well understood, and storage systems have been developed over the years. For example, hydrogen is often used as a reactant in laboratory research or in manufacturing. In these applications, compressed hydrogen gas is deployed in appropriate size cylinders at pressures up to about 200–300 bar. In addition, when hydrogen is used in a unique and critical capability (e.g., as a rocket propellant for launching spacecraft), then cryogenic liquid hydrogen (LH_2) is used and stored in cryogenic containment vessels as part of the launch vehicle. On the other hand, whenever hydrogen is to be used in a role that would either replace existing technologies or serve a novel function, the current storage approaches generally must be heavily adapted, or new ones developed, to meet the different operating parameters and constraints.

Chapter 1 describes the increasing concerns over the energetic, political, and environmental sustainability of our current energy supplies. Hydrogen has often been proposed [1] as an attractive candidate for the primary fuel carrier for vehicles and in numerous other applications that currently utilize batteries or generators powered by the combustion of fossil fuels (e.g., backup or remote power sources, mobile electronics, power tools, etc.). As described in Chapter 2, the combination of hydrogen with both internal combustion engines (ICEs) and fuel cell power systems offers the potential of matching or even extending the performance capabilities of the current options, providing that appropriate and cost-effective hydrogen storage solutions are also developed.

Figure 3.1 lists the various physical states of hydrogen (solid, liquid, gas, chemical compound) that can, in principle, be used for hydrogen storage. In the broadest sense, hydrogen can be contained either as a diatomic molecule (i.e., H_2) via physical constraints (i.e., in some kind of vessel) or as monatomic hydrogen (i.e., H atom) reacted and bonded with other elements in the form of chemical compounds or materials. Ideally, these hydrogen storage materials would be "reversible." By reversible, we mean that hydrogen gas can be directly used to reestablish the original hydrogen storage material after the original (stored) hydrogen is used. Reversibility typically implies conditions of temperature and pressure such that the material can be regenerated "on-board" the vehicle. These are termed *on-board-reversible* materials. Another set of the chemical compounds in Figure 3.1 is "irreversible" in that regeneration of the initial H-containing phase or compound cannot be achieved on-board the vehicle or "in place" for a nonvehicular application. Reversibility of these materials can only be achieved by excessive pressure/temperature conditions [2] requiring removal of the material from the vehicle, followed by processing via an industrial-scale process. Such materials are considered "off-board-reversible" materials. This book discusses recent advancements in both on-board-reversible and off-board-reversible hydrogen storage materials.

There is a long history of the synthesis and characterization of materials with large hydrogen contents that theoretically could act as hydrogen storage media [3–18]. These investigations have substantially intensified over the past decade as a greater emphasis has been placed on hydrogen-powered vehicles as a way of reducing greenhouse gas (GHG) emissions. The automobile application is a particularly challenging setting for the use of various hydrogen storage approaches due to the volume and weight constraints. A number of chapters in this book give extensive descriptions of the scientific and engineering investigation of materials being explored and considered as hydrogen storage media.

In this chapter, we give a broad overview of the types of hydrogen storage approaches that have been investigated in the past and the interesting applications that have needed hydrogen storage. We cover some of the key generic material properties and characteristics having an impact on the configuration and performance of the various types of hydrogen storage systems discussed and focus on materials that can directly supply molecular hydrogen (i.e., H_2) for immediate usage. This book does not consider those materials requiring additional processing to release molecular H_2. Examples of systems not covered include those that store and release ammonia (NH_3) for downstream decomposition to form H_2 or the reformation of hydrocarbons (e.g., gasoline) to produce hydrogen.

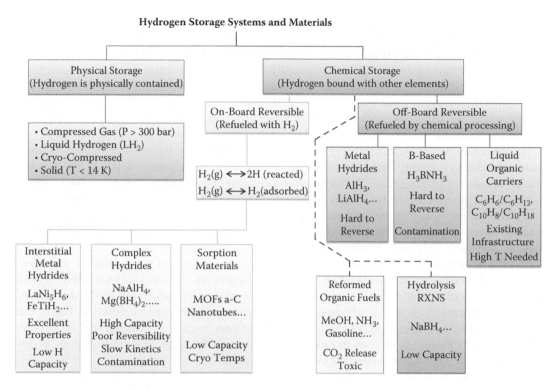

FIGURE 3.1 (See color insert.)
Overview of hydrogen storage systems and materials.

We begin with those storage systems that physically contain molecular hydrogen as a gas, cryogenic liquid, or cryogenic solid to provide historical perspectives on the roles, advantages, and limitations of each configuration. In Chapter 4, Guillaume Petitpas and Salvador Aceves describe the development of high-pressure vessels (both gaseous and cryogenic) for storing hydrogen from a more thermodynamic perspective. In fact, compressed gas and cryogenic liquid have been the most common methods historically of storing molecular hydrogen, and they continue to be the dominant choices due to the existing infrastructures and commercial availability. For decades, the development of chemical- and material-based storage methods has been explored to overcome the fundamental limitations of the physical storage methods, primarily volumetric density, gas pressure requirements, or the necessary cryogenic conditions. In many cases, these hydrogen storage materials also provided additional capabilities for specialized or niche applications.

Compressed Gas Hydrogen Storage

In the Editor's Introduction, Klebanoff has discussed some of the general properties of hydrogen. Hydrogen is a gas at essentially all normal use and storage temperatures. Hydrogen is the lightest of all elements with a very low normal density of 0.09 g/L at 288 K and 1.0 bar. With a lower heating value (LHV) of 120.9 kJ/g, hydrogen has the highest specific energy of any known fuel, making it highly applicable to weight-sensitive applications such as aircraft and spacecraft. Liquid hydrogen has a normal boiling temperature of about 20 K and a critical temperature of approximately 33 K, above which LH_2 cannot be formed through the application of pressure. To overcome the low volumetric density of normal hydrogen gas, it is usually compressed to high pressures. Merchant hydrogen gas is typically delivered at 150 to 300 bar [19]. Hydrogen has been stored and transported as a compressed gas in metal cylinders since the late 19th century, when the British used wrought-iron metal cylinders weighing in excess of 500 kg for transporting hydrogen to inflate balloons during military expeditions across Asia and Africa [20].

Hydrogen exhibits significant deviation from ideal gas law behavior at elevated pressure, and the gas density increases much more slowly than the pressure [1]. At 350 bar, the molar volume of hydrogen is 22% larger than predicted by the ideal behavior due to intermolecular repulsion of the hydrogen molecules. As a result, it takes more pressure-volume work to compress it to a desired density as the pressure increases. Storing gas at high pressures requires robust pressure vessels, which incur significant weight and cost penalties. In addition, repeated cycling from low to high pressure stresses the materials used in hydrogen tanks. The design, manufacture, transport, and use of gas cylinders and pressure vessels are typically regulated by government agencies. The regulations often require compliance with specific design, manufacture, or use codes and standards developed by organizations such as the International Organization for Standardization (ISO), the Compressed Gas Association (CGA), and the American Society of Mechanical Engineers (ASME). Chris Sloane discusses in much more detail in Chapter 12 the hydrogen "codes and standards" that have developed over the past years.

Historically, there have been four standard types of cylinders developed and used for the transport and storage of hydrogen: Type I, all-metal cylinders; Type II, hoop-wrapped composite cylinders; Type III, fully wrapped composite cylinders with metallic liners; and Type IV, fully wrapped composite cylinders with non-load-bearing nonmetallic liners [1]. Type I steel or aluminum cylinders are the most common type found in use for merchant hydrogen delivery and storage. Metals selected for these vessels must not permit hydrogen permeation or be subject to hydrogen embrittlement, especially when their use involves extensive pressure or temperature cycling. The cylinders are designed for a maximum working pressure, with the minimum wall thickness determined by the metal's yield and tensile strength [1]. Because the mass of the metal is substantial, the mass of hydrogen stored is typically only about 1% of the cylinder mass and will drop to less than 1% at pressures of 350 bar and higher as the tank walls need to be thicker to hold back the pressure.

For automotive applications [21], weight and volume constraints make Type I cylinders impractical. However, they are suitable for many stationary applications in laboratory or manufacturing facilities as well as some types of specialty and utility vehicles (such as forklifts), for which excess weight is not a concern. In Type III and IV cylinders, thin, lightweight metal or nonmetallic liners, respectively, are wrapped by a fiber/epoxy matrix. Aluminum metal is commonly used for the liner as it cannot be embrittled by hydrogen. The fiber wrapping supplies the strength to contain the high-pressure gas, while the liner primarily acts as a gas permeation barrier. When high-tensile-strength carbon fiber is used to provide the strength for 350- and 700-bar cylinders, recent analyses [22] indicated that hydrogen system capacities of 5.9% and 4.7% by weight, respectively, are achievable. Although the carbon fiber wrappings do reduce the weight of the cylinders dramatically from the all-metal variety, they add significant cost. This cost is currently projected to be nearly three-quarters of the storage vessel cost, taking into account high-volume manufacturing methodology [22].

The most severe limitation of compressed gas storage systems (especially for any transportation application) is the overall volume occupied by the tank itself. The density of H_2 gas at room temperature is 23 g/L and 39 g/L at 350 bar and 700 bar, respectively. Thus, to store 5 kg of hydrogen on-board a light-duty hydrogen-powered vehicle, minimum volumes of 217 L and 128 L are needed just to accommodate the gas volumes at 350 bar and 700 bar, respectively. In reality, the volume must be even greater since the additional volumes of the cylinders and balance-of-plant components (i.e., supply lines, valves, pressure regulators, sensors, etc.) are required. The 5 kg of hydrogen is the estimated quantity a fuel-cell-powered vehicle needs for a 300-mile driving range [19].

The use of higher pressures will severely impact weight and cost of the vessel as well as decrease efficiency for filling the tank. Nevertheless, types III and IV storage systems have been extensively used in demonstration assessments of prototype automobiles due mainly to their commercial availability, manageable cost, and straightforward means of installation into the vehicles and refilling at the test facilities [23, 24]. To fit the compressed H_2 storage systems into these vehicles with minimal impact on the passenger spaces, they are often separated into multiple cylinders located in trunks, under seats, or in spare wheel compartments. Compressed gas tanks have also been used in larger demonstration vehicles, including trucks, buses, and even a switch locomotive [25–27]. Type III tanks are being used in the Fuel Cell Mobile Light discussed in Chapter 2.

While the compressed gas tanks are put into the beds of pickup trucks, they are often located just below roofs of buses (i.e., above and outside the passenger compartment) as well as in the fuel cell hybrid locomotive as shown in Figure 3.2. To operate the nominal 250-kW PEM (proton exchange membrane) fuel cell in the locomotive, 70 kg of H_2 gas was stored in 14 carbon-wrapped tanks as indicated in Figure 3.2b [26]. During demonstration tests in an urban railyard, the mean time between required refueling of the locomotive was reported to be 11 h [27].

Cryogenic Hydrogen Storage Systems

The most obvious first approach to increasing the volumetric storage density of hydrogen is to change the physical state of hydrogen from gas to liquid. James Dewar first liquefied molecular hydrogen in 1898. During the first half of the 20th century only relatively small quantities of liquid hydrogen (LH_2) were produced and stored, mainly to conduct and support fundamental research activities. Commercial-scale production of LH_2 was stimulated in the 1950s by the need of very large quantities of liquid hydrogen for the launching of spacecraft [28, 29]. Extensive efforts were implemented to enhance the exothermic conversion rate of ortho-H_2 to para-H_2 to greater than 95% during the liquefaction process [28, 29] to contain the costs of hydrogen liquefaction and minimize evaporative losses of this low-boiling-temperature (i.e., ~20 K) cryogenic liquid.

The international development of launch vehicles containing one or more LH_2 and liquid oxygen (LO_2) stages has led to facilities being constructed at locations in several countries to produce, store, and transfer the large quantities of liquid hydrogen consumed during launches of spacecraft for military, communications, and space exploration missions [30]. As an example, the three engines on the NASA Space Shuttle orbiters burned up to 230,000 kg of hydrogen during each launch. Figure 3.3 shows the Space Shuttle vehicle

(a)

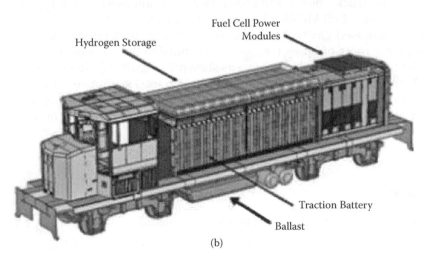

(b)

FIGURE 3.2
Hydrogen fuel cell locomotive designed and built by Vehicle Projects, Inc. from Reference [26]: (a) Photograph
and (b) diagram showing hydrogen storage in 14 cylinders of compressed gas.

with the orbital and the external LH_2/LO_2 storage tank, which is the only component of the
launch system that is not reused.

The density of liquid hydrogen at its 20 K boiling point is 71 g/L, which is nearly twice
that of 700-bar compressed gas at room temperature. Still, LH_2 is itself a low-density liquid.
For context, the density of water is 1000 g/L, and the mass density of hydrogen in water
is 111 g hydrogen per liter of water. The storage of LH_2 requires highly specialized and
sophisticated vessels that can minimize heat transfers from thermal conduction, thermal
convection, and thermal radiation while exhibiting mechanical robustness throughout
their operational life. The cryogenic storage vessels must also safely manage the release
of the evaporated gas (i.e., boil-off) [20, 28, 31] due to heating from residual thermal leaks
or the ortho-para conversion. Evaporative loss is exacerbated by the very low enthalpy of

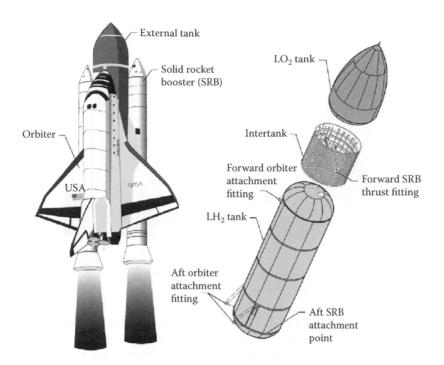

FIGURE 3.3
Schematic of Space Shuttle spacecraft highlighting external tank that contains the liquid hydrogen and oxygen storage systems supplying these propellants to the three main engines in the *Orbiter*. The solid rocket boosters are also shown. (From Bowman, R.C., Jr. 2006. Roles of hydrogen in space explorations. In *Hydrogen in Matter*, ed. G.R. Myneni, B. Hjorvarsson, 175–199. New York: American Institute of Physics. With permission of American Institute of Physics.)

vaporization ΔH_{vap} of LH_2. LH_2 has a ΔH_{vap} value of only 0.92 kJ/mole, 9.2 times less than that of liquid natural gas (LNG), whose $\Delta H_{vap} = 8.5 \dfrac{kJ}{mole}$. In other words, if you look at LH_2 cross-eyed, it will evaporate.

Nearly all LH_2 storage vessels use metallic double-walled containers that are evacuated and contain multiple layers of alternating metallic and thermally insulated polymeric or glass films to reduce heat leaks to the cryogenic fluid via convection and radiation. Designs and materials used to construct the containers and all of the components are chosen to minimize thermal conduction. Klell [20] stated that most current LH_2 tanks have evaporation losses between 0.3% and 3% per day. One innovative vessel design used circulated liquid air formed when liquid hydrogen is delivered from the inner vessel to cool its wall and can extend storage times by up to a factor of about 4 [20]. Furthermore, the addition of activated cryocoolers within the LH_2 storage vessel is being developed for zero boil-off (i.e., no H_2 gas venting) for long-duration space flights [32]. This last approach is probably not an economical solution for most terrestrial applications, such as storage for hydrogen-powered vehicles.

The first space-qualified liquid hydrogen and oxygen storage vessels that operated in near-zero gravity environments were developed for the NASA Apollo missions [30]. These vessels held the reactants for the alkaline fuel cells that provided the electrical power in the Apollo Command/Service Modules (CSM) and generated potable water as a by-product. These tanks supplied H_2 and O_2 at 4.1 bar to the fuel cells, which were operated at nominally 483 K, forming extremely pure water saturated with hydrogen gas. An Apollo LH_2

FIGURE 3.4
Photograph of a liquid hydrogen storage vessel from the NASA Apollo Command/Service Modules (CSMs) showing the external thermal insulating blanket.

storage vessel is shown in Figure 3.4. There were three separate vessels of each cryogen; the total stored quantity of LH_2 was 36.5 kg at liftoff. During a normal lunar mission, the fuel cells would supply about 650 kWh of energy with a generation rate for potable water of about 0.54 kg/h [30].

More recently, the Space Shuttle orbiters used liquid hydrogen and oxygen to provide electrical power and potable water during these missions [30]. In this case, the LH_2 and LO_2 were both stored in vacuum-insulated spherical tanks under the payload bay liner of the orbiters. The hydrogen vessels were constructed of aluminum alloy 2219 with an outside diameter of 1.16 m and total volume of 606 L. Each of these tanks has an empty mass of 98 kg and stores about 42 kg hydrogen at an initial temperature of 22 K. H_2 gas is supplied to the alkaline fuel cell power plants at pressures between 13.8 and 15.4 bar [30].

Liquid H_2 storage has remained the primary option for proposed aviation systems due to its much greater effective storage capacity per volume and mass compared to compressed gas [28, 33, 34]. As described in Chapter 2, aviation interest in hydrogen drove the development of hydrogen-burning gas turbines starting in the 1950s. Covering work up to about 1990, Peschka [28] gives a detailed description of the potential and issues of hydrogen as an aviation fuel. Included in this review are brief descriptions of flight tests performed in the United States and former Soviet Union on jet engines and aircraft running on hydrogen with LH_2 storage. More recent assessments and conceptual designs are still being explored to incorporate LH_2 storage tanks for hydrogen-fueled aircraft [33, 34].

Starting in the early 1970s, liquid hydrogen storage was used in several prototype ground vehicles operating with ICEs. Development took place in the United States, Japan, and Germany, where the most sustained efforts were by the BMW Corporation [31]. The designs and configurations of these LH_2 storage vessels were adapted from aerospace technology but had been significantly modified for accommodation in buses and passenger cars [31]. The capacities of these tanks were mostly between 7 and 11 kg of hydrogen, with the maximum pressures usually below 10 bar. Numerous refinements were incorporated over the years in designs and construction methods to lengthen storage dormancy by reducing liquid boil-off as well as to facilitate safe filling of the tank with minimal loss of the cryogenic hydrogen [31, 35].

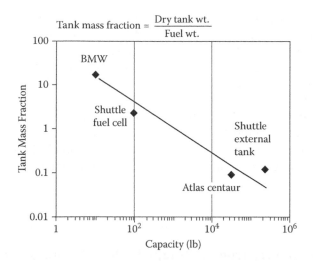

FIGURE 3.5
Plot of tank mass fraction vs. LH_2 capacity for different applications. (Courtesy Russ Jones, Boeing, private communication to L. E. Klebanoff.)

One clear concept emerging from these diverse historic uses of LH_2 is the following: If you are going to store LH_2, it is better to store a lot of it. Figure 3.5 gives a plot of tank mass fraction, defined as (Dry tank weight)/(Fuel mass), for a number of these different applications involving LH_2 [36]. Ideally, the tank mass fraction would be zero. Since the mass of the spherical tank in an LH_2 storage system varies as the square of the radius of the storage tank, whereas the mass of stored fuel varies as the cube of the radius, the weight penalty associated with the storage tank drops significantly as the total stored quantity of LH_2 increases. This approximately explains the tank mass fraction variation shown in Figure 3.5.

An approach initiated at the Lawrence Livermore National Laboratory (LLNL) combines high pressure and cryogenic storage vessels to increase gravimetric and volumetric capacities while extending storage times before significant venting of the boil-off gas [37]. Over the past several years, LLNL along with BMW have worked to improve the performance of these "cryocompressed" hydrogen storage vessels so that they could be viable for use in passenger cars [37, 38]. Independent systems and cost analyses [39, 40] recently projected that currently configured cryocompressed vessels have the largest storage capacities along with competitive costs when compared to the properties and status of other systems now available for passenger vehicles. The properties of these cryocompressed hydrogen storage systems is examined further in Chapter 4 by Guillaume Petitpas and Salvador Aceves.

Since the density of solid molecular hydrogen exceeds 85 g/L below the 14 K triple point (see Chapter 9 of Reference 28), both solid and slush hydrogen (a two-phase cryogenic fluid) have been considered for propellant storage in space and aviation systems [1, 20, 28]. However, the numerous technical challenges in producing and thermally isolating solid and slush hydrogen have kept experimental investigations to the research laboratory up to the present.

On the other hand, hydrogen stored as a solid has been successfully utilized as a cryogen to cool infrared sensors and optics to below 12 K in two space missions [30, 41]. The first use of solid hydrogen as a cryogen was for cooling the long-wavelength infrared (LWIR) sensors in the Spatial Infrared Imaging Telescope III (SPIRIT III) sensor systems [42] of the

Mid-Course Space Experiment (MSX). The SPIRIT III cryogen tank [41, 42] contained 80 kg of solid hydrogen at about 9 K in a volume of 944 L that also had a 1.7% dense aluminum foam structure for heat conduction to the tank walls. The MSX instrument was launched in 1996 to test a variety of multispectral imaging technologies [42]. The SPIRIT III instruments were successfully controlled by the solid hydrogen cryogen to 10.5 K for nearly 11 months before all the hydrogen was depleted. In 2009, NASA launched the Wide-Field Infrared Survey Explorer (WISE) satellite to generate a survey of the entire sky in the 3.5- to 23-μm infrared region while in Earth orbit [43]. A two-stage solid hydrogen cryostat [43] cooled the detectors and optics below 15 K for approximately 10 months before loss of its solid coolant. For more information on the WISE flight mission, see Reference 44.

On-Board Reversible Hydrogen Storage Materials and Systems

So far, the discussion has focused on hydrogen storage associated with the physical states of molecular hydrogen, namely, gas, liquid, and solid. Another "physical state" is atomic hydrogen bound with other elements in chemical compounds. We alluded to this when we described water as having a hydrogen storage mass density of 111 g hydrogen per liter. Water is actually a good hydrogen storage material, although it is too stable to be used in mobile applications. Fortunately, it turns out that hydrogen forms some remarkable compounds with other elements that do have application for reversible high-capacity hydrogen storage.

Interstitial Metal Hydrides

Hydrogen is very reactive with numerous metals and alloys as well as forming chemical compounds or complexes with most other elements [3]. A few examples that are potential candidates for hydrogen storage are shown in Figure 3.1. Metal hydrides (MH_x) can be classified in terms of the hydrogen bonding to the metal (i.e., metallic, ionic, or covalent with complexes being intermediate cases). For most metals, intermetallic compounds, and alloys, hydrogen is bound in the interstitial sites in a metallic state with usually minor distortions of the generally stable H-free alloy lattice [7, 8, 12]. Many practical MH_x materials are denoted as AB_y that involve a combination of elements A that are strongly exothermic hydrogen absorbers (e.g., Mg, Ti, Zr, La, etc.) with elements B that are either endothermic or very weakly exothermic hydrogen absorbers (e.g., Ni, Fe, Co, Mn, etc.), and y is in the range 0–5. There have been many studies on the formation, properties, and applications of these AB_yH_x materials over the past 40+ years, as summarized in several books [1, 4, 6, 15, 18] and a number of review articles [8–10, 12–14, 45, 46]. Ben Chao and Lennie Klebanoff in Chapter 5 give an updated review of the hydrogen storage properties of these "interstitial metal hydrides."

Listed in Table 3.1 are applications for which metal hydrides have been seriously considered or are currently used. Various key properties along with candidate MH_x systems are given in Table 3.1 for each application along with references where further descriptions of these hydrides and systems as well as the requirements and limitations imposed in specific applications can be found. Nickel metal hydride (Ni-MH) batteries [5, 6, 53] are now well-established commodities in the international marketplace, and several niche devices

TABLE 3.1

Energy Storage and Conversion Applications for Metal Hydrides

Application	Desired MH_x Attributes	Candidate Metal Alloys	References
Stationary fuel storage (e.g., backup and remote power, etc.)	$P_d \sim$ 1–10 bar, very low cost, uses waste heat, H capacity > 2 wt%, safety	TiFe, V alloys, Mg alloys, AB_2 alloys	4, 5, 45–48
Portable and mobile fuel storage (e.g., power and communication equipment, light carts, generators, etc.)	$P_d \sim$ 1–10 bar, compactness, low cost, uses air cooling or waste heat, fast kinetics, H capacity > 2 wt%, safety, durability	AB_5, AB_2, AB, alanates	1, 4–6, 45
Passenger vehicle fuel storage (internal combustion/fuel cells)	H capacity > 5 wt%, cost, $P_d \sim$ 1–10 bar, uses waste heat, fast kinetics, durability during cycling, safety, contamination	AB_2, Mg alloys, alanates (TiFe, AB_5 used in the past, but H capacities too low)	4, 11, 15, 45, 49–51
Specialty and utility vehicles (e.g., forklifts, tow tractors, scooters, boats, submarines)	$P_d \sim$ 1–10 bar, compactness, low cost, uses waste heat, H capacity > 1 wt%, fast kinetics, safety, durability	AB_5, AB_2, AB, A alloys	1, 4, 8, 52
Electrodes/Ni-MH batteries	Cost, reversible energy capacity, power density, activation, P_d < 1 bar	AB_5, AB_2, AB	5, 6, 53
Chemical heat pumps and refrigerators	Very fast kinetics, cost, H capacity, $P_d \sim$ 1–5 bar, uses waste heat	AB_5, AB_2, AB	4, 6, 54
Purification, chemical separation, and isotope separation (fusion energy)	Kinetics, activation, impurity contamination, reaction efficiency, stability, durability, safety	Pd, V alloys, Zr alloys (AB_2, AB)	4, 6
Reversible gettering (vacuum)	Very low pressure, kinetics, pumping speed, activation, durability	U, Zr alloys (AB_2, AB, $A_xB_yO_z$)	4, 6
Gas gap thermal switches	P_d < 0.05 bar, fast kinetics, low power (~10 mW), durability during temperature cycling, contamination, reliability	ZrNi, U, Zr-alloys (AB_2, $A_xB_yO_z$)	55, 56
Compressors (up to ~500 bar) for liquefaction or filling high-pressure gas storage tanks	Thermal efficiency (i.e., high $\Delta P/\Delta T$ ratio), fast kinetics, cycling stability, safety, cost	V alloys, AB_5, AB_2, AB	4, 6
Sorption cryocoolers (space flight applications)	Fast kinetics, cycling stability, constant P_a absorption plateau, power, reliability	$LaNi_{4.8}Sn_{0.2}$, V alloys, AB_5, AB_2	30, 57, 58

Source: Updated from Bowman, R.C., Jr., B. Fultz. 2002. Metal hydrides I: hydrogen storage and other gas-phase applications. *MRS Bull.* **27**(9):688–693. (With permission of Cambridge University Press.)

Note: P_d is desorption pressure.

(i.e., getters, purifiers, portable storage vessels, etc.) using hydrides have been developed and are commercially available [4–6, 45, 46]. Prototypes and demonstrations of numerous specialty, utility, and passenger hydrogen-powered vehicles have used hydrides for on-board storage [1, 4–6, 8, 45, 46, 49] as well as for chemical heat pumps and refrigerators [4, 6, 54]. However, the limited gravimetric capacities (i.e., less than 5 wt%) for all known interstitial and many complex hydrides (e.g., $NaAlH_4$) limit their potential [11, 21, 50, 51] for general-purpose passenger vehicles. There have been successful development and deployment of hydride systems on at least one military application (e.g., storage in fuel-cell-powered submarines [52]), gas gap heat switches [55, 56], and sorption cryocoolers [57, 58] for space flight missions.

Many of the intermetallic compounds and solid-solution alloys can readily absorb and desorb hydrogen gas around room temperature over the pressure range of 1–100 bars. Unfortunately, these hydrides can reversibly store only 1–3 wt% H_2, which is not considered [19] large enough for most passenger vehicles. However, these capacities are fully suitable for stationary storage systems [45, 46] as well as various types of mobile applications, including utility/service vehicles such as forklifts and tow tractors [6, 7], submarines [52], canal boats [59], scooters [60], and so on.

One significant advantage of interstitial metal hydride storage is the ability to refuel it using readily available low-pressure (~150 bar) sources of bottled merchant hydrogen, eliminating the need for a hydrogen compressor to generate high pressures. Another significant advantage of these "low-temperature" hydrides is the relative simplicity of using the waste heat from an ICE or a proton exchange membrane (PEM) fuel cell to discharge the hydrogen gas from the storage beds via a heat exchange fluid [5, 7]. The transfer of waste heat from a hydrogen energy conversion device to the metal hydride storage medium is an example of the balance of plant thermal management. Any practical engineered hydrogen storage system using metal hydrides must bring these thermal management issues under control, providing heat to the metal hydride for hydrogen release and removing heat from the engineered system when the spent material is recharged with hydrogen.

One example of a hydride storage system is shown in Figure 3.6; two Type III (i.e., carbon-wrapped metal) vessels containing company proprietary AB_2 alloy were mounted on the underside of a hybrid Prius model 2005 automobile by Ovonic Hydrogen Solutions [61]. These hydride tanks could reversibly store up to 3.6 kg hydrogen with waste heat from an ICE used during discharge. On the other hand, some ionic hydrides (e.g., MgH_2 and LiH) are capable of storing 7–12 wt% H_2, but they must first be heated above about 600 K to release gas at pressure of about 1 bar. These desorption temperatures are much higher than the approximately 350 K waste heat normally available from a PEM fuel cell or ICE.

FIGURE 3.6
Two metal hydride storage tanks mounted on the underside of a modified 2005 Prius hybrid passenger car by the Ovonics Corporation. (Photo reproduced with permission from Reference [61a].)

Complex Metal Hydrides

Over the past decade, substantial progress has been made with "complex metal hydrides" [13] containing mixed ionic-covalent bonding that can reversibly store more than 4 wt% H_2, often with operating temperatures below about 400 K. Development challenges presented by the complex metal hydrides have included improving the slow reaction kinetics of these materials, as well as decreasing the high heats of desorption, which require desorption temperatures above those available from ICE or PEM fuel cells. These complex metal hydrides are reviewed in Chapter 6 by Vitalie Stavila, Lennie Klebanoff, John Vajo, and Ping Chen.

The most common engineering solution to the problem of too high a desorption temperature is to catalytically oxidize (i.e., burn) a portion of the hydrogen stored in the bed. A recent concept [62] for a storage system based on this approach for catalyzed $NaAlH_4$ is shown in Figure 3.7; a 12-kW combustor is needed. Thermal modeling analyses [50, 62] of alanate storage beds indicated that about 25–40% of the total hydrogen content would be consumed during H_2 discharge, which greatly decreases both the effective gravimetric and volumetric capacities. Hence, the advantages of $NaAlH_4$ and most other complex hydrides are clearly negated when much of the hydrogen must be consumed to provide the heat for desorbing the gas. This motivates the desire to find complex metal hydrides with an enthalpy of hydrogen release that is consistent with the temperature of the fuel cell waste heat, typically about 85°C. This requirement is described further in Chapter 6. One possible alternative would be the development and incorporation of fuel cells that operate

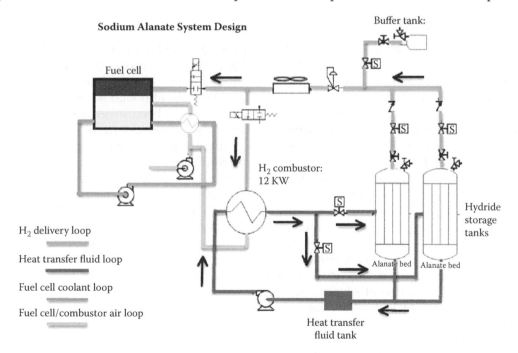

FIGURE 3.7 (See color insert.)
Schematic configuration for a reversible hydrogen storage system in fuel-cell-powered passenger cars that uses $NaAlH_4$ as the storage medium. (Figure from Pasini, J.M., B.A. van Hassel, D.A. Mosher, and M.J. Veenstra. 2012. System modeling methodology and analyses for materials-based hydrogen storage. *Int. J. Hydrogen Energy* **37**:2874–2884 with permission of Elsevier.)

at higher temperatures (i.e., >400 K) to provide the necessary waste heat for desorption from the complex hydride [63].

Heat transfer during both hydrogen desorption and absorption involving any metal hydride is a major engineering issue due to their heats of reactions and the low thermal conductivities of the materials stemming from their fine-powder nature [64]. The material engineering properties of selected hydrogen storage materials are discussed by Daniel Dedrick in Chapter 9. Over the years, there have been numerous bed designs and configurations in successful attempts to provide sufficient thermal management of these systems. Figure 3.8 provides a diagrammatic summary [65] for most of the system designs and components that have been proposed.

Many variants of these configurations have been described in the hydrogen storage literature [54, 57, 64–67]. There have always been major conflicts in designing any hydride storage vessel in an effort to reconcile heat transfer management, minimal mass and volume, mechanical and structural stability, and so on. Some impacts of the performance of a nearly full-scale (i.e., up to 3 kg H_2 capacity) storage system using sodium alanate have been experimentally investigated, revealing that both reaction kinetics and thermal management issues are involved but can be brought under control [68]. The complications that arise in balancing contradictory factors against the operational requirements imposed on the hydride storage systems to meet performance requirements are addressed more thoroughly in Chapter 10 by Terry Johnson and Pierre Bénard [69].

Adsorption Materials

The easily reversible adsorption of H_2 molecules by solids with large surface areas (i.e., usually much greater than 1000 m^2/g) and very highly porous structures offers another type of hydrogen storage method [1, 11, 15]. Most attention has been focused on activated carbons (a-C) and metal organic framework (MOF) compounds, although nanocarbons, zeolites, microporous polymers, and other materials have also been investigated over the past two decades [70–72]. Since adsorption bonding arises from relatively weak van der Waals attractions of the H_2 molecules to the sorbent surfaces, the binding energies are usually about 2–10 kJ/mole. This limits significant storage capacities at ambient temperatures requiring the sorbent materials to be cooled to cryogenic temperatures (i.e., usually ~100 K or lower). Extensive measurements have established that H_2 adsorption capacities increase approximately linearly with pressure, yielding maximum contents of about 5–6 wt% and about 7–7.5 wt% for a-C and MOFs, respectively, at 77 K and pressures of 50–150 bar [70, 73, 74]. Since these materials generally have very rapid kinetics for adsorption as well as other potentially attractive properties for hydrogen storage, several systems have been proposed and their thermal performance parameters analyzed for vehicular applications [75–78]. Recent advances in the science of hydrogen adsorption materials are given in Chapter 7 by Justin Purewal and Channing Ahn.

Figure 3.9 shows one viable configuration for a cryogenic adsorption storage vessel based on a high-performance activated carbon that has been evaluated by Ahluwalia and Peng [77]. Although the heat of reaction between hydrogen and this carbon is low, aluminum foam is required for effective thermal management during both the adsorption and the desorption processes, and a large temperature change is necessary to provide sufficient storage capacity [77]. Furthermore, the use of sufficient liquid nitrogen to cool the carbon beds from ambient or discharge temperatures and during charging was shown to be ineffective for the overall energy balance [77]. Direct filling with liquid hydrogen was shown to be a much better option, although this clearly requires LH_2 to be available at the filling station.

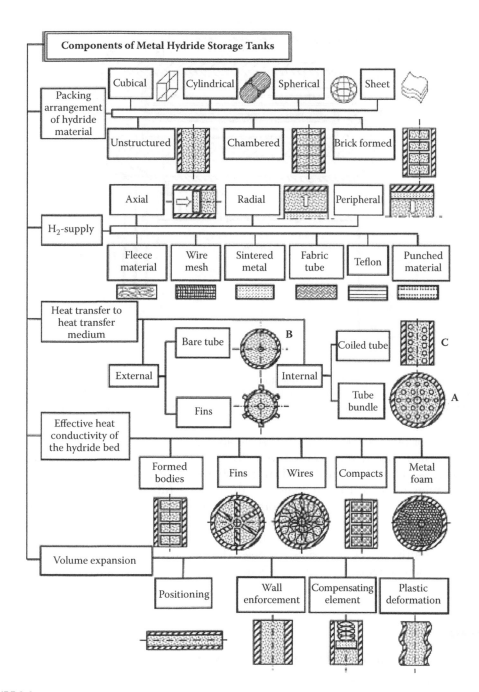

FIGURE 3.8
Design configurations and options for components in metal hydride storage vessels. (From Ranong, C.N., M. Hohne, J. Franzen, J. Hapke, G. Fleg, M. Dornheim, N. Eigen, J.M. Bellosta von Colbe, O. Metz. 2009. Concept, design, and manufacture of a prototype hydrogen storage tank based on sodium alanate. *Chem. Eng. Technol.* **32**:1154–1163 with permission of John Wiley and Sons.)

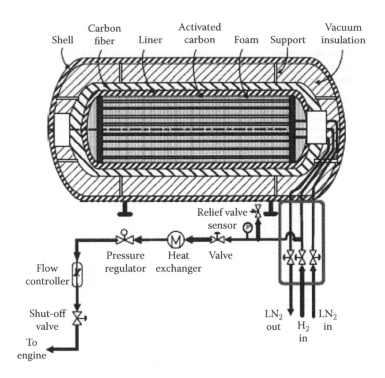

FIGURE 3.9

Schematic of a complete system for the reversible storage of hydrogen via adsorption on activated carbon. (Figure reproduced with permission from Ahluwalia, R.K. and J.K. Peng. 2009. Automotive hydrogen storage system using cryo-adsorption on activated carbon. *Int. J. Hydrogen Energy* **34**, 5476–5487 with permission of Elsevier.)

To the best of our knowledge, no full-scale adsorption storage system has been fabricated and tested in a demonstration vehicle. However, there have been smaller-scale prototype vessels built and evaluated in the laboratory to validate thermal simulations of the behavior of these carbon materials during the filling and discharge processes [79–82]. The thermal models for these storage beds were based on the measured hydrogen isotherms for the sorbents and gave reasonable representations of the observed behaviors during the tests. New sorbents with much greater storage capacities and greater compaction than currently known materials are needed to improve the performance levels from adsorption-based systems. The situation is described more fully in Chapters 7 and 10.

Off-Board Reversible Hydrogen Storage Materials and Systems

The final general hydrogen storage category identified in Figure 3.1 involves materials where hydrogen has formed sufficiently strong chemical bonds such that the material after the release of hydrogen cannot be regenerated on-board the vehicle using H_2 gas at "reasonable" pressures and temperatures [2, 9, 11]. The on-board reformation of hydrocarbons, alcohols, and ammonia creates primarily gaseous by-products (e.g., CO_2, H_2O, N_2), which are inconvenient to collect and reprocess into the original hydrogen-storing molecule. Hence, refueling the storage vessel requires replacement by fresh fuel produced from new outside resources, much like we do currently with gasoline. Since there is also simultaneous release of the detrimental greenhouse gas CO_2 during the creation of hydrogen via the reformation of the carbon-based fuels, there are immediate environmental impacts using this approach to hydrogen storage. Consequently, reformation is not discussed any further in the present review. In contrast, there are several classes of compounds and types of chemical reactions that do produce condensed phase products following hydrogen release. These spent materials are generally available to regenerate the original fuel on the addition of hydrogen, albeit typically needing considerably complex and energy-intensive processing that cannot be performed within the storage containers themselves. These last systems are addressed in this section.

Liquid Organic Hydrogen Carriers

There are a number of liquid cyclic hydrocarbons that react reversibly with H_2 gas at pressures of about 100 bar or lower, albeit at rather elevated temperatures of 500–600 K in the presence of suitable catalysts [83–87]. Examples of these liquid hydrogen carriers include benzene/cyclohexane (7.1 wt%), toluene/methylcyclohexane (6.1 wt%), naphthalene/decalin (7.2 wt%), and several substituted carbazoles [71, 86, 87]. Because these moderately high-capacity liquids can be transported and handled as conventional fuels for filling the storage tanks of vehicles and other storage systems in remote or distance locations, they have often been proposed as attractive options in a hydrogen fuel infrastructure [1, 71, 84–86].

The development of practical systems using these liquid organics has been hindered by the need for much more efficient and robust catalysts during both the hydrogenation and the dehydrogenation reactions [86]. This results in burning significant portions of the released hydrogen to maintain the reactor chambers at temperatures high enough to continue the dehydrogenation reaction [71]. Maintenance of an elevated temperature is also required to avoid solidification of the reaction products [71, 86, 87] so that they can be separately stored and removed from the vehicle for regeneration [84–86]. There have not been any demonstrations of complete hydrogen storage systems using any of the candidate liquid organic hydrogen carriers, although several concepts of stations and distribution systems for fueling vehicles have been published [83–86]. The scientific advances in developing liquid organic carriers of hydrogen are reviewed by Guido Pez and Alan Cooper in Chapter 8.

Hydrolysis Reactions

The hydrolysis reactions between some metals (e.g., Al [88, 89]), elemental hydrides (e.g., LiH, MgH_2, CaH_2 [14, 90, 91]), complex hydrides (e.g., $LiAlH_4$, $NaBH_4$, etc. [14, 90, 91]), boranes (e.g., NH_3BH_3 [92, 93] and NaB_3H_8 [94]), and several Si-based compounds [95–98] release hydrogen gas when placed in contact with either gaseous or liquid water that has sufficient purity to act as viable storage medium. This behavior has been known for decades. For example, the hydrolysis of CaH_2 in portable canisters was used to inflate balloons for meteorological and military applications since before World War II [90]. Many concepts based on these hydrolysis reactions combine the functions of storage, production, and delivery into a single system in which the reactants are kept separate until hydrogen is released. Since hydrogen can be supplied at low pressures and in an ambient thermal environment, devices based on hydrolysis have been proposed for a wide range of portable and remote applications for fuel cells [99–101] where power requirements range from below 1.0 W to over about 10 kW.

Diverse approaches for initiating and controlling the highly exothermic reactions with water have been reported in the patent and research literature [88, 90, 93, 98, 102, 103]. While hydrolysis reactions can take place in both acidic and basic media, the importance of including appropriate catalysts and additives has been widely recognized, and they have been incorporated into nearly all these chemical hydrogen storage/production concepts [88, 90, 91, 95, 101, 103–105]. The hydrolysis of aluminum powders has been suggested for various systems [88, 89], including submarines powered by fuel cells [106]; however, since all the hydrogen must come from the water reactant alone, reactions with metal hydrides would provide larger yields since both components contribute hydrogen [14, 90, 91]. A very interesting approach has been proposed [101] that combines thermal and hydrolysis decomposition of alane (AlH_3) with $NaH/NaAlH_4$ mixtures to enhance the yield of gaseous hydrogen. However, the issues of regenerating the extremely stable aluminum oxides/hydroxides remain.

Within the past decade, several organizations have built and tested laboratory prototypes and demonstration hydrogen storage/generation systems based on various hydrolysis reactions [90, 96–98, 102, 103, 106]. A few companies, including Millenium Cell, Incorporated [107], Jadoo [108], and SiGNa Chemistry [96, 97], have offered commercial portable systems for specialty or military applications. A nominal device to operate portable 300-W fuel cell power sources is shown in Figure 3.10, which identifies the major components. Hydrogen is generated by the reaction

$$2NaSi \text{ (solid)} + 5\ H_2O \text{ (liquid)} \rightarrow 5\ H_2 \text{ (gas)} + Na_2Si_2O_5 \text{ (aqueous)} \qquad (3.1)$$

which in principle has a gravimetric storage capacity of 5.2 wt%. However, its practical capacity is substantially reduced by the additional water required along with the storage vessels and all the other components that are necessary.

The catalyzed hydrolysis of sodium borohydride ($NaBH_4$) is by far the most common candidate [90, 91, 102–105, 109] for hydrogen storage and generation based on the idealized reaction

$$NaBH_4 \text{ (solution)} + 2H_2O \rightarrow 4H_2 \text{ (gas)} + NaBO_2 \text{ (aqueous)} + \sim 300 \text{ kJ} \qquad (3.2)$$

For a 30 wt% $NaBH_4$ solution, 6.7 wt% hydrogen gas is available, neglecting the contributions from the catalysts, excess water to maintain the borate by-products in solution, and the containment and control components. Note this is a highly exothermic reaction requiring significant heat management when there are multiple kilograms of reactants and fast hydrogen release rates.

The system developed by Millennium Cell [107, 110, 111] that integrates hydrolysis of $NaBH_4$ solutions with fuel cell power sources is shown in Figure 3.11. This approach was demonstrated for a variety of applications, including the modification and testing of a Chrysler Natrium passenger vehicle with a nominal 50-kW fuel cell power system [110, 111]. Following a technical review [112] held by the Department of Energy (DOE) in 2007, it was concluded that the configuration shown in Figure 3.11 had numerous limitations when considered for passenger vehicles. Factors that jeopardize the process include (1) the unproven single-tank bladder system (which is based on a single-tank design separating fresh and spent fuel via a movable diaphragm); (2) the requirement for large amounts of excess water on-board the vehicle; and (3) issues dealing with the precipitation of the $NaBO_2$ product. In addition, there were concerns about the practicality of using a 30% solution of $NaBH_4$ at ambient temperature, which is near the solubility limit (and may be above it depending on temperature) [113]. Whether $NaBO_2$ precipitation can be inhibited during the process of generating hydrogen remains unclear.

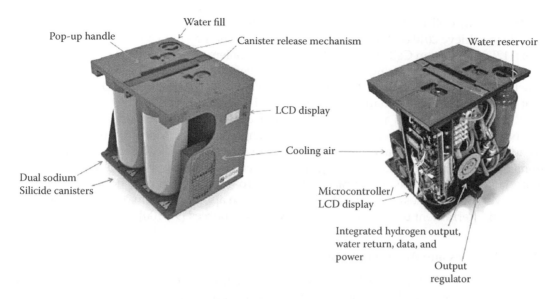

FIGURE 3.10
A portable hydrogen storage/generation system based on the hydrolysis of sodium silicide (NaSi) that was developed by SiGNa Chemistry Incorporated for operation with a 300-W PEM fuel cell power system. (Reproduced from Wallace, A., M. Lefenfeld. 2010. NaSi and Na-SG powder hydrogen fuel cells. Proceedings of DOE Annual Merit Review. http://www.hydrogen.energy.gov/pdfs/review10/st055_lefenfeld_2010_p_web. pdf; and Wallace, A., M. Lefenfeld. 2010. DOE annual progress report. http://www.hydrogen.energy.gov/pdfs/ progress10/iv_h_1_lefenfeld.pdf.)

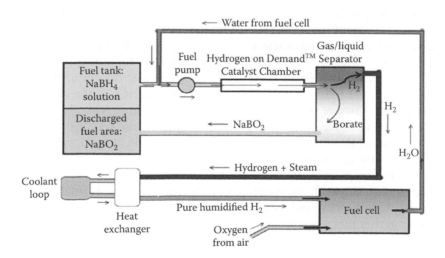

FIGURE 3.11
Schematic concept for integrated $NaBH_4$ hydrolysis storage and fuel cell systems that were developed by Millennium Cell, Incorporated. (Figure from Wu, Y. 2005. Development of advance chemical hydrogen storage and generation system. Proceedings of 2005 DOE Hydrogen Annual Merit Review. http://www.hydrogen.energy.gov/pdfs/review05/stp_10_wu.pdf; Wu, Y., R.M. Mohring. 2003. Sodium borohydride for hydrogen storage. *Prepr. Pap.-Am Chem. Soc., Div. Fuel Chem.* **48**:940; Wu, Y. 2003. Hydrogen storage via sodium borohydride—current status, barriers, and R&D roadmap. Presented at the GCEP, Stanford University. http://www. stanford.edu/group/gcep/pdfs/hydrogen_workshop/Wu.pdf.)

Millennium Cell essentially concluded that the solution-based $NaBH_4$ approach was unable to achieve 2010 on-board capacity targets specified by the DOE for passenger vehicles [112]. Millennium Cell also reported that the problem of accumulating a solid product was a significant engineering issue that had not been addressed adequately, and that no practical engineering solution has been identified. Finally, Millennium Cell pointed out that the hydrogen cost remains well above the DOE targets [112] with this system. The DOE consequently issued a "no-go" decision that ceased any further funding for developing $NaBH_4$-hydrolysis concepts for use in passenger vehicles. More recent independent comparison of vehicle storage systems has clearly confirmed the limitations of $NaBH_4$ solutions relative to all the other options [114]. However, hydrolysis of $NaBH_4$ and many other materials remains viable and attractive for numerous portable power devices when fuel cell power levels are less than 10 kW, especially those applications involving remote locations and intermittent usage, such as standby power sources [99, 100].

Metal Hydrides and Ammonia Borane

Catalyzed thermal decomposition of compounds with hydrogen contents greater than about 10 wt% has often been suggested as hydrogen storage candidates even when these reactions cannot be readily reversed [2, 11, 71, 91, 93, 114–120]. Examples of those materials releasing hydrogen during endothermic reactions are AlH_3 and $LiAlH_4$, while NH_3BH_3, amidoboranes, and numerous other B-N-H compounds [117, 120] produce hydrogen from exothermic decompositions. Although hydrogen release from these materials can be rapid and substantial, providing sufficiently active catalysts are available [114, 116, 118, 120], the creation of highly stable products (e.g., Al, BN, etc.) usually leads to challenging and costly schemes for regeneration. The intense research and development (R&D) directed to understanding the hydrogen storage potential of these metal hydride and B-containing materials and developing methods to regenerate them are reviewed by Jason Graetz, Lennie Klebanoff, David Wolstenholme, and Sean McGrady in Chapter 8.

Detailed systems engineering assessments [114, 118, 119] have been recently performed that address both the on-board and off-board performance for storage systems based on the endothermic decomposition of AlH_3 and the exothermic decomposition of NH_3BH_3. A major issue with NH_3BH_3 and most related B-N-H materials is the concurrent formation of copious quantities of contaminant gaseous species (e.g., NH_3, B_2H_6, borazine, etc.) along with hydrogen during decomposition. Not only are these contaminants detrimental to the performance of the fuel cells (poisoning the fuel cell catalyst), and greatly impact the recyclability of the decomposition products, but they are also toxic to humans. While various approaches ranging from selective catalysts to encapsulation of nanosize particles in benign hosts have shown to be effective in reducing impurity formation, their control will require improvements to both the reaction chemistry and engineering design of the chemical hydrogen storage systems.

Concluding Remarks

Summarizing, we have reviewed various materials and methods that have been developed over the past several decades for storing hydrogen for a variety of applications. Hydrogen storage via physical containment of molecular hydrogen (gas, solid, liquid) as well as different types of chemical phases and reactions (on-board reversible, off-board reversible) can be used in portable or mobile devices, stationary power sources, different modes of transportation, and aerospace technologies. The benefits and limitations for each storage approach were identified, and examples of established storage systems or prototypes produced for feasibility and functionality demonstrations were given when possible. The benefits, issues, and limitations for each storage approach were briefly related to the requirements imposed in several representative applications. Our hope is that this review provides a solid introduction to the chapters that follow.

Acknowledgments

We express our appreciation to Drs. Ned T. Stetson and Gregory L. Olson for insightful discussions and contributions relating to hydrogen storage technology. This work was partially supported by the U.S. Department of Energy, Office of Energy Efficiency and Renewable Energy.

References

1. Züttel, A., M. Hirscher, K. Yvon, et al. 2008. Hydrogen storage. In *Hydrogen as a Future Energy Carrier*, ed. A. Züttel, A. Borgschulte, L. Schlapbach, Chapter 6, 165–264. Weinheim, Germany: Wiley-VCH.
2. Aardahl, C.L., S.D. Rassat. 2009. Overview of systems considerations for on-board chemical hydrogen storage. *Int. J. Hydrogen Energy* 34:6676–6683.
3. Mueller, W.M., J.P. Blackledge, G.G. Libowitz. 1968. *Metal Hydrides*. New York: Academic Press.
4. Sandrock, G., S. Suda, L. Schlapbach. 1992. Applications. In *Hydrogen in Intermetallic Compounds II*, ed. L. Schlapbach. Topics in Applied Physics 67:197–258. Berlin: Springer-Verlag.
5. Sandrock, G. 1995. Applications of hydrides. In *Hydrogen Energy System—2 Production and Utilization of Hydrogen and Future Aspects*, ed. Y. Yurum, NATO ASI Series E 295:253. Amsterdam: Kluwer.
6. Dantzer, P. 1997. Metal-hydride technology: a critical review. In *Hydrogen in Metals III*, ed. H. Wipf, 279–340. Berlin: Springer.
7. Sandrock, G. 2003. Hydride storage. In *Handbook of Fuel Cells—Fundamentals, Technology and Applications*, ed. W. Vielstich, H.A. Gasteiger, A. Lamm, 3(2):101–112. Chichester, UK: Wiley.
8. Bowman, R.C., Jr., B. Fultz. 2002. Metal hydrides I: hydrogen storage and other gas-phase applications. *MRS Bull.* 27(9):688–693.
9. Felderhoff, M., C. Weidenthaler, R. von Helmolt, U. Eberle. 2007. Hydrogen storage: the remaining scientific and technological challenges. *Phys. Chem. Chem. Phys.* 9:2643–2653.
10. Sakintuna, B., F. Lamari-Darkrim, M. Hirscher. 2007. Metal hydride materials for solid hydrogen storage: a review. *Int. J. Hydrogen Energy* 32:1121–1140.
11. Yang, J., A. Sudik, C. Wolverton, D.J. Siegel. 2010. High capacity hydrogen storage materials: attributes for automotive applications and techniques for materials discovery. *Chem. Soc. Rev.* 39:656–675. DOI: 10.1039/b802882f.
12. Sandrock, G. 1999. A panoramic overview of hydrogen storage alloys from a gas reaction point of view. *J. Alloys Compd.* 293–295:877–888.
13. Orimo, S.-I., Y. Nakamori, J.R. Eliseo, A. Züttel, C.M. Jensen. 2007. Complex hydrides for hydrogen storage. *Chem. Rev.* 107:4111–4132.

14. Schüth, F., B. Bogdanović, M. Felderhoff. 2004. Light metal hydrides and complex hydrides for hydrogen storage. *Chem. Commun.* **October 21**:2249–2258.

15. Varin, R.A., T. Czujko, Z.S. Wronski. 2009. *Nanomaterials for Solid State Hydrogen Storage.* New York: Springer.

16. Grochala, W., P.P. Edwards. 2004. Thermal decomposition of the non-interstitial hydrides for the storage and production of hydrogen. *Chem. Rev.* **104**:1283–1315.

17. Ross, D.K. 2006. Hydrogen storage: the major technological barrier to the development of hydrogen fuel cell cars. *Vacuum* **80**:1084–1089.

18. Hirscher, M. 2010. *Handbook of Hydrogen Storage.* Weinheim, Germany: Wiley-VCH.

19. Satyapal, S., J. Petrovic, J., Read, C., Thomas, G., and Ordaz, G. 2007. The U.S. Department of Energy's National Hydrogen Storage Project: progress towards meeting hydrogen-powered vehicle requirements. *Catal. Today* **120**:246–256.

20. Klell, M. 2010. Storage of hydrogen in the pure form. In *Handbook of Hydrogen Storage*, ed. M. Hirscher, 1–37. Weinheim, Germany: Wiley-VCH.

21. Irani, R.S. 2002. Hydrogen storage: high-pressure gas containment. *MRS Bull.* **27**(9):680–682.

22. Hua, T.Q., R.K. Ahluwalia, J.-K. Peng, M. Kromer, S. Lasher, K. McKenney, K. Law, J. Sinha. 2011. Technical assessment of compressed hydrogen storage tank systems for automotive applications. *Int. J. Hydrogen Energy* **36**:3037–3049.

23. Wipke, K., S. Sprik, J. Jurtz, H. Thomas, J. Garbak. 2008. FCV learning demonstration: project midpoint status and first-generation vehicle results. *World Electric Vehicle J.* **2**(3):4–17. http://www.nrel.gov/hydrogen/pdfs/ja-560-45468.pdf.

24. Wipke, K., S. Sprik, J. Kurtz, J. Garbak. 2009. Field experience with fuel cell vehicles. In *Handbook of Fuel Cells—Fundamentals, Technology and Applications*, ed. W. Vielstich, H. Yokokawa, H.A. Gasteiger. *Volume 6: Advances in Electrocatalyst, Materials, Diagnostics and Durability*, Chapter 60. New York: Wiley. http://www.nrel.gov/hydrogen/pdfs/43589.pdf.

25. Miller, A.R., K.S. Hess, D.L. Barnes, T.L. Erickson. 2007. System design of a large fuel cell hybrid locomotive. *J. Power Sources* **173**:935–942.

26. Barnes, D.L., A.R. Miller. 2007. Fuel cell prototype locomotive. DOE Hydrogen Program FY2007 Annual Progress Report, pp. 1083–1085. http://www.hydrogen.energy.gov/pdfs/progress07/vi_a_9_barnes.pdf.

27. Hess, K.A. 2010. Demonstration of a hydrogen fuel cell locomotive. Presentation at the American Public Transportation Association (APTA) 2010 Rail Conference, Vancouver, BC, Canada. http://www.apta.com/mc/past/2010/2010rail/Presentations/EET-Hess-rail10.pdf.

28. Peschka, W. 1992. *Liquid Hydrogen—Fuel of the Future.* Vienna, Austria: Springer-Verlag.

29. Kinard, G.E. 1998. The commercial use of liquid hydrogen over the last 40 years. In *Proceedings of the International Cryogenic Engineering Conference*, Vol. ICEC-17, ed. D. Dew-Hughes, R.G. Scurlock, J.H.P. Watson, 39–44. Bristol, UK: Institute of Physics.

30. Bowman, R.C., Jr. 2006. Roles of hydrogen in space explorations. In *Hydrogen in Matter*, ed. G.R. Myneni, B. Hjorvarsson, 175–199. New York: American Institute of Physics.

31. Wolf, J. 2002. Liquid-hydrogen technology for vehicles. *MRS Bull.* **27**(9):684–687.

32. Haberbusch, M.S., Nguygen, C.T., Stochl, R.J., Hui, T.Y. 2010. Development of No-Vent™ liquid hydrogen storage system for space applications. *Cryogenics* **50**:541–548.

33. Haglind, F., A. Hasselrot, R. Singh. 2006 Potential of reducing the environmental impact of aviation by using hydrogen. Part I: background, prospects and challenges. *Aeronaut. J.* **110**:533–540.

34. Verstraeta, D., P. Hendrick, P. Pilidis, K. Ramsden. 2010. Hydrogen fuel tanks for subsonic transport aircraft. *Int. J. Hydrogen Energy* **35**:11085–11098.

35. Krainz, G., G. Bartlok, P. Bodner, P. Casapicola, Ch. Doeller, F. Hofmeister, E. Neubacher, A. Zieger. 2004. Development of automotive liquid hydrogen storage systems. *Adv. Cryogenic Eng.* **49**:35–40.

36. Plot courtesy of Russ Jones, Boeing, private communication with L.E. Klebanoff, December 6, 2005.

37. Aceves, S.M., G.D. Berry, J. Martinez-Frias, F. Espinsoa-Loza. 2006. Vehicular storage of hydrogen in insulated pressure vessels. *Int. J. Hydrogen Energy* **36**:2274–2283.
38. Aceves, S.M., F. Espinosa-Loa, E. Ledesma-Orozco, T.O. Ross, A.H. Weisberg, T.C. Brunner, O. Kircher. 2010. High-density automotive hydrogen storage with cryogenic capable pressure vessels. *Int. J. Hydrogen Energy* **35**:1219–1226.
39. Ahluwalia, R.K., J.K. Peng. 2008. Dynamics of cryogenic hydrogen storage in insulated pressure vessels for automotive applications. *Int. J. Hydrogen Energy* **33**(17):4622–4633.
40. Ahluwalia, R.K., T.Q. Hua, J.-K. Peng, S. Lasher, K. McKenney, J. Sinha, M. Gardiner. 2010. Technical assessment of cryo-compressed hydrogen storage tank systems for automotive applications. *Int. J. Hydrogen Energy* **35**:4171–4184.
41. Donabedian, M. 2003. Stored solid cryogen systems. In *Spacecraft Thermal Control, Handbook—Volume II: Cryogenics*, ed. M. Donabedian, 31–39. El Segundo, CA: Aerospace Press.
42. Bartschi, B.Y., D.E. Morse, T.L. Woolston. 1996. The Spatial Infrared Imaging Telescope III. *John Hopkins APL Technical Digest* **17**:215–225.
43. Liu, F., R. Cutri, G. Greanias, V. Duval, P. Eisenhardt, J. Elwell, I. Heinrichsen, J. Howard, W. Irace, A. Mainzer, A. Razzaghi, D. Royer, E.L. Wright. 2008. Development of the Wide-field Infrared Survey Explorer (WISE) Mission. *Proc. SPIE* **7017**:70170M.1–70170M.12.
44. WISE Mission home page. http://wise.ssl.berkeley.edu/.
45. Bernauer, O. 1988. Metal hydride technology. *Int. J. Hydrogen Energy* **13**:181–190.
46. Lynch, F.E. 1990. Metal hydride practical applications. *J. Less Common Met.* **172–174**:943–958.
47. Gray, E. MacA., C.J. Webb, J. Andrews, B. Shabani, P.J. Tsai, S.L.I. Chan. 2011. Hydrogen storage for off-grid power supply. *Int. J. Hydrogen Energy* **36**:654–663.
48. Bielmann, M., U.F. Bogt, M. Zimmermann, A. Zuttel. 2011. Seasonal energy storage system based on hydrogen for self sufficient living. *J. Power Sources* **196**:4054–4060.
49. Topler, J., K. Feucht. 1989. Results of a test fleet with metal hydride motor cars. *Z. Physk. Chem. N.F.* **164**:1451–1461.
50. Ahluwalia, R.K. 2007. Sodium alanate hydrogen storage system for automotive fuel cells. *Int. J. Hydrogen Energy* **32**:1251–1261.
51. Mori, D., K. Hirose. 2009. Recent challenges of hydrogen storage technologies for fuel cell vehicles. *Int. J. Hydrogen Energy* **34**:4569–4574.
52. Pommer, H., P. Hauschildt, R. Teppner, W. Hartung. 2006. Air-independent propulsion system for submarines. *ThyssenKrupp Techforum* **1**:64–69.
53. Joubert, J.-M., M. Latroche, A. Percheron-Guegan. 2002. Metallic hydrides II: materials for electrochemical storage. *MRS Bull.* **27**(9):694–698.
54. Muthukumar, P.M., M. Groll. 2010. Metal hydride based heating and cooling systems; a review. *Int. J. Hydrogen Energy* **35**:8816–8829.
55. Prina, M., J.G. Kulleck, R.C. Bowman, Jr. 2002. Assessment of Zr-V-Fe getter alloy for Gas-Gap heat switches. *J. Alloys Compd.* **330–332**:886–891.
56. Reiter, J.W., P.B. Karlmann, R.C. Bowman, Jr., M. Prina. 2007. Performance and degradation of gas gap heat switches in hydride compressor beds. *J. Alloys Compd.* **446–447**:713–717.
57. Bowman, R.C., Jr. 2003. Development of metal hydride beds for sorption cryocoolers in space applications. *J. Alloys Compd.* **356–357**:789–793.
58. Pearson, D., R. Bowman, M. Prina, P. Wilson. 2007. The Planck sorption cooler: using metal hydrides to produce 20 K. *J. Alloys Compd.* **446–447**:718–722.
59. Bevan, A.I., A. Züttel, D. Book, I.R. Harris. 2011. Performance of a metal hydride store on the "Ross Barlow" hydrogen powered canal boat. *Faraday Discuss.* **151**:353–367.
60. Shang, J.L., B.G. Pollet. 2010. Hydrogen fuel cell hybrid scooter (HFCHS) with plug-in features on Birmingham campus. *Int. J. Hydrogen Energy* **35**:12709–12715.
61. (a) Young, R.C., Y. Li, J. Giedzinsski, B. Chao, V. Myasnkkov, S.R. Ovshinsky. 2006. Recent advances of metal hydride hydrogen ICE vehicles and dispensing systems. Presented at 17th Annual National Hydrogen Association Conference. Long Beach, CA. March 2006. (b) Chao, B.S., R.C. Young, V. Myasnikov, Y. Li, B. Huang, F. Gingl, P.D. Ferro, V. Sobolev, S.R. Ovshinsky. 2004. Recent advances in solid hydrogen storage systems. *Mater. Res. Soc. Symp. Proc.* **801**:BB1.4.1.

62. Pasini, J.M., B.A. van Hassel, D.A. Mosher, M.J. Veenstra. 2012. System modeling methodology and analyses for materials-based hydrogen storage. *Int. J. Hydrogen Energy* **37**:2874–2884.

63. Pfeifer, P., C. Wall, O. Jensen, H. Hahn, M. Fichtner. 2009. Thermal coupling of a high temperature PEM fuel cell with a complex hydride tank. *Int. J. Hydrogen Energy* **34**:3457–3466.

64. Zhang, J., T.S. Fisher, P.V. Ramachandran, J.P. Gore, I. Mudawar. 2005. A review of heat transfer issues in hydrogen storage technologies. *J. Heat Transfer* **127**:1391–1399.

65. Ranong, C.N., M. Hohne, J. Franzen, J. Hapke, G. Fleg, M. Dornheim, N. Eigen, J.M. Bellosta von Colbe, O. Metz. 2009. Concept, design, and manufacture of a prototype hydrogen storage tank based on sodium alanate. *Chem. Eng. Technol.* **32**:1154–1163.

66. Botzung, M., S. Chaudourne, O. Gillia, C. Perret, M. Latroche, A. Percheron-Guegan, P. Marty. 2008. Simulation and experimental validation of a hydrogen storage tank with metal hydrides. *Int. J. Hydrogen Energy* **33**:98–104.

67. Visaria, M., I. Mudaward, T. Pourpoint, S. Kumar. 2010. Study of heat transfer and kinetics parameters influencing the design of heat exchangers for hydrogen storage in high-pressure metal hydrides. *Int. J. Heat Mass Transfer* **53**:2229–2239.

68. Johnson, T., Jorgensen, S., and Dedrick, D. 2011. Performance of a full-scale hydrogen-storage tank based on complex hydrides. *Faraday Discuss.* **151**:327–352.

69. See Chapter 10 for in-depth discussions of thermal design and analyses as well as structural aspects of hydrogen storage systems.

70. Thomas, K.M. 2007. Hydrogen adsorption and storage on porous materials. *Catalysis Today* **120**:389–398.

71. Eberle, U., M. Felderhoff, F. Schüth. 2009. Chemical and physical solutions for hydrogen storage. *Angew. Chem. Int. Ed.* **48**:6608–6630.

72. Panella, B., M. Hirscher. 2010. Physisorption in porous materials. In *Handbook of Hydrogen Storage*, ed. M. Hirscher, 39–62. Weinheim, Germany: Wiley-VCH.

73. Meisner, G.P., Q. Hu. 2009. High surface area microporous carbon materials for cryogenic hydrogen storage synthesized using new template-based and activation-based approaches. *Nanotechology* **20**:204023.

74. Fierro, V., A. Szczurek, C. Zlotea, J.F. Mareche, M.T. Izquierdo, A. Albiniak, M. Latroche, G. Furdin, A. Celzard. 2010. Experimental evidence of an upper limit for hydrogen storage at 77 K on activated carbons. *Carbon* **48**:1902–1911.

75. Vasiliev, L.L., L.E. Kanonchik, A.G. Kulakov, V.A. Babenko. 2007. Hydrogen storage system based on novel carbon materials and heat pipe heat exchanger. *Int. J. Therm. Sci.* **46**:914–925.

76. Weinberger, B., F.D. Lamari. 2009. High pressure cryo-storage of hydrogen by adsorption at 77 K and up to 50 MPa. *Int. J. Hydrogen Energy* **34**:3058–3064.

77. Ahluwalia, R.K., J.K. Peng. 2009. Automotive hydrogen storage system using cryo-adsorption on activated carbon. *Int. J. Hydrogen Energy* **34**:5476–5487.

78. Ghosh, I., S. Naskar, S.S. Bandyopadhyay. 2010. Cryosorption storage of gaseous hydrogen for vehicular application—a conceptual design. *Int. J. Hydrogen Energy* **35**:161–168.

79. Hermosilla-Lara, G., G. Momen, P.H. Marty, B. Le Neindre, K. Hassoni. 2007. Hydrogen storage by adsorption on activated carbon: investigation of the thermal effects during the charging process. *Int. J. Hydrogen Energy* **32**:1542–1553.

80. Richard, M.-A., D. Cossement, P.-A. Chandonia, R. Chahine, D. Mori, K. Hirose. 2009. Preliminary evaluation of the performance of an adsorption-based hydrogen storage system. *AICHE J.* **55**:2985–2996.

81. Paggiaro, R., P. Benard, W. Polifke. 2010. Cryo-adsorptive hydrogen storage on activated carbon. I: Thermodynamic analysis of adsorption vessels and comparison with liquid and compressed gas hydrogen storage. *Int. J. Hydrogen Energy* **35**:638–647.

82. Paggiaro, R., F. Michl, P. Benard, W. Polifke. 2010. Cryo-adsorptive hydrogen storage on activated carbon. II: Investigation of the thermal effects during filling at cryogenic temperatures. *Int. J. Hydrogen Energy* **35**:648–659.

83. Hodoshima, S., H. Arai, Y. Saito. 2003. Liquid-film-type catalytic decalin dehydrogeno-aromatization for long-term storage and long-distance transportation of hydrogen. *Int. J. Hydrogen Energy* 28:197–204.

84. (a) Hodoshima, S., H. Arai, S. Takaiwa, Y. Saito. 2003. Catalytic decalin dehydrogenation/naphthalene hydrogenation pair as a hydrogen source for fuel-cell vehicle. *Int. J. Hydrogen Energy* 28:1255–1262. (b) Pez, G.P., A.R. Scott, A.C. Cooper, H. Cheng. US Patent. Appl. Publ. 2004/0223907Al.

85. Okada, Y., E. Sadaki, E. Watanabe, S. Hyodo, H. Nishijima. 2006. Development of dehydrogenation catalyst for hydrogen generation in organic chemical hydride method. *Int. J. Hydrogen Energy* 31:1348–1356.

86. Crabtree, R.H. 2008. Hydrogen storage in liquid organic heterocycles. *Energy Environ. Sci.* 1:134–138.

87. Eblagon, K.M., D. Rentsch, O. Friedrichs, A. Remhof, A. Zuettel, A.J. Ramirez-Cuesta, S.C. Tsang. 2010. Hydrogenation of 9-ethylcarbazole as a prototype of a liquid hydrogen carrier. *Int. J. Hydrogen Energy* 35:11609–11621.

88. Petrovic, J., G. Thomas. 2010. Reaction of aluminum with water to produce hydrogen. DOE report. http://www1.eere.energy.gov/hydrogenandfuelcells/pdfs/aluminum_water_hydrogen.pdf.

89. Ziebarth, J.T., J.M. Woodall, R.A. Kramer, G. Choi. 2011 Liquid phase-enabled reaction of Al-Ga and Al-Ga-In-Sn alloys with water. *Int. J. Hydrogen Energy* 36: 5271–5279.

90. Kong, V.C.Y., F.R. Foulkes, D.W. Kirk, J.T. Hinatsu. 1999. Development of hydrogen storage for fuel cell generators. I: Hydrogen generation using hydrolysis hydrides. *Int. J. Hydrogen Energy* 24:665–675.

91. Marrero-Alfonso, E.Y., A.M. Beaird, T.A. Davis, M.A. Matthews. 2009. Hydrogen generation from chemical hydrides. *Ind. Eng. Chem. Res.* 48:3703–3712.

92. Yan, J.-M., X.-B. Zhang, S. Han, H. Shioyama, Q. Xu. 2008. Iron-nanoparticle-catalyzed hydrolytic dehydrogenation of ammonia borane for chemical hydrogen storage. *Angew. Chem. Int. Ed.* 47:2287–2289.

93. Diwan, M., D. Hanna, A. Varma. 2010. Method to release hydrogen from ammonia borane for portable fuel cell applications. *Int. J. Hydrogen Energy* 35:577–584.

94. Huang, Z., X. Chen, T. Yisgedu, J.-C. Zhao, S.G. Shore. 2011. High-capacity hydrogen release through hydrolysis of NaB_3H_8. *Int. J. Hydrogen Energy* 36:7038–7042.

95. Dye, J.L., K.D. Cram, S.A. Urbin, M.Y. Redko, J.E. Jackson, M. Lefenfeld. 2005. Alkali metals plus silica gel: powerful reducing agents and convenient hydrogen sources. *J. Am. Chem. Soc.* 127:9338–9339.

96. Wallace, A., M. Lefenfeld. 2010. NaSi and Na-SG powder hydrogen fuel cells. *Proceedings of DOE Annual Merit Review.* http://www.hydrogen.energy.gov/pdfs/review10/st055_lefenfeld_2010_p_web.pdf.

97. Wallace, A., M. Lefenfeld. 2010. DOE annual progress report. http://www.hydrogen.energy.gov/pdfs/progress10/iv_h_1_lefenfeld.pdf.

98. Kim, Y., C.W. Yoon, W.-S. Han, S.O. Kang, S.W. Nam. 2012. A portable power-pack fueled by carbonsilane-based chemical hydrides. *Int. J. Hydrogen Energy* 37:3319–3327.

99. Friedrich, K.A., F.N. Buchi. 2008. Fuel cells using hydrogen. In *Hydrogen as a Future Energy Carrier*, ed. A. Zuettel, A. Borgschulte, L. Schlapbach, Chapter 8.1, p. 356. Weinheim, Germany: Wiley-VCH.

100. McWhorter, S., C. Read, G. Ordaz, N. Stetson. 2011. Materials-based hydrogen storage: attributes for near-term, early market PEM fuel cells. *Curr. Opin. Sol. St. Mater. Sci.* 15:29–38.

101. Teprovich, J.A., T. Motyka, R. Zidan. 2012. Hydrogen system using novel additives to catalyze hydrogen release from the hydrolysis of alane and activated aluminum. *Int. J. Hydrogen Energy* 37:1594–1603.

102. Zhu, L., V. Swaminathan, B. Gurau, R.I. Masel, M.A. Shannon. 2009. An onboard hydrogen generation method based on hydrides and water recovery for micro-fuel cells. *J. Power Sources* 192:556–561.

103. Kojima, Y., K.-I. Suzuki, K. Fukumoto, Y. Kawai, M. Kimbara, H. Nakanishi, S. Matsumoto. 2004. Development of 10 kW-scale hydrogen generator using chemical hydride. *J. Power Sources* 125:22–26.

104. Muir, S.S., X. Yao. 2011. Progress in sodium borohydride as a hydrogen storage material: development of hydrolysis catalysts and reaction systems. *Int. J. Hydrogen Energy* **36**:5983–5997.
105. Retnamma, R., A.Q. Novais, C.M. Rangel. 2011. Kinetics of hydrolysis of sodium borohydride for hydrogen production in fuel cell applications: a review. *Int. J. Hydrogen Energy* **36**:9772–9790.
106. Nikiforov, B.V., A.V. Chigarev. 2011. Problems of designing fuel cell power plants for submarines. *Int. J. Hydrogen Energy* **36**:1226–1229.
107. Wu, Y. 2005. Development of advance chemical hydrogen storage and generation system. *Proceedings of 2005 DOE Hydrogen Annual Merit Review*. http://www.hydrogen.energy.gov/pdfs/review05/stp_10_wu.pdf.
108. Wallace, A. 2007. Advances in chemical hydride based PEM fuel cells for portable power applications. Presented 2007 Joint Service Power Conference. http://www.dtic.mil/ndia/2007power/NDIARegency/Thur/Session18JadooJSPEWallace/JadooJointServicePower2007_VERSION_2.pdf.
109. Santos, D.M.F., C.A.C. Sequeira. 2011. Sodium borohydride as a fuel for the future. *Renewable Sustainable Energy Rev.* **15**:3980–4001.
110. Wu, Y., R.M. Mohring. 2003. Sodium borohydride for hydrogen storage. *Prepr. Pap.-Am Chem. Soc., Div. Fuel Chem.* **48**:940.
111. Wu, Y. 2003. Hydrogen storage via sodium borohydride—current status, barriers, and R&D roadmap. Presented at the GCEP, Stanford University. http://www.stanford.edu/group/gcep/pdfs/hydrogen_workshop/Wu.pdf.
112. Go/no-go recommendation for sodium borohydride for on-board vehicular hydrogen storage. An independent review of the U.S. Department of Energy Hydrogen Program, November 2007. Publication NREL/MP-150-42220. http://www1.eere.energy.gov/hydrogenandfuelcells/pdfs/42220.pdf.
113. Shang, Y., R. Chen. 2006. Hydrogen storage via the hydrolysis of $NaBH_4$ basic solution: optimization of $NaBH_4$ concentration. *Energy Fuels* **20**:2142–2148.
114. Ahluwalia, R.K., T.Q. Hua, J.K. Peng. 2012. On-board and off-board performance of hydrogen storage options for light-duty vehicles. *Int. J. Hydrogen Energy* **37**:2891–2910.
115. Stephens, F.H., V. Pons, R.T. Baker. 2007. Ammonia–borane: the hydrogen source par excellence? *Dalton Trans.* **7**:2613–2626.
116. Smythe, N.C., J.C. Gordon. 2010. Ammonia borane as a hydrogen carrier: dehydrogenation and regeneration. *Eur. J. Inorg. Chem.* **2010**:509–521.
117. Staubitz, A., A.P..M. Robertson, I. Manners. 2010. Ammonia-borane and related compounds as dihydrogen sources. *Chem. Rev.* **110**:4079–4124.
118. Ahluwalia, R.K., J.K. Peng, T.Q. Hua. 2011. Hydrogen release from ammonia borane dissolved in an ionic liquid. *Int. J. Hydrogen Energy* **36**:15689–15697.
119. Devarakonda, M., K. Brooks, E. Rönnebro, S. Rassat. 2012. Systems modeling, simulation and material operating requirements for chemical hydride based hydrogen storage. *Int. J. Hydrogen Energy* **37**:2779–2793.
120. Chu, Y.S., P. Chen, G. Wu, Z. Xiong. 2011. Development of amidoboranes for hydrogen storage. *Chem. Commun.* **47**:5116–5129.

4

Hydrogen Storage in Pressure Vessels: Liquid, Cryogenic, and Compressed Gas

Guillaume Petitpas and Salvador Aceves

CONTENTS

Introduction 91
Density 92
Capital Cost 95
Compression, Liquefaction, Delivery, and Dispensing 97
Dormancy 98
Refueling 100
Safety 105
Conclusions 106
Acknowledgments 107
References 107

Introduction

This chapter describes and compares the different options for hydrogen storage in pressure vessels. Compressed (ambient temperature and high pressure), liquid (very low temperature and low pressure), and cryogenic pressure vessels (low temperature and high pressure) are all described and analyzed from a thermodynamic point of view, in which key performance figures: Weight and volume performance, capital cost, fuel delivery and dispensing, dormancy, refueling, and safety are shown to be controlled by thermodynamic functions such as density, internal energy, exergy, enthalpy, and entropy.

Unlike other fuels, hydrogen (H_2) can be generated and consumed without producing carbon dioxide (CO_2). This creates both significant engineering challenges and unsurpassed ecological advantages for H_2 as a fuel while enabling an inexhaustible (closed) global fuel cycle based on the cleanest, most abundant, natural, and elementary substances: H_2, O_2, and H_2O. If generated using light, heat, or electrical energy from solar, wind, fission, or (future) fusion power sources, H_2 becomes a versatile, storable, and universal carbonless energy carrier, a necessary element for future global energy systems aimed at being free of air and water pollution, CO_2, and other greenhouse gases [1]. This has been discussed in Chapter 1.

The physical and chemical properties of H_2 make its use superior to fossil fuels. H_2 is a simple nontoxic molecule that generates power cleanly and efficiently, even silently and without combustion if desired. However, widespread use of H_2 has been challenging due to its low density. Whether it is compressed to extreme pressures, cooled to cryogenic temperatures, or bonded to metals or sorbent structures, H_2 stores little energy per unit of volume relative to conventional (hydrocarbon) fuels. Therefore, storing enough H_2 fuel on-board a vehicle to achieve sufficient (500+ km) driving range in a compact, lightweight, rapidly refuelable, and cost-effective system remains the predominant technical barrier limiting widespread use of H_2 automobiles. Other issues (e.g., hydrogen supply) will likely be addressed during the early deployment of hydrogen infrastructure.

This chapter focuses on H_2 storage in pressure vessels. Depending on pressure and temperature operating regimes, three common approaches exist to H_2 pressure vessel storage:

1. Compressed gas vessels operating at high pressure (typically 350–700 bar) and near ambient temperature; typically made of carbon fiber composite wound around an aluminum (Type III vessels) or polymer (Type IV vessels) liner [2, 3].

2. Liquid hydrogen (LH_2) vessels for low pressure (~6 bar) and temperatures below the H_2 critical point (20–30 K); consist of an inner stainless steel vessel surrounded by vacuum insulation and an outer metallic vacuum jacket [4].

3. Cryogenic pressure vessels, sometimes called "cryocompressed" vessels, compatible with both high pressure (350 bar) and cryogenic temperatures down to LH_2 (20 K) and comprising a Type III aluminum-composite vessel surrounded by a vacuum space and an outer metallic vacuum jacket [5].

Building on the introduction to these systems given in Chapter 3 by Bob Bowman and Lennie Klebanoff, in this chapter we analyze these three approaches from a thermodynamic point of view, focusing on how the pressure vessel key performance features (density, weight, volume, cost, refueling, dormancy, safety, environmental performance) are controlled by thermodynamic properties such as pressure, internal energy, enthalpy, and exergy.

Density

System hydrogen storage density is the most important parameter in H_2 vessels. High storage density enables long-range vehicles necessary for practical carbonless transportation. Increasing the storage density also increases the H_2 weight fraction (kilograms H_2/kilograms vessel and accessories) and reduces cost (dollars/kilogram H_2) because compact vessels demand less material (metal and composite) for manufacture. Compact vessels also fit better within available spaces in the vehicle, leaving more space for passengers or cargo.

Figure 4.1 shows the gravimetric density of H_2 gas (grams/liter, solid lines) as a function of temperature and pressure. Focusing for a moment on the 300 K line, it can be observed that the density of ambient temperature compressed gas is a relatively low 50 g/L, even when pressurized to 1000 bar. For context, note that the density of saturated LH_2 at 1 bar is 70.7 g/L, and gasoline has the energy equivalent density of 264 g H_2/L. In addition to this, curvature in the 300 K line indicates that gaseous H_2 is not an ideal gas.

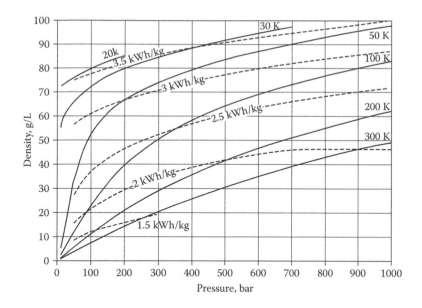

FIGURE 4.1
Calculated results for temperature (solid lines) and exergy (dashed lines) as a function of H_2 pressure and density. Exergy does not include ortho-para H_2 conversion, which can be important for H_2 stored more than a few days at temperatures below 60 K. The ideal work of converting normal (75% ortho) LH_2 to pure para-LH_2 is 0.65 kWh/kg.

It is less compressible as the pressure increases and becomes more so at high pressures. At 350 bar, the molar volume of hydrogen is about 20% larger than that predicted by the ideal gas law due to intermolecular repulsion of the hydrogen molecules, as described in Chapter 3. Nonetheless, significant increases in H_2 volumetric density can be achieved via compression.

Aside from pressurization, much potential exists to increase storage density by cooling. Very cold gaseous H_2 (also called cryogenic hydrogen) at 100 K reaches a density of 50 g/L at a moderate pressure of 300 bar. The density of LH_2 is 70.7 g/L when saturated at 1 bar at 20.3 K. Densities above 90 g/L (~30% denser than LH_2) can be obtained with very cold pressurized H_2, although refueling at such high density has not been demonstrated at vehicle scale. Solid hydrogen storage was described in Chapter 3 for specialized spacecraft systems.

Increasing H_2 density is typically a productive endeavor. While the previous discussion considers only the density associated with the physical state of hydrogen (gas, liquid), it is the *system* storage density that ultimately matters since thick vessel walls and thermal insulation occupy volume otherwise available for H_2 storage. The strength of practical materials and the performance of thermal insulation therefore limit ultimate vessel storage system density. Beyond the tank itself, other storage system hardware, such as valves and manifolds, contribute to the total system volume.

The net impact of H_2 real gas behavior, vessel wall strength, and thermal insulation is best illustrated by modeling the amount of H_2 that can be stored in a volume of fixed dimensions as shown in Figure 4.2. Consider a long (2 m), slender (35-cm outer diameter) pressure vessel, similar to the dimensions preferred by manufacturers due to efficient packaging along the central tunnel. The fixed total outer cylindrical vessel volume is 170 L, assuming ellipsoidal heads. Internal vessel volume is a function of design pressure since

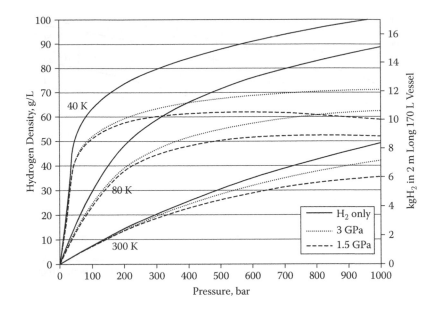

FIGURE 4.2

Predictions for H_2 storage capacity vs. pressure for idealized pressure vessels of varying wall stress at ambient (300 K) and cryogenic (40 and 80 K) temperatures. All vessels shown are cylindrical with ellipsoidal end caps and identical external volumes (170 L) and dimensions (2-m length, 35-cm diameter). Vessel wall thickness calculations include thick-wall effects. Three vessels are shown for each temperature case: 3.0-GPa wall strength, 1.5-GPa wall strength, and H_2 only (amount stored considering zero wall thickness). Safety factor is 2.25. Analysis assumed negligible liner thickness. Cryogenic vessels had 1.5 cm vacuum insulation. Corresponding H_2 storage system densities are shown on the left axis.

wall thickness increases as pressure increases to maintain stresses below a maximum (ultimate strength divided by 2.25 safety factor). For cryogenic vessels, vacuum insulation (assumed 1.5 cm thick from a recent prototype [5]) occupies 29 L, which has to be deducted from the total volume.

We assume two vessels made of materials with two different inherent wall strengths: 1.5 GPa, similar to today's advanced composite materials; and 3 GPa, which may be obtained in the future with improved materials. Figure 4.2 assumes no vessel liner or a thin vessel liner. For any design pressure, a material with higher strength allows vessels with thinner walls, leading to a greater available inner volume. Increasing the inner volume increases the available hydrogen storage volume contained within the fixed external volume, leading to greater volumetric efficiency. We call the ratio of inner volume to external volume the *geometric volume efficiency*.

Compressed H_2 vessels made of high-strength materials (dotted lines in Figure 4.2) achieve 90–95% geometric volumetric efficiency at pressures up to 350 bar at ambient temperature. Above 350 bar, as thicker walls become necessary, the inner volume available to store gas goes down, and the geometric volumetric efficiency and hydrogen storage capacity decrease. For example, a 700-bar, 170-L (external volume) pressure vessel made of 1.5-GPa material with a safety factor of 2.25 is predicted to store only 5.3 kg of H_2 instead of the theoretical 6.6 kg H_2 due to the reduced inner volume available. The same 170-L vessel made of materials with twice the strength (3.0 GPa) would have thinner walls and could store 6 kg H_2 at 700 bar.

The disadvantage of low-strength wall materials is made worse by the real-gas nonlinear relationship between pressure and H_2 density. Referring to Figure 4.2, let us consider the situation at 700-bar storage at 300 K. For H_2 gas only, with a volume equal to the external volume of 170 L, the calculations show we can store about 7 kg of H_2. If we go to a real storage vessel with wall strength of 3.0 GPa, we take up some inner volume with the tank wall. At 3.0 GPa, we can only store about 6 kg of hydrogen in the reduced inner volume at 300 K and 700 bar. If we want to get back to 7 kg, we would have to increase the pressure to about 900 bar at 300 K for the 3.0-GPa case. Alternatively, if we went with a 1.5-GPa wall material, the vessel wall would be even thicker still, and instead of storing 7 kg of H_2 (gas only) at 700 bar and 300 K, we could now only store about 5 kg of H_2. It is impossible to recover the 7 kg H_2 storage capacity of a 1.5-GPa vessel by raising pressure.

At higher pressures (1000 bar), an ultimate H_2 storage density is reached as H_2 becomes more incompressible and vessel walls must be ever thicker. These two factors ultimately limit maximum H_2 storage density to pressures of approximately 10% of vessel wall material strength and 60% geometric volumetric efficiency, whereas maximum *economic* compressed H_2 pressures are likely to occur at approximately 5% of vessel wall material strength.

In summary, system storage density of compressed gas vessels is limited by the nonideal properties of H_2 and the strength of available vessel materials. Although a maximum system density can be reached at a pressure about 10% of the wall strength (e.g., 1500 bar for a 1.5-GPa composite material), it is unlikely that pressure vessels will operate beyond about 5% of ultimate strength because only marginal density gains are obtained at considerable increase in material cost.

Compressed gas and LH_2 vessels operate at (nearly) constant temperature. Cryogenic pressure vessels, sometimes called cryocompressed vessels, do not. They continuously change temperature, depending on how the hydrogen-powered car is driven. The cryocompressed tanks cool down if the vehicle is driven, and they warm up if the vehicle is parked for a long time. Figure 4.2 therefore considers two cryogenic temperatures: 40 and 80 K. Hydrogen at these temperatures deviates from ideal gas behavior at very low pressure (~100 bar), where the lines start exhibiting considerable nonlinear behavior. As a result, while cryogenic H_2 density increases to over 100 g/L (at 40 K and 1000 bar), system H_2 density for real vessels with walls reaches its maximum at much lower pressures than for ambient temperature gas: 300–700 bar for 1.5-GPa vessels and slightly over 1000 bar for 3.0-GPa vessels. While there is little density gain in increasing cryogenic vessel pressure rating beyond 350 bar, there is a dormancy gain that may justify the additional expense, as discussed further in this chapter.

Capital Cost

Storage system cost is another key parameter for development of practical H_2 vehicles. Considering that liquid hydrocarbon (gasoline and diesel) tanks are very inexpensive (~$100), it is anticipated that H_2 vessels will represent the largest cost premium vs. today's vehicles, even after large-scale introduction of H_2 vehicles.

As shown in Figure 4.3 [6], Argonne National Laboratory and TIAX have conducted detailed cost modeling of H_2 storage vessels (capacity 5.6 kg H_2) when introduced in large

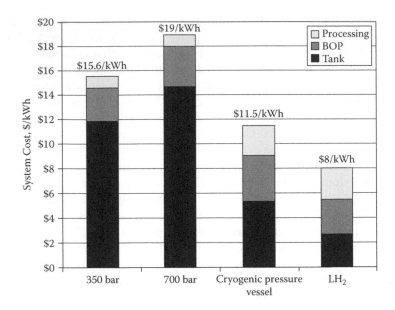

FIGURE 4.3
Argonne National Laboratory estimates of on-board storage system cost as a function of usable H_2 capacity when fabricated in large scale (500,000/yr [6]). The figure shows *system* cost (vessel, accessories, and processing) in dollars per kilowatt hour (33.3 kWh = 1 kg H_2) for compressed gas vessels at 350 and 700 bar; cryogenic pressure vessels; and LH_2 tanks, for 5.6 kg available H_2 capacity. The DOE cost target of $4/kWh [7] has thus far proven out of reach for available technologies.

scale (500,000 vehicles per year). The relatively low density of compressed H_2 demands large vessels made of expensive materials (metal and composite) that contribute about 80% of the high cost of compressed gas storage: $15.3/kWh for 350-bar and $18.6/kWh for 700-bar vessels. This cost difference is mainly due to the reduced compressibility of H_2 at the higher pressure. Balance of plant (valves, regulator) and processing represent the remaining 20%. Total cost for a 5.6-kg H_2 vessel is $2800–3500 (1 kg H_2 = 33.33 kWh). These values are about half the current costs for small-scale production of $1000/kg H_2, or $33/kWh.

The higher density of cryogenic H_2 reduces vessel size and cost. LH_2 vessels are the least-expensive option because low-pressure operation reduces material expense (no composites required). Cryogenic pressure vessels demand thick composite walls for high-pressure operation and therefore cost more than LH_2 tanks, although still considerably less than compressed gas vessels at $11.3/kWh due to the reduced cryocompressed pressure of about 300 bar. Cryogenic storage systems demand vacuum insulation, cryogenic valves that reduce heat entry, cryogenic regulators that do not freeze, and pressure, temperature, or level sensors to determine state of charge. Therefore, balance of plant and processing are larger fractions of the total cost, and vessel costs are only 30% of the total for LH_2 tanks and 50% for cryogenic pressure vessels.

The cost structure of cryogenic storage, where the vessel contributes 50% or less of the total cost, favors applicability to high-capacity systems. While compressed gas system cost scales almost linearly with kilograms H_2, cryogenic pressure vessel capacity can be almost doubled (from 5.6 to 10.4 kg H_2) while increasing system cost by only $700 [8] (system cost decreasing from $11.5/kWh to $8/kWh). Sensitivity to environmental heat transfer (dormancy and evaporative losses) is also less challenging for larger cryogenic vessels due to reduced area-to-volume ratio, as shown in Figure 3.5 of Chapter 3 for LH_2 storage systems. Cryogenic vessels may therefore be the best choice for large vehicles with low fuel economy or small vehicles demanding a long driving range.

Compression, Liquefaction, Delivery, and Dispensing

Vessel capital cost is an important parameter for achieving widespread commercialization of H_2 vehicles. However, the key parameter for the consumer is life-cycle cost. This includes refueling cost in addition to vessel capital cost.

Alternatives for H_2 delivery and dispensing are similar to those used for vehicle storage: liquefaction and compression [9]. These two paths can be thermodynamically analyzed in terms of exergy. Defined as the minimum theoretical work necessary for compressing and cooling H_2 from a reference state (300 K and 1 bar) to any state (T, p), exergy is therefore an indication of the energy necessary for H_2 densification.

Going back to Figure 4.1, the dashed lines show exergy as a function of pressure and density (or temperature). The figure shows that H_2 compression is exergetically inexpensive compared to cooling: Any level of densification from 10 to 100 g/L is achieved with minimum possible exergy by maximizing pressurization and minimizing cooling. To better appreciate this, select any density (e.g., 40 g/L) and observe how exergy drops monotonically as we move toward higher temperatures. H_2 liquefaction is especially expensive (3.92–3.27 kWh/kg depending on whether ortho-para H_2 conversion is included or not [10]). This is reflected in reality by the very large liquefaction energy (7–13.4 kWh/kg) vs. compression energy (1.5–2 kWh/kg theoretical and 3–5 kWh/kg in practice). The energetic disadvantage of liquefaction also makes it more expensive and more polluting when using today's average grid electricity [9].

LH_2 has, however, a virtue that largely mitigates the high cost of liquefaction: high density at low pressure. Being dense, LH_2 can be dispensed in inexpensive low-pressure tanker trucks that carry about 10 times as much H_2 as tube trailers for compressed H_2 (4000 vs. 550 kg), reducing driver and truck costs and therefore reducing delivery cost ($0.50/kg vs. $1.50–$2/kg for compressed gas) [9].

Finally, LH_2 dispensing is less expensive than compressed gas dispensing. Following the 3.5-kWh/kg exergy line in Figure 4.1, we can see that 100 g/L H_2 at 1000 bar and 43 K has the same exergy as 70.7 g/L LH_2 at 1 bar and 20 K. It would therefore be theoretically possible to densify LH_2 to 100 g/L without work input. While impractical in practice, this points to the thermodynamic advantage of LH_2 pumps (discussed further in this chapter) vs. pressurizing gaseous H_2. Aside from demanding considerable work input (2 kWh/kg exergy for 1000 bar and 50 g/L), compressed H_2 dispensing is expensive because it demands compressors, a cascade charging system, and, for 700-bar delivery, a booster compressor and a chiller [9, 11].

In summary, the total fueling cost is comparable for compressed gas, LH_2, and cryocompressed options because the high cost and energy of liquefaction is mitigated by LH_2's ease of delivery and dispensing. Cost difference between these delivery alternatives under typical scenarios is less than $0.50 per kilogram of hydrogen when delivering to midsize (400 kg H_2/day) stations [9]. Considering that capital cost is lowest for cryogenic vessels (Figure 4.3) and refueling costs are similar, life-cycle costs, including capital and refueling costs, are lowest for LH_2 and cryogenic pressure vessels [9]. A discussion of wells-to-wheels costs for several different types of hydrogen storage can be found in Reference 9.

Dormancy

Minimum life-cycle cost, potential for large capacity, and compact size make LH_2 and cryogenic pressure vessels a compelling option. Evaporative losses are, however, a key challenge. Operating at cryogenic temperatures, these systems are exposed to continued environmental heat transfer. While high-performance vacuum insulation can go a long way toward reducing heat transfer, even a few watts can result in evaporative losses when a vehicle is not frequently operated. The dormancy (period of inactivity before a vessel releases H_2 to reduce pressure buildup) is an important parameter for cryogenic storage acceptability.

Dormancy can be calculated from the first law of thermodynamics [12] and the properties of H_2 [13] and can be illustrated with a diagram of H_2 thermodynamic properties (Figure 4.4) to simplify visualization and graphical calculation of H_2 vessel dormancy. A full discussion can be found in Reference 5.

Figure 4.4 is a simplified version of a more complete figure from Reference 5. Figure 4.4 uses axes of specific internal energy and density instead of more traditional temperature and pressure. An appropriate choice of scales in Figure 4.4 radically simplifies dormancy

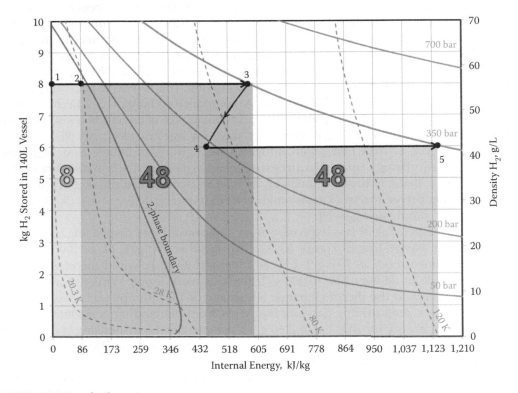

FIGURE 4.4 (See color insert.)
Phase diagram for H_2 showing density (right vertical axis) and internal energy (horizontal axis), with lines for constant pressure (solid) and temperature (dashed). Curves for entropy are not shown. The left axis shows the mass of H_2 contained in a vessel with 140-L internal volume, which would store 10 kg of LH_2 at 20 K and 1 bar. The figure also shows points to areas representing dormancy (in watt-days) of conventional LH_2 tanks (light gray) and cryogenic pressure vessels (all shadings). Thermodynamic states that are "inside" (to the left of) the 2-phase boundary line consist of two phases; whereas states to the right of this line consist of one supercritical phase.

calculations. The grid scale in the specific internal energy (horizontal) axis is set at 86.4 kJ/kg H_2, which converts to 1 watt-day/kg H_2 (1 day = 86,400 s). The grid scale in the vertical axis represents 1 kg H_2. Therefore, the area of a grid square represents 1 watt-day of heating. The total change in internal energy (in watt-days) can be easily calculated by counting the squares under the curve representing changing the state of hydrogen from one point to another in Figure 4.4. Dormancy (days) is calculated by dividing the internal energy change (in watt-days) by the rate of heat transfer (in watts).

A dormancy calculation begins by identifying the initial thermodynamic state in Figure 4.4 of the H_2 contained in the vessel. As an illustration, consider a parked automobile with a LH_2 tank with 140-L internal volume and 6 bar maximum working pressure, which is 80% full with 8 kg LH_2 at 20 K and 1 bar (point 1 in Figure 4.4). As the vehicle is parked (not using H_2), heat transfer warms the H_2, increasing both its temperature and pressure. Dormancy ends in this case when the pressure reaches 6 bar (point 2), when H_2 venting or vehicle driving becomes necessary to maintain pressure within the vessel rating. Total heat absorbed during this process from point 1 to point 2 can be calculated by counting the number of squares (8 watt-days) in the area marked by the light gray shaded area under the 1-2 line. Dormancy can then be calculated by dividing 8 watt-days by the heat transfer rate. If one has a heat leak rate of 4 W total, then the dormancy would be 2 days before loss of hydrogen.

Figure 4.4 illustrates the dramatic increase in dormancy of automobiles with cryogenic pressure vessels. An auto initially filled with 8 kg LH_2 at 1 bar and 20 K can remain parked until the pressure reaches 350 bar (point 3 in the figure) without venting any H_2. Counting squares under the line joining point 1 and point 3 (two shaded regions), we obtain 8 + 48 = 56 watt-days, *seven times* greater thermal endurance than a conventional LH_2 tank.

Furthermore, unlike conventional LH_2 vessels, cryogenic pressure vessels dramatically extend dormancy as the vehicle is driven. For example, if the parked vehicle is driven when the H_2 is at point 3 (Figure 4.4), consuming 2 kg of H_2 fuel, the remaining H_2 in the vessel expands and cools following a constant entropy line from point 3 to a point of lower hydrogen density, point 4. If the vehicle were not driven further, the system would progress from point 4 to point 5, extending the vessel thermal endurance by an additional 48 watt-days. Eventually, loss would occur at point 5 as the internal pressure exceeded the 350-bar pressure limit of the vessel. In this way, periodic driving substantially extends dormancy, essentially eliminating fuel evaporation for even very moderate driving patterns.

Figure 4.4 is somewhat conservative because it neglects secondary effects such as vessel heat capacity and heat potentially absorbed by conversion between the two states of nuclear spin arrangement (i.e., *para*-H_2 conversion to *ortho*-H_2) of H_2 molecules. Both of these effects lower the internal energy of the stored hydrogen, thereby increasing the dormancy of the vessels. However, these effects are most significant only for warmer temperatures ($T > 77$ K). Both effects are negligible at the very low (i.e., 20–30 K) temperatures at which LH_2 tanks operate.

From Figure 4.4, it is clear that insulation performance can be traded off vs. vessel pressure rating: Dormancy can be increased by either improving insulation (reducing heat leaks) or strengthening the vessel (raising the vent pressure). This trade-off is better illustrated in the predictions of Figure 4.5, which shows contour lines of dormancy vs. heat transfer rate (*x*-axis) and vessel-rated pressure (*y*-axis) for an initially full vessel (140 L internal volume and 10 kg LH_2 at 1 bar). The figure shows that a pressure vessel may reach acceptable dormancy (for example, 5 days) through either high-performance insulation (~3-W heat leak rate) of an inexpensive low-pressure (~167-bar) vessel or, alternatively, using low-performance insulation (~17-W heat leak rate) of a high-pressure (700-bar) vessel.

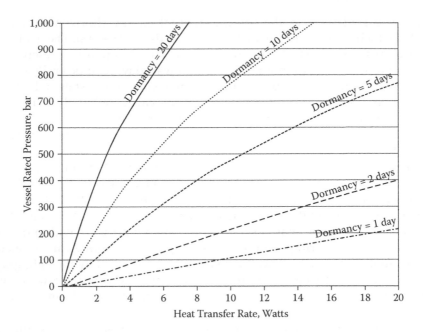

FIGURE 4.5
Contour lines of dormancy vs. heat transfer rate (*x*-axis) and vessel-rated pressure (*y*-axis) calculated for a 140-L vessel initially full with 10 kg LH_2 saturated at 1 bar.

The optimum design point depends on the relative cost of vessel materials and insulation, as well as the particular mission requirements. Even larger heat transfer rates would be allowable in continuously driven vehicles (e.g., taxis, buses) since driving increases dormancy, as shown in Figure 4.4, and the allowable dormancy can be as short as 1 day in such fleet vehicles. Strong vessels (>500-bar rated pressure) deliver 1 day of dormancy at heat leak rates under 50 W, possibly allowing use of inexpensive powder or foam insulation.

It should also be noted that Figure 4.5 assumes a completely full vessel at LH_2 density—the best case for low-pressure vessels and the worst case for high-pressure vessels. Low-pressure vessels (<10 bar) have maximum dormancy when full because the thermal inertia of the evaporating LH_2 slows the pressure rise. On the other hand, cryogenic pressure vessels gain most of their dormancy from containing the H_2 as it heats up and therefore have the longest dormancy at low fill levels, at which the vessel can heat up more before reaching the rated pressure. Dormancy is infinite when ambient temperature H_2 at rated pressure is denser than the cryogenic H_2 stored in the vessel.

Refueling

Rapid, high-density refueling is key for H_2 storage acceptability. Chapter 10 discusses how refueling of solid-state hydrogen storage is limited by the thermodynamics of the processes of heat generation and heat removal. Pressure vessel refueling is also greatly affected by thermodynamics, which determines ultimate H_2 storage density and therefore vehicle driving range. We discuss the thermodynamics of both compressed gas and LH_2 storage, beginning with compressed hydrogen storage.

Typically, before one fills an empty compressed hydrogen vessel with H_2, one needs to know the following: (1) How hot will the hydrogen vessel get as we are compressing hydrogen gas within it? (2) What will the final pressure be? and (3) How much hydrogen will be in the vessel? Pressure vessel refueling can be modeled with the first law of thermodynamics for open systems [12]. To illustrate the thermodynamic implications of vessel fill processes, we start by considering a simplified case: a large hydrogen station tank storing H_2 at temperature T_i and pressure p_i. This tank fills a relatively small compressed gas storage vessel on-board a vehicle at the station.

As a first approximation, we consider an initially empty vehicle vessel with negligible thermal mass and negligible heat transfer to/from the environment. Under these assumptions, the first law of thermodynamics simplifies to $u_f = h_i$, where u_f is the specific internal energy of the hydrogen inside the vehicle vessel, and h_i is the specific enthalpy of the H_2 flowing at the station vessel. The enthalpy h_i is calculated at the hydrogen station's tank conditions (p_i, T_i), assumed constant due to its large relative size. From thermodynamics, $h_i = u_i + p_i v_i$. The term $p_i v_i$, frequently named flow work, explains the heating that occurs when gases are forced into a vessel.

Figure 4.6 shows how the density of hydrogen in the vehicle vessel changes during the fill and how the temperature of the gas increases due to the compression. In this modeling study, the large storage tank at the hydrogen filling station is assumed to have an initial pressure p_i of 700 bar, and with varying assumed initial temperatures T_i of 100, 200, and 300 K. Figure 4.6 shows that compressing hydrogen in the vehicle vessel to the 700 bar limit heats the hydrogen in all cases, but by smaller amounts if the initial station tank temperature T_i is lower. If the station hydrogen is initially at 300 K, the final delivered gas temperature in the vessel is 460 K, an increase of 160 K. This higher temperature limits the density of the delivered hydrogen to only 28 g/L (at 700 bar) in the vehicle's vessel (dashed line at 300 K). Alternatively, if we lower the temperature of hydrogen in the station tank to 200 K, we are able to deliver 38 g/L into the vehicle at a final temperature of 310 K at the

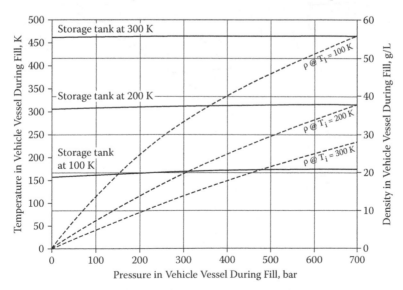

FIGURE 4.6
Model results for automotive vessel temperature and density during the fill process as a function of pressure when filled from a large hydrogen station tank at $p_i = 700$ bar and $T_i = 100, 200,$ and 300 K, assuming an initially empty vehicle vessel with negligible thermal mass and no environmental heat transfer.

700 bar pressure limit. If we are able to store large quantities of 100 K hydrogen gas at the station, then filling the vehicle vessel leads to a final vessel temperature of 175 K (only a 75 K temperature rise), and the final density of hydrogen in the vehicle's storage vessel is 56 g/L, consistent with the lower gas temperature.

Although there is a jump in temperature from station tank to vehicle vessel, the temperature inside the vehicle vessel, once established, is nearly constant during the fill process, changing only slightly due to H_2's nonideal behavior, especially at lower temperature. It is worth pointing out that H_2 heats up as it flows into the vessel regardless of the value of the Joule-Thomson coefficient (negative at ambient temperature, near zero at 200 K, and positive at 100 K): Vessel fill processes are not isenthalpic and therefore not controlled by the Joule-Thomson coefficient. Flow work plays the key role in describing the process. In addition, heating of the hydrogen gas with compression can lead to degradation of the internal surfaces of the tank if the temperature gets too high.

Results in Figure 4.6 can be generalized by varying p_i and T_i over broad ranges. Figure 4.7 predicts vehicle vessel temperature (labeled in K), and vehicle vessel fill density (solid lines labeled in g/L), at the end of the refueling process, when pressure equilibrium is reached with the hydrogen station tank ($p = p_i$), for an initially empty vehicle vessel with negligible thermal mass, for any combination of H_2 station tank pressure and temperature (p_i, T_i). As an example, assume that the H_2 station tank contains H_2 at $T_i = 100$ K and $p_i = 300$ bar. From the figure, the vehicle vessel at the end of the fill process (when both vessels equilibrate at 300 bar) would be at 150 K and about 40 g/L.

Figure 4.7 once again shows the considerable heating that occurs during the fill process. For an ideal gas with constant specific heat, heating is constant ($T = \gamma T_i$, where γ is the

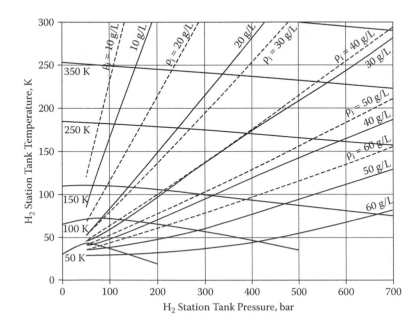

FIGURE 4.7
Model calculations for vehicle vessel temperature (labeled in K) and fill density (solid lines labeled in g/L), at the end of the refueling process, when pressure equilibrium is reached with the H_2 station tank ($p = p_i$), for an initially empty vehicle vessel with negligible thermal mass, for any combination of station tank pressure and temperature (p_i, T_i). The figure also shows (dashed lines) the density of the hydrogen in the station tank at (p_i, T_i). The difference between dashed lines and solid density lines represents the density losses due to flow work.

specific heat ratio c_p/c_v), and therefore temperature lines are fairly horizontal at low pressures and high temperatures. At higher pressures, heating increases because H_2 becomes increasingly incompressible ($pv > nRT$). Minimum heating occurs at low temperature and pressure, where H_2's compressibility factor drops as H_2 approaches the liquid phase. This is observed by the 50 K blue line reaching a local maximum at about 50 bar.

In particular, looking at $T_i = 300$ K (top of Figure 4.7), we can see the substantial heating that occurs during compressed gas filling. H_2 heats up to 440–464 K when filled to 350–700 bar (off the top of the chart in Figure 4.7), reducing storage capacity by 38% and demanding 48–55% overpressurization to recover the density that would be obtained during isothermal filling. Fortunately, in practice, pressure vessels absorb some of this thermal energy, so warming to only about 350 K is observed [11]. This reduces storage capacity by only 14–12% (at 350–700 bar), and the storage capacity can be restored with 17% overpressure. Overpressure effects vary little with filling pressure between 350 and 1000 bar and are lessened because H_2 is easier to compress as it warms.

In addition to fill density and temperature, Figure 4.7 also shows H_2 station tank hydrogen density ρ_i at p_i, T_i (dashed lines). This would be the fill density if no flow work existed, or if the vehicle vessel were somehow cooled down to the station tank temperature during fill. Density losses during vessel fill are represented in Figure 4.7 by the difference between the dashed and the solid density lines. Flow work therefore has a large impact on delivered density. On average, fill density would be approximately 1.4 times higher without flow work. The fact that this average factor is almost identical to the specific heat ratio γ is no surprise since γ determines fill heating, at least when H_2 behaves ideally and specific heats remain constant.

The results of Figures 4.6 and 4.7 characterize fill processes involving compressed hydrogen gas for both a hydrogen station storage tank and the vehicle hydrogen storage vessel being filled. However, cryogenic vessels are more likely to be filled with an LH_2 pump coupled to a large station LH_2 Dewar. Considering that pressurizing LH_2 is exergetically inexpensive (Figure 4.1), LH_2 pumping enables rapid compression and efficient densification with low evaporation (less than 3%) in commercially available systems. Evaporated LH_2 is recycled into the Dewar, avoiding losses and repressurizing the Dewar after LH_2 extraction.

Predicting LH_2 pump refueling is similar to predicting compressed gas vessel refueling, and the same equations dominate the process. However, delivery conditions are functions of LH_2 pump performance details, and these remain unpublished. We therefore assume, as a first approximation, isentropic LH_2 pumping. While real pumps will not reach this level of performance, final vessel density may be reasonably well approximated due to the low exergetic cost of LH_2 pumping. We assume that the large LH_2 hydrogen station Dewar is saturated at 3 bar and 24.6 K, filling a vehicle pressure vessel to 700 bar at various temperatures.

Modeling results in Figure 4.8 show final vehicle H_2 density after refueling (when the vehicle drives away from the fueling station) as a function of vehicle H_2 density before refueling (when the vehicle drives into the fueling station) for multiple initial vehicle vessel temperatures: 30, 50, 100, 200, and 300 K. Two sets of lines are shown: isentropic pump with negligible pressure vessel thermal mass (dashed) and isentropic pump including the vessel thermal mass and assuming thermal equilibrium between vessel and H_2 (solid lines). We assume vessel refueling to 700 bar. Note the ordinate of Figure 4.8 begins at 30 g/L.

Looking first at the dashed lines, we see that all fills from empty end at very high density, nearly 90 g/L, because the initial vessel temperature does not matter if the vehicle pressure vessel has negligible thermal mass. Comparing dashed lines vs. solid lines in Figure 4.8 shows that nonnegligible vessel thermal mass plays a large role in fill density for warm (200–300 K), empty vessels, refueling from the LH_2 pump at the station. These

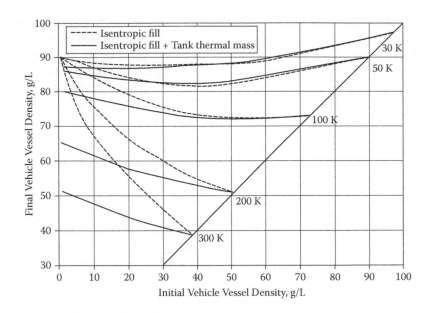

FIGURE 4.8
Model predictions for H_2 density after refueling with an isentropic LH_2 pump as a function of H_2 density before refueling for multiple initial vessel temperatures (30, 50, 100, 200, and 300 K). Two sets of lines are shown: isentropic pump with negligible pressure vessel thermal mass (dashed) and isentropic pump including the vessel thermal mass and assuming thermal equilibrium between vessel and H_2. The black diagonal line in the right of the figure gives the initial vehicle hydrogen density if the vehicle had a full-capacity charge of hydrogen at 700 bar at the temperatures indicated.

realistic vessels cool only partially during refueling, driving away pressurized (700 bar) at relatively low density due to their elevated temperature. A few rapid drive-refuel cycles would be necessary to reach low temperature and full density.

At very low starting temperatures (30–50 K), the heat absorbed by the vessel is relatively small because the starting temperature is only 10–30 K different from that of LH_2. As a result, the difference between the solid and dashed lines for 30–50 K starting temperatures is much smaller, and high-density refueling is achieved regardless of the initial fill level. In the limit of an initially empty vessel with zero thermal mass, the fill density is 91.45 g/L, obtained from isentropic compression of 3-bar saturated LH_2. All dashed lines approach this density for an initially empty vessel. Even higher densities (>95 g/L) are obtained when refueling very cold (30 K), nearly full vessels, for which the pump is essentially compressing low-entropy H_2 already in the vessel.

The black diagonal line in the right of Figure 4.8 gives the initial vehicle hydrogen density if the vehicle had a full capacity charge of hydrogen at 700 bar at the temperatures indicated. Since Figure 4.8 assumes refueling to 700 bar, the black line is a limit where no refueling is possible, and density after refueling equals density before refueling. Driving is necessary before refueling to reduce pressure and cool the vessel.

In summary, the low exergetic cost of pumping LH_2 may lead to rapid and efficient refueling at high densities. Flow work is also lowest near the liquid phase (Figure 4.7), minimizing heating and density losses during vessel fill. Future experiments will reveal how closely the isentropic pump models real pump performance. However, the density of the fueling is reduced by warm tanks with nonzero thermal mass.

Safety

Safety, both real and perceived, is an often-raised criticism of pressurized H_2 storage. However, the safety risks of storing compressed or cryogenic fluids are not a simple function of pressure. The overall safety of pressure vessels can be counterintuitive, for although vessel wall strength and impact resistance increase directly with storage pressure, the maximum mechanical energy released by sudden expansion (e.g., in a vessel rupture) of the stored gas (H_2) does not. Here, we consider some of the thermodynamic aspects of safety associated with compressed gas, LH_2, and cryocompressed approaches. The "codes and standards" that have developed around compressed gas hydrogen storage technology are described in Chapter 12 by Chris Sloane.

Thermodynamically, the mechanical energy released during a sudden expansion can be calculated as $W = \int pdv$, where p is pressure and dv is a differential volume expansion [14]. Considering that sudden expansions are often rapid enough to be adiabatic, the first law of thermodynamics reduces to $\Delta U = W$. Therefore, the mechanical energy equals the adiabatic change in internal energy when the gas expands from storage pressure to ambient pressure while cooling down considerably in the process.

The most dramatic and perhaps counterintuitive result from integrating the expansion energy equation is the radically lower theoretical burst energy of cold H_2. Figure 4.9 shows the theoretical maximum specific mechanical energy released by a sudden adiabatic expansion to atmospheric pressure (e.g., in a vessel rupture) of high-pressure H_2 gas from three temperatures (60, 150, and 300 K). H_2 stored at 70 bar and 300 K will release a

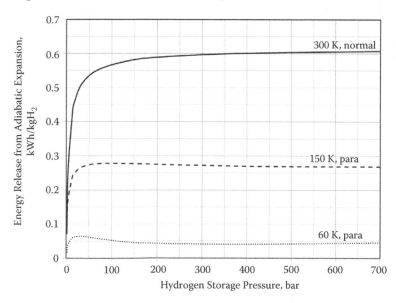

FIGURE 4.9
Calculated results for the maximum mechanical energy (per kilogram of H_2) released on instantaneous expansion of H_2 gas (e.g., from a pressure vessel) as a function of initial storage pressure at 60, 150, and 300 K. For comparison, note that the chemical energy content of H_2 is 33.3 kWh/kg. The figure reports normal-H_2 expansion energy at 300 K and para-H_2 expansion energy at 60 and 150 K. In the last two cases, the real adiabatic expansion energies are slightly lower because the para-H_2 content is less than 100% before expansion (between 99.8% and equilibrium, depending on the previous use of the storage and the actual ortho-to-para H_2 transition rate).

maximum mechanical energy of 0.55 kWh/kg H_2 if suddenly (i.e., adiabatically) expanded to atmospheric pressure (cooling substantially in the process). Counterintuitively, this maximum *specific* energy release increases only slightly with much higher H_2 pressures. Raising vessel pressure to 700 bar (1000% increase from 70 bar) increases maximum specific mechanical energy release by only 10%, while shrinking vessel volume and strengthening (thickening) vessel walls many times over. We emphasize here that these numbers are for *specific energy release* (kilowatt hours/kilograms H_2). To the extent that higher-pressure vessels of equal internal volume hold a larger number of kilograms of hydrogen, more total mechanical energy release can be realized from higher-pressure vessels.

Cooling down the H_2, on the other hand, considerably reduces specific expansion energy (by nearly an order of magnitude between 300 and 60 K). In addition to this, the specific expansion energy of cold H_2 (150–60 K) decreases slightly as the pressure increases between 100 and 700 bar due to nonideal gas behavior. The low burst energy and high H_2 storage density of cryogenic temperatures combine synergistically, permitting smaller vessels, which can be better packaged on-board to withstand automobile collisions. The vacuum jacket surrounding a cryogenic pressure vessel offers a second layer of protection, eliminating environmental impacts over the life of the pressure vessel. Vacuum jacketing also provides expansion volume to mitigate shocks from H_2 release [15].

Cryogenic vessels avoid the fast fill heating and potential overpressures typical of compressed gas storage ambient temperature vessels, consequently operating at higher safety factors, especially as driving the automobile cools the remaining H_2 fuel and reduces average H_2 pressures further over typical driving and refueling cycles. Finally, due to the high storage density of cryogenic H_2 and the potential for relatively low refueling pressure, the number and amplitude of pressure peaks in a cryogenic pressure vessel can be lower than in ambient high-pressure vessels, reducing thermal cycling wear. Of course, ambient pressure hydrogen is completely harmless if it comes into contact with humans (save for simple asphyxiation, which is not so simple if you are the one being asphyxiated), whereas severe frostbite can result from exposure of skin to LH_2. However, reliable and safe procedures have been developed for the handling of LH_2 based on the decades-long experience of NASA.

Conclusions

This chapter describes the thermodynamics of H_2 storage in pressure vessels: compressed, cryogenic, and liquid. In addition to pressure, it has been shown that temperature is an extra degree of freedom for designing compact and cost-effective storage vessels. The chapter describes weight and volume performance and vessel cost as a function of H_2 density; compression and liquefaction are analyzed in terms of exergy; delivery and dispensing are sensitive to density and exergy; cryogenic dormancy is fully characterized through internal energy and entropy; density losses during refueling are due to flow work (enthalpy minus internal energy); and energy release during vessel failure can be calculated from internal energy differences between H_2 inside the vessel and H_2 adiabatically expanded to ambient pressure. In summary, most issues, advantages, and future potential of pressure vessel storage can be defined in terms of thermodynamics.

Acknowledgments

This project was funded by the Department of Energy, Office of Hydrogen and Fuel Cell Technologies—Ned Stetson and Scott Weil, technology development managers. This work was performed under the auspices of the U.S. Department of Energy by Lawrence Livermore National Laboratory under Contract DE-AC52-07NA27344.

References

1. Berry, G.D., Aceves, S.M., The case for hydrogen in a carbon constrained world, *J. Energy Res. Technol.* **127**, 89–94 (2005).
2. von Helmolt, R., Eberle, U., Fuel cell vehicles: status 2007, *J. Power Sources* **165**, 833–843 (2007).
3. Ciancia, A., Pede, G., Brighigna, M., Perrone, V., Compressed hydrogen fuelled vehicles: reasons of a choice and developments in ENEA, *Int. J. Hydrogen Energy* **21**, 397–406 (1996).
4. Amaseder, F., Krainz, G., Liquid hydrogen storage systems developed and manufactured for the first time for customer cars, SAE Paper 2006-01-0432, 2006. SAE International, Warrendale, PA.
5. Aceves, S.M., Espinosa-Loza, F., Ladesma-Orozco, E., Ross, T.O., Weisberg, A.H., Brunner, T.C., et al., High-density automotive hydrogen storage with cryogenic capable pressure vessels, *Int. J. Hydrogen Energy* **35**(3), 1219–1226 (2010).
6. Ahluwalia, R.K., Hua, T.Q., Peng, J.K., On-board and off-board performance of hydrogen storage options for light-duty vehicles, *Int. J. Hydrogen Energy* **37**, 2891 (2012).
7. Department of Energy, Office of Energy Efficiency and Renewable Energy Hydrogen, Fuel cells and infrastructure technologies program multi-year research, development and demonstration plan. http://www.eere.energy.gov/hydrogenandfuelcells/mypp.
8. Ahluwalia, R.K., Hua, T.Q., Peng, J.K., Lasher, S., McKenney, K., Sinha, J., et al., Technical assessment of cryo-compressed hydrogen storage tank systems for automotive applications, *Int. J. Hydrogen Energy* **35**, 4171 (2010).
9. Paster, M., Ahluwalia, R.K., Berry, G., Elgowainy, A., Lasher, S., McKenney K., Hydrogen storage technology options for fuel cell vehicles: well-to-wheel costs, energy efficiencies, and greenhouse gas emissions, *Int. J. Hydrogen Energy* **36**, 14534 (2011).
10. Peschka, W., *Liquid Hydrogen, Fuel of the Future.* Vienna, Austria: Springer-Verlag, 1992.
11. Maus, S., Hapke, J., Ranong, C.N., Wüchner, E., Friedlmeier, G., Wenger, D., Filling procedure for vehicles with compressed hydrogen tanks, *Int. J. Hydrogen Energy* **33**, 4612–4621 (2008).
12. Van Wylen, G.J., Sonntag, R.E., *Fundamentals of Classical Thermodynamics.* New York: Wiley, 1978.
13. Lemmon, E.W., McLinden, M.O., Huber, M.L., REFPROP: NIST reference fluid thermodynamic and transport properties. Gaithersburg, MD: National Institute of Standards and Technology, 2004. NIST Standard reference database 23, version 7.1.
14. Mannan, S., ed. Explosion. In *Lee's Loss Prevention in the Process Industries.* 3rd edition, Volume 2, Chapter 17. Burlington, MA: Elsevier Butterwoth-Heinmann, 2005.
15. Petitpas, G., Aceves, S.M., Modeling of sudden hydrogen expansion from cryogenic pressure vessel failure, Proceedings of the International Conference on Hydrogen Safety, San Francisco, CA, 2011.

5

Hydrogen Storage in Interstitial Metal Hydrides

Ben Chao and Lennie Klebanoff

CONTENTS

Introduction 109
Reversible Metal Hydride Materials 110
AB-Type Intermetallic Interstitial Compounds 113
AB$_5$-Type Intermetallic Compounds 115
AB$_2$-Type Intermetallic Compounds 117
V-Ti-Cr-Based Solid Solution BCC Alloys 123
Applications of Interstitial Metal Hydrides 126
References.. 128

Introduction

Guillaume Petitpas and Salvador Aceves discussed in detail in Chapter 4 the physical limits to the hydrogen storage density achievable using compressed hydrogen gas at various temperatures as well as LH$_2$. Considering only the hydrogen itself (and ignoring the tankage and balance of plant), one can compress gaseous hydrogen at 300 K to a density of 39 g/L using 700-bar pressures. If one cools the gas to 100 K, the gas intrinsically becomes denser, and it requires only 300-bar pressure to densify it to 50 g/L. Liquid hydrogen LH$_2$ at 20 K has a density of 70.7 g/L. Isentropically compressing LH$_2$ at 3 bar can increase the density into the solid hydrogen range of about 90 g/L. More details can be found in Chapter 4.

So, one can see that the low-density nature of hydrogen, in all of its pure states, allows storage density to at most 90 g/L and considerably less when one takes into account the necessary tankage. The cryogenic options offer considerably more density than the room temperature compression of hydrogen gas. However, LH$_2$ has, up to now, been nowhere near as available as hydrogen from compressed gas hydrogen stations and merchant compressed gas cylinders. As a result, there has historically been interest in finding another means of storing hydrogen that combines the near-ambient-temperature character of compressed gas storage with the higher density and lower-pressure attributes of LH$_2$ and cryo-compressed storage.

This third option is storing hydrogen in a chemical compound that ideally releases and reabsorbs hydrogen like a sponge, with little inducement for either process. This was alluded to in Chapter 3, where it was noted that the mass density of hydrogen in water (H$_2$O) is 111 g/L at room temperature and ambient pressure. It turns out water is a lousy sponge for hydrogen because the H-O bond is sufficiently strong (~426 kJ/mol) that water does not readily release H atoms that would eventually form molecular H$_2$. However, there exist a number of compounds with even higher volumetric hydrogen density than water that *do* act as excellent sponges for hydrogen. Storing hydrogen in the solid-state hydride form not only holds a volumetric advantage over compressed and liquid hydrogen states but also can potentially offer several additional key features. These features include low-pressure operation, compactness, safety, full reversibility, tailorable delivery pressure, excellent absorption/desorption kinetics, modular design for easy scalability, and long cycle life.

As given by Figure 3.1 in Chapter 3, research on solid-phase hydrogen storage systems has focused on "on-board-reversible materials," by which the spent material remains on-board the vehicle and is refueled with molecular hydrogen, and "off-board-reversible materials," for which the rehydrogenation requires removal of the material off the vehicle followed by industrial processing. The on-board-reversible materials include interstitial metal hydrides, complex hydrides, and sorption materials. The interstitial and complex hydrides involve chemical bonding between the hydrogen and elements in the storage material, whereas "sorption materials" involve hydrogen physically absorbed on materials with high surface densities, such as various forms of carbon and the metal organic frameworks (MOFs) and their derivatives.

Activated carbon is a good example of a sorption material in which molecular hydrogen is adsorbed on the carbon surface by the weakly bonded van der Waals force. The storage capacity of physical adsorption is dramatically increased with reduced temperature and increased pressure, a regime called *cryoadsorption* in which the hydrogen gas forms a condensed form on the substrate at the temperature of liquid nitrogen (77 K) [1]. The development of sorption materials for hydrogen storage is comprehensively examined by Channing Ahn and Justin Purewal in Chapter 7.

Metal hydrides, in which atomic hydrogen is chemically bonded to the host elements, is an example of chemical hydrogen storage. Depending on the nature of chemical bonding and its bond strength, the hydride formation process can be either reversible or irreversible. The nature of the metal-hydrogen bonds can be classified into three types: metallic, ionic, and covalent. Listing in order of decreasing level of reversibility, we can order the types of bonding as metallic > ionic > covalent in terms of reversibility of hydride formation. For example, the metallic-bonded $TiFeH_2$, $LaNi_5H_6$, and $TiMn_2H_3$ are all reversible. Similarly, the ionic-bonded lithium, calcium, and magnesium hydrides (LiH, CaH_2, and MgH_2) are also reversible but with more difficulty. However, the covalent bonds formed between hydrogen and metals and metalloids of elements in group IB to VB of the periodic table lead to chemical compounds that are irreversible, such as CH_4 (methane) and C_8H_{18} (octane).

A special class of metal hydrides, called *complex metal hydrides*, consists of compounds with a mixed ionic and covalent character. For example, the material $LiAlH_4$ consists of Li^+ cations bound to $(AlH_4)^-$ anions, yet within the $(AlH_4)^-$ moiety there is substantial covalency. As a class, these complex metal hydrides often have reversibility problems introduced by their partial covalent nature. The complex metal hydrides are addressed by Vitalie Stavila, Lennie Klebanoff, John Vajo, and Ping Chen in Chapter 6. We continue here with a discussion of the interstitial metal hydrides.

Reversible Metal Hydride Materials

As briefly reviewed in Chapter 3, metal hydrides are a broad class of materials that undergo a reversible reaction with hydrogen. The reaction is written as

$$M_{solid} + x/2\ H_{2\ gas} \Leftrightarrow MH_{x\ solid} + Heat,\ \Delta H = \text{enthalpy of formation} < 0 \qquad (5.1)$$

where M is the metal element or alloy to be hydrogenated, MH_x is the metal hydride, and ΔH is the enthalpy of formation of the metal hydride. The forward charging reaction is exothermic ($\Delta H < 0$), where heat is liberated, whereas the reverse discharge reaction is endothermic ($\Delta H > 0$), requiring the supply of heat to release hydrogen. Without adding heat to metal hydride MH_x during hydrogen release, the temperature at the metal hydride bed would drop as heat was scavenged from the surroundings.

The reaction of gaseous hydrogen with a metal or alloy is illustrated in Figure 5.1 by a pressure-composition-temperature curve, commonly referred to as PCT curve. At a given temperature, when gaseous hydrogen is introduced into a vessel containing a pure metal or alloy, molecular hydrogen first dissociates on the material surface into atomic hydrogen. The atomic hydrogen then dissolves in the crystal lattice of the metal alloy or intermetallic compound to form a solute solution prior to reaction with the metal. This is usually designated as the α phase. Further increases in hydrogen pressure above the substrate leads to more dissolved hydrogen, and the pressure in the vessel remains relatively flat. The chemically bonded hydride (MH_x) phase, or the β phase, starts to grow. As more hydrogen is introduced above the sample, the β phase continues to grow, and in an ideal scenario, the hydrogen pressure remains at a flat plateau value as more hydrogen is added.

The β-phase formation is a hydrogen nucleation and growth mechanism resulting in expansion of the lattice. Depending on the absorbed hydrogen content, the volume expansion can be up to 30–40%, a large expansion indeed. The α and β phases coexist in the plateau pressure region. As more hydrogen is added, the β phase will continue to grow until atomic H reaches its solubility limit, at which point the β phase is completed. After the solubility limit, further additions of hydrogen cannot be absorbed by the sample, and the hydrogen pressure will rise sharply in the test volume.

Reversing the process above, when the applied hydrogen pressure decreases below the equilibrium value for the complete β phase, the metal hydride MH_x starts to release atomic hydrogen, which diffuses to the surface of the metal alloy or intermetallic compound to

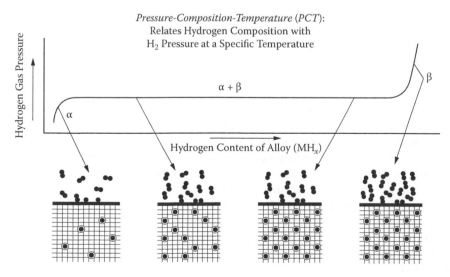

FIGURE 5.1
A PCT curve illustrates the hydride formation mechanism on metal hydride.

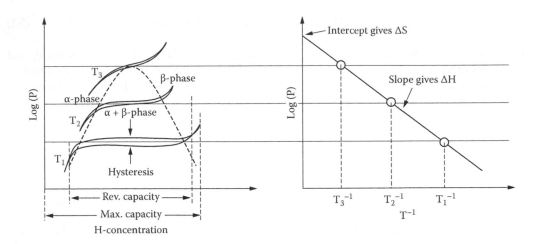

FIGURE 5.2
PCT absorption/desorption curves conducted at three temperatures (left) and the corresponding van't Hoff plot (right).

form molecular hydrogen. The H_2 thus formed desorbs off the material, and the lattice starts to contract back toward its original state.

It is really remarkable that the lattice of the metal could expand/contract in this way during the hydrogen absorption/desorption reaction. The hydrogen-metal system can be considered as a gaseous analogy to a water sponge. Metal crystals contain interstices that will absorb and desorb hydrogen atoms just as sponges have pores that will absorb and desorb drops of water. However, unlike the purely physical action of water in the sponge, the hydrogenation and dehydrogenation reactions of interstitial metal hydrides are true reversible chemical reactions.

The thermodynamic aspects of hydride formation from gaseous hydrogen can be derived from the PCT curves, as displayed in Figure 5.2. It consists of a set of PCT curves at three temperatures. The plateau or equilibrium pressure depends strongly on the temperature. The enthalpy of formation ΔH and the entropy of formation ΔS of a metal hydride can be derived from the PCT curves by means of the van't Hoff equation written for the hydrogenation reaction:

$$\ln P = \Delta H/RT - \Delta S/R \qquad (5.2)$$

Plotting $\ln P$ vs. $1/T$ yields a line with slope $\Delta H/R$ and y-axis intercept of $-\Delta S/R$. Since ΔH is typically negative for hydrogenation, the plot slopes downward with increasing $1/T$.

For the interstitial metal hydrides, the entropy of formation ΔS is determined largely by the change in entropy of the hydrogen itself as it transforms from molecular H_2 into dissolved atomic hydrogen. A ΔS value of -120 J/K mol H_2 is typical for many of the all metal hydrogen systems. However, it is shown in Chapter 6 that complex metal hydrides can have dramatically different values of ΔS that are important for determining the equilibrium pressure as described by Equation 5.2. The enthalpy term characterizes the stability or bond strength of the metal-hydrogen bond. To reach an equilibrium pressure of 1 bar at 298 K [$\ln P = 0$ in Equation 5.2], ΔH needs to be -35.8 kJ/mol H_2 assuming a ΔS value of -120 J/K mol H_2.

We shall soon see that many interstitial metal hydrides have enthalpy barriers for hydrogen release on the order of 30 kJ/mol H_2. For context, the bond dissociation energy of molecular hydrogen is 436 kJ/mol H_2, and the H-O bond strength in water is about 426 kJ/mol [2]. So, we can see that it takes about an order of magnitude less energy to release atomic hydrogen (on a molar basis) from the average interstitial metal hydride than it does to release atomic hydrogen from typical covalent chemical bonds.

The absorption and desorption reactions of metallic hydrides are surprisingly easy and simple. Since there exists no distinct activation energy required in forming metallic hydrides, the absorption and desorption reactions readily occur spontaneously. The intrinsic material kinetics do not limit the practical hydride/dehydride cycle times. The main constraints at the storage system level actually lie on (1) how to engineer the heat management and (2) how to prevent the metal hydride surfaces from becoming contaminated.

The majority of the binary metal (single-element) hydrides have large negative enthalpy of formation ΔH values, meaning too stable metal-hydrogen bonds that would require higher heat to break the bonds (>150°C). These elements are mostly in the IA, IIA, IIIA, IVA, and VA groups in the periodic table (see inside back cover), such as lithium, magnesium, sodium, potassium, calcium, titanium, vanadium, yttrium, zirconium, niobium, rare earth elements, and so on [3] and are not suitable for the hydrogen storage applications. The enthalpy of formation can be significantly reduced for several metal hydrides based on the intermetallic compound groups [4–9].

This chapter focuses on the reversible intermetallic compound materials that serve as the host matrix to store and release hydrogen in and out of their interstitial sites. They are called the *interstitial metal hydride* materials, which can be conveniently denoted by AB_X where A represents the hydride formers, B are the nonhydriding metals, and x is the atomic ratio of B/A, that is, AB, A_2B, AB_2, AB_3, A_2B_7, and AB_5. Among many AB_X intermetallic compounds, AB, AB_2, and AB_5 systems have been extensively studied due to their reasonable hydrogen storage capacity, ease of synthesis by the conventional cast methods, flexibility in tailoring the thermodynamic properties, good absorption/desorption kinetics, and cycle life.

AB-Type Intermetallic Interstitial Compounds

AB-type intermetallic compounds exhibit the CsCl simple cubic structure (space group $Pm\bar{3}m$), where titanium and zirconium generally occupy the A sites and iron, manganese, nickel, aluminum, and other third-row transition metals occupy the B sites. ZrNi was the first reported AB-type intermetallic compound to absorb hydrogen in a reversible way in 1958 [10]. Other AB-type compounds include TiFe, TiCo, TiNi, and ZrCo. They are able to absorb hydrogen with a mole fraction level of H/M = 1–1.5 (M: metals). Recall that in water, the mole fraction of hydrogen to oxygen is 2.0.

Historically, the major interest in the AB-type materials was due to their hydrogen storage capacity, low specific weight, and reasonable cost. Iron titanium compound, TiFe, is the most recognized AB type in the group. PCT measurements of TiFe at 25°C are shown in Figure 5.3. TiFe has the cubic lattice constant of 0.2976 nm and has good thermodynamic properties (ΔH, ΔS), such that hydrogen can absorb and desorb from the material

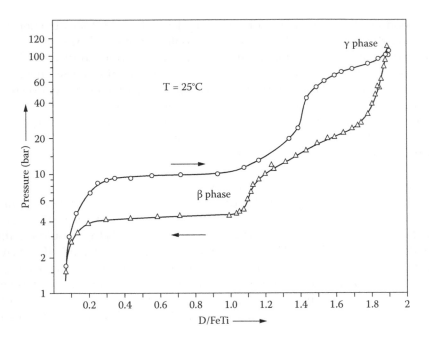

FIGURE 5.3
PCT absorption/desorption curves of FeTi deuteration forming TiFeD (β phase) and TiFeD₂ (γ phase) at 25°C.
(From J. Schefer, P. Fischer, W. Halg, F. Stucki, L. Schlapback, and A.F. Andresen, *Mater. Res. Bull.* 14(10), 1281
(1979) with permission of Pergamon.)

at ambient conditions [6]. Deuteration of TiFe has been examined by neutron diffraction,
which showed that the D first occupies the octahedral sites (among two iron and four
titanium atoms), which causes a distortion from the original simple cubic structure to
the orthorhombic β-phase structure (space group $P222_1$) with TiFeD. Further deuteration
forms the monoclinic γ phase (space group $P2/m$): TiFeD₂ [11]. Two plateau pressures are
observed on the PCT absorption/desorption curves at 25°C with desorption pressure of
4.3 and about 10 bars correlated to TiFeD/TiFeH (β phase) and TiFeD₂/TiFeH₂ (γ phase),
respectively. The volume expansion is about 11% for β phase and about 18% for γ phase,
with the corresponding enthalpy values of –28.1 and –30.6 kJ/mol H₂, respectively [6].

A fully activated TiFe stoichiometric material can reach the hydrogen capacity of 1.5 wt%
[12]. In other words, 105.7 g of TiFeH₂ (1 mol of TiFeH₂) contains 1.58 g of H, corresponding
to 0.78 mol of H₂. Although the molar fraction of hydrogen to TiFe is quite good, the poor
weight percent is entirely due to the use of relatively heavy Ti and Fe. Chapter 6 explores
the more recent efforts to use much lighter elements, such as Li, Al, and B, to increase this
wt% figure.

Due to its high sensitivity to surface contamination by gaseous impurities, it is hard for
TiFe to achieve full activation [13, 14]. However, partial substitution of iron by manganese
and nickel can improve its hydrogenation reactivity [6, 12, 15]. In addition, manganese substitution changes the PCT curve to a sloped absorption/desorption pressure curve with
significantly reduced absorption/desorption hysteresis and still manages to maintain the
hydrogen storage capacity [12].

The phenomenon of hysteresis was shown schematically in Figure 5.2 and displayed in practical data for TiFe in Figure 5.3. Nearly all metal hydrides show this phenomenon, which can be stated as follows: The pressure existing over the alloy when it is being hydrided is larger than the pressure that exists over the alloy *at the same elemental composition* when the hydride alloy is being decomposed. Another way to look at this is that for the same elemental states of the sample (H content or H/M ratio), there are two different possible values for the "equilibrium" plateau pressure. Clearly, the elemental composition given by the H/M ratio or the H content does not fully specify the true state of the material.

A full discussion of this hysteresis phenomenon is beyond the scope of this chapter. A discussion of the phenomenon with a review of historical explanations has been given by Flanagan and Clewley [16]. It turns out that when a metal is hydrided, a volume expansion occurs that is accompanied by dislocations in the lattice (also called plastic deformation [17]). These dislocations take up free energy, changing the thermodynamics of the process. At a given point on the hydriding plateau (the top plateau in Figure 5.3), the pressure over the alloy is the pressure required to reach that composition in the presence of some number of lattice dislocations. When the hydride material is decomposed, some of these dislocations are annihilated as the material contracts, producing fewer (but nonzero) lattice distortions at any given composition. As a result, neither of the plateaus corresponds to the true equilibrium in the absence of lattice dislocations, which is in-between the two observed plateaus.

The existence of hysteresis affects practical systems because the free energy that goes into the dislocations is "lost" as heat to the surroundings when the dislocations are annihilated on decomposition. In a sense, it represents a loss in thermal efficiency. More details can be found in the discussion of Reference 16 and references therein.

AB$_5$-Type Intermetallic Compounds

The AB$_5$-type metal hydride is based on the CaCu$_5$ hexagonal structure (*P6/mmm* space group). The LaNi$_5$ metal hydride family has been one of the most frequently studied AB$_5$-type metal hydride materials [18–21]. The first report of LaNi$_5$ hydrogen absorption properties was published by Van Vucht et al. in 1970 [18]. LaNi$_5$ metal hydride material can readily and reversibly achieve the hydrogen capacity of 1.4 wt% [i.e., LaNi$_5$H$_6$ where H/M (La + 5Ni) = 1] at room temperature under an equilibrium hydrogen pressure of 2 bars (30 psig). Figure 5.4 shows the AB$_5$-type structure, which contains a variety of tetrahedral and octahedral interstice sites.

Hydrogen occupancies depend on the composition of the alloys, with the hydrogen atoms preferring to occupy tetrahedral interstice sites formed by A and B atoms (i.e., A$_2$B$_2$, B$_4$, and AB$_3$) per formula [22–24]. These tetrahedral sites are shown in Figure 5.4. Neutron-scattering investigation by Fisher et al. reported the hydrogen occupies both tetrahedral sites of 6*d* (D2) and 3*c* (D1) in LaNi$_5$D$_6$ alloy (*P3lm* space group) [25–27]. Unit cell volume expansion in the range of 25% was observed for LaNi$_5$D$_x$ (x = 6–6.7) [24, 25].

Even though LaNi$_5$ has remarkable properties as hydrogen storage material, its use as an anode electrode material for nickel metal hydride (NiMH) batteries was not suitable due

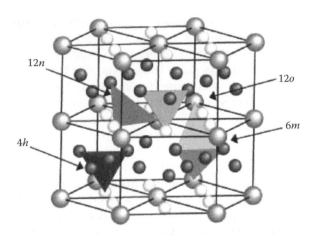

FIGURE 5.4

Schematic diagram of the AB$_5$-type hydride crystal structure: A atoms indicated by large balls; B atoms by small balls. The sites occupied by hydrogen are shown in the form of tetrahedrons: $6m$ (A$_2$B$_2$), $12n$ (AB$_3$), $12o$ (AB$_3$), and $4h$ (B$_4$) in Wyckoff description. (From F. Cuevas, J.-M. Joubert, M. Latroche, and A. Percheron-Guegan, *Appl. Phys.* A72, 225 (2001) with permission of Springer-Verlag.)

to two constraints: too high equilibrium pressure at ambient conditions and too short a cycle life [28].

In the NiMH battery application, the hydrogen storage material absorbs and releases atomic hydrogen as the battery charges and discharges, respectively. However, in the battery application, formation of molecular H$_2$ is undesired and suppressed.

Many researchers began to modify the LaNi$_5$ alloy by substituting for either La on the A sites or Ni on the B sites or both. La can be replaced with other rare earth metals, such as Ce, Pr, Nd, or even a mixture of these metals (also called mischmetals [Mm]) forming a solid solution, as well as by Zr, and Mg [29–31]. The original intention of Mm replacement was only aimed to reduce the cost of MH electrodes [31]. Surprisingly, using Mm led to a better cycle life [32]. Partial substitution of Ni by a combination of the metals, such as Mn, Fe, Co, Ni, Cu, and Al, has significantly improved the alloy properties for battery application, such as lower plateau pressure, better high rate discharge, low-temperature power performance, and long cycle stability [33–43]. Based on the early work by Achard [44], a linear relation was observed between the unit cell volume of the intermetallic compound and the logarithm of the plateau pressure. Figure 5.5 depicts a group of alloys for various replacements on the La and Ni sites [22].

The importance of Figure 5.5 is that it demonstrates the tuning of the plateau pressure over orders of magnitude by changing the interstitial metal alloy composition. We shall show a bit later that it is important to match the plateau pressure with the needs of the particular application.

Currently, MmNi$_5$-type metal hydride chemistry (La$_{10.5}$Ce$_{4.3}$Pr$_{0.5}$Nd$_{1.4}$Ni$_{60.0}$Co$_{12.7}$Mn$_{5.9}$Al$_{4.7}$) widely used for consumer batteries and the hybrid electric vehicle (HEV) exhibits the usable reversible capacity of about 1.2 wt% [29]. The high-nickel-containing MmNi$_5$-type metal hydride alloys are the key to meet the high-power demand for HEVs. This special feature does not benefit from much hydrogen storage. A maximum 1.5 wt% hydrogen storage capacity in this alloy family is the main constraint for using these materials beyond the battery application. Further research does not expect to significantly improve the hydrogen storage capacity of the AB$_5$-type metal hydride alloys family.

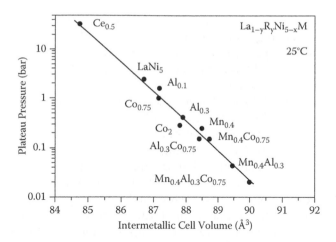

FIGURE 5.5

Linear relationship of the plateau pressure as a function of the intermetallic cell volume for various $La_{1-y}R_yNi_{5-x}M_x$ compounds at room temperature. (Reproduced from F. Cuevas, J.-M. Joubert, M. Latroche, and A. Percheron-Guegan, *Appl. Phys.* A72, 225 (2001) with permission of Springer-Verlag.)

AB_2-Type Intermetallic Compounds

Three crystal structures have been observed to associate with the stoichiometric AB_2-type intermetallic compounds: $MgCu_2$ (cubic) and $MgZn_2$ and $MgNi_2$ (hexagonal). These three structures can be described as different stacking sequences of layers involving A and B atoms in their respective close-packed structure classes. The schematic diagrams of three Laves structures are displayed in Figure 5.6 below and in Color Figure 5.6 in the color insert, which shows more clearly the A and B atom locations..

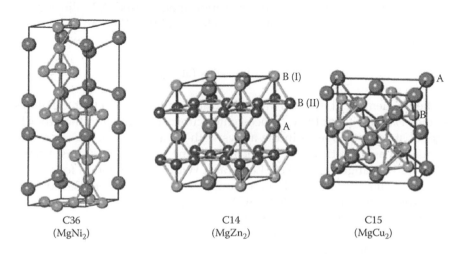

C36
($MgNi_2$)

C14
($MgZn_2$)

C15
($MgCu_2$)

FIGURE 5.6 (See color insert.)

Schematics of three Laves structures in AB_2-type intermetallic compounds.

MgCu$_2$-type cubic structure (space group $Fd\bar{3}m$) is also called C15 after the Strukturbericht designation with the stacking sequence of a-b-c-a-b-c. MgZn$_2$ and MgNi$_2$ types have the same hexagonal structure (space group $P6_3/mmc$) and are called C14 and C36, respectively. The former has a close-packed layer sequence of a-b-a-b, and the latter has an intergrowth packing sequence of a-b-a-c-a-b-a-c.

Transition-metal-based AB$_2$-type intermetallic compounds have higher hydrogen storage capacities than the rare-earth-based AB$_5$-type alloy family due to use of relatively lighter transition metals. Binary compounds like ZrV$_2$ (C15), ZrMn$_2$ and TiMn$_2$ (C14), and ZrCr$_2$ and TiCr$_2$ (can be either C14 or C15 depending on the synthesis conditions) can absorb an appreciable amount of hydrogen, more than the rare-earth-based AB$_5$ alloys, for example, ZrMn$_2$H$_{3.6}$ and ZrV$_2$H$_{5.2}$ (where H/M > 1) [45, 46]. However, these binary compounds, like the magnesium-based AB$_2$ compounds, are too stable in the hydride phases (strong metal-hydrogen bonds), which require higher temperature (>250°C) to release hydrogen. In addition, the compounds cannot be used for battery applications due to their poor corrosion resistance properties and poor cycle stability.

Ovshinsky and his team introduced the multielement, multiphase, and compositionally disordered alloy design concept to lead the way to the successful commercialization of the NiMH battery technology [47, 48]. For transition-metal-based AB$_2$-type intermetallic compounds, zirconium and titanium are the main elements on the A sites. Small amounts of yttrium, hafnium, niobium, molybdenum, tantalum, and tungsten have also been added on the A sites. A broad range of elements can be used on the B sites. The most common elements are vanadium, chromium, manganese, iron, cobalt, nickel, copper, zinc, and aluminum [47–50]. For battery applications, zirconium and nickel are designed to be the dominant elements on the A sites and B sites, respectively. In terms of gas-phase hydrogen storage application, titanium and manganese are, respectively, the key elements on the A sites and B sites.

The stability of the transition-metal-based AB$_2$ Laves structures are generally influenced by the size factor principles and the electron concentration factor e/a [51, 52]. The electron concentration factor e/a is defined as the average number of electrons per atom outside the closed electron shells of the component atoms [52]. Electron concentration factors of the key elements generally used for the transition-metal-based AB$_2$ alloys are listed in Figure 5.7. The average electron concentration factor e/a of the alloy can be directly derived from the following equation:

$$(e/a)_{\text{alloy}} = \Sigma_i \left[(e/a)_i \times (\text{at}\%)_i \right] \tag{5.3}$$

where i is the ith element. It is found (also illustrated in Figure 5.7) that the alloys having higher average e/a (>7.1) favor the C15 face-center-cubic structure. On the other hand, the C14 hexagonal structure appears in the alloys with e/a around 6.5. The atomic size ratio (R_A/R_B) is defined as the ratio of the atomic radii of the two neutral atoms A and B and is ideally 1.225, with a range of 1.05–1.68. Laves and Witte's investigation found that for the Mg-based binary alloy systems, with increasing valence electron concentration, the structure moved from MgCu$_2$ (C15) to MgNi$_2$ (C36) to MgZn$_2$ (C14) [51]. The works by Bardos et al. [53] and Watson and Bennett [54] were successfully extended to correlate the average electron concentration and d-band hole number to the crystal structure in the transition-metal-based AB$_2$ Laves structures.

The alloys having average e/a between 6.5 and 7.1 are mixtures of C14 and C15 structures [51, 55–58]. When e/a is below 5.5, then the alloys display a body-centered-cubic (BCC)

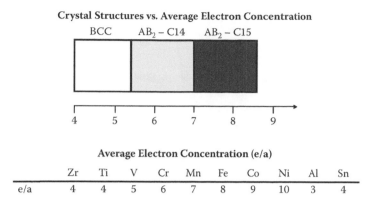

FIGURE 5.7

Crystal structure of transition-metal-based AB_2-type alloy vs. average electron concentration factor (e/a).

phase. Interestingly, since the C36-type Laves structure, so much like the C14 structure, has not been carefully identified by x-ray diffraction (XRD) observed in the transition-metal-based AB_2-type alloys.

For the transition-metal-based AB_2-type alloys, the C14 and C15 Laves structures (see color insert Figure 5.6), each A atom (red balls) (i.e., either zirconium or titanium, the hydride formers) is coordinated with 12 B atoms (such as vanadium, chromium, manganese, iron, cobalt, nickel, copper, zinc, aluminum, etc.) in a predominantly icosahedra configuration and surrounded tetrahedrally by 4 A atoms. In the case of the C15 cubic structure, B atoms (green balls) form the tetrahedral unit B_4, which are linked together by sharing a vertex to form a three-dimensional network. For the C14 hexagonal structure, tetrahedral units are built from two crystallographically distinct B(I) and B(II) atoms (green and blue balls) and are connected by B(I) vertex sharing and by triangle B(II)-B(II)-B(II) face sharing. The ratio of the B(I) and B(II) sites is 1:3. Preliminary neutron diffraction investigation suggests that in the C14 structure, zirconium and titanium are randomly mixed at the A sites, nickel atoms prefer to occupy the B(I) sites [59], and vanadium, chromium, manganese, and the remaining nickel are randomly distributed in the B(II) sites. For the C15 cubic structure, zirconium and titanium randomly occupy A sites, and all B sites are randomly mixed by the vanadium, chromium, manganese, and nickel elements [60].

Both C14 and C15 Laves structures contain a total of 17 tetrahedrally coordinated sites per AB_2 formula [27, 61]. In the C14 structure, there are three distinct types of tetrahedral sites available for hydrogen, that is, 12 [A_2B_2] sites, 4 [AB_3] sites, and 1 [B_4] site. Considering the size constraints proposed for the hydrogen occupation, namely, that the separation between H sites should be greater than 0.21 nm and the diameter of the site should be greater than 0.08 nm [62], only 6 tetrahedral sites per formula are practically available for hydrogen, that is, AB_2H_6. In other words, two hydrogen atoms for each metal atom (H/M = 2) is the maximum hydrogen capacity (~3.8 wt%) that would be expected in the AB_2-type intermetallic compounds.

Most studies indicated that the larger [A_2B_2] sites are occupied first, followed by the [AB_3] sites, and the smallest [B_4] sites remain empty [27, 46]. Figure 5.8 (see also color Figure 5.8) is a schematic diagram of the C14 structure showing two [A_2B_2] tetrahedral sites 24l and 12k that are preferentially occupied by hydrogen [63]. Hydrogen reversible capacity of 2.4 wt% has been reported for the transition-metal-based AB_2-type multielement and multiphase (Zr, Ti)(V, Cr, Mn, Fe, Al)$_2$ alloys [64].

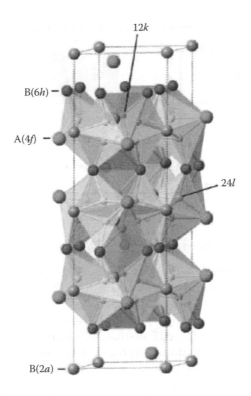

FIGURE 5.8 (See color insert.)
A schematic diagram of C14 structure indicating the two preferred [A_2B_2] sites for hydrogen. (Reproduced from
I. Levin, V. Krayzman, C. Chiu, K.-W. Moon, and L.A. Bendersky, *Acta Mater.* 60, 645 (2012) with permission of
Elsevier.)

Transition-metal-based AB_2-type alloys synthesized by the conventional induction- and
arc-melt cast techniques typically consist of a mixture of C14 or C15 main hydrogen storage
phases along with secondary phases of Zr_7Ni_{10}, Zr_9Ni_{11}, and a TiNi- or ZrNi-like B2-related
structure [55, 56, 65–71]. Most diffraction peaks associated with the C15 structure overlap
with those of the C14 structure. An example is shown in Figure 5.9, which shows the XRD
patterns of the as-prepared and hydride MF139Z alloy, $Zr_{27}Ti_9V_5Cr_5Mn_{16}Ni_{36}$ (e/a = 6.91).
In the 2θ range of 27–53° shown in Figure 5.9a, the (220), (311), and (222) peaks of the C15
structure totally overlap to the (110), (112), and (004) peaks of the C14 structure, respectively. In addition, the C14 structure has more allowed peaks in the same range. The result
of Rietveld crystal structure refinement indicates that the alloy consists of 63% C14 and
37% C15. The lattice constants are a_o = 0.498 nm and c_o = 0.812 nm for C14 and a_o = 0.703 nm
for C15. The peaks marked with the asterisk are due to the minor Zr_7Ni_{10}, Zr_9Ni_{11}, and
ZrO_2 phases.

The characteristics of a C14/C15 mixture are much more evident in the hydrided state
(see Figure 5.9b). All major peaks are observed to shift to lower 2θ values resulting from
the hydrogen occupied in the interstitial sites causing the lattice expansions. Peaks indicated by the arrows labeled with C14 are related to the C14 structure in the hydride state.
The expanded lattice constants of the C14 structure are centered at a_o = 0.532 nm and
c_o = 0.868 nm, equivalent to about 22% expansion in volume from the as-prepared alloy.
Peaks indicated by the arrows labeled with C15 are due to the hydride C15 structure.
Its expanded lattice constant is centered at a_o = 0.746 nm, which accounts for a 19% volume

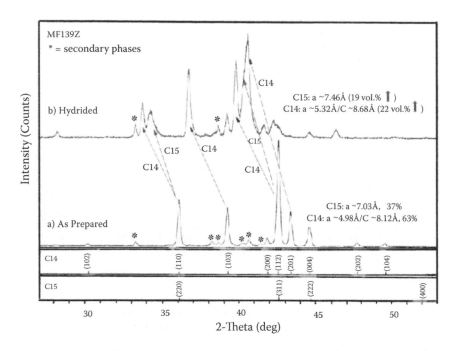

MF139Z
* = secondary phases

b) Hydrided

C14

C15: a ~7.46Å (19 vol.% ↑)
C14: a ~5.32Å/C ~8.68Å (22 vol.% ↑)

C15   C14

C14   C15

C14   C14

C15: a ~7.03Å, 37%
C14: a ~4.98Å/C ~8.12Å, 63%

a) As Prepared

Intensity (Counts)

C14   (102)   (110)   (103)   (200) (112) (201)   (004)   (202)   (104)

C15   (220)   (311)   (222)   (400)

2-Theta (deg)

30   35   40   45   50

FIGURE 5.9

XRD patterns of MF139Z alloy of (a) as-prepared and (b) hydrided. (*): minor phases. (Reproduced with permission from Reference [55].)

expansion. It is interesting to point out that the C14 structure seems to exhibit larger lattice expansion than the C15 structure in this alloy.

One of the favorable features associated with the metal hydride alloys is the flexibility of the equilibrium plateau pressure produced by varying the alloy composition. Among the types of alloys discussed, the transition-metal-based AB_2-type alloy family is the most flexible and versatile. Increase in the fraction of larger atomic size elements in the A or B sites and enriching the alloy in A elements are two common ways to increase the unit cell volume. Similar to the AB_5-type intermetallic compounds (Figure 5.5), increasing the unit cell volume in this alloy family also correlates well to the decrease in the alloy desorption plateau pressure. Figure 5.10 shows the pressure-temperature relationship of three alloys. At 20°C, the desorption plateau pressure varies from 30 psia (OV694) to 120 psia (OV679) to 250 psia (OV610) with the corresponding heat of formation of −30.4, −25.8, and −24.0 kJ/mol H_2, respectively. The reversible hydrogen capacities of these alloys are in the range of about 2.3 wt% [72].

The criteria that are used to select an appropriate alloy for a particular application are (1) the highest reversible capacity, (2) the type of waste heat available, (3) a required delivery pressure, and (4) the maximum charging pressure available at the site of the application. In the selection of a particular metal hydride alloy for on-board vehicle application, the alloy has to perform well, utilizing waste heat that is available from either a fuel cell or an internal combustion engine (ICE) vehicle. The temperatures of the cooling loop from the engine or the fuel cell stack are similar, in a range of 50–85°C, only varying in its flow rate. In this case, the alloy is limited to transition-metal-based metal hydride materials. Due to the cold start demand, a relatively high plateau pressure alloy is required at low temperatures. Transition-metal-based AB_2-type alloys have been demonstrated to be the most flexible

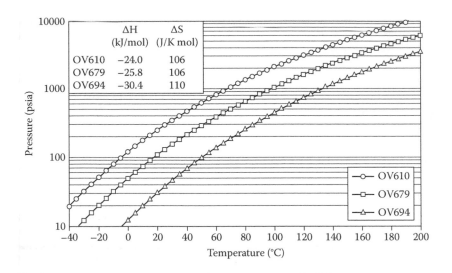

FIGURE 5.10
Pressure-temperature relationship of three proprietary transition-metal-based AB_2-type alloys. (From B.S. Chao, R.C. Young, V. Myasnikov, Y. Li, B. Huang, F. Gingl, P.D. Ferro, V. Sobolev, and S.R. Ovshinsky, *Mater. Res. Soc. Symp. Proc.* 801-BB1.4, 27 (2003) with permission of Cambridge University Press.)

and versatile materials to be able to meet the high desorption plateau pressure requirement at the approximately 85°C waste heat temperature. As shown in Figure 5.10, OV610 has the plateau pressure of about 50 psia at –20°C and is more suitable for the on-board hydride storage vessel application [73]. However, an alloy with higher desorption plateau pressure is needed if the cold start requirement is at –40°C.

On-site hydrogen generation by an electrolyzer combined with the hydrogen bulk storage system is an attractive hydrogen stationary application. Typical deliverable pressure from an electrolyzer is in the range of 150–200 psig. OV694 can meet the low-pressure charging for a bulk hydrogen storage system [72]. The ability to charge the material at the relatively low pressure available from merchant hydrogen cylinders, about 2000 psi (138 bar), is a major advantage for hydrogen storage using these interstitial metal hydrides. In many cases, merchant hydrogen is all that could be considered available. In the case of the Fuel Cell Mobile Light discussed in Chapter 2, the commercial systems will be configured to be filled from both merchant hydrogen cylinders or a hydrogen filling station.

Among all potential applications for a solid metal hydride system, portable power products provide an early opportunity for commercialization in the areas of small backup power and emergency and uninterruptible power supply (UPS) systems. For most UPS systems, portable power hydrogen canisters can be easily integrated with an air-cooled proton exchange membrane (PEM) fuel cell. The canisters generally contain only a passive heat exchange system. Waste heat from the fuel cell stacks, in the form of air heated to between 25°C and 60°C (depending on fuel cell power level), is used to assist the discharge of hydrogen from the canisters. Canisters are required to deliver at least 10 psig pressure at 5°C, and the maximum pressure must be less than 250 psig at 30°C [72, 74]. In Figure 5.10, OV694 is ruled out for the application because it cannot sustain the required deliverable flow rate at 5°C due to the lack of heat transferred by hot air from the fuel cell stack to the canister. OV610 has too high of a plateau pressure, and this alloy cannot be charged with 250 psig at 25°C. OV679 alloy is capable of meeting all the technical requirements for the portable power canister application.

V-Ti-Cr-Based Solid Solution BCC Alloys

Apart from the AB-, AB_2-, and AB_5-type alloys, solid solution alloys based on vanadium-titanium-chromium (V-Ti-Cr) with BCC structure exhibit higher hydrogen absorption storage capacity. The alloys represent a group of hydrogen storage materials containing the main elements of titanium, vanadium, and chromium incorporated with additional elements as modifiers, including zirconium, manganese, iron, cobalt, nickel, copper, and zinc. Other modifiers of aluminum, boron, molybdenum, tantalum, yttrium, magnesium, rare earth elements, hafnium, and tungsten have also been studied [75–94]. To ensure the hydrogen storage capacity advantage, the sum of the modifiers in the BCC alloys is generally no more than about 15 at%. The lattice constant and unit cell volume of the V-Ti-Cr-based BCC solid solution alloys are in the range of 0.300–0.305 nm (space group $Im\bar{3}m$) and 27.00–28.35×10^{-3} nm³, respectively. A set of representative XRD patterns recorded from two Ti-V-Cr-Mn BCC alloys is displayed in Figure 5.11.

The theoretical hydrogen absorption capacity of this family is about 3.85 wt% (i.e., H/M = 2; M: metal) at a modest hydrogen pressure and at room temperature. More than 2.2 wt% of desorption hydrogen capacity has been reported [76, 78, 84–86, 90, 92]. The V-Ti-Cr-based BCC solid solution alloys generally consist of two pressure plateaus in the PCT isotherms corresponding to two phase transition regions: a solid solution BCC-type hydride, where hydrogen occupies the octahedral sites, and a face-centered-cubic (FCC)-type hydride, where hydrogen resides in the tetrahedral sites, as illustrated in Figure 5.12. The reduced number of bonding partners (four) for hydrogen in the FCC tetrahedral sites naturally leads to lower stabilization of the interstitial atomic hydrogen, permitting greater ease of release compared to the BCC octahedral sites, where H sees six bonding partners.

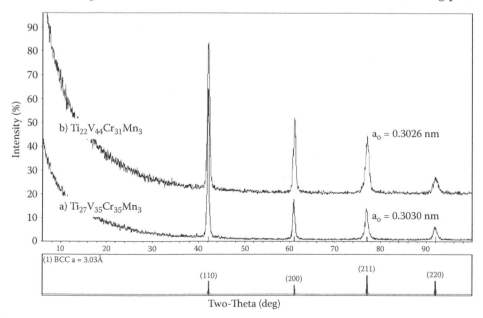

FIGURE 5.11
XRD patterns of two BCC alloys: (a) $Ti_{27}V_{35}Cr_{35}Mn_3$; (b) $Ti_{22}V_{44}Cr_{31}Mn_3$. (From B. Chao, B. Huang, and F. Gingl, TARDEC-AES/ECD milestone reports, Contract number FA8222-05-D-0001-0044-01 (October 2008, March and September 2009) with permission.)

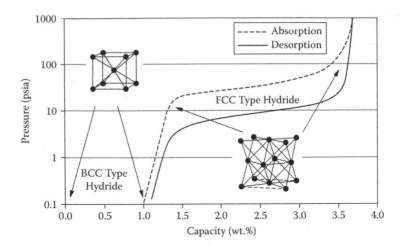

FIGURE 5.12
Schematic view of structure transformation from BCC hydride to FCC hydride in the PCT study of V-Ti-Cr BCC alloy at 25°C. (From B. Chao, B. Huang, and F. Gingl, TARDEC-AES/ECD milestone reports, Contract number FA8222-05-D-0001-0044-01 (October 2008, March and September 2009) with permission.)

Figure 5.12 shows the first plateau is below 0.1 psia, near vacuum level at 25°C. In the hydriding process, the V-Ti-Cr-based alloy family exhibits the structure transformation from the as-cast BCC structure to a body-centered-tetragonal (BCT) structure (space group *I4/mmm*) correlated to the monohydride state and to the FCC structured (space group *Fd3m*) dihydride state. Volume expansion of the first stage between the BCC base structure and the BCT structure is in the range of about 15–17%. Smaller expansion of the monohydride might be due to the larger octahedral interstitial sites. Close to 35–40% expansion in volume is realized for the dihydride state [82, 83]. The hydrogen atoms that mostly occupy the octahedral interstice sites in BCC structure require higher temperature (>250°C) to release from the metal alloy host matrix; otherwise, they are trapped in the alloy lattice. Typically, it is estimated that 1.0–1.3 wt% of hydrogen is not reversible at room temperature conditions. Only those hydrogen atoms residing at the FCC tetrahedral interstice sites are reversible at ambient temperature.

The arc melting technique instead of the conventional induction casting method is used to synthesize the V-Ti-Cr-based BCC solid solution alloys. Arc melting minimizes the contamination from the crucible resulting from the high melting temperatures (>1600°C) of the raw materials associated with the induction casting method. Those alloys that are rich in chromium or titanium or containing less than 50 at% vanadium exhibit relatively sloped PCT plateau curves that require an additional annealing process at temperatures of about 1300°C or higher to flatten the absorption/desorption plateau pressures [76, 92, 95] (e.g., see Figure 5.13). High manganese content in the alloys promotes the secondary C14 Laves phase [76] more prominently in the alloys when the total amount of three main elements (vanadium/titanium/chromium) is less than 75 at%. Based on the atomic radius argument, increasing titanium or vanadium content in the alloys correlates to a larger BCC lattice constant and a lower PCT absorption/desorption plateau pressure [90]. On the other hand, the vanadium-rich BCC alloys (>50 at% vanadium) generally exhibit much flatter PCT plateau curves [90–92], similar to the one shown in Figure 5.14.

In general, a higher desorption temperature can increase the desorption capacity of the alloy by assisting extra hydrogen atoms to release from the metal alloy host matrix, as

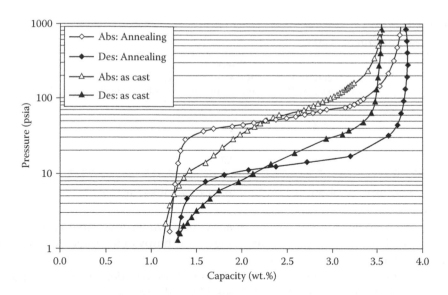

FIGURE 5.13
PCT curves of $Ti_{24}V_{41}Cr_{31}Mn_3$ as cast and annealed at 20°C. (From B. Chao, B. Huang, and F. Gingl, TARDEC-AES/ECD milestone reports, Contract number FA8222-05-D-0001-0044-01 (October 2008, March and September 2009) with permission.)

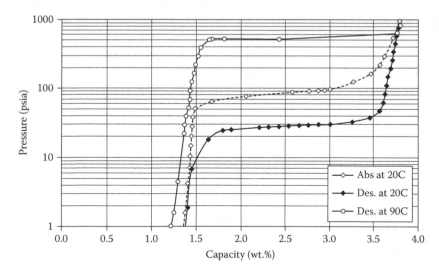

FIGURE 5.14
PCT curves of $V_{80}Ti_{6.8}Cr_{12}Mn_1Ni_{0.2}$ alloy: absorption and desorption at 20°C and desorption at 90°C. (From B. Chao, B. Huang, and F. Gingl, TARDEC-AES/ECD milestone reports, Contract number FA8222-05-D-0001-0044-01 (October 2008, March and September 2009) with permission.)

shown in Figure 5.14. It includes a pair of PCT absorption/desorption curves at 20°C and an additional desorption curve at 90°C. The desorption hydrogen capacity increases from 2.41 wt% (solid diamond) at 20°C to 2.63 wt% (open circle) at 90°C. Desorption plateau pressure is noticeably increased from 30 psia at 20°C to 515 psia at 90°C, which is consistent with what you would expect from the van't Hoff relationship (Equation 5.2). Raising the desorption temperature from 20°C to 90°C promotes additional hydrogen release from the

TABLE 5.1

Summary: Interstitial Metal Hydrides

Type	Theoretical Capacity (wt%)	Observed Reversible Capacity (wt%)	Volume Expansion in Hydride (%)	$|\Delta H|$ (kJ/mol H$_2$)
AB	2.8	1.9	~18	28–35
AB$_2$	~3.8	2.4	18–30	22–35
AB$_5$	1.6	1.6	15–25	24–35
V-Ti-Cr-based BCC	3.8	3.0	~40	28–35

alloy lattice, resulting in about 0.2 wt% gain in reversible capacity for the V-Ti-Cr-based BCC alloys.

Table 5.1 summarizes the hydrogen capacity, observable reversible capacity, volume expansion in the hydride, and the possible range of enthalpy of formation of the four interstitial metal hydride families discussed.

Even though the V-Ti-Cr-based BCC solid solution alloys have demonstrated higher hydrogen storage capacity than other interstitial metal hydride groups, more research is needed to overcome some disadvantages associated with these alloys before they can be widely adopted for practical application. The challenges are to reduce the absorption/desorption pressure hysteresis, to reduce the hydrogen trapping in the octahedral coordinated sites at room temperature, to have appropriate thermodynamic properties and not sacrificing the hydrogen desorption capacity, to reduce material cost associated with the expensive vanadium raw materials, and to improve the tolerance of gaseous impurities resulting in better cycle life. Right now, transition-metal-based AB$_2$-type alloys are excellent hydrogen storage candidates for many practical applications due to their facile reversibility, excellent kinetics, tailorability of plateau pressure (heat of formation), good cycle life, and ease of synthesis.

Applications of Interstitial Metal Hydrides

Chapter 3 gave a brief history of some of the historic applications of hydrogen storage, including those using interstitial metal hydrides. Here, we provide a bit more of the application history. The first hydride storage unit was introduced in 1976 by the Billings Energy Company [96]. The unit was filled with TiFe or Ti(Fe,Mn) hydride alloy. It had no internal heat exchanger and had a 0.22 kg hydrogen storage capacity. Subsequently, several other companies [97–101] began marketing their small storage containers with storage capacities between 2.5 g and 228 g of hydrogen. Only a limited number of units have been made since 1985 because of relatively weak market demand.

In 1974, Brookhaven National Lab built the first prototype large stationary hydride storage unit for a New Jersey Public Service Electric and Gas electric peak shaving experiment [102, 103]. The system had a total weight of 564 kg, of which 400 kg were Ti(Fe,Mn)-type hydride alloy (1.6 wt%). The system storage capacity was 6.4 kg H$_2$. Since 1974, several additional large stationary storage units have been built [104–106], demonstrating the technical viability of large-scale hydrogen storage systems.

Metal hydride on-board storage received much attention in the 1970s and early 1980s [107]. In 1969, K. C. Hoffman presented the first Society of Automotive Engineers (SAE) paper with the concept of using metal hydrides as a fuel source for vehicle propulsion [108]. Many design challenges remained unresolved, and more development appeared to be needed to achieve practicality [109, 110]. The quest of using hydrogen as a transportation fuel and the research required developing a better metal hydride on-board system declined and reached the lowest point in the early 1980s.

Upon entering the 21st century, the urgency of energy security and the awareness of climate change resulted in the reinitiation of national and worldwide hydrogen programs [111]. In 2000, Texaco Ovonic Hydrogen Systems LLC, a 50-50 joint venture between Energy Conversion Devices, Inc. (ECD Ovonics), and a unit of ChevronTexaco Corporation was formed to develop and advance solid-state metal hydride hydrogen storage systems for portable power, stationary, and on-board applications [72]. The transition-metal-based AB_2-type alloy family was selected for the hydrogen storage media in the solid hydrogen storage canisters and prototype vessels. A prototype fiber-wrapped hybrid vessel equipped with a proprietary heat exchanger and metal hydride powder containment system was developed and demonstrated on a converted Prius vehicle to run with hydrogen [73]. A picture of the storage system comprised of two vessels (each with a nominal capacity of 1.8 kg hydrogen) was shown in Figure 3.6 of Chapter 3.

A larger vessel incorporated with the same alloy under 1500 psig charging pressure can deliver a nominal hydrogen capacity of 3 kg [72, 73]. The corresponding gravimetric and volumetric densities of this vessel are 1.58 wt% and 60 g/L, respectively. The material's intrinsic gravimetric and volumetric efficiencies are 2.1 wt% and 110 g/L, respectively. Thus, we see that the engineering of the material into a practical storage tank with the necessary heat exchange hardware led to a 25% penalty applied to the gravimetric efficiency and a 45% penalty applied to the volumetric efficiency.

The system gravimetric density was below the target specified by the U.S. Department of Energy (U.S. DOE) light duty vehicle hydrogen storage goal for 2010. The system volumetric efficiency actually exceeded the U.S. DOE 2010 goal but is below the 2015 target [111]. A driving range of 137 miles was accomplished at 55 mph speed with the vessel filled by 1200 psig to about 2.7 kg H_2. Tailpipe emissions demonstrated from the vehicle incorporated with the prototype metal hydride vessel met the SULEV (super ultra low emissions vehicle) standard [73].

In addition to the Prius demonstration system, Sandia National Laboratories (SNL) and General Motors Corporation (GM) developed automotive-scale hydrogen storage systems based on metal hydrides [112]. This is discussed in more detail in Chapter 10 by Terry Johnson and Pierre Bénard. In addition, D. Mori and his team at Toyota Motors Corporation have developed a high-pressure metal hydride and gaseous hybrid vessel (rated at 350 bars) using AB_2 Laves phase Ti-Cr-Mn alloy exhibiting high disassociation pressure ($\Delta H < -20$ kJ/mol H_2) for a fuel cell vehicle [113]. The alloy showed a desorption hydrogen capacity of 1.9 wt% at 298 K (25°C). The vehicle, equipped with four high-pressure hybrid tanks (total volume of 180 L) having a total on-board hydrogen capacity of 7.3 kg, demonstrated the driving range of 700 km (437.5 miles), which is about 2.5 times longer than that of a 350-bar compressed vessel system with the same volume. The volumetric density is derived to be 40.6 g/L.

Apart from the light-duty transportation sector, the metal hydride family is potentially easier to be adopted in the application areas that are less sensitive to weight constraints, such as mechanical handling equipment (MHE), forklift, and stationary backup power.

In summary, the interstitial metal hydrides are compound materials that display remarkably versatile properties for hydrogen storage. Their properties can be tuned by changing the composition of both the hydride formers A and nonhydriding metals B in the AB_x formula to yield materials with a good plateau hydrogen pressure to suit most applications. A variety of alloys can be considered from the AB family (e.g., TiFe), AB_5 (e.g., $LaNi_5$), and AB_2 (e.g., ZrV_2) as well as the V-Ti-Cr family of solid solution BCC alloys. Their bonding in the varying interstitial sites in these materials was discussed from both crystallographic and electronic points of view. As a result of these properties, the interstitial metal hydrides have been demonstrated in diverse applications, such as utility-scale energy storage, hydrogen-powered vehicles, as well as in hydrogen storage for fuel cell material equipment (MHE) and forklifts. The major area for improvement in the metal hydrides lies with improving their gravimetric storage capacity beyond the 3.0 wt% set by the V-Ti-Cr family of BCC alloys, while it is hoped maintaining their otherwise-excellent hydrogen storage properties. Complex metal hydrides for on-board-reversible hydrogen storage with improved gravimetric storage density are the subject of the next chapter.

References

1. J.S. Noh, R.K. Agarwal, and J.A. Schwarz, *Int. J. Hydrogen Energy* 12, 693 (1987).
2. S.J. Blanksby and G.B. Ellison, *Acc. Chem. Res.* 36, 255 (2003).
3. H.H. Van Mal, *Philips Research Reports*, Supplement No. 1, Philips Research Laboratory, Eindhoven, Netherlands, 1976.
4. G.G. Libowitz, H.F. Hayes, and T.R.P Gibbs, *J. Phys. Chem.* 76, 62 (1958).
5. J.J. Reilly and R.H. Wiswall, *J. Inorg. Chem.* 7, 2254 (1968).
6. J.J. Reilly and R.H. Wiswall, *J. Inorg. Chem.* 13, 218 (1974).
7. J.J. Reilly, in *Hydrides for Energy Storage*, edited by A.F. Andersen and A.J. Maeland, Pergamon Press, New York, 301 (1978).
8. J.H.N. Van Vucht, F.A. Kuijpers, and H.C.A.M. Brunning, *Philips Res. Rep.* 25, 133 (1970).
9. R. Wiswall, in *Hydrogen in Metals*, edited by G. Alefeld and J. Volkl, Springer-Verlag, Berlin, Chapter 5 (1978).
10. G.G. Libowitz, H.F. Hayes, and T.R.P. Gibb, Jr., *J. Phys. Chem.* 62, 76 (1958).
11. J. Schefer, P. Fischer, W. Halg, F. Stucki, L. Schlapback, and A.F. Andresen, *Mater. Res. Bull.* 14(10), 1281 (1979).
12. J.J. Reilly and J.R. Johnson, Proceedings First World Hydrogen Energy Conference, Miami, Florida (1976).
13. G.D. Sandrock and P.D. Goodell, *J. Less-Common Metals* 104, 159 (1984).
14. L. Schlapback, *Hydrogen in Intermetallic Compounds II*, edited by L. Schlapback (Topics in Applied Physics), Springer, Berlin, 63, 76 (1992).
15. G.D. Sandrock, *J. Alloys Compd.* 293–295, 877 (1999).
16. T.B. Flanagan and J.D. Clewley, *J. Less Common Metals* 83, 127 (1982).
17. N.A. Scholtus and W.K. Hall, *J. Chem. Phys.* 39, 868 (1963).
18. J.H.N. Van Vucht, F.A. Kuijpers, and H.C.A.M. Brunning, *Philips Res. Rep.* 25, 133 (1970).
19. H.H. Van Mal, K.H.J. Buschow, and A.R. Miedema, *J. Less-Common Metals* 35, 65 (1974).
20. L. Schlapback, *J. Less-Common Metals* 73, 145 (1980).
21. G.D. Sandrock, in *Proceedings Second World Hydrogen Energy Conference, Zurich*, edited by T.N. Veziroglu and W. Seifritz, Pergamon Press, New York, 3, 1625 (1978).
22. F. Cuevas, J.-M. Joubert, M. Latroche, and A. Percheron-Guegan, *Appl. Phys.* A72, 225 (2001).

23. A. Percheron-Gregan, C. Lartigue, and J.-C. Achard, *J. Less-Common Metals* 74, 1 (1980).

24. M. Latroche, J. Rodriguez-Carvajal, A. Percheron-Guegan, and F. Bouree-Vigneron, *J. Alloys Compd.* 218, 64 (1995).

25. P. Fischer, A. Furrer, G. Busch, and L. Schlapbach, *Helv. Phys. Acta* 50, 421 (1977).

26. P. Fisher, W. Halg, L. Schlapbach, and Th. Von Wadkirch, *Helv. Phys. Acta* 51, 4 (1978)

27. K. Yvon and P. Fischer, *Hydrogen in Intermetallic Compounds I*, edited by L. Schlapbach (Topics in Applied Physics), Springer, Berlin, 63, 87 (1988).

28. H.F. Bittner and C.C. Badcock, *J. Electrochem. Soc.* 130, 193C (1983).

29. H. Ogawa, M. Ikoma, H. Kawano, and I. Matsumoto, *J. Power Sources* 12, 393 (1988).

30. T. Sakai, H. Miyamura, N. Kuriyama, A. Kato, K. Oguro, and H. Ishikawa, *J. Electrochem. Soc.* 137, 795 (1990).

31. C. Folonari, F. Iemmi, F. Manfredi, and A. Rolle, *J. Less-Common Metals* 74, 371 (1980).

32. G.D. Adzic, J.R. Johnson, J.J. Reilly, J. McBreen, and S. Mukerjee, *J. Electrochem. Soc.* 142, 3429 (1995).

33. J.J.G. Williams, *Philips J. Res.* 39, 1 (1984).

34. P.H.E. Notten, R.E.F. Einerhand, and J.L.C. Daams, *J. Alloys Compd.* 210, 221 and 233 (1994).

35. Z.-P. Li, Y.-Q. Lei, C.-P. Chen, J. Wu, and Q.-D. Wang, *J. Less-Common Metals* 172–174, 1256 (1991).

36. T. Sakai, K. Oguro, H. Miyamura, N. Kuriyama, A. Kato, and H. Ishikawa, *J. Less-Common Metals* 161, 193 (1990).

37. S. Wakao and Y. Yonemura, *J. Less-Common Metals* 89, 481 (1983).

38. T. Sakai, H. Miyamura, N. Kuriyama, A. Kato, K. Oguro, and H. Ishikawa, *J. Less-Common Metals* 159, 127 (1990).

39. Z. Li, D. Yan, and S. Suda, *Res. Rep. Kagakuin Univ.* 75, 113 (1993).

40. T. Hara, N. Yasucha, Y. Takeuchi, T. Sakai, and A. Uchiyama, *J. Electrochem. Soc.* 140, 2450 (1993).

41. T. Sakai, H. Yashinaga, H. Miyamura, N. Kuriyama, and H. Ishikawa, *J. Alloys Compd.* 180, 37 (1992).

42. J. Choi and C.-N. Park, *J. Alloys Compd.* 217, 25 (1995).

43. F. Meli, A. Zuttel, and L. Schlapback, *J. Alloys Compd.* 190, 17 (1992) and 202, 81 (1993).

44. J.-C. Achard, A. Perccheron-Guegan, H. Diaz, F. Briaucourt, and F. Demany, *Proc. 2nd Int. Congr. Hydrogen Metals* (Paris 1977), p. 1E12.

45. J.J. Didisheim, K. Yvon, D. Shaltiel, and P. Fisher, *Solid State Commun.* 31, 47 (1979).

46. D. Fruchart, A. Rouaut, C.B. Shoemaker, and D.P. Shoemaker, *J. Less-Common Metals* 73, 363 (1980).

47. K. Sapru, B. Reichman, A. Reger, and S.R. Ovshinsky, U.S. Patent 4, 623, 597 (1986).

48. S.R. Ovshinsky, M.A. Fetcenko, and J. Ross, *Science* 260, 176 (1993).

49. J. Huot, E. Akiba, T. Ogura, and Y. Ishido, *J. Alloys Compd.* 218, 101 (1995).

50. J. Huot, E. Akiba, and H. Iba, *J. Alloys Compd.* 228, 181 (1995).

51. F. Laves and H. Witte, *Metallwirschaft* 14, 645 (1935) and 15, 840 (1936).

52. J.H. Zhu, P.K. Liaw, and C.T. Liu, *Mater. Sci. Eng.* A239–240, 260 (1997).

53. D.I. Bardos, K.P. Gupta, and P.A. Beck, *Trans. Met. Soc. AIME* 221, 1087 (1961).

54. R.E. Watson and L.H. Bennett, *Acta Metall.* 32, 477 and 491 (1984).

55. B.S. Chao, R.C. Young, S.R. Ovshinsky, D.A. Pawlik, B. Huang, J.S. Im, and B.C. Chakoumakos, *Mater. Res. Soc. Symp. Proc.* 575, 193 (2000).

56. K. Young, J. Nei, T. Ouchi, and M.A. Fetcenko, *J. Alloys Compd.* 509, 2277 (2011).

57. O. Bernauer, J. Topler, D. Noreus, R. Hempelmann, and D. Richter, *Int. J. Hydrogen Energy* 14, 187 (1989).

58. K. Young, T. Ouchi, B. Huang, J. Nei, and M.A. Fetcenko, *J. Alloys Compd.* 501, 236 (2010).

59. M. Bououdina, J.L. Soubeyrous, D. Fruchar, and P. de Rango, *J. Alloys Compd.* 257, 82 (1997).

60. M. Yoshida and E. Akiba, *J. Alloys Compd.* 224, 121 (1995).

61. D. Shaltiel, I. Jacob, and D. Davidov, *J. Less-Common Metals* 53, 117 (1977).

62. D.G. Westalke, *J. Alloys Compd.* 90, 251 (1983).

63. I. Levin, V. Krayzman, C. Chiu, K.-W. Moon, and L.A. Bendersky, *Acta Mater.* 60, 645 (2012).

64. B.S. Chao, B. Huang, and F. Gingl, DLA Quarterly Technical Report (4th Quarter), Contract number N00164-07-C-6967 (June 2008).

65. J.M. Joubert, M. Latroche, and A. Percheron-Guegan, *J. Alloys Compd.* 231, 494 (1995).

66. Q.A. Zhang, Y.Q. Lei, X.G. Yang, K. Ren, and Q.D. Wang, *J. Alloys Compd.* 292, 236 (1999).
67. J.C. Sun, S. Li, and S.J. Ji, *J. Alloys Compd.* 404–406, 687 (2005).
68. F.C. Ruiz, E.B. Castro, S.G. Real, H.A. Peretti, A. Visintin, and W.E. Triaca, *Int. J. Hydrogen Energy* 33, 3576 (2008).
69. W.J. Boettinger, D.E. Newbury, K. Wang, L.A. Bendersky, C. Chiu, U.R. Kattner, K. Young, and B. Chao, *Metall. Mater. Trans.* A41, 2033 (2010).
70. L.A. Bendersky, K. Wang, W.J. Boettinger, D.E. Newbury, K. Young, and B. Chao, *Metall. Mater. Trans.* A41, 1891 (2010).
71. K. Young, T. Ouchi, B. Huang, B. Chao, M.A. Fetcenko, L.A. Bendersky, K. Wang, and C. Chiu, *J. Alloys Compd.* 506, 841 (2010).
72. B.S. Chao, R.C. Young, V. Myasnikov, Y. Li, B. Huang, F. Gingl, P.D. Ferro, V. Sobolev, and S.R. Ovshinsky, *Mater. Res. Soc. Symp. Proc.* 801-BB1.4, 27 (2003).
73. R.C. Young, B.S. Chao, Y. Li, V. Myasnikov, B. Huang, and S.R. Ovshinsky, *SAE Int. 04 Annu.* 606 (2004)
74. B.S. Chao, P.D. Ferro, M. Zelinsky, and N. Stetson, Proceeding of 2004 NHA Conference. Washington, DC (2004).
75. H. Iba and E. Akiba, *J. Alloys Compd.* 231, 508 (1995).
76. E. Akiba and H. Iba, *Intermetallics* 6, 461 (1998).
77. H. Iba and E. Akiba, *J. Alloys Compd.* 253–254, 21 (1997).
78. E. Akiba and M. Odada, *MRS Bullet.* 27(9), 699 (2002).
79. S. Ono, K. Nomura, and Y. Ikeda, *J. Less-Common Metals* 72, 159 (1980).
80. A.J. Maeland, G.G. Libowitz, and J.F. Lynch, *J. Less-Common Metals* 104, 133 and 361 (1984).
81. M. Tsukahara, K. Takahashi, T. Mishima, A. Isomura, and T. Sakai, *J. Alloys Compd.* 243, 133 (1995).
82. Y. Fukai, in *The Metal-Hydrogen Systems*, Springer Series in Materials Science, Vol. 21, Springer, Berlin, Chapter 3 (1993).
83. Y. Nakamura and E. Akiba, *J. Alloys Compd.* 345, 175 (2002).
84. K. Sapru, Z. Tan, M. Bazzi, S. Ramachandran, and S. Ovshinsky, U.S. Patent 6,616,891 B1 (2003).
85. B. Huang and S. Ovshinsky, U.S. Patent 7,108,757 B2 (2006).
86. K. Young, M. Fetcenko, T. Ouchi, J. Im, S. Ovshinsky, F. Li, and M. Reinhout, U.S. Patent 7,344,676 B2 (2008).
87. S.W. Cho, C.S. Han, C.N. Park, and E. Akiba, *J. Alloys Compd.* 288, 294 (1999).
88. M. Kada, T. Kuriiwa, T. Tamura, H. Takamura, and A. Kamegawa, *J. Alloys Compd.* 330–332, 511 (2002).
89. T. Tamura, T. Kazumi, A. Kamegawa, H. Takdamura, and M. Kada, *J. Alloys Compd.* 356–357, 505 (2003).
90. Y. Yan, Y. Chen, H. Liang, X. Zhou, C. Wu, M. Tao, and L. Ping, *J. Alloys Compd.* 454, 427 (2008).
91. J.J. Reilly and R.H. Wiswall, *J. Inorg. Chem.* 9, 1678 (1970).
92. B. Chao, B. Huang, and F. Gingl, TARDEC-AES/ECD milestone reports, Contract number FA8222-05-D-0001-0044-01 (October 2008, March and September 2009).
93. M.V. Lototsky, V.A. Yartys, and I.Y. Zavaliy, *J. Alloys Compd.* 404–406, 421 (2005).
94. C. Wu, X. Zheng, Y. Chen, M. Tao, G. Tong, and J. Zhou, *Int. J. Hydrogen Energy* 35(15), 8130 (2010).
95. H. Iba, Ph.D. thesis, Tohoku University, Sendai, Japan (1997).
96. R.M. Hartley (Ed.), *Hydrogen Progress*, Billings Energy Corp., Provo. UT, 2nd quarter (1977).
97. Ergenics, Inc., 247 Margaret King Avenue, Ringwood, NJ 07456 USA.
98. Hydrogen Consultants, Inc. (now Hydrogen Components, Inc.), 12420 N. Dumont Way, Littleton, CO 80125 USA.
99. J.C. Mccue, *J. Less-Common Metals* 74, 333 (1980).
100. HWT Gesellschaft fur Hydrid–und Wasserstofftechnik mbH, Postfach 100827, 4330 Mulheim a.d., Ruhr, FRG.
101. Suzuki Shokan Co., 3-1 Kojimachi, Chiyoda-ku, Tokyo 102, Japan.
102. G. Strickland, J.J. Reilly, and R.H. Wiswall, Jr., *Proc. Hydrogen Economy Miami Energy (THEME) Conf.* (University of Miami, Coral Gables, FL), S4-9–S4-21 (1974).

103. J.M. Burger, P.A. Lewis, R.J. Isler, F.J. Salzano, and J.M. King, Jr., *Proc. 9th Intersoc. Energy Conversion Eng. Conf.* (ASME, New York), 428–434 (1974).

104. N. Baker, L. Huston, F. Lynch, L. Olavson, and G. Sandrock, Final Phase I Report for U.S. Bureau of Mines Contract H0202034, Eimco Mining Machinery International, Salt Lake City, UT, 127–129 (1981).

105. G. Sattler, *J. Power Sources* 71, 144 (1998) and 86, 61 (2000).

106. A. Reller, Solide TAtsachen Schaffen, ENET news, Energy Switzerland (March 2004).

107. See, for example, *Hydrogen Power*, by L.O. Williams, Pergamon Press, New York (1980); P.P. Turillon, *Proc. 4th World Hydrogen Energy Conf.*, 1289 (1982).

108. K.C. Hoffman, W.E. Wische, R.H. Wiswall, J.J. Reilly, T.V. Sheehan, and C.H. Waide, SAE paper 690232, presented at the International Automotive Engineering Conference, Detroit, USA (1969).

109. F.E. Lynch and E. Snape, *Alternative Energy Sources*, Vol. 3, edited by T.N. Veziorglu, Hemisphere, Washington, DC, 1479 (1978).

110. G. Strickland, *Alternative Energy Sources*, Vol. 8, edited by T.N. Veziorglu, Hemisphere, Washington, DC, 3699 (1978).

111. See, for example, the DOE Web site, http://www.eere.energy.gov/hydrogenandfuelcells/.

112. T. Johnson, S. Jorgensen, and D. Dedrick, *Faraday Discuss.* 151, 327 (2011).

113. D. Mori and K. Hirose, *Int. J. Hydrogen Energy* 34(10), 4569 (2009).

6

Development of On-Board Reversible Complex Metal Hydrides for Hydrogen Storage

Vitalie Stavila, Lennie Klebanoff, John Vajo, and Ping Chen

CONTENTS

Introduction 134
Classes of Complex Metal Hydrides 138
   Metal Alanates 138
   Synthesis of Metal Alanates 138
   Metal Alanate Crystal Structures 139
      $LiAlH_4$ 139
      $NaAlH_4$ 140
      $KAlH_4$ 140
      $Mg(AlH_4)_2$ 140
      $Ca(AlH_4)_2$ 141
   Hydrogen Storage Properties of Alanates 142
      $LiAlH_4$ 143
      $NaAlH_4$ 143
      $KAlH_4$ 145
      $Mg(AlH_4)_2$ 146
      $Ca(AlH_4)_2$ 146
      Na_2LiAlH_6 and $LiMg(AlH_4)_3$ 147
Metal Borohydrides 147
   Synthesis of Metal Borohydrides 148
   Crystal Structures of Metal Borohydrides 149
      $LiBH_4$ 149
      $NaBH_4$ 150
      $Mg(BH_4)_2$ 150
      $Ca(BH_4)_2$ 150
   Hydrogen Storage Properties of Metal Borohydrides 150
      $LiBH_4$ 150
      $NaBH_4$ 153
      $Mg(BH_4)_2$ 153
      $Ca(BH_4)_2$ 154
      Other Metal Borohydrides 155
Amides, Imides, Nitrides 157
   Li_3N 157
   $LiNH_2$ Modified with Mg 161
   LiMgN 166
Mixed-Anion Complex Metal Hydrides 169

Destabilized Complex Metal Hydrides 172
Theoretical Prediction of Complex Metal Hydride Materials 179
Nanoscale Complex Metal Hydrides 185
Summary and Outlook 198
    Destabilized Materials 198
    Nanoconfinement 199
    Kinetics of Solid-State Reactions 200
    Effect of Additives on the Rates of Solid-State Reactions 200
    Borohydrides 201
References 201

Introduction

Complex metal hydrides represent a class of compounds composed of metal cations (typically group I and II elements) and "complex" hydrogen-containing anions such as alanates (AlH_4^-), borohydrides (BH_4^-), and amides (NH_2^-). The IUPAC (International Union of Pure and Applied Chemistry) recommended names for the AlH_4^- and BH_4^- salts are tetrahydroaluminates and tetrahydroboronates, although these names are seldom used in the literature. Unlike the interstitial metal hydrides discussed in Chapter 5 by Ben Chao and Lennie Klebanoff, complex metal hydrides display hydrogen atoms covalently bound to Al, B, and N. Complex metal hydrides are of interest for hydrogen storage applications due to their light weight and high hydrogen content [1–4]. They release molecular hydrogen either by heating or by a chemical reaction, such as hydrolysis. In fact, many complex hydrides release hydrogen in the presence of water or aqueous solutions. However, such reactions are quite exothermic and are not easily reversible. Here, we consider explorations of complex metal hydrides with the goal of finding a reversible hydrogen storage system with higher gravimetric density than the interstitial hydride materials but retaining the other attractive features of fast kinetics, favorable thermodynamics, and release of very pure hydrogen gas. We seek a material that can release hydrogen and reabsorb it without having to remove the material or the tank from the vehicle or piece of equipment powered by hydrogen. In other words, we seek an "on-board-reversible" complex hydride material.

Chemists have used complex hydrides for almost a century in organic syntheses involving reduction of esters, carboxylic acids, and organic amides. It was not until the mid-1990s that complex metal hydrides were considered for hydrogen storage applications. The pioneering work of Bogdanovic and Schwickardi [5] charged the field when they demonstrated that H_2 can be stored reversibly in sodium alanate ($NaAlH_4$) doped with titanium [5]. Titanium addition has proven beneficial because it lowers the decomposition temperature of $NaAlH_4$ but, more importantly, promotes full reversibility. This was a remarkable breakthrough and demonstrated for the first time that H_2 release from a complex metal hydride could be reversible. Ti-catalyzed sodium alanate remains one of the better *reversible* complex metal hydrides known.

Chapter 5 presented the van't Hoff expression for the formation of metal hydrides and how the thermodynamic changes in enthalpy ΔH and entropy ΔS are derived from pressure-composition-temperature (PCT) studies. As discussed in Chapter 5, for hydrogen storage applications, we seek a ΔH of dehydrogenation near 40 kJ/mol, assuming

ΔS in the range of about 120 J/K mol H_2. For a hydride with an equilibrium pressure of 1 atm, a 10-kJ/mol H_2 variation in ΔH results in about an 80 K change in the decomposition temperature [6]. In general, higher values of ΔH suggest higher stability of the complex metal hydride, while lower ΔH values suggest lower stability. A majority of complex metal hydrides require heat to release H_2, but there are also some important exceptions [7–14].

It is important to consider the kinetics of hydrogen absorption/desorption for complex metal hydride reactions. Solid-state reactions involving H_2 often suffer from high kinetic barriers required for diffusion. The dependence of the rate constant k on the reaction temperature (in kelvin) and activation energy E_a is given by the Arrhenius equation,

$$k = Ae^{-E_a/RT}$$

where A is the preexponential factor. The activation energy originates from the barrier associated with bond breaking in the transition state of the potential energy surface between the reactants and products of a reaction. Higher temperatures typically result in accelerated H_2 release reaction rates.

Many reactions involving metal hydrides occur in the presence of catalysts. The role of the catalysts is to accelerate a dehydrogenation or rehydrogenation reaction without modifying ΔG (standard Gibbs energy change) of the reaction or being consumed in the process. The catalysts do participate in the reaction and lower the activation energy for various processes; however, they are regenerated as the reaction proceeds to completion. Often, the role of the catalysts in reactions involving complex metal hydrides is to form activated species for rapid H_2 release or aid the dissociation of H_2 at the gas/solid interphase and accelerate diffusion of atomic H. To extract values of the E_a, measurements are made of the variation of the rate constant k with temperature, as follows:

$$\ln k = -\frac{E_a}{RT} + \ln A$$

Plotting $\ln k$ vs. $1/T$ gives a line of slope $-E_a/R$, with the preexponential factor obtainable from the intercept. With $R = 8.314$ J·mol^{-1}·K^{-1}, the units of activation energy are joules per mole. The E_a values for hydrogen release from complex metal hydrides are typically ≥ 100 kJ·mol^{-1}, but catalysts can significantly reduce the activation energy (to tens of kilojoules per mole) [11–18], as discussed in the following sections.

Although many metal hydrides display high volumetric and gravimetric hydrogen densities, for true commercial viability, the complex metal hydrides need to satisfy many performance requirements. The requirements are especially stringent for light-duty hydrogen-powered vehicles with fuel cells or internal combustion engines (ICEs) [19]. Complex metal hydrides have high gravimetric and volumetric capacities and tunable thermodynamics, which can allow reversible H_2 storage without removing the material from the H_2 tank. On-board reversibility is one of the most challenging requirements driving recent complex hydride materials discovery [11, 13, 16, 17, 19–26].

To help focus and guide scientific research and development (R&D) in the field, the U.S. Department of Energy (DOE) developed a Hydrogen Storage Multi-Year Research Development and Demonstration Plan (MRDDP) that identifies barriers to using solid-state hydrogen storage for hydrogen-powered cars. Some of these barriers are as follows:

A. **Cost.** We need low-cost materials and components for hydrogen storage systems, as well as low-cost, high-volume manufacturing methods.

B. **Weight and Volume.** Materials and components are needed that allow compact, lightweight, hydrogen storage systems while enabling greater than a 300-mile range in all light-duty vehicle platforms. Reducing the weight and volume of thermal management components is required concurrently.

C. **Efficiency.** The energy required to get hydrogen in and out of the material is an issue for reversible solid-state materials. Thermal management for charging and releasing hydrogen from the storage system needs to be optimized to increase overall efficiency.

D. **Durability.** Materials and components must enable a hydrogen storage system with 1500-cycle lifetime, with tolerance to fuel contaminants.

E. **Refueling Time.** There is a need to develop hydrogen storage systems with a refueling time of less than 3 min for 5 kg of hydrogen. Thermal management during refueling is a critical issue for metal hydrides that release heat on hydrogenation.

The U.S. DOE, in consultation with the FreedomCAR and Fuel Partnership Program, released an evolving set of technical targets for practical light-duty vehicle applications of hydrogen storage media. Since the DOE recently completed a 5-year funding cycle in which a great deal of hydrogen storage research was performed, we list the DOE targets for hydrogen storage systems, revised in 2009, in Table 6.1 [27].

While all of the DOE system targets are important for a properly functioning vehicular hydrogen tank, we emphasize that complex metal hydrides aim to improve on the one deficiency of the interstitial metal hydrides, namely, *gravimetric capacity*. Therefore, the DOE targets for gravimetric capacity are particularly important. For example, the 2017 system gravimetric target, indicating the mass of hydrogen stored per mass of the entire hydrogen storage system (including hydrogen storage material, tankage, and necessary plumbing), is 5.5 wt%. Assuming a 50% weight penalty arising from the necessary system hardware, the actual material's hydrogen storage capacity needs to be about 11 wt% or higher to satisfy the 2017 targets. This is much larger than than the 2–3 wt% gravimetric density provided by the interstitial metal hydride materials.

Another critical requirement specified by the DOE is that of *material reversibility*. The requirement for reversibility is implicit in the DOE requirement for cycle lifetime (2017 target: 1500 cycles). This is a challenging requirement from a materials perspective, and experience has shown that reversibility is especially difficult for the higher-capacity complex hydride materials. From a practical perspective, the reversibility would need to be in excess of 99.9% per cycle for a commercial storage system.

A third system consideration involves the *thermodynamic requirements*. It is desirable to be able to use the waste heat from a fuel cell operating at 70–85°C to drive off hydrogen from the metal hydride. Beyond the practical engineering issue of collecting and transporting fuel cell waste heat, if the material requires a high temperature to liberate hydrogen, then energy efficiency drops dramatically. In a sense, the hydrogen storage material needs to be "metastable." The material should be stable enough to store hydrogen near room temperature, yet be sufficiently unstable that only a modest amount of additional heat is required to liberate hydrogen completely and quickly. The high H_2 desorption temperatures of most of the complex metal hydrides arise from thermodynamic limitations. However, even a material with good thermodynamics can require high desorption temperatures if the material has lousy kinetics [11–18].

TABLE 6.1

U.S. DOE Targets for On-Board Hydrogen Storage Systems

Storage Parameter	Units	2010	2017	Ultimate
System gravimetric capacity: Usable, specific energy from H_2 (net useful energy/max system mass)	kWh/kg (kg H_2/kg system)	1.5 (0.045)	1.8 (0.055)	2.5 (0.075)
System volumetric capacity: Usable energy density from H_2 (net useful energy/max system volume)	kWh/L (kg H_2/L system)	0.9 (0.028)	1.3 (0.040)	2.3 (0.070)
Storage system cost	$/kWh net ($/kg H_2)	6 (200)	4 (133)	2 (67)
Fuel cost	$/gge at pump	3-7	2-4	2-4
Durability/operability				
Operating ambient temperature	°C	–30/50 (sun)	–30/60 (sun)	–40/60 (sun)
Min/max delivery temperature	°C	–40/85	–40/85	–40/85
Cycle life (1/4 tank to full)	Cycles	1000	1500	1500
Minimum delivery pressure from tank	bar (abs)	5 FC/35 ICE	4 FC/35 ICE	3 FC/35 ICE
Maximum delivery pressure from tank	bar (abs)	12 FC/100 ICE	12 FC/100 ICE	12 FC/100 ICE
On-board efficiency	%	90	90	90
"Well" to power plant efficiency	%	60	60	60
Charging/discharging rates				
System fill time (5 kg)	min	4.2	3.3	2.5
Minimum full flow rate	(g/s)/kW	0.02	0.02	0.02
Start time to full flow (20°C)	s	5	5	5
Start time to full flow (–20°C)	s	15	15	15
Transient response 10–90% and 90–0%	s	0.75	0.75	0.75
Fuel purity (H_2 from storage)	% H_2	SAE J2719 and ISO 14687-2 (99.97% dry basis)		
Environmental health and safety				
Permeation and leakage	Scc/h	Meets or exceeds applicable standards		
Toxicity	—	0.1	0.05	0.05
Safety	—			
Loss of usable H_2	(g/h)/kg H_2 stored			

Note: FC, fuel cell ICE, internal combustion engine.

A fourth material property considers *material stability and volatilization,* which is not explicitly called out in the DOE targets but is implicit in the requirements for cycle lifetime and hydrogen purity (Table 6.1). Ideally, it is preferred that the hydrogen storage material liberates only hydrogen when heated and does not release volatile and reactive components such as NH_3, BH_3, or other gas-phase components. This requirement serves two purposes: preserving the fuel cell catalysts (which become poisoned by reactive impurities in the hydrogen gas stream) and maintaining the storage material's physical integrity. If the storage material loses some of its components by volatilization as the material is heated, that is bad.

For fuel cell systems, the contamination target levels are less than 10 ppb sulfur, 1 ppm carbon monoxide, 100 ppm carbon dioxide, 1 ppm ammonia, and less than 100 ppm nonmethane hydrocarbons on a C-1 basis. Furthermore, oxygen, nitrogen, and argon must not exceed 2%. A H_2 ICE engine would have less-stringent requirements for the purity of the hydrogen stream.

Finally, *material kinetics* is critically important and forms the basis of the DOE targets for fuel-dispensing rate and hydrogen discharge. Although the H_2 desorption rates are critical, perhaps even more challenging are the material's kinetics associated with rehydrogenation. In analogy with the current refueling operation of automobiles, a storage material must be capable of receiving a 5-kg charge of hydrogen in about 3.3 min for a 5-kg hydrogen charge (2017 target).

In the following sections, we discuss the current state of R&D in complex metal hydride research. We focus on the general synthetic methods used to prepare them, their observed crystallographic structures, and finally their hydrogen storage properties.

Classes of Complex Metal Hydrides

Metal Alanates

Metal alanates, also known as "metal aluminum hydrides," contain metal cations such as Li^+ and tetrahedral $(AlH_4)^-$ anions or $(AlH_6)^{3-}$. Most of the group I and II metal alanates are white powders, which are air and moisture sensitive. Lithium alanate ($LiAlH_4$) and sodium alanate ($NaAlH_4$) are commercially available and used to synthesize other materials in the alanate family. They have been the subject of chemical interest for more than 60 years.

Synthesis of Metal Alanates

$LiAlH_4$ was first synthesized in 1947 by Finholt et al. [28] according to the following route:

$$4LiH + AlCl_3 \text{ (ether)} \rightarrow LiAlH_4 + 3LiCl \tag{6.1}$$

Another wet chemical approach is to use the metathesis reaction of $NaAlH_4$ and $LiCl$ in tetrahydrofuran (THF) to synthesize $LiAlH_4$ according to the following reaction [29]:

$$NaAlH_4 + LiCl \rightarrow LiAlH_4 + NaCl \tag{6.2}$$

A similar protocol was described for the wet chemical synthesis of $Mg(AlH_4)_2$ [30]. Magnesium alanate $Mg(AlH_4)_2$ can also be synthesized by a metathesis reaction of $NaAlH_4$ and MgH_2 [31]:

$$2NaAlH_4 + MgH_2 \rightarrow Mg(AlH_4)_2 + 2NaH \tag{6.3}$$

Direct hydrogenation of aluminum metal in the presence of LiH to form $LiAlH_4$ is achieved in the presence of catalytic amounts of $TiCl_3$ using dry THF as a solvent [29, 32].

Ashby and coworkers found that $NaAlH_4$ is formed in almost quantitative yield by reacting aluminum with either sodium hydride or sodium metal and applying 5000-psi H_2 pressure for 4 h in THF [29]. Solvent-free synthesis is also possible; however, higher hydrogen pressures are typically required. $KAlH_4$ was reported to form in good yield from KH and Al at 175-bar H_2 pressure on heating [33]. The same procedure is used to synthesize $NaAlH_4$ from NaH and Al at 175-bar H_2 pressure and 270°C:

$$2MH + 2Al + 3H_2 \rightarrow 2MAlH_4 \qquad (M = Li, Na, K) \tag{6.4}$$

Morioka et al. [34] reported that in contrast to $NaAlH_4$, bulk mixtures of KH and Al can be hydrogenated to $KAlH_4$ under a much smaller H_2 pressure of 10 bar between 250°C and 340°C without the need of a catalyst. An alternative route for the synthesis of alkali-metal alanates that does not require hydrogen gas was proposed by Dymova et al. and involves ball milling mixtures of the corresponding alkali-metal hydrides MH and aluminum hydride, AlH_3 [35]:

$$MH + AlH_3 \rightarrow MAlH_4 \qquad (M = Li, Na) \tag{6.5}$$

Summarizing, the alanates can be synthesized by straightforward wet synthetic approaches and solid-state routes in quantitative yield. The procedures do require glovebox equipment for their handling as all of the metal alanates are extremely air sensitive.

One of the attractive aspects of complex alanate materials is the diversity of compositions in the alanate family of materials, with associated variability in their hydrogen storage properties. For example, mixed metal alanates $M'M''(AlH_4)_n$ can be isolated using both solid-state and solution-based approaches. $LiNa_2AlH_6$ can be synthesized by reacting $LiAlH_4$ with 2 equivalents of NaH in toluene [36] or, alternatively, by ball milling a mixture of LiH, NaH, and $NaAlH_4$ [37]. LiK_2AlH_6 is synthesized from ball-milled $LiAlH_4$ and KH in a 1:2 molar ratio on heating to 320–330°C under 100- to 700-bar H_2 pressure [38]. Similarly, NaK_2AlH_6 is synthesized by ball milling $NaAlH_4$ and KH followed by annealing at 150°C under 100-bar H_2 pressure. The mixed Li-Mg alanate $LiMg(AlH_4)_3$ can be prepared from $LiAlH_4$ and $MgCl_2$ [39] or by solid-state metathesis between the two reagents in a 3:1 molar ratio [40, 41]:

$$3LiAlH_4 + MgCl_2 \rightarrow LiMg(AlH_4)_3 + 2LiCl \tag{6.6}$$

Other synthetic approaches to metal alanates exist; however, they have received less interest due to difficult purification or low yield [2, 16].

Metal Alanate Crystal Structures

R&D in complex-hydride hydrogen storage materials involves a lot of crystal structure determination. Theoretical predictions for the thermodynamics and reactive properties of these materials require accurate crystal structures as input, putting ever more emphasis on crystal structures. X-ray diffraction (XRD) is one of the chief methods used to determine the geometric structures in a complex metal hydride or even to determine if a solid-state reaction takes place. Neutron diffraction is useful as well and has the added benefit of being sensitive to hydrogen. We now give a discussion of the crystal structures of the alanate materials. Figure 6.1 displays and compares the crystal structures of both $LiAlH_4$ and $NaAlH_4$.

LiAlH$_4$

The crystal structure of $LiAlH_4$ was determined by XRD to be in the space group $P2_1/c$ from single-crystal data [42]. Each lithium atom is surrounded by five hydrogen atoms from the AlH_4^- tetrahedra with a wide spread of Li-H distances between 1.88 and 2.16 Å. Hauback et al. [43] used a combination of neutron and XRD techniques to determine accurate values

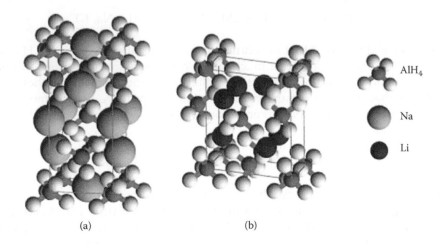

FIGURE 6.1
Crystal structure of the tetragonal NaAlH$_4$ (a) and monoclinic LiAlH$_4$ (b). (Reprinted from Z. Ma, M.Y. Chou, *Journal of Alloys and Compounds*, 479 (2009) 678–683 with permission of Elsevier.)

for bond lengths and angles in the deuterated version LiAlD$_4$. Li atoms are bonded to five deuterium atoms in slightly distorted trigonal bipyramidal coordination, as shown in Figure 6.1 [44]. The Al-D distances are in the range of 1.603–1.633 Å at room temperature.

NaAlH$_4$

The crystal structure of NaAlH$_4$ has been solved from single-crystal XRD studies [44, 45] to be in the $I4_1/a$ tetragonal space group. Each Na atom is surrounded by eight AlH$_4^-$ groups in a square antiprismatic geometry. Although these studies elucidated most of the structural details, some inconsistencies remained due to small size of the H atoms, which makes the determination of their coordinates problematic. Hauback et al. examined the structure of the deuterated version NaAlD$_4$ and found Al-D distances of 1.627(2) Å at 8 K and 1.626(2) Å at 295 K [45]. The experimentally observed crystal structures of NaAlH$_4$ and LiAlH$_4$ agree well with the first-principles calculations by Ma et al. [44] (Figure 6.1).

KAlH$_4$

The deuterated version KAlD$_4$ has a BaSO$_4$-type structure and crystallizes in the *Pnma* space group with each of the potassium atoms surrounded by 10 D atoms from seven different AlD$_4^-$ tetrahedra. The D-Al-D angles are within 106.2° to 114.6°, which is close to the ideal tetrahedral angle 109.5°. The minimum K-D distance of 2.596 Å is longer than in the corresponding Li-D and Na-D distances in LiAlH$_4$ and NaAlH$_4$.

Mg(AlH$_4$)$_2$

Mg(AlH$_4$)$_2$ crystallizes in the hexagonal space group $P3m1$ [46], as determined by XRD and neutron diffraction studies. The Al-H distances range from 1.606 to 1.634 Å at 295 K and 8 K and from 1.561 to 1.672 Å at 295 K. As can be seen from Figure 6.2, the sheetlike structure contains Mg atoms surrounded by tetrahedral AlH$_4^-$ groups in a slightly distorted octahedral MgH$_6$ geometry.

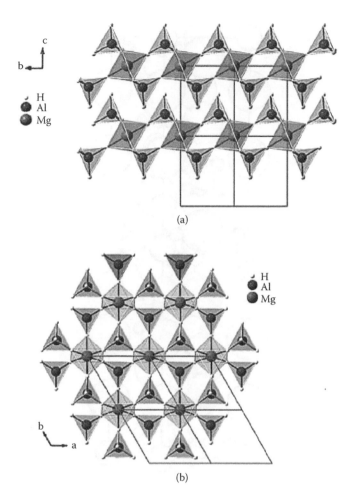

(a)

(b)

FIGURE 6.2
Crystal structure of the tetragonal $Mg(AlH_4)_2$. (Reprinted from A. Fossdal, H.W. Brinks, M. Fichtner, B.C. Hauback, *Journal of Alloys and Compounds*, 387 (2005) 47–51 with permission of Elsevier.)

$Ca(AlH_4)_2$

Fichtner et al. [47] reported the crystal structure of the THF adduct $Ca(AlH_4)_2 \cdot 4THF$ with two formulas per monoclinic unit cell $P2_1/n$. Attempts to solve the crystal structure of the solvent-free $Ca(AlH_4)_2$ have been unsuccessful. Two groups recently predicted that calcium alanate crystallizes in the octahedral space group *Pbca* [48, 49].

The mixed-cation materials are of interest because they offer the potential to tune hydrogen storage properties with variations in composition. Recent studies have shown that $LiNa_2AlH_6$ crystallizes in the cubic space group $Fm\overline{3}m$ [a = 7.38484(5) Å], displaying a perovskite structure with Li and Al atoms in octahedral positions [36, 50]. NaK_2AlH_6 also crystallizes in the $Fm\overline{3}m$ space group, although with a larger unit cell [a = 8.118(1) Å] [51]. The $[AlH_6]^{3-}$ octahedra display a close packing arrangement with potassium ions in the octahedral interstices. The experimentally determined cubic elpasolite-type structure is in

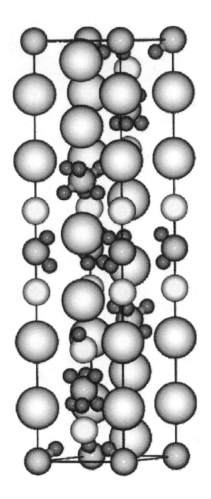

FIGURE 6.3
Crystal structure of LiK$_2$AlH$_6$. (Reprinted from E. Rönnebro, E.H. Majzoub, *Journal of Physical Chemistry B*, 110 (2006) 25686–25691 with permission of Elsevier.)

accordance with the theoretical predictions by Løvvik and Swang [52]. Figure 6.3 displays the crystal structure of LiK$_2$AlH$_6$. Graetz et al. reported that LiK$_2$AlH$_6$ crystallizes in the same cubic $Fm\bar{3}m$ space group [53]. A recent study by Rönnebro and Majzoub [38] suggested that LiK$_2$AlH$_6$ forms a rhombohedral structure within the space group $R\bar{3}m$. The Li and K cation sites are mutually exclusive, and no cation mixing is observed. The deuterated version of LiMgAlH$_6$, LiMgAlD$_6$, was found to crystallize in the trigonal space group $P321$, with hexacoordinated Li and Mg atoms and [AlD$_6$]$^{3-}$ octahedra [54].

Hydrogen Storage Properties of Alanates

The alanates LiAlH$_4$ and NaAlH$_4$ have good theoretical hydrogen gravimetric capacities, 10.6 wt% and 7.4 wt%, respectively. In the following, we present a review of the dehydrogenation and rehydrogenation reactions of these and other alanate systems. The history of these materials is that it is challenging to access the full theoretical hydrogen capacity.

LiAlH$_4$

Davis et al. [55] reported the first thermodynamic parameters for LiAlH$_4$, suggesting a ΔH_f value of 119.4 kJ/mol at 25°C. Subsequent studies allowed a more complete characterization, including an accurate value for the free energy of formation [56, 57]. In 1952, Garner and Haycock [58] proposed that the decomposition of LiAlH$_4$ proceeds via a stable intermediate LiAlH$_2$:

$$LiAlH_4 \rightarrow LiAlH_2 + H_2 \tag{6.7}$$

However, with the isolation of Li$_3$AlH$_6$ by Ehrlich et al. [59], it became obvious that this is the actual intermediate formed during the decomposition of LiAlH$_4$ [60]. Balema et al. demonstrated that LiAlH$_4$ is transformed under mild conditions into Li$_3$AlH$_6$ in the presence of catalytic amounts of Fe or TiCl$_4$ [61–63]. The reaction is almost quantitative and occurs at room temperature under ball-milling conditions.

Jang et al. [64] compiled the existing experimental thermodynamic parameters for LiAlH$_4$, Li$_3$AlH$_6$, and LiH and calculated the enthalpy and Gibbs energy change for the rehydrogenation reaction:

$$Li_3AlH_6 + 2Al + 3H_2 \rightarrow 3LiAlH_4 \tag{6.8}$$

There is experimental evidence [65, 66] that H$_2$ pressures in excess of 1000 bar are required for Reaction 6.8 to proceed, which makes the hydrogenation impractical, rendering LiAlH$_4$ irreversible. However, the excellent hydrogen desorption properties of LiAlH$_4$ have motivated efforts to find "off-board" methods of rehydrogenation. Such methods are discussed in more detail in Chapter 8.

Mikheeva and Arkhipov [60] suggested that the full decomposition pathway for LiAlH$_4$ follows the following sequence:

$$3LiAlH_4 \rightarrow Li_3AlH_6 + 2Al + 3H_2 \qquad T = 160{-}180°C;\ \Delta H = -10\ \text{kJ/mol}\ H_2 \tag{6.9}$$

$$2Li_3AlH_6 \rightarrow 6LiH + 2Al + 3H_2 \qquad T = 180{-}225°C;\ \Delta H = 25\ \text{kJ/mol}\ H_2 \tag{6.10}$$

$$2LiH \rightarrow Li + H_2 \qquad T \geq 400°C;\ \Delta H = 140\ \text{kJ/mol}\ H_2 \tag{6.11}$$

The last sequence 2LiH $\rightarrow$ Li + H$_2$ has far too high a desorption temperature to be useful for hydrogen storage. Thus, the effective gravimetric density for LiAlH$_4$ arises from the first two steps and is limited to about 7 wt%.

NaAlH$_4$

The hydrogen release from bulk NaAlH$_4$ is preceded by melting, which occurs at 183°C. The decomposition proceeds via stable intermediates, Na$_3$AlH$_6$ and NaH, which are stable up to 220°C and 400°C, respectively [67, 68]. The three-step decomposition process is as follows [68]:

$$3NaAlH_4 \rightarrow Na_3AlH_6 + 2Al + 3H_2 \qquad T = 180{-}190°C;\ \Delta H = 37\ \text{kJ/mol}\ H_2 \tag{6.12}$$

$$Na_3AlH_6 \rightarrow 3NaH + Al + 3/2H_2 \qquad T = 190{-}225°C;\ \Delta H = 47\ \text{kJ/mol}\ H_2 \tag{6.13}$$

$$NaH \rightarrow Na + ½H_2 \qquad T \geq 400°C;\ \Delta H = 120\ \text{kJ/mol}\ H_2 \tag{6.14}$$

Based on the stoichiometry of these reactions, the first step corresponds to 3.7 wt% H, the second 1.9 wt% H, while the last step accounts for 1.8 wt% H. As was the case with $LiAlH_4$, the last step is not useful for hydrogen storage due to the excessively high temperatures, so the viable gravimetric capacity of $NaAlH_4$ is 5.6 wt% and, due to kinetic limitations, is more like 4–5 wt%.

In the seminal work by Bogdanovic and Schwickardi [5], Ti addition to solid $NaAlH_4$ was found to dramatically increase the rate of hydrogenation, thereby making the material reversible. Catalyzed $NaAlH_4$ releases hydrogen for temperatures as low as 150°C (compared to 180–190°C in bulk), while the rehydrogenation reaction was completed in 5 h at 170°C and H_2 pressure of 152 bar. The reversibility after the first cycle was 4.2 wt% H_2, after the second it was 3.8 wt% H_2, and it dropped further to 3.1 wt% H_2 after 31 dehydrogenation/hydrogenation cycles [5]. Mechanochemical approaches [68–75] were found to incrementally improve transition metal doping of $NaAlH_4$, leading to improved properties. Jensen et al. [65, 66] achieved superior cycling behavior in titanium- and zirconium-doped $NaAlH_4$ by mechanically mixing the components. Interestingly, Ti was superior to Zr for dehydriding $NaAlH_4$ into Na_3AlH_6, while Zr was superior to Ti for dehydriding Na_3AlH_6 into NaH and Al. The authors suggested that the presence of both Ti and Zr is beneficial and achieved 4 wt% or more cyclable hydrogen capacity with the onset of H_2 release below 100°C [65]. Mechanical milling efficiently introduced the dopants in a single step. Moreover, the particle size of the hydride is also reduced, contributing to enhanced kinetics of the reaction [66–68]. In fact, ball milling $NaAlH_4$ alone was shown to improve its dehydriding kinetics [69].

Brinks et al. [70] performed synchrotron x-ray and neutron diffraction studies on sodium alanate samples doped with 2 mol% Ti additives. After initial ball milling, no Ti-containing phase was detected. However, after several cycles of dehydrogenation/rehydrogenation, a *fcc* solid solution with the approximate composition $Al_{0.93}Ti_{0.07}$ was observed. Kuba et al. investigated the effect of Ti additives on $NaAlH_4$ using electron paramagnetic resonance (EPR) spectroscopy [71] and detected several Ti(III) and Ti(0) species present. $TiCl_3$ was shown to be reduced more readily to Ti(0) compared to TiF_3. An important conclusion from the EPR studies was that the enhancement in H_2 storage properties in Ti-doped $NaAlH_4$ is from a minority Ti phase, while most of the titanium is in a resting state.

Solid-state nuclear magnetic resonance (NMR) studies revealed that Ti doping perturbs the mobility of AlH_4^- groups and promotes the mobility of H atoms. Anelastic spectroscopy measurements confirmed the presence of highly mobile hydrogen species [72]. Heating Ti-doped $NaAlH_4$ to 436 K resulted in a thermally activated relaxation process, suggesting the formation of a point defect with high mobility. Evidence that Ti doping affects the Al-H bonds in sodium alanate was obtained by infrared (IR) spectroscopy [73]. On doping, the Al-H asymmetric stretching mode is shifted by 15 cm^{-1} to higher frequencies, while H-Al-H asymmetric bending is shifted by 20 cm^{-1} to lower frequencies.

Theoretical calculations predict that Ti can substitute for Na in the structure; however, it only stays on the surface rather than diffusing into bulk [74–76]. Chaudhuri et al. computed the reverse reaction of formation of $NaAlH_4$ in the presence of Ti [77, 78]. A pure Al surface has a very low affinity for H_2. However, the studies suggested a specific local environment of two adjacent Ti atoms is responsible for the rehydrogenation reaction. The hydrogen atoms diffuse into the bulk to form AlH_3 (alane) species. This is consistent with experiments on hydrogen/deuterium scrambling, showing D_2 and H_2 mix to form HD on the surface of Ti-catalyzed $NaAlH_4$. The reaction does not occur when Ti is not present, so this is a direct proof that H_2 splitting occurs with direct participation of Ti species [79]. Extended x-ray absorption fine structure (EXAFS) data suggest that dehydrogenated samples contain a highly disordered distribution of Ti-Al distances with no long-range order beyond the second coordination sphere [77].

Density functional theory (DFT) calculations on possible Ti arrangements on the Al(001) surface identified low-barrier sites in which the incipient doped surface-H_2 adduct's highest occupied molecular orbital (HOMO) incorporates the sigma antibonding molecular orbital of hydrogen, allowing the transfer of charge density from the surface to dissociate the molecular hydrogen. Once AlH_3 forms, it reacts instantly with NaH to form $NaAlH_4$ without the need of a catalyst or hydrogen overpressure [77]. DFT calculations indicate that other metals, such as Zr, Fe, and V, also have catalytic effects on H_2 splitting; however, Ti is still one of the best catalysts [77, 78].

Interestingly, the hydrogen storage properties of other alanates can usually be enhanced using similar dopants as in the case of sodium alanate. Doping $LiAlH_4$ and Li_3AlH_6 by ball milling with $TiCl_3/AlCl_3$ results in a change in the dehydriding/rehydriding behavior and induces some reversibility [80, 81]. The apparent activation energies for dehydriding $LiAlH_4$ and Li_3AlH_6 were found to be 42.6 and 54.8 kJ/mol H_2, respectively, values that are almost two times lower compared to the undoped materials. X-ray photoelectron spectroscopy (XPS) revealed that the reversibility and the accelerated H_2 release are related to the distribution in the hydride matrix of Ti-Al clusters and Ti catalyst particles.

$KAlH_4$

Potassium alanate has received considerably less interest compared to the Li and Na counterparts, mainly because of lower theoretical hydrogen capacity (5.7 wt%). Hydrogen release from bulk $KAlH_4$ was studied by NMR, and the intermediate species were proposed to be AlH_3 and KH [82, 83]. However, subsequent work by Morioka et al. [34] suggested that $KAlH_4$ undergoes similar thermal decomposition as in the case of its lighter counterparts, $LiAlH_4$ and $NaAlH_4$:

$$3KAlH_4 \rightarrow K_3AlH_6 + 2Al + 3H_2 \quad \text{2.9 wt% H; } T = 300\text{–}330°C \tag{6.15}$$

$$K_3AlH_6 \rightarrow 3KH + Al + 3/2H_2 \quad \text{1.4 wt% H; } T = 340\text{–}370°C \tag{6.16}$$

$$KH \rightarrow K + ½ H_2 \quad \text{1.4 wt% H; } T \geq 420°C \tag{6.17}$$

This mechanism was also considered in ab initio calculations of $KAlH_4$ [84], which predicted that the enthalpies of Reactions 6.15 and 6.16 are 55 and 70 kJ/mol H_2, respectively, consistent with relatively high decomposition temperatures found experimentally. The decomposition of $KAlH_4$ doped with $TiCl_3$ resulted in a decrease in the decomposition temperature necessary to transform $KAlH_4$ into K_3AlH_6 by 50°C (Reaction 6.15), yet the decomposition of K_3AlH_6 proceeds at the same temperatures as in undoped material [85]. Decomposition of $KAlH_4$ was monitored with [39]K and [27]Al magic angle spinning (MAS) NMR [82, 83], and evidence for the appearance of $[AlH_6]^{3-}$ units was found. Rehydrogenation of KH and Al under 9-bar pressure starts at 200°C and ends at 340°C with a reversible capacity of 3–4 wt% hydrogen [34].

Mg(AlH₄)₂

Magnesium alanate has one of the highest H capacities (9.3 wt% H) of the metal alanates. Fichtner et al. proposed a three-step decomposition process for $Mg(AlH_4)_2$ [31, 86]:

$$Mg(AlH_4)_2 \rightarrow MgH_2 + 2Al + 3H_2 \qquad 7.0 \text{ wt\% H; } 110\text{--}200°C \qquad (6.18)$$

$$MgH_2 \rightarrow Mg + H_2 \qquad 2.3 \text{ wt\% H; } 240\text{--}380°C \qquad (6.19)$$

$$2Al + Mg \rightarrow \tfrac{1}{2} Mg_2Al_3 + \tfrac{1}{2}Al \qquad T \geq 400°C \qquad (6.20)$$

Reaction 6.18 is exothermic, which makes the prospects of rehydrogenating "spent" $Mg(AlH_4)_2$ problematic. More recent in situ synchrotron XRD studies, however, suggested a more complex mechanism of $Mg(AlH_4)_2$ decomposition [87]:

$$(1 - x/2)Mg(AlH_4)_2 \rightarrow (1 - (3/2)x)MgH_2 + 2Al(1 - x/2)Mg(x/2) + (3 + (3/2)x)H_2 \quad (6.21)$$

$$(1 - (3/2)x)MgH_2 + 2Al(1 - x/2)Mg(x/2) \rightarrow Al_xMg_y + (1 - (3/2)x)H_2 \qquad (6.22)$$

Titanium doping accelerates the kinetics of H_2 release from $Mg(AlH_4)_2$, with most of the gaseous H_2 evolved at 150°C. Other studies [88, 89] confirmed the accelerated kinetics of H_2 release; however, no reversibility was found for $Mg(AlH_4)_2$ under a variety of studied conditions, suggesting the rehydrogenation is thermodynamically unfavorable.

Ca(AlH₄)₂

Calcium alanate has received considerably less interest compared to lithium, sodium, or magnesium alanate, mainly due to its lower H capacity. The decomposition of $Ca(AlH_4)_2$ can be represented according to the following sequence of reactions:

$$Ca(AlH_4)_2 \rightarrow CaAlH_5 + Al + 3/2H_2 \qquad 3.0 \text{ wt\% H} \qquad (6.23)$$

$$CaAlH_5 \rightarrow CaH_2 + Al + 3/2H_2 \qquad 3.0 \text{ wt\% H} \qquad (6.24)$$

$$CaH_2 \rightarrow Ca + H_2 \qquad 2.0 \text{ wt\% H} \qquad (6.25)$$

Reaction 6.23 is slightly exothermic, and the enthalpy change was estimated to be –7 kJ/mol H_2 [90]. In contrast, $CalAlH_5$ decomposes endothermically, and the enthalpy of reaction is predicted to be a more reasonable 28 kJ/mol H_2. A detailed differential scanning calorimetry (DSC) investigation revealed that the first step of decomposition of $Ca(AlH_4)_2$ occurs at 127°C, whereas the second endothermic reaction occurs at 250°C. The third step (decomposition of CaH_2) is highly endothermic (172 kJ/mol H_2) and occurs at temperatures too high for reversible H_2 storage ($t \geq 700°C$).

Recent first-principles DFT calculations by Wolverton and Ozolins suggested that doping $Ca(AlH_4)_2$ with two equivalents of Al metal changes the thermodynamics of H_2 release considerably, and the resulting enthalpy of the reaction

$$CaH_2 + Al \rightarrow CaAl_2 + H_2 \qquad (6.26)$$

was predicted to be 72 kJ/mol H_2 [48], a significant reduction from 172 kJ/mol H_2 for pure CaH_2, but still too high for reversible hydrogen storage applications. Recently, Iosub et al. proposed that the decomposition of $Ca(AlH_4)_2$ proceeds via a stable hexaaluminumhydride complex, $Ca_3(AlH_6)_2$ [91].

Na_2LiAlH_6 and $LiMg(AlH_4)_3$

Hydrogen desorption from Na_2LiAlH_6 was studied by PCT measurements, and the total H_2 capacity achieved was 3.2 wt% [37]. A doped version of Na_2LiAlH_6 released only 3.0 wt% H_2 on thermal decomposition, too low for practical applications. In contrast, $LiMg(AlH_4)_3$, which contains almost 10 wt% H, has one of the highest capacity among metal alanates. The thermal decomposition of $LiMg(AlH_4)_3$ proceeds as follows [40]:

$$LiMg(AlH_4)_3 \rightarrow LiMgAlH_6 + 2Al + 3H_2 \qquad \text{3.0 wt\% H; 100–130°C} \qquad (6.27)$$

$$LiMgAlH_6 \rightarrow LiH + MgH_2 + Al + 3/2H_2 \qquad \text{3.0 wt\% H; 150–180°C} \qquad (6.28)$$

However, DSC studies revealed that the first step of H_2 release (Reaction 6.27) is exothermic, with a $\Delta H = -15.1$ kJ/mol H_2 [41], making this hydride less attractive for reversible H_2 storage applications.

Metal alanates remain one of the most studied classes of complex metal hydrides, and although their reversible cycling capacity is insufficient to satisfy the technical targets presented in Table 6.1, some of their properties are quite remarkable, making them suitable for other hydrogen storage applications, for example, supporting the use of fuel cells in construction and materials-handling equipment. For instance, $NaAlH_4$ has an equilibrium pressure of 1 bar at 35°C for the first decomposition step, releasing up to 3.7 wt% H_2. This reaction has a desirable ΔH value of 37 kJ/mol H_2. For a H_2 tank, one important problem is the heat management, so lower values of ΔH^0 are preferred.

Metal Borohydrides

Metal borohydrides, also known as tetrahydroborates, contain metal cations and tetrahedral BH_4^- anions. Group I and II metal borohydrides are white powders that are air and moisture sensitive. A number of metal borohydrides are commercially available, including $LiBH_4$, $NaBH_4$, KBH_4, $Mg(BH_4)_2$, and $Ca(BH_4)_2$. Lithium and sodium borohydrides are used to synthesize transition metal borohydrides through metathesis reactions and have been used in organic synthesis for decades for various redox processes.

The interest in borohydrides as hydrogen storage media originated from the work of Züttel et al., who investigated hydrogen storage properties of $LiBH_4$ in the early 2000s [92–94]. The continued interest of borohydrides and borohydride-based materials is due to their high gravimetric and volumetric hydrogen densities. In analogy with the alanates, we first discuss the syntheses of borohydrides and then give details about their structural properties. Finally, we describe the hydrogen storage properties of the most promising borohydride materials.

Synthesis of Metal Borohydrides

$LiBH_4$ was first synthesized in 1940 by Schlesinger and Brown [95], according to the following scheme:

$$LiEt + B_2H_6 \rightarrow LiBH_4 + BEtH_2 \tag{6.29}$$

Subsequently, they proposed substituting ethyllithium with the more accessible lithium hydride:

$$LiH + 1/2B_2H_6 \rightarrow LiBH_4 \tag{6.30}$$

Alternatively, $LiBH_4$ can be obtained in good yield by direct reaction of Li metal and elemental boron on heating under hydrogen pressure (30–150 bar) [96, 97]:

$$Li + B + 2H_2 \rightarrow LiBH_4 \tag{6.31}$$

Analogously, $NaBH_4$ can be synthesized from NaEt using the route of Reaction 6.29 or by reacting NaH with trimethyl borate [98]:

$$4NaH + B(OMe)_3 \rightarrow NaBH_4 + 3NaOMe \tag{6.32}$$

Reaction 6.32 represents one of two large-scale industrial processes for sodium borohydride production, with reaction temperatures of 250–280°C. The second industrial method involves the reaction of less-expensive borosilicate glass with Na and hydrogen at high temperatures (400–500°C):

$$Na_2B_4O_7 + 7\,SiO_2 + 16Na + 8H_2 \rightarrow 4NaBH_4 + 7Na_2SiO_3 \tag{6.33}$$

Currently, millions of kilograms of sodium borohydride are produced annually using the two approaches mentioned, making it by far the most manufactured complex metal hydride [99].

Other borohydrides can be synthesized using metathesis reactions of alkali metal borohydrides MBH_4 (M = Li, Na, K) with other metal salts, such as transition metal halides. For example, zinc(II) borohydride can be synthesized using the reaction of $ZnCl_2$ and various metal borohydrides [100–103]:

$$ZnCl_2 + 2MBH_4 \rightarrow Zn(BH_4)_2 + 2MCl \text{ (M = Li, Na, K)} \tag{6.34}$$

Zanella et al. proposed two approaches to synthesize $Mg(BH_4)_2$ from dibutyl magnesium [104]:

$$3Mg(C_4H_9)_2 + 2Al(BH_4)_3 \rightarrow 3Mg(BH_4)_2 + 2Al(C_4H_9)_3 \tag{6.35}$$

$$3Mg(C_4H_9)_2 + 8BH_3\cdot SMe_2 \rightarrow 3Mg(BH_4)_2\cdot 2SMe_2 + 2B(C_4H_9)_3\cdot SMe_2 \tag{6.36}$$

Both Reaction 6.35 and Reaction 6.36 may be suitable for the synthesis of other metal borohydrides when the corresponding alkylmetal derivatives are readily available. A nice compilation of the available synthetic approaches toward metal borohydrides was published by Hagemann and Cerny [105].

Crystal Structures of Metal Borohydrides

Complex metal borohydride crystal structures have been extensively studied in the past several years. Here, we provide a review for the most prominent materials in this class.

LiBH₄

The crystal structure of $LiBH_4$ was first determined by Harris and Meibohm [106] to be in the orthorhombic space group *Pcmn*. The unit cell contains four $LiBH_4$ molecules, with each Li^+ ion surrounded by four BH_4^- tetrahedra. Soulie et al. redetermined the structure and found that the actual space group for the room temperature orthorhombic phase is *Pnma* [107]. The tetrahedra BH_4^- anions are aligned along two orthogonal directions and are severely distorted with respect to bond lengths [d(B-H) = 1.04–1.28 Å] and angles (H-B-H = 85.1–120.1°). On heating to 408 K, there is a phase transition, and the structure undergoes a first-order transition and becomes hexagonal, space group *P6mc*. In the high-temperature (HT) phase, the BH_4^- tetrahedra align along the *c*-axis and become more symmetric [d(B-H) = 1.27–1.29 Å] and (H-B-H = 106.4–112.4°). Züttel et al. also employed synchrotron radiation to study the phase transition in $LiBH_4$ and confirmed the identity of the low-temperature (LT) and high-temperature (HT) phases [108]. The two structures (Figure 6.4) are in good agreement with first-principles theoretical predictions [109, 110].

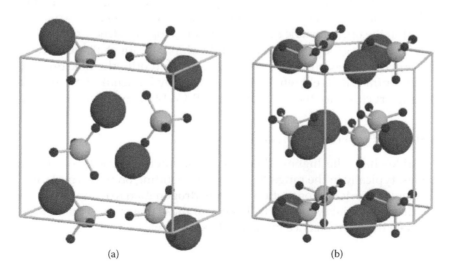

(a)　　　　　　　　　　(b)

FIGURE 6.4
Crystal structures of $LiBH_4$ in (a) orthorhombic phase at room temperature and (b) hexagonal phase at high temperature. Large, midsize, and small spheres represent Li, B, and H atoms, respectively. (Reprinted from K. Miwa, N. Ohba, S.-I. Towata, Y. Nakamori, S.-I. Orimo, *Physical Review B*, 69 (2004) 245120; S.C. Abrahams, J. Kalnajs, *Journal of Chemical Physics*, 22 (1954) 434–436 with the permission of American Physical Society.)

NaBH₄

Abrahams and Kalnajs investigated $NaBH_4$ using powder XRD and determined the parameters of the unit cell [111]. A more recent neutron diffraction study revealed that $NaBH_4$ crystallizes in a NaCl-type structure with sodium and boron atoms arranged in a body-centered tetragonal fashion [112]. It was reported that the deuterated version of $NaBH_4$, $NaBD_4$, belongs to the $P\bar{4}21c$ space group [13].

Mg(BH₄)₂

The crystal structures of the LT and HT phases of $Mg(BH_4)_2$ are shown in Figure 6.5. Konoplev et al. suggested $Mg(BH_4)_2$ exists in two crystal modifications, a LT tetragonal phase and a HT cubic phase [113]. More recent theoretical predictions suggested lower symmetry for both phases [114–118]. Neutron and powder XRD studies on the LT phase suggested the symmetry is hexagonal, space group $P6_1$ [117, 119]. Filinchuk et al. published a revised structure for the LT phase in the $P6_122$ setting with four almost-ideal BH_4^- tetrahedral groups forming a dodecahedral MgH_8 coordination around each Mg atom. The H-H distances between two BH_4^- groups were remarkably short (2.18–2.28 Å) [120]. When heated to 453 K, the LT phase is transformed into an orthorhombic structure within the space group $Fddd$ [119]. It was found [16] that $Mg(BH_4)_2$ underwent an irreversible structural transition at 3.35 GPa, which did not match any of the first-principles predicted monoclinic $P2/c$ [118], orthorhombic $Pmc2_1$ [114], or tetragonal $I\bar{4}m1$ [121] structures.

Ca(BH₄)₂

Kedrova et al. reported the unit cell of $Ca(BH_4)_2$ in 1977 [122]; however, no attempts to solve the structure were made. A number of attempts to predict the ground crystal structure were made by various groups [123–125]. Miwa et al. refined the crystal structure of the LT α-$Ca(BH_4)_2$ within the orthorhombic $Fddd$ space group [126]. The structure contains Ca^{2+} ions coordinated by six BH_4^- groups in an octahedral fashion. Riktor et al. found two HT modifications, β-$Ca(BH_4)_2$ and γ-$Ca(BH_4)_2$ having tetragonal and orthorhombic unit cells, respectively [127]. In contrast, Rietveld refinements reported by Filinchuk et al. suggest $F2dd$ and $P\bar{4}$, respectively, structures for the α-$Ca(BH_4)_2$ and β-$Ca(BH_4)_2$ modifications, respectively.

Hydrogen Storage Properties of Metal Borohydrides

The borohydrides have the highest theoretical gravimetric capacity of all the complex metal hydrides, typically in the range 10–14 wt%, depending on the formulation. We present next a review of the dehydrogenation and rehydrogenation reactions of the most prominent borohydride material systems. The history of these materials is that it is challenging to rehydrogenate the material once hydrogen has been released.

LiBH₄

Fedneva and coworkers studied the thermal behavior of $LiBH_4$ by differential thermal analysis (DTA); they reported three endothermic peaks and assigned them to the following processes [128]: (1) polymorphic phase transition between 108°C and 112°C; (2) melting of $LiBH_4$ at 268–286°C; and (3) decomposition with hydrogen evolution at 483–492°C. Later

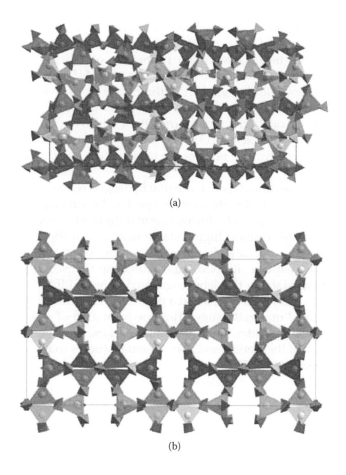

FIGURE 6.5
Crystal structure of low- (a) and high-temperature (b) Mg(BH$_4$)$_2$ phases. (Reprinted from J.-H. Her, P.W. Stephens, Y. Gao, G.L. Soloveichik, J. Rijssenbeek, M. Andrus, J.-C. Zhao, *Acta Crystallographica Section B*, 63 (2007) 561–568 with permission of IUCr.)

studies of the decomposition of LiBH$_4$ in a hydrogen atmosphere up to 10 bar suggested two possible decomposition reactions [129]:

$$LiBH_4 \rightarrow Li + B + 2H_2 \tag{6.37}$$

$$LiBH_4 \rightarrow LiH + B + 3/2H_2 \tag{6.38}$$

Züttel et al. proposed that the decomposition of LiBH$_4$ occurs through the formation of a LiBH$_2$ intermediate species [92, 94]. Ball milling LiBH$_4$ with SiO$_2$ in a 1:3 ratio resulted in a significant decrease in H$_2$ desorption temperature with the onset at 523 K and total capacity of 9 wt% hydrogen [94]. The product of the dehydrogenation reaction of LiBH$_4$ can be rehydrogenated at 873 K under 35-MPa H$_2$ pressure for 12 h [130]. Vajo et al. reported that a 2:1 mixture of LiH and MgB$_2$ can absorb 7.84 wt% hydrogen at 350°C under 100-bar H$_2$ pressure to form LiBH$_4$ and MgH$_2$ [131]. The absorption was performed in the presence of 3 mol% TiCl$_3$.

The kinetics of hydrogen desorption from $LiBH_4$ can be significantly enhanced in the presence of additives. Au et al. performed a comprehensive study and verified the effect of various additives on the H_2 desorption properties of $LiBH_4$ [132–134]. The experimental results indicated that additives, such as Mg, Al, MgH_2, AlH_3, CaH_2, $TiCl_3$, TiF_3, TiO_2, ZrO_2, V_2O_3, and SnO_2, added via ball milling have a positive effect and reduce the dehydrogenation temperature of $LiBH_4$, but some other dopants, such as C, Ni, In, Ca, and NaH, increase its desorption temperature. Other additives were examined for their promotion of kinetics of H_2 release [135–140]. However, so far the effect is limited to the dehydrogenation reaction, and none of the tested additives/dopants rendered $LiBH_4$ fully reversible.

There is compelling experimental evidence that hydrogen release from $LiBH_4$ is complex and involves one or several stable intermediate species. Theoretical predictions indicated that $Li_2B_{12}H_{12}$ is a stable intermediate during thermal decomposition of $LiBH_4$ [141]. *Closo*-dodecahydrododecaborates contain highly stable isocahedral $[B_{12}H_{12}]^{2-}$ cages, which are extremely stable due to the aromatic character of bonding in the cage. This prediction was verified experimentally with ex situ NMR and Raman spectroscopy [142, 143]. Figure 6.6 shows comparison NMR data for the dehydrogenated states of $LiBH_4$, $Mg(BH_4)_2$, $(Ca(AlH_4)_2$ + $6LiBH_4)$, and $(ScCl_3 + 3LiBH_4)$. It seems that a peak similar to that of $K_2B_{12}H_{12}$ is present in all these dehydrogenated materials, suggesting the formation of a $[B_{12}H_{12}]^{2-}$ intermediate is a general issue for borohydrides. The appearance of these relatively unknown $M_xB_{12}H_{12}$ stimulated a great deal of theoretical and experimental studies to reveal the structure and reactivity of these species [144–152].

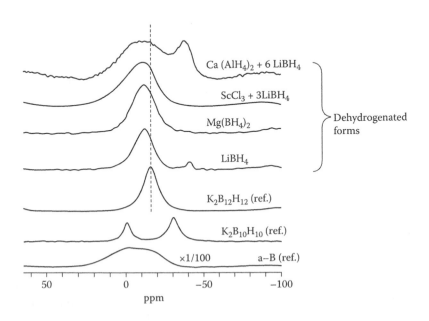

FIGURE 6.6
^{11}B NMR data for the calibration compounds B, $K_2B_{10}H_{10}$, and $K_2B_{12}H_{12}$. Also shown are NMR data for the dehydrogenated states of $LiBH_4$, $Mg(BH_4)_2$, $(Ca(AlH_4)_2 + 6LiBH_4)$, and $(ScCl_3 + 3LiBH_4)$. (Reprinted from S.-J. Hwang, R.C. Bowman, Jr., J.W. Reiter, J. Rijssenbeek, G.L. Soloveichik, J.-C. Zhao, H. Kabbour, C.C. Ahn, *Journal of Physical Chemistry C*, 112 (2008) 3164–3169 with permission of American Chemical Society.)

$NaBH_4$

Bulk sodium borohydride has a high desorption temperature and is mainly considered a chemical hydride in off-board regeneration schemes by catalytic dehydrogenation of its aqueous solutions. See Chapter 3 for a fuller discussion of this process. Although bulk sodium borohydride is considered an irreversible H_2 storage material, a report by Ngene and coworkers found that carbon-nanoconfined $NaBH_4$ displayed partial reversibility at 325°C under 60-bar H_2 [153].

$Mg(BH_4)_2$

Magnesium borohydride is attractive because of its exceptionally high theoretical hydrogen weight percentage, 14.9%. $Mg(BH_4)_2$ was first synthesized by Konoplev [154]. The H_2 desorption from $Mg(BH_4)_2$ was observed above 320°C, with ΔH about 53 kJ/mol H_2. However, almost nothing was known about either the reaction pathway or if the material was reversible. Zhao et al. used in situ XRD to directly correlate $Mg(BH_4)_2$ crystal structure with temperature and hydrogen release [119, 155, 156]. The XRD studies revealed that the material undergoes a phase transition from an LT form to an HT form. The residual gas analysis (RGA) data confirmed that this structural phase transition occurs before H_2 is released. As in the case of $LiBH_4$, a stable *closo*-dodecahydrododecaborate intermediate $MgB_{12}H_{12}$ was observed [155, 157] once hydrogen was liberated from the sample. From these initial studies, an attractive ΔH for hydrogen desorption of about 40 kJ/mol H_2 was observed, lower than Konoplev's result, indicating a decent chance for reversibility under reasonable conditions if no kinetic limitations existed [158].

The overall reaction sequence for hydrogen release from $Mg(BH_4)_2$ is shown in Figure 6.7 and can be written as

$$6Mg(BH_4)_2 \rightarrow 5MgH_2 + Mg(B_{12}H_{12}) + 13H_2 \tag{6.39}$$

$$5MgH_2 \rightarrow 5Mg + 5H_2 \tag{6.40}$$

$$5Mg + Mg(B_{12}H_{12}) \rightarrow 6MgB_2 + 6H_2 \tag{6.41}$$

Thus, hydrogen is first released from $Mg(BH_4)_2$ to form MgH_2 and $Mg(B_{12}H_{12})$. MgH_2 then releases H_2, followed by eventual release at high temperature from the highly stable $Mg(B_{12}H_{12})$ material. A full 12 wt% of hydrogen is released from the material, and the low ΔH of desorption indicates facile hydrogenation of MgB_2 should be possible.

The reversibility of $Mg(BH_4)_2$ was first demonstrated by Severa et al. [159], in which the end product of dehydrogenation from $Mg(BH_4)_2$, namely, MgB_2, was subjected to hydrogenation at high pressures (950 bar) and high temperatures (400°C) using the high-pressure hydrogenation cell at Sandia National Laboratories. The results of the experiments showed that $Mg(BH_4)_2$ could in fact be produced by the hydrogen exposure, and that the $Mg(BH_4)_2$ thus produced released more than 11 wt% of hydrogen when heated. MAS ^{11}B NMR spectroscopy demonstrated that $Mg(BH_4)_2$ is the major product of the reaction, with small amounts of amorphous $MgB_{12}H_{12}$ comprising less than 5% of the product mixture. Summarizing, it has been shown that in the case of $Mg(BH_4)_2$, 12 wt% of hydrogen could be released, and that the end product (MgB_2) could then be rehydrogenated all the way back to $Mg(BH_4)_2$ according to the following reaction:

$$MgB_2 + 4H_2 \rightarrow Mg(BH_4)_2 \tag{6.42}$$

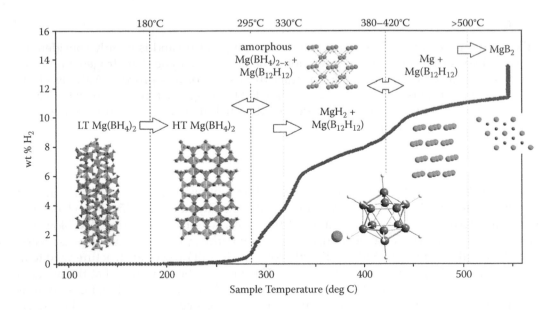

FIGURE 6.7
Overall reaction sequence for the release of hydrogen from Mg(BH$_4$)$_2$ [155]. (Based on the work of G.L. Soloveichik, Y. Gao, J. Rijssenbeek, M. Andrus, S. Kniajanski, R.C. Bowman, Jr., S.-J. Hwan, J.-C. Zhao, *International Journal of Hydrogen Energy*, 34 (2009) 916–928.)

The 12 wt% reversible hydrogen capacity for Mg(BH$_4$)$_2$ makes it a record-breaking reversible H$_2$ storage material. Although the process conditions of pressure and temperature for both dehydrogenation and hydrogenation are higher than desired, Mg(BH$_4$)$_2$ has one of the highest reversible capacities for any metal hydride material discovered thus far. It has been demonstrated that the dehydrogenation temperature can be significantly reduced using a mixed ScF$_3$/TiCl$_3$ additive, with most of the H$_2$ released below 300°C [160, 161]. Mg(BH$_4$)$_2$ remains one of the most interesting compounds for H$_2$ storage due to the high theoretical weight percentage H (14.9%), good ΔH (40 kJ/mol H$_2$) for hydrogen desorption, and demonstrated reversibility at high temperature and pressure. We believe that if the kinetic limitations in this material can be overcome, then the thermodynamics of the system would allow facile hydrogen release and reversibility.

Ca(BH$_4$)$_2$

Calcium borohydride received considerable attention due to its high H content (9.6%) and early reports of high reversible capacity. Ozolins et al. used first-principles methods to predict the enthalpies for the following two decomposition pathways for Ca(BH$_4$)$_2$:

$$Ca(BH_4)_2 \rightarrow 2/3CaH_2 + 1/3CaB_6 + 10/3H_2 \qquad (6.43)$$

$$Ca(BH_4)_2 \rightarrow 5/6CaH_2 + 1/6CaB_{12}H_{12} + 13/6H_2 \qquad (6.44)$$

The two reactions were predicted to have almost identical enthalpies, about 40 kJ/mol H$_2$ [144]. A somewhat lower value of 32 kJ/mol H$_2$ for the enthalpy of Reaction 6.43 was reported from calculations by Miwa et al. [126]. While the thermodynamic characteristics of the two reactions seem to be similar, the amounts of hydrogen released differed significantly in Reactions 6.43 and 6.44, corresponding to 9.63% and 6.26% wt% hydrogen, respectively.

It has to be noted that the amount of hydrogen released in Reaction 6.43 can be achieved at about 380°C in pure $Ca(BH_4)_2$ [162] and at slightly lower temperatures when various additives are used [163]. At higher temperatures, Kim et al. [164] found that about 8.3% mass fraction of hydrogen was released in the presence of NbF_5 and suggested that CaB_6 and NbB_2 might be among the decomposition products. The reported hydrogen content for this reaction was close to the H_2 released in a hypothetical Reaction 6.45, 8.67%:

$$Ca(BH_4)_2 \rightarrow Ca(B_2H_2) + 3H_2 \tag{6.45}$$

Calcium borohydride represents a rare example of a partially reversible BH_4-containing material. The reverse reaction of 6.43 has been shown to occur in moderate yield at 440°C and 70 MPa when CaH_2 and CaB_6 reacted in the presence of $TiCl_3/Pd$ [125]. Since the $[B_{12}H_{12}]^{2-}$ salts are known to display remarkable thermal and chemical stability [165], if $CaB_{12}H_{12}$ is present among the decomposition products, it may severely limit reversibility. Indeed, solvent-free $CaB_{12}H_{12}$ displays high thermal stability and starts desorbing H_2 only above 600°C [152]. Although early indications suggested that the system was at least partially reversible, subsequent studies revealed that with cycling, the reversibility decreased. The cycling behavior of $Ca(BH_4)_2$ revealed a steady decrease in hydrogen capacity. Coupled with ^{11}B NMR studies, it became apparent that accumulation of $CaB_{12}H_{12}$ species is responsible for the loss in reversibility.

Other Metal Borohydrides

In addition to group I and II borohydrides [7, 158, 166–168], a number of transition metal borohydrides were also considered for hydrogen storage applications [9, 13, 169]. Many trivalent and tetravalent borohydrides have high hydrogen content (10 wt% and higher) and typically decompose at lower temperatures compared to main group counterparts. However, some of the transition metal borohydrides release significant amounts of diborane and other toxic boron hydrides on decomposition. For example, $CuBH_4$, $Mn(BH_4)_2$, and $Zn(BH_4)_2$ decompose by releasing large amounts of diborane [30, 102, 103, 167, 168, 170–172]:

$$CuBH_4 \rightarrow CuH + 1/2B_2H_6 \tag{6.46}$$

$$M(BH_4)_2 \rightarrow M + B_2H_6 + H_2 \ (M = Mn, Zn) \tag{6.47}$$

Some additives were found to reduce the amount of diborane formed; however, it is extremely challenging to fully suppress diborane formation from some of the borohydrides mentioned [102, 170].

Trivalent and tetravalent borohydrides $Al(BH_4)_3$, $Ti(BH_4)_3$, $Zr(BH_4)_4$, and $Hf(BH_4)_4$ are volatile and can be distilled or sublimed under vacuum. It was reported that $M(BH_4)_x$ compounds decompose slowly even at room temperature to release hydrogen. On heating, they typically release B_2H_6, as in the case of $Al(BH_4)_3$ [167, 172]:

$$2Al(BH_4)_3 \rightarrow Al_2B_4H_{18} + B_2H_6 \tag{6.48}$$

Mixed-metal borohydrides offer some interesting advantages over the corresponding monometallic compounds. Thus, $Li_2Zr(BH_4)_6$ decomposes below 100°C with no detectable diborane release. Other mixed-cation borohydrides have been considered for hydrogen storage applications, including $LiCa(BH_4)_3$ [173]; $LiSc(BH_4)_4$ [174, 175]; $NaSc(BH_4)_4$ [176]; $KSc(BH_4)_4$ [177]; $LiZn(BH_4)_3$ [178]; $NaZn(BH_4)_3$ [178, 179]; $LiZn_2(BH_4)_5$ [178–180]; $NaZn_2(BH_4)_5$ [178, 179]; $Li_4Al_3(BH_4)_{13}$ [181]; $Mn_xMg_{1-x}(BH_4)_2$ [182]. Unfortunately, none of the mixed cation borohydride materials known to date can be rehydrogenated, and the materials therefore are not reversible [176, 177, 181, 183, 184].

The thermodynamic stabilities of metal borohydrides have been predicted from first-principles calculations [118, 185]. The results indicated different degrees of ionic bond character in $M(BH_4)_x$ compounds, where $x = 1$ to 4. Nakamori et al. compiled all the results and deduced a linear dependency between the enthalpy of formation of a metal borohydride ΔH and the Pauling electronegativity χ_P [118]:

$$\Delta H \text{ (borohydride)} = 248.7 \, \chi_P - 390.8 \text{ (in kJ/mol } BH_4)$$

Figure 6.8 displays the obtained fit for representative borohydrides with the metal ion in various oxidation states. As a rule, the thermal desorption temperatures for the $M(BH_4)_x$ compounds decrease with increasing χ_P. Interestingly, the same correlation between observed desorption temperatures and the average Pauling electronegativity exists for mixed-metal borohydrides [167, 186].

Summarizing, the borohydrides are readily synthesized by now-standard techniques. Their crystal structures are rather complicated but amenable to XRD analysis. A very broad variation in the properties of the borohydrides (desorption temperature, tendency to release B_2H_6, reversibility) is observed. Closoborane intermediates $[B_{12}H_{12}]^{2-}$ were found to be important intermediates in metal borohydride dehydrogenation reactions, tying up hydrogen in the stable closoborane structure. Perhaps the biggest challenge is to rehydrogenate the spent material, making the borohydride reversible.

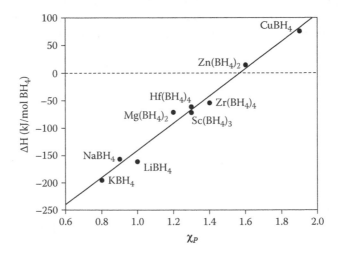

FIGURE 6.8
Enthalpy of formation of metal borohydrides as a function of of the Pauling electronegativity [167]. (Reprinted from Y. Nakamori, H.W. Li, K. Kikuchi, M. Aoki, K. Miwa, S. Towata, S. Orimo, *Journal of Alloys and Compounds*, 446 (2007) 296–300; H.W. Li, S. Orimo, Y. Nakamori, K. Miwa, N. Ohba, S. Towata, A. Züttel, *Journal of Alloys and Compounds*, 446 (2007) 315–318 with permission of Elsevier.)

Amides, Imides, Nitrides

Of the light elements that can support chemistry with hydrogen, the element nitrogen has received a great deal of attention. The initial starting point for these N-related studies was the compound Li_3N and its hydrogenated cousin $LiNH_2$. These compounds form the foundation for hydrogen storage using amides ($-NH_2$), imides ($-NH$), and nitrides ($-N$).

Li_3N has been investigated for at least 100 years [187] for a variety of chemical and electrical properties. Much of the modern work since the 1970s has been directed to understanding the geometric structure [188, 189] and ionic conductivity [190] of this material. The ionic conductivity was particularly of interest. Li_3N is an example of a relatively straightforward binary alkali metal nitride. Furthermore, it is a fast lithium ion conductor that has made it a potential solid-state electrolyte for lithium-based batteries [191]. More recently, the hydrogen uptake properties of Li_3N has been investigated as the basis for a new route to reversible hydrogen storage materials.

Li_3N

Li_3N is now commercially available [190, 192]. Historically, it was typically made by reacting molten Li [193, 194] or a melt of Li-Na [188] with gaseous nitrogen at 100–650°C [193, 194] and a pressure of about 500 torr [189]. The procedures yield a crystalline solid of α-phase Li_3N.

The geometric structure of Li_3N has been investigated by a variety of neutron-based [188] and x-ray-based techniques [189, 195], and it has been found that there are two Li_3N phases: α- and β-Li_3N. The α-Li_3N is stable at room temperature and ambient pressure and belongs to the space group $P6/mmm$ with lattice parameters of $a = b = 3.648$ Å, $c = 3.875$ Å. At a pressure of 15 kbar and at 300 K, the loosely packed α-Li_3N transforms to β-Li_3N (Na_3As type) of a hexagonal D46h structure, belonging to the space group $P63/mmc$ with $a = b = 3.552$ Å, $c = 6.311$ Å [195]. The α-Li_3N is a layered hexagonal structure with alternating layers of nominal composition Li_2N connected by a layer of Li atoms, as shown in Figure 6.9. As discussed by Rabenau [189], α-Li_3N has a highly symmetric structure in which the nitrogen ions are surrounded by eight lithium ions: six Li ions within the plane of the nitrogen ions, a seventh Li ion underneath, and an eighth above, as shown in Figure 6.9. The nominal composition of the nitrogen-containing layers is Li_2N, and they are separated by dilutely populated lithium layers.

The interaction between hydrogen and Li_3N was first investigated over 100 years ago by Dafert and Miklauz [187]. The modern interest in using this compound as hydrogen storage material was ignited by the pioneering study of Chen and coworkers in 2002 [194]. Chen observed that Li_3N uptook a significant amount of hydrogen, and the resulting material (proposed to be $LiNH_2 + 2LiH$) could also desorb H_2, giving a reversible hydrogen storage system operating below about 250°C. The sample weight variations observed with hydrogen absorption and desorption from Li_3N are shown in Figure 6.10 [194].

It can be seen from Figure 6.10 that a small amount of hydrogen absorption occurs at a very low temperature of 100°C, with a much faster uptake seen at about 200°C. In this original study [194], it was seen that a total increase of about 9.3 wt% hydrogen can be achieved if the sample is kept at 255°C for half an hour. It was also observed that substantial loadings of H in Li_3N could also be achieved for temperatures about 200°C, but a much longer absorption time was required.

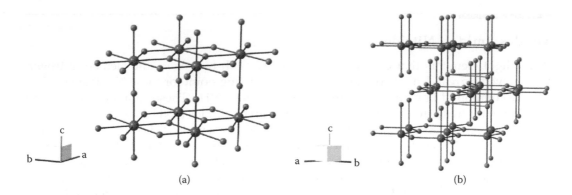

FIGURE 6.9
Crystal structures of (a) α-Li₃N and (b) β-Li₃N. (From W. Li, G. Wu, C.M. Araujo, R.H. Scheicher, A. Blomqvist, R. Ahuja, Z. Xiong, Y. Feng, P. Chen, *Energy & Environmental Science*, 3 (2010) 1524–1530. With permission.)

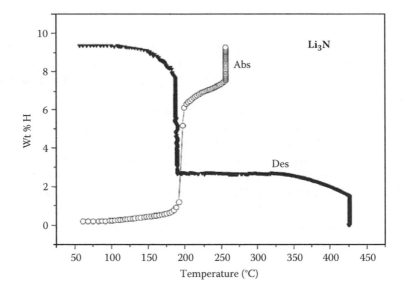

FIGURE 6.10
Uptake of H_2 by Li_3N and subsequent desorption from the hydrogenated product. (Figure reproduced with permission from P. Chen, Z. Xiong, J. Luo, J. Lin, K.L. Tan, *Nature*, 420 (2002) 302–304 with permission of Nature Publishing Group.)

Once the sample has been hydrogenated in this way, desorption of hydrogen from the H-Li_3N complex can be seen (Figure 6.10), making Li_3N the basis for a reversible hydrogen storage material. Figure 6.10 shows that heating the hydrogenated product to about 200°C releases about 6 wt% hydrogen. Further heating the sample to temperatures greater than 320°C releases the remaining 3 wt% hydrogen [194]. Equilibrium measurements are required to assess the pressures attainable for the H-Li_3N hydrogen storage material. The first equilibrium pressure-composition measurements for hydrogen storage on Li_3N are shown in Figure 6.11 from the work of Chen et al. [194].

It can be seen from Figure 6.11 that the maximal molar ratio of H to Li_3N unit is 3.6. A formula unit of $Li_3NH_{3.6}$ corresponds to a hydrogenated Li_3N state of mass percent 9.4% hydrogen. Figures 6.11b and 6.11c also show that there exist two plateaus in the PCT

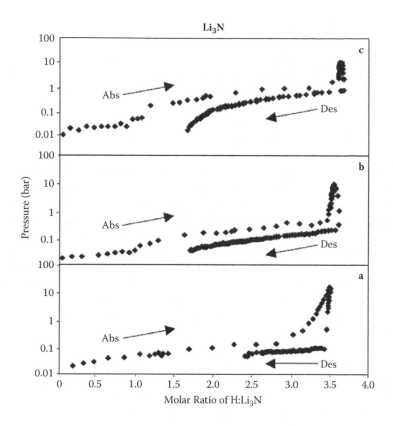

FIGURE 6.11
Pressure-composition-temperature (PCT) measurements of hydrogen absorption (abs) and desorption (des) from Li_3N. (a) for measurement at 195°C; (b) for 230°C; and (c) for 255°C. The *x*-axis is the molar ratio of H atoms to the molecular Li_3N unit. (Figure reproduced from P. Chen, Z. Xiong, J. Luo, J. Lin, K.L. Tan, *Nature*, 420 (2002) 302–304 with permission of Nature Publishing Group.)

measurements. The first one (with $0 < H{:}Li_3N < 1.1$) has an equilibrium pressure below about 0.07 bar for temperatures 255°C and below. This low pressure, far lower than the approximately 3- to 5-bar pressure needed to feed a proton exchange membrane (PEM) fuel cell [196], suggests absorption/desorption to a very stable $H{-}Li_3N$ phase. The second plateau with $1.4 < H{:}Li_3N < 3$ has a higher equilibrium pressure of about 0.8 bar at 255°C. A reaction sequence consistent with the gravimetric (Figure 6.10) and PCT (Figure 6.11) measurements, in addition to structural studies of the reaction intermediates by XRD, was put forth by Chen et al. as follows [194]:

$$Li_3N + H_2 \rightarrow Li_2NH + LiH \tag{6.49}$$

$$Li_2NH + LiH + H_2 \rightarrow LiNH_2 + 2LiH \qquad \Delta H = -66.1 \text{ kJ/mol } H_2 \tag{6.50}$$

As described by Equation 6.49, the hydrogenation of Li_3N proceeds first to the "imide" Li_2NH and LiH. Further reaction of Li_2NH and H_2 (Equation 6.50) leads to lithium amide $LiNH_2$ and additional LiH. The reaction schemes 6.49 and 6.50 involve 4 H atoms ($2H_2$) hydrogenating a single molecular unit of Li_3N to form lithium amide and lithium hydride (i.e., 4 hydrogen atoms for every Li_3N unit). In this way, the "theoretical" hydrogen capacity of the overall reaction scheme is 10.4 wt%, in good agreement with the total H mass observed in the gravimetric studies of Figure 6.10 and the H:Li$_3$N mass derived from the PCT measurements (Figure 6.11). This reaction sequence is also consistent with XRD studies of the reaction and the known XRD patterns of Li$_3$N, LiNH$_2$, and Li$_2$NH. The overall enthalpy for the hydrogenation of Li$_3$N to LiNH$_2$ + LiH is calculated to be –161 kJ/mol. Experimental assessments of Reaction 6.50 suggest an enthalpy of hydrogenation of the imide Li$_2$NH to be –66 kJ/mol H$_2$ [194].

Summarizing this initial work [194], a new materials basis for lightweight and reversible hydrogen storage material was found based on Li$_3$N. Under various conditions of treatment, a hydrogen capacity of about 9–10 wt% H was observed, full reversibility was achieved, and the required temperatures, although somewhat high from 200°C to about 320°C, were closer to the desired 100°C temperature range than many other observed systems. The temperatures required to affect Reaction 6.50 are less than that needed to affect Reaction 6.49, making the conversion of LiNH$_2$ + LiH to Li$_2$NH + H$_2$ a more favorable process for hydrogen storage reactions, albeit with lower hydrogen capacity. The equilibrium plateau pressure for this reaction would need to be significantly improved if the LiNH$_2$/2LiH system were to be used in fueling PEM fuel cell systems. Interestingly, these materials were observed to be kinetically fast in H$_2$ absorption. For example, for a 500-mg sample of Li$_3$N, a substantial amount of hydrogen could be absorbed at 255°C in 10 min at 30-bar hydrogen pressure [194].

Subsequent studies after the original paper by Chen et al. and others have sought to confirm and understand the properties of the Li$_3$N ↔ LiNH$_2$ + 2LiH system (or stoichiometric variants thereof) [197–199]. Most of these have started from the hydrogenated side, namely, mixtures of LiNH$_2$ and 2LiH for which the LiNH$_2$ and LiH samples were mechanically ball milled together to form starting material. In an interesting study, Chen speculated [199] that one of the driving forces for the reaction between LiNH$_2$ and 2LiH was a strong (H$^{δ+}$)-(H$^{δ-}$) interaction since hydrogen has a positive character in LiNH$_2$ and negative character in LiH. If this were the dominant mechanistic driver for the reaction, then substituting 2LiD in the reaction with LiNH$_2$ would lead to a predominance of HD in the gas phase of the dehydrogenation reaction. The majority of the product turned out to be H$_2$; however, there was a substantial amount of HD produced [199]. It is unclear to what extent isotopic exchange beyond the initial reaction of LiNH$_2$ with 2LiD is responsible for the observed HD signal.

Although ammonia liberation as a side reaction of reaction schemes 6.49 and 6.50 was an original concern [194], Ichikawa et al. [198] carefully examined ammonia release, as LiNH$_2$ can decompose according to Reaction 6.51, shown in Figure 6.12:

$$2\text{LiNH}_2 \rightarrow \text{Li}_2\text{NH} + \text{NH}_3 \ \Delta H = 84.1 \text{ kJ/mol NH}_3 \quad\quad (6.51)$$

Pure LiNH$_2$ releases NH$_3$ starting at low levels at about 200°C and then increasing markedly at about 350°C. Ammonia release from the LiNH$_2$/LiH system would be a major concern for practical hydrogen storage systems for two reasons. First, nitrogen release represents an irreversible loss of the storage material that cannot be recovered. For practical matters, if the ammonia content in the hydrogen stream is about 200 ppm or higher, one would have a 10% loss after 1000 cycles, which needs improvement. A tighter requirement on the ammonia content arises from the need to keep PEM fuel cells from being degraded by NH$_3$. This more stringent requirement requires that the NH$_3$ levels in the released H$_2$ be 0.1 ppm or less [27]. Ichikawa et al. found [198] that heating mixtures of LiNH$_2$ and LiH tended to release reduced levels of NH$_3$ along with hydrogen, as shown in Figure 6.13.

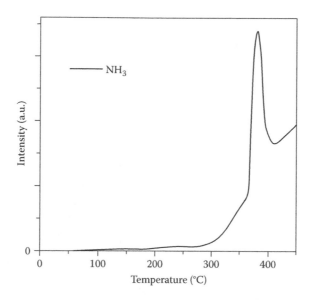

FIGURE 6.12

Thermal desorption of ammonia from pure $LiNH_2$, with constant heating rate of 5°C/min. (Figure reproduced from T. Ichikawa, S. Isobe, N. Hanada, H. Fujii, *Journal of Alloys and Compounds*, 365 (2004) 271–276 with permission of Elsevier BV.)

It is evident from Figure 6.13 that ammonia release accompanies hydrogen release from all these samples. Comparing the top panel (1:1 $LiNH_2LiH$) and the middle panel (1:2 $LiNH_2LiH$), one can see that increasing the relative amount of LiH decreases ammonia release. Similarly, comparing the top panel to the bottom panel shows that ball milling further reduces ammonia emission. Thus, intimate contact between the starting $LiNH_2$ and LiH would seem to be important in reducing NH_3 emission. Comparing the temperature scales of Figures 6.12 and 6.13, it would seem that hydrogen release precedes ammonia release at any given temperature, but there is always (perhaps trace) levels of ammonia present in the hydrogen stream released. These authors found [198] that introducing 1 mol% of $TiCl_3$ into mixtures of 1:1 $LiNH_2/LiH$ essentially eliminated NH_3 release. Presumably, Ti catalyzes the reaction between $LiNH_2$ and LiH, dramatically increasing the rate of hydrogen release at temperatures of about 220°C, such that all the $LiNH_2$ is converted to hydrogen product before significant amounts of $LiNH_2$ can decompose to NH_3. Lamb and coworkers have conducted some interesting work showing that adding molecular N_2 to the H_2 gas stream acts to enhance the reversibility of Li_3N [200].

$LiNH_2$ Modified with Mg

Researchers discovered quickly that improvements to the base material $LiNH_2$ + LiH could be affected by adding magnesium to the system [199, 201–208]. While other additives to the system, such as Ca [204] and Na [207], have been examined, Mg seems to be the most potent in reducing the thermodynamic requirements. This is an example of "thermodynamic destabilization," discussed more fully later in this chapter. Nakamori and Orimo [193] suggested that the high ionicity (and therefore bond strength) of the Li^+-NH_2^- interaction could be reduced by replacing lithium with an element that is more electronegative. Magnesium is more electronegative, so it was thought that introducing

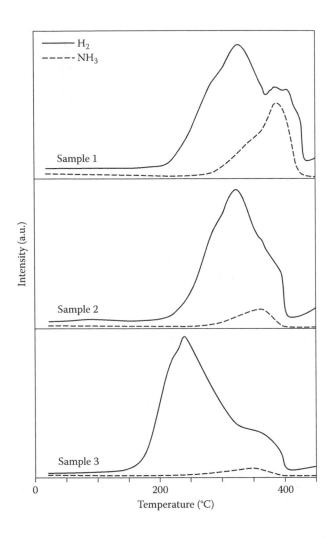

FIGURE 6.13

Mass-resolved thermal desorption spectra of hydrogen (solid line) and ammonia (dashed line) release from mixtures of $LiNH_2/LiH$ in various ratios (sample 1 = hand-mixed 1:1 $LiNH_2LiH$, sample 2 = hand-mixed 1:2 $LiNH_2LiH$, sample 3 = ball-milled 1:1 $LiNH_2LiH$). (Figure reproduced from T. Ichikawa, S. Isobe, N. Hanada, H. Fujii, *Journal of Alloys and Compounds*, 365 (2004) 271–276 with permission of Elsevier BV.)

$Mg(NH_2)_2$ to the reaction mixture would lead to lowered desorption temperatures. In fact, these authors observed [193] that thermal desorption of hydrogen from MNH_2 where M = 90% Li with 10% Mg does indeed occur at about 50°C lower temperature than for $LiNH_2$ alone. These notions were supported by first-principles calculations of the electronic structure of $LiNH_2$ conducted by Orimo and coworkers [209]. Xiong et al. [204] and Leng and coworkers [202] demonstrated the importance of Mg for destabilizing $LiNH_2$.

Luo [206] also found that replacement of LiH with MgH_2 would effectively destabilize $LiNH_2$ since MgH_2 is less stable than LiH. The formation enthalpy of MgH_2 is –74 kJ/mol, whereas that for LiH is –90 kJ/mol. In these studies, large amounts of Mg were introduced into the system, with stoichiometry equivalent to $2LiNH_2 + MgH_2$. A PCT measurement from the material [201] is shown in Figure 6.14.

As indicated in Figure 6.14, hydrogen can be desorbed from $2LiNH_2 + MgH_2$ at 220°C with a pressure of about 40 bar. This is a dramatic increase in plateau pressure compared to

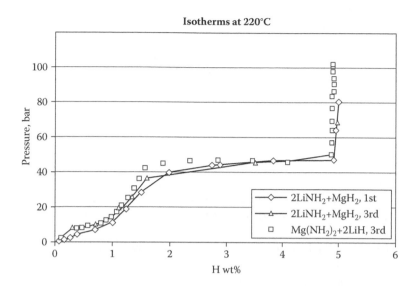

FIGURE 6.14
PCT measurement of 2LiNH$_2$ + MgH$_2$, recorded at 220°C. Diamond data are for the first cycle of 2LiNH$_2$ + MgH$_2$. The triangles represent the third cycle of 2LiNH$_2$ + MgH$_2$. The squares represent PCT data collected for a sample of Mg(NH$_2$)$_2$ + 2LiH on its third cycle. (Reproduced from W.F. Luo, S. Sickafoose, *Journal of Alloys and Compounds*, 407 (2006) 274–281 with permission of Elsevier.)

the LiNH$_2$-LiH system (<0.1 bar at 200°C) with no Mg present and at the same temperature. At these temperatures, 5 wt% hydrogen could be reversibly desorbed and then reabsorbed onto the material. Interestingly, the PCT measurement is essentially identical to that obtained from Mg(NH$_2$)$_2$ + 2LiH [198]. This observation will have important implications for the overall reaction mechanism for the 2LiNH$_2$ + MgH$_2$ system (discussed further in this chapter). From this early work [206], the initial hydrogen desorption/absorption process from 2LiNH$_2$ + MgH$_2$ was assigned to be

$$2\text{LiNH}_2 + \text{MgH}_2 \leftrightarrow \text{Li}_2\text{Mg(NH)}_2 + 2\text{H}_2 \qquad (6.52)$$

Subsequent XRD and Fourier transform infrared (FTIR) measurements by Luo and Sickafoose [201] revealed that while this overall reaction process was correct, there were complications in the reaction scheme that did not allow one to recover 2LiNH$_2$ + MgH$_2$ on cycling. The results indicated that the mechanism discussed next is actually operative for the (2LiNH$_2$ + MgH$_2$) system.

On heating, the starting material (2LiNH$_2$ + MgH$_2$) undergoes an irreversible transformation to Mg(NH$_2$)$_2$ + 2LiH. This is the "hydrogenated" material that exists as this material undergoes hydrogen desorption/adsorption cycling. When (Mg(NH$_2$)$_2$ + 2LiH) is heated further, hydrogen is released to form Li$_2$Mg(NH)$_2$ + 2H$_2$. The Li$_2$Mg(NH)$_2$ material belongs to a class of N-H materials called *imides*. Generally, imides are stable compounds and do not readily release the remaining hydrogen in the molecule. Thus, the final product produced by heating is Li$_2$Mg(NH)$_2$, and the material system is limited to about 5 wt% hydrogen. When Li$_2$Mg(NH)$_2$ is hydrogenated, the intermediate material formed is Mg(NH$_2$)$_2$ + 2LiH, and *not* the original lithium amide (2LiNH$_2$ + MgH$_2$). As a result, the system cycles between Mg(NH$_2$)$_2$ + 2LiH and Li$_2$Mg(NH)$_2$ as shown in Figure 6.15. The presence of Mg(NH$_2$)$_2$ + 2LiH is also confirmed by the fact that PCT measurements of an authentic sample of Mg(NH$_2$)$_2$ + 2LiH are identical to that obtained using 2LiNH$_2$ + MgH$_2$

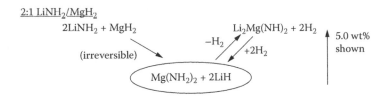

FIGURE 6.15
Pathways for hydrogen release from ($2LiNH_2 + MgH_2$).

as the starting point, as shown in Figure 6.14. Subsequent DSC measurements have shown that the initial reaction $2LiNH_2 + MgH_2 \rightarrow Mg(NH_2)_2 + LiH$ is in fact exothermic [210]. The exothermicity of this step makes it difficult to hydrogenate back to the original $2LiNH_2 + MgH_2$ material.

Recent studies [211–213] have explored the cyclic dehydrogenation/hydrogenation in Figure 6.15 in more detail and confirmed the overall picture of Figure 6.15. The study by Rijssenbeek and coworkers [211] is particularly noteworthy for the powerful array of in situ characterization techniques brought to bear on the reaction, including in situ synchrotron XRD and RGA. Geometric structures for the species $Li_2Mg(NH)_2$ were determined in this study [211]. Although the results of studies [211–213] generally support the reaction pathway of Figure 6.15, intermediates such as $Li_2Mg_2(NH)_3$ [212] and $Li_4Mg_3(NH_2)_2(NH)$ [213] were reported, with some diversity in the observations depending on the starting compositions and conditions.

Figure 6.16 shows that the material $2LiNH_2 + MgH_2$ shows excellent cyclability [214]. This material was cycled 264 times, albeit with a 23% loss of the storage capacity. This is the largest number of cyclic absorption/desorptions ever recorded for a complex metal hydride. The cause of this cycling-induced capacity loss was not determined.

One might expect that, just like ammonia release from $LiNH_2$ can cause ammonia contamination of the released hydrogen stream, so perhaps ammonia release from $Mg(NH_2)_2$ might also cause difficulties with ammonia contamination. Ammonia is seen to be released from $Mg(NH_2)_2$ at a temperature of 375°C [215]. This was examined in the work of Luo, who found that the ammonia contamination in the hydrogen stream for $2LiNH_2 + MgH_2$ was about 200–700 ppm, varying over a desorption temperature range of 180–240°C [216]. This would have to be reduced for the $2LiNH_2 + MgH_2$ material to be used in a practical hydrogen storage system for use with a PEM fuel cell or hydrogen ICE.

By this time, the $2LiNH_2 + MgH_2$ system was also being referred to as $Mg(NH_2)_2 + 2LiH$ due to the realities depicted in Figure 6.15, in which $Mg(NH_2)_2 + 2LiH$ is produced after the first hydrogenation cycle. Although this system has shown excellent reversibility, and a good reversible hydrogen capacity of 5 wt%, it still requires too high an operating temperature, and the kinetics is slower than desired. A major improvement in the performance of the material was made by Wang et al. [217], who found that introducing 3 mol% of KH dramatically improved the dehydrogenation and absorption kinetics of the material. The original report by Wang is shown in Figure 6.17. It can be seen that the peak in hydrogen desorption shifts from 186°C to 132°C with modification by KH. Figure 6.17 also shows little NH_3 being evolved, although this would require more sensitive investigation. Wang et al. also reported significant improvements in the hydrogenation of the potassium-modified imide phase [217].

Subsequent work by Luo et al. [210] confirmed the efficacy of the KH modification and reported dramatic increases in the absorption kinetics, shown in Figure 6.18. It is believed

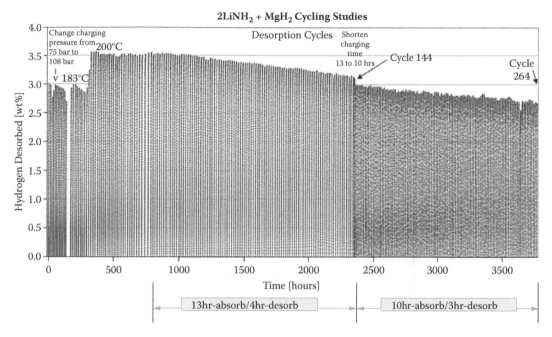

FIGURE 6.16
Cycling studies of (2LiNH$_2$ + MgH$_2$). (Figure reproduced from L.E. Klebanoff, Metal hydride center of excellence, presentation at the 2006 DOE Hydrogen Program Annual Merit Review, p. 17. http://www.hydrogen.energy.gov/annual_review06_storage.html#metal, (2006).)

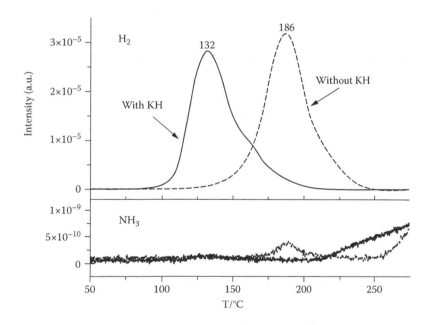

FIGURE 6.17
Thermal desorption studies examining H$_2$ and NH$_3$ release from pristine Mg(NH$_2$)$_2$ + 2LiH (dashed line) and the same material modified with 3 mol% KH (solid line). (Figure reproduced from J. Wang, T. Liu, G. Wu, W. Li, Y. Liu, C.M. Araujo, R.H. Scheicher, A. Blomqvist, R. Ahuja, Z. Xiong, P. Yang, M. Gao, H. Pan, P. Chen, *Angewandte Chemie-International Edition*, 48 (2009) 5828–5832 with permission of Wiley-VCH.)

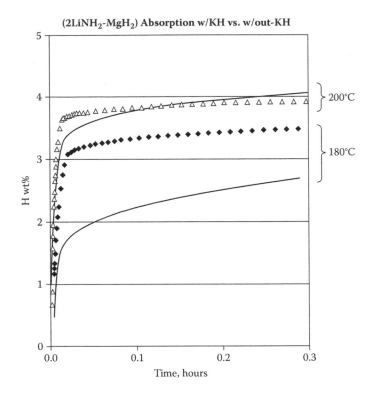

FIGURE 6.18
Increased absorption kinetics for the $2LiNH_2/MgH_2$ system observed with the incorporation of 4 mol% KH. Symbols are for catalyzed samples, line for uncatalyzed. Absorption is at 115 bar at the two temperatures indicated. (Figure reproduced from W. Luo, V. Stavila, L.E. Klebanoff, *International Journal of Hydrogen Energy*, 37 (2012) 6646–6652 with permission of International Association of Hydrogen Energy.)

that part of the efficacy of KH can be attributed to the ability of potassium to bind with nitrogen in both the amide and imide phases [217].

The vastly improved kinetics and reversibility allowed measurements leading to a van't Hoff plot of the system. The values of ΔH and ΔS for $(2LiNH_2 + MgH_2)$ dehydrogenation are 42.0 kJ/mol H_2 [217] and 99 J/K-mol H_2 [210], respectively. These can be compared with the corresponding values of ΔH and ΔS for hydrogen release from $LaNi_5H_6$, which are 30.2 kJ/mol H_2 and 104 J/K-mol H_2, respectively.

LiMgN

As indicated, the $(2LiNH_2 + MgH_2)$ system possesses limited hydrogen storage capacities. However, experience has shown that when the molar ratio of reactants changes, new reactions between lithium amide and simple or complex metal hydrides can take place, opening up new reaction sequences with higher ultimate hydrogen production. So, changes in the initial stoichiometry of mixed components can really perform as a different hydrogen storage system. Lu and coworkers [218] have investigated the hydrogen desorption/adsorption properties of 1:1 $LiNH_2 + MgH_2$. It turns out that this stoichiometry allows the formation of the fully dehydrogenated product LiMgN on desorption of H_2, as shown by the following reaction for 1:1 $MgH_2/LiNH_2$:

$$MgH_2 + LiNH_2 \rightarrow LiMgN + 2H_2 \qquad (6.53)$$

This system was actually predicted theoretically by Alapati et al. [219] to be of interest due to the very attractive theoretical desorption enthalpy of $\Delta H = 32$ kJ/mol H_2, with a hydrogen capacity of 8.2 wt%. The high predicted weight percentage of hydrogen is due to the full dehydrogenation all the way to LiMgN, bypassing the undesired imide intermediate. Because Reaction 6.53 represents the complete dehydrogenation of the system, LiMgN would be an important candidate material for hydrogen storage if the dehydrogenation takes place at low temperature and if it were shown to be reversible.

To demonstrate the feasibility of Reaction 6.53 experimentally, a mixture of MgH_2 and $LiNH_2$ with a molar ratio of 1:1 was prepared by ball milling for 24 h [218]. Figure 6.19 shows the thermogravimetric analysis (TGA) profile of the as-milled $MgH_2/LiNH_2$ mixture as the sample is heated. The reaction starts at about 120°C with the weight loss accelerated at about 200°C. The total weight loss was 8.1 wt% of the initial weight after the sample was held at 220°C for 20 min. Assuming all the weight losses were due to the release of hydrogen, the dehydrogenation process can be considered complete, confirming the theoretically predicted Reaction 6.53. DSC measurements of the hydrogen release were carried out [218]. The results indicated that the dehydrogenation was completed in a one-step reaction, with a ΔH of 33.5 kJ/mol H_2, which is very close to the theoretically predicted reaction enthalpy of 32 kJ/mol H_2 reported by Alapati et al. [219]. XRD studies of the reaction showed the presence of LiMgN in the fully dehydrogenated state, confirming Reaction 6.53.

Of course, the material should be reversible to be most useful for hydrogen storage. Although the material could be reversed without catalysts, the reaction was slow and

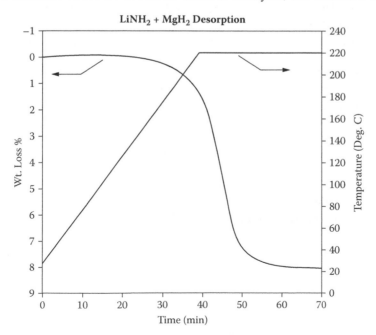

FIGURE 6.19
Thermogravimetric analysis (TGA) data for desorption from ball-milled ($LiNH_2 + MgH_2$). (Figure reproduced from J. Lu, Z.Z.G. Fang, Y.J. Choi, H.Y. Sohn, *Journal of Physical Chemistry C*, 111 (2007) 12129–12134 with permission of American Chemical Society.)

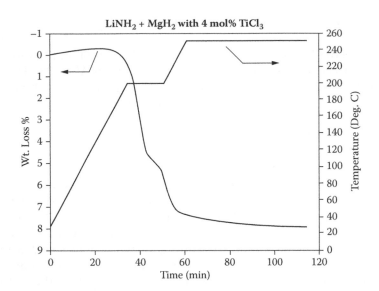

FIGURE 6.20
TGA curves for TiCl$_3$-doped LiMgN after hydrogenation at 160°C under 2000 psi H$_2$ for 6 h. (Figure reproduced from J. Lu, Z.Z.G. Fang, Y.J. Choi, H.Y. Sohn, *Journal of Physical Chemistry C*, 111 (2007) 12129–12134 with permission of American Chemical Society.)

incomplete. It was found [218] that addition of 4 mol% TiCl$_3$ to the reaction mixture significantly aided the absorption kinetics. The LiMgN/Ti sample was hydrogenated at 137.9 bar and 160°C for 6 h. The sample was then heated to release H$_2$, leading to the hydrogen desorption shown in Figure 6.20. Figure 6.20 shows that TiCl$_3$-doped LiMgN had gained about 8.0 wt% hydrogen from the rehydrogenation process and starts to resemble the original desorption curve of Figure 6.19.

Extensive FTIR and XRD measurement support the assignment that on hydrogenation, LiMgN converts to LiH + ½Mg(NH$_2$)$_2$ + ½MgH$_2$, as depicted in Figure 6.21. The assignment of the imide Li$_2$Mg(NH)$_2$ in this reaction as an intermediate is speculative as its detection as an intermediate in the cycling has not been confirmed, although the TGA data do show a multistep behavior. More recent theoretical studies by Akbarzadeh et al. have suggested that the conversion of LiNH$_2$ + MgH$_2$ to LiMgN likely involves multiple steps [220].

The nature of the reaction pathway to LiMgN is clearly complex, and the system needs more work. It seems reasonable to assume that the presence of excess MgH$_2$ allows the further dehydrogenation of the imide Li$_2$Mg(NH)$_2$ to the fully dehydrogenated product LiMgN, as shown in Figure 6.20. Lu et al. [221] have reported that the procedures used

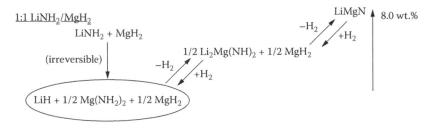

FIGURE 6.21
Proposed mechanism for hydrogen absorption/desorption involving 1:1 LiNH$_2$/MgH$_2$.

to ball mill the starting materials $LiNH_2$ and MgH_2 can have a dramatic influence on the reactions seen. Low-energy ball milling for 96 h, followed by subsequent dehydrogenation, led solely to LiMgN. The same milling process used for only 12 h led, on dehydrogenation, to a mixture of LiMgN and Mg_3N_2. This variability may account for the diversity of products seen in this reaction in other studies [222–225].

The work on the 1:1 $LiNH_2/MgH_2$ material, while revealing the complexity of the system, confirmed in a general way the earlier theoretical predictions [219] and in fact revealed a reversible N-based system that could be cycled up to about 8 wt%. In addition, the dehydrogenation and hydrogenation temperatures are in the range 160–220°C, which is higher than desired but in the realm of possibility for use with PEM fuel cells or hydrogen ICE technology. The yield on cycling is rather poor, and substantial hydrogen storage capacity is lost with each cycle [222, 223]. The kinetics are slow, but responsive to additives. Although incorporation of $TiCl_3$ improved the kinetics, significant improvements in desorption and absorption kinetics are needed for this material to be a viable hydrogen storage system. A nice review of the Li-Mg-N-H system can be found in the work of Liang and coworkers [8].

Mixed-Anion Complex Metal Hydrides

Similar to the concept of tuning the stability of metal borohydrides by using metal ions of appropriate electronegativity, one can try to tune the stability by using mixed-anion phases. One such system was proposed by Aoki et al. [226] and Pinkerton et al [227] in 2005 based on a 1:2 $LiBH_4/LiNH_2$ mixture [226]:

$$LiBH_4 + 2LiNH_2 \rightarrow Li_3BN_2 + 4H_2 \tag{6.54}$$

First-principles calculations predicted an enthalpy of reaction of 23 kJ/mol H_2 for Reaction 6.54. However, experimental evidence suggested that lithium borohydride and lithium amide react on heating to form new compounds, depending on the relative ratio. The first compound in the Li-B-N-H system was identified as $Li_4(BH_4)(NH_2)_3$ by Pinkerton et al. [227] and later by Chater et al. [228]. $Li_4(BH_4)(NH_2)_3$ forms a *bcc* cubic lattice with discrete BH_4^- and NH_2^- groups surrounding Li^+ cations [228, 229]. Reacting $LiBH_4$ and $LiNH_2$ in a 1:1 molar ratio results in a different phase, $LiBNH_6$ [227, 230–232]. $LiBNH_6$ contains two symmetry-independent Li^+ cations: One is surrounded by three BH_4^- and one NH_2^- group, while the other is coordinated by one BH_4^- and three NH_2^- groups. These structures are shown in Figure 6.22. $Li_4BN_3H_{10}$ and Li_2BNH_6 have high H contents of 11.9 wt% and 13.3 wt% respectively and release H_2 at relatively low temperatures. However, at least the initial dehydrogenation steps are exothermic; thus, the materials are not fully reversible.

An interesting concept of a "self-catalyzing" hydrogen storage system was introduced by Yang et al. [233]. The ternary composite phase $2LiNH_2/LiBH_4/MgH_2$ exhibits a self-catalyzing reaction pathway that results in faster H_2 desorption, lower desorption temperatures, and suppression of volatiles other than H_2 (NH_3, diborane) compared to the constituent compounds. The enhanced properties arise from the incorporation of an ionic liquid phase ($Li_4BH_3H_{10}$) and from ancillary reaction seeding of a reversible H_2 storage reaction. The reaction mechanism is complex and involves several steps, as determined

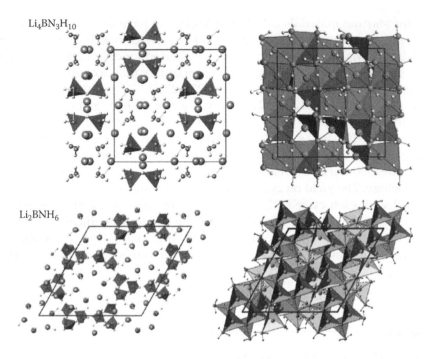

FIGURE 6.22 (See color insert.)
Crystal structure of $Li_4BN_3H_{10}$ and Li_2BNH_6. (Reprinted from H. Wu, W. Zhou, T.J. Udovic, J.J. Rush, T. Yildirim, *Chemistry of Materials*, 20 (2008) 1245–1247 with permission of American Chemical Society.)

from temperature-programmed description (TPD) and in situ XRD studies shown in Figure 6.23.

On heating to 100°C, $Li_4BN_3H_{10}$ melts and reacts with MgH_2 to form $Mg(NH_2)_2$ and $LiBH_4$:

$$2Li_4BN_3H_{10} + 3MgH_2 \rightarrow 3Mg(NH_2)_2 + 2LiBH_4 + 6LiH \qquad (6.55)$$

As the temperature is increased above 100°C, hydrogen starts desorbing according to the following reaction:

$$2Li_4BN_3H_{10} + 3MgH_2 \rightarrow 3Li_2Mg(NH)_2 + 2LiBH_4 + 6H_2 \qquad (6.56)$$

Reaction 6.55 serves to directly catalyze the subsequent reversible reaction between $Mg(NH_2)_2$ and LiH that occurs between approximately 190 and 230°C:

$$Mg(NH_2)_2 + LiH \rightarrow Li_2Mg(NH)_2 + 2H_2 \qquad (6.57)$$

The ternary composite is thus self-catalyzed because one Reaction 6.56 produces nuclei of $Li_2Mg(NH)_2$ in situ, which accelerates the overall hydrogen desorption for Reaction 6.57. At temperatures above 350°C the decomposition of $Li_2Mg(NH)_2$ and $LiBH_4$ occurs to form various crystalline products, including Li_3BN_2, Mg_3N_2:

$$3Li_2Mg(NH)_2 + 2LiBH_4 \rightarrow 2Li_3BN_2 + Mg_3N_2 + 2LiH + 6H_2 \qquad (6.58)$$

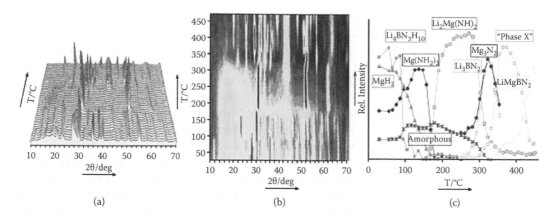

FIGURE 6.23
(a) Raw PXRD data for the ternary composition as a function of temperature (25–450°C). (b) The corresponding two-dimensional contour plot derived from the raw patterns in Figure 6.23a. (c) Plot of the relative amounts of individual phases as a function of temperature. (Reprinted from J. Yang, A. Sudik, D.J. Siegel, D. Halliday, A. Drews, R.O. Carter, III, C. Wolverton, G.J. Lewis, J.W.A. Sachtler, J.J. Low, S.A. Faheem, D.A. Lesch, V. Ozolins, *Angewandte Chemie-International Edition*, 47 (2008) 882–887 with permission of Wiley.)

Several studies were directed toward reversible H_2 storage in the $LiBH_4$-$LiNH_2$-MgH_2 system; however, since some of the reaction steps are believed to be exothermic, only 3–4 wt% hydrogen can be cycled under practical conditions of temperature and H_2 pressure [233–240]. After the initial publications on the $LiBH_4$-$LiNH_2$ systems, a number of binary and ternary compositions have been under investigation, including $LiBH_4$-$NaNH_2$ [241]; $LiBH_4$-$Mg(NH_2)_2$ [242]; $Ca(BH_4)_2$-$LiNH_2$ [243, 244]; $Ca(BH_4)_2$-$Mg(NH_2)_2$ [245]; $Ca(BH_4)_2$-$Ca(NH_2)_2$ [245]; and $LiBH_4$-$LiNH_2$-LiH [246].

In addition to borohydride-amide materials, several alanate-amide systems have high hydrogen content and were investigated for H_2 storage applications. Lu and Fang [247] found that hand-milled mixtures of $LiAlH_4$ and $LiNH_2$ in a 2:1 molar ratio start desorbing H_2 as low as 80°C and release 8.1 wt% H_2 on heating to 310°C:

$$2LiAlH_4 + LiNH_2 \rightarrow Li_2NH + LiH + 2Al + 4H_2 \qquad (6.59)$$

The most likely explanation for the significant reduction in desorption temperature is that the hydridic-protic interactions promote H_2 formation. Ball milling together the lithium alanate and amide turned out to be counterproductive since large amounts of H_2 form during the milling process [248, 249]. After 60 min of continuous milling, about half the H capacity was lost. A 1:1 ball-milled mixture of $LiAlH_4$ and $LiNH_2$ was reported to release up to 8 wt% H_2 on heating [250]; however, during the milling there was evidence some H_2 was lost, resulting in Li_3AlH_6. Kojima et al. used directly a mixture of Li_3AlH_6 and $LiNH_2$ in a 1:2 molar ratio and observed a decomposition according to the following reaction [251]:

$$2Li_3AlH_6 + 4LiNH_2 \rightarrow Li_3AlN_2 + 2Li_2NH + 3LiH + Al + 15/2H_2 \qquad (6.60)$$

The measured H_2 capacity was 6.9 wt%, slightly lower compared to the full theoretical capacity of 7.6 wt% H_2 based on Equation 6.60, with the difference in H_2 probably lost during ball milling. Interestingly, the Li-Al-N-H system is partially reversible: The undoped material absorbs 1–2 wt% H_2, while in the presence of a nano-Ni catalyst the reversible capacity is increased to 3–4 wt% [251].

Mao and coworkers investigated hydrogen storage properties of ball-milled $LiAlH_4$-$LiBH_4$ composites in a 1:2 molar ratio [252]. Small amounts of TiF_3 were found to exhibit catalytic effects on hydriding/dehydriding kinetics. XRD patterns of the dehydrogenated samples revealed AlB_2 and LiAl phases in the fully desorbed samples. PCT data suggest that the reaction enthalpy was decreased from 74 kJ/mol H_2 for the pure $LiBH_4$ to 60.4 kJ/mol H_2 in the TiF_3-doped $LiAlH_4$–$LiBH_4$ system. The dehydrogenation products of the TiF_3-doped $LiAlH_4$–$LiBH_4$ sample absorbed 4.78 wt% of hydrogen after 14 h at 600°C and under 40 bar of hydrogen. Powder XRD studies confirmed the presence of $LiBH_4$ in the rehydrogenated sample. Other transition metal halides also were reported to enhance the kinetics of H_2 release from the $LiAlH_4$–$LiBH_4$ system [253]. Ball milling of $LiBH_4$-$NaAlH_4$ in a 1:1 molar ratio resulted in a metathesis reaction with formation of sodium borohydride, which then decomposed into amorphous boron and NaH; however, the reaction required temperatures in excess of 340°C [254].

In addition to mixed anionic species, metal salt adducts with neutral molecules, such as ammonia, have also been investigated. Soloveichik et al. synthesized the ammonia adduct $Mg(BH_4)_2(NH_3)_2$ by reacting $Mg(BH_4)_2$ with NH_3 [255]. The stoichiometric $Mg(BH_4)_2(NH_3)_2$ released mostly hydrogen on heating with only traces of NH_3 present. Presumably, the dihydrogen bonds between the NH_3 group and the BH_4 group significantly facilitated the hydrogen desorption, reducing the onset of H_2 release from the ~270°C of $Mg(BH_4)_2$ to about 100°C for $Mg(BH_4)_2(NH_3)_2$. However, only partial reversibility was observed after H_2 desorption. The lithium and calcium versions, that is, $LiBH_4 \cdot NH_3$ [256] and $Ca(BH_4)_2(NH_3)_2$, respectively, were also synthesized [257] and released H_2 with only traces of NH_3 when decomposed in a closed system. However, in vacuum or in an open system, the materials decomposed to regenerate pure $LiBH_4$ and $Ca(BH_4)_2$. It is reasonable to deduce that borohydrides with cations that bind NH_3 more strongly would be more likely to release H_2 without NH_3 impurity.

The enhancement in H_2 desorption properties of borohydride-ammonia compounds is believed to occur through enhanced hydridic-protic interactions between BH_4^- and NH_3 groups. Zheng and coworkers reported the release of 17.8 wt% of hydrogen from the cocatalyzed lithium borohydride ammoniate $Li(NH_3)_{4/3}BH_4$, which contains an equivalent number of protic and hydridic hydrogen atoms. To prevent ammonia loss and allow sufficient time for it to react with BH_4^- groups, the decomposition was performed in the temperature range of 135–250°C in a closed vessel. The dehydrogenation occurred via a two-step process with the intermediates $Li_4BN_3H_{10}$ and $LiBH_4$, and the final solid residue was a mixture of BN and Li_3BN_2 [258].

Destabilized Complex Metal Hydrides

We have seen that many of the complex anionic materials are thermodynamically challenged due to a mixture of covalent and ionic interactions. Within moieties such as $[BH_4]^-$ and $[AlH_4]^-$, the bonding is primarily covalent (with dipole character), with strong ionic bonds between these anions and their counterions, such as Na^+, Li^+, and Mg^{2+}. As a result, these materials can bind hydrogen too tightly. For example, MgH_2 contains 7.7 wt% hydrogen, but it requires a rather high temperature of 275°C to have 1 bar equilibrium pressure above the material. Starting with the pioneering work of Reilly and Wiswall [259], it has been found that additives can in some cases modify the thermodynamics of a metal hydride for the better, a modification that has come to be known as *thermodynamic destabilization*. We have already seen an example of destabilization with MgH_2 improving the thermodynamic properties of $LiNH_2$. Here, we discuss the phenomenon in much greater detail.

Reilly and Wiswall [259] observed that the hydrogen storage properties of MgH_2 were significantly changed by introducing copper. Whereas Mg forms two intermetallic compounds with Cu, namely, Mg_2Cu and $MgCu_2$ [259], only Mg_2Cu shows any hydrogen storage potential. Reilly et al. found that a reversible (albeit low weight percentage) hydrogen storage behavior was observed with Mg_2Cu via the reaction sequence

$$3MgH_2 + MgCu_2 \leftrightarrow 2Mg_2Cu + 3H_2 \tag{6.61}$$

Whereas pure MgH_2 has an enthalpy ΔH (at 298 K) for hydrogen desorption of 78.2 kJ/mol H_2 [259], the ΔH for the forward direction of Reaction 6.61 was observed to be 72.8 kJ/mol H_2 [259]. This is a small, but definite, drop in the thermodynamic requirements for hydrogen release. A similar result was found for the Ni analog of Reaction 6.61 [260],

$$Mg_2NiH_4 \leftrightarrow Mg_2Ni + 2H_2 \tag{6.62}$$

For Reaction 6.62, a ΔH value of 64.4 kJ/mol H_2 was found for the hydrogen release. This is a significant drop in the enthalpy of dehydrogenation for the MgH_2 system with added Ni. A similar enhancement has been seen by Zaluska et al. [261] with Al additive, in which case the added aluminum, leading to a "final state" Mg/Al alloy, was also expected to improve the thermal conductivity of the material [261].

How are these additives able to reduce the enthalpy of dehydrogenation? Vajo and coworkers introduced the term *destabilization* to describe the phenomenon, schematically shown in Figure 6.24 [262]. AH_2 represents a generic metal hydride, for example, MgH_2. Typically, in the absence of additives, a metal hydride must release a mole of hydrogen to form gaseous H_2 and the dehydrogenated product A. This requires a typically large

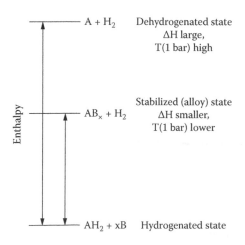

FIGURE 6.24
A generalized enthalpy diagram illustrating destabilization of a metal hydride AH_2 with an additive B, forming the alternative final alloy AB_x on dehydrogenation.

enthalpy change, which requires a high equilibrium temperature to establish 1-bar pressure over the material. However, if an additive xB is added, then this component B (Cu or Ni in the examples) can lead to the production of an intermediate alloy AB_x, which can be reached with a lower enthalpy of reaction. Note that there are still 2 equivalents of hydrogen being released, only with the "final state" being a lower energy alloy AB_x.

The term *destabilization* can be a bit confusing. It is important to note that the original material AH_2 is not directly modified [262]. The A-H bonds are not in the "initial state" being changed in any way. Rather, the overall reaction sequence is changing, such that the liberation of hydrogen occurs with a smaller enthalpic change. In this way, the additive makes the initial state (AH_2 + B) more unstable with respect to decomposition via hydrogen release and therefore "destabilized." The (AH_2 + B) mixture may be considered as a "destabilized chemical system" because destabilization is occurring on a system level, not for any individual component.

Destabilization of binary metal hydrides can be achieved with nonmetallic additives as well. LiH is a binary hydride with high weight capacity (12.5 wt%), but it is much too stable to be a hydrogen storage material, requiring 910°C to achieve an equilibrium pressure of 1 bar [263]. Because of its stability, LiH is not a viable hydrogen storage material. Silicon is known to bind well with Li, forming a series of Li_xSi alloys and therefore might significantly destabilize LiH with respect to dehydrogenation. It is also a fairly light element, which it is hoped would not too dramatically decrease the gravimetric storage density of the material. Figure 6.25 shows the dehydrogenation of a mixture with stoichiometry 4LiH + Si from the work of Reference 263. Figure 6.25 also shows desorption from pure LiH for comparison.

The LiH/Si mixture releases hydrogen beginning at approximately 270°C and reaches 0.8 wt% at 450°C. This capacity was limited by the H_2 pressure because dehydrogenation was conducted in a closed volume. After dehydrogenation under dynamic vacuum, XRD

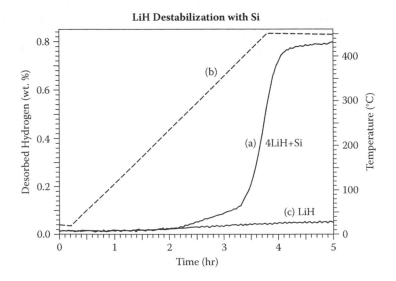

FIGURE 6.25

Dehydrogenation in the LiH/Si system from milled LiH and 4LiH + Si. Curves (a) and (b) show the weight percent and temperature, respectively, during heating 4LiH + Si at 2°C/min to 450°C. Curve (c) shows the weight percent hydrogen during heating of pure LiH. As expected, no hydrogen desorption occurred for pure LiH in this temperature range. (Reprinted with permission from J.J. Vajo, F. Mertens, C.C. Ahn, R.C. Bowman, Jr., B. Fultz, *Journal of Physical Chemistry B*, 108 (2004) 13977–13983. Copyright 2004, American Chemical Society.)

data showed that the final dehydrogenated product was not the expected Li_4Si (a known silicide phase), but rather a silicide phase with composition $Li_{2.3}Si$. This suggested that an initial mixture of 2.5LiH + Si would be more appropriate, and cycle at about 5.0 wt% hydrogen, which was in fact observed [263]. Characterization of the equilibrium behavior showed that the mixture 2.5LiH + Si lowered the enthalpy of hydrogen desorption from 190 kJ/mol H_2 (pure LiH) to 120 kJ/mol H_2 and increased the equilibrium hydrogen pressure (at ~500–560°C) by a factor of over 10^4 [263].

While the 2.5LiH + Si system still operated at too high a temperature to be a viable hydrogen storage material, the study showed the power of the destabilization strategy to alter the thermodynamics of hydrogen release from metal hydrides, in this case with a light nonmetallic element. Silicon was also found to destabilize MgH_2, leading to a four-fold increase in the desorbed hydrogen pressure at 300°C. However, attempts to rehydrogenate Mg_2Si were unsuccessful, presumably due to kinetic limitations [263].

Vajo et al. [131] also demonstrated for the first time that complex anionic materials such as $LiBH_4$ can be destabilized. $LiBH_4$ is known to have a high gravimetric capacity (18 wt% theoretical), releasing about 13.6 wt% according to the reaction $LiBH_4 \rightarrow LiH + B + 3/2H_2$ [131]. The ΔH for the hydrogen desorption was estimated to be 67 kJ/mol H_2, and if reversible, pure $LiBH_4$ would produce 1 bar of H_2 pressure at $T \sim 400°C$ [131]. Subsequent measurements [264] indicated a ΔH for desorption of 74 kJ/mol H_2, producing a 1-bar pressure of hydrogen at about 370°C. A nice account of the hydrogen storage chemistry of $LiBH_4$ can be found in the recent review by Li et al. [10]. Vajo et al. showed [131] that $LiBH_4$ can be reversibly cycled with a reduced reaction enthalpy by the addition of stoichiometric amounts of MgH_2, as described by Reaction 6.63:

$$LiBH_4 + \tfrac{1}{2}MgH_2 \leftrightarrow LiH + \tfrac{1}{2}MgB_2 + 2H_2 \qquad (6.63)$$

The hydrogen storage capacity of Reaction 6.63 is 11.4 wt%. The addition of magnesium stabilizes the final state MgB_2, thereby destabilizing the ($LiBH_4 + \tfrac{1}{2}MgH_2$) mixture with respect to decomposition yielding hydrogen. Figure 6.26 summarizes the hydrogen desorption/absorption studies from Vajo et al. [131] in the form of a van't Hoff plot, using plateau pressures measured at a composition of 4 wt% hydrogen.

A linear fit to the data at 315–400°C (1.7–1.48 abscissa values in Figure 6.26) indicates a dehydrogenation enthalpy of 40.5 kJ/mol H_2 and an equilibrium pressure of 1 bar at 225°C. Comparing with the values for $LiBH_4$ alone [131, 264], addition of MgH_2 reduced the ΔH of hydrogen desorption by 33.5 kJ/mol H_2 and increased the equilibrium pressure by approximately one order of magnitude, still yielding a material cycling at about 10 wt% hydrogen.

Interestingly, the equilibrium pressure behavior for the $LiBH_4 + 1/2MgH_2$ system crosses the curve for MgH_2/Mg at about 360°C (1000/T = 1.57). At temperatures below 360°C (larger 1000/T), the equilibrium pressures are greater than those for pure MgH_2. Thus, in addition to $LiBH_4$ being destabilized, the MgH_2 was also destabilized in this temperature region. For temperatures lower than 360°C, the combined $LiBH_4 + 1/2MgH_2$ system has equilibrium pressures higher than either individual component. The properties of a number of systems of the general formula $LiBH_4/MgX$ are summarized in Reference 26.

We described previously that Reilly and Wiswall [260] examined the hydrogen storage properties of Reaction 6.62: $Mg_2NiH_4 \leftrightarrow Mg_2Ni + 2H_2$. Orimo and Fujii [265] conducted an extensive set of studies examining the materials science and hydrogen storage properties of a number of hydrides in the Mg-Ni family. Vajo et al. pushed the explorations further

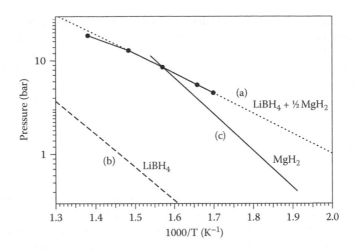

FIGURE 6.26
Van't Hoff plots for destabilized LiBH$_4$ + ½MgH$_2$, pure LiBH$_4$, and MgH$_2$. (a) Equilibrium pressures obtained from absorption isotherms at 4 wt%. (b) An estimate of the behavior for dehydrogenation of LiBH$_4$ to LiH + B. (c) Equilibrium pressure for MgH$_2$/Mg. (Reprinted with permission from J.J. Vajo, S.L. Skeith, F. Mertens, *Journal of Physical Chemistry B*, 109 (2005) 3719–3722. Copyright 2005 American Chemical Society.)

by examining the combined system LiBH$_4$/Mg$_2$NiH$_4$. The LiBH$_4$/Mg$_2$NiH$_4$ system was studied in detail [266] because of its remarkable features, which include full reversibility, reaction through a direct LT kinetic pathway, formation of a unique ternary boride phase, and low reaction enthalpy coupled with low entropy. We give a more complete description of this interesting system in the following as it reveals in many ways the full power of the destabilization approach and points the way to possible future courses of study. Dehydrogenation of a mixture of 4LiBH$_4$ + 5Mg$_2$NiH$_4$ is shown in Figure 6.27.

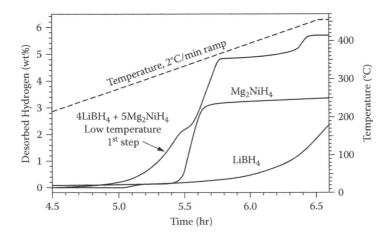

FIGURE 6.27
Dehydrogenation of 4LiBH$_4$ + 5Mg$_2$NiH$_4$, Mg$_2$NiH$_4$, and LiBH$_4$. Dehydrogenation was conducted using a 2°C/min temperature ramp in 4 bar of H$_2$ for 4LiBH$_4$ + 5Mg$_2$NiH$_4$ and Mg$_2$NiH$_4$ and (initial) vacuum for LiBH$_4$. The small desorption step for LiBH$_4$ (at 5.1 h) occurs at the melting point (~285°C). Thus, the first step for the 4LiBH$_4$ + 5Mg$_2$NiH$_4$ mixture begins below the melting point for LiBH$_4$. (Reproduced from J.J. Vajo, W. Li, P. Liu, *Chemical Communications*, 46 (2010) 6687–6689 with permission of the Royal Society of Chemistry.)

The reaction occurs in three steps. The first step is consistent with the reaction given by

$$4LiBH_4 + 5Mg_2NiH_4 \leftrightarrow 2MgNi_{2.5}B_2 + 4LiH + 8MgH_2 + 8H_2 \quad (6.64)$$

This reaction step releases 2.6 wt% hydrogen. As shown in Figure 6.26, dehydrogenation occurs at temperatures lower than the dehydrogenation temperature for either pure $LiBH_4$ or pure Mg_2NiH_4. This is a critically important feature as it indicates that a new hydrogen release pathway has been opened by mixing the $LiBH_4$ and Mg_2NiH_4 materials. This pathway likely involves direct reaction between $LiBH_4$ and Mg_2NiH_4, rather than H_2 release from either of the individual components. The reaction begins at temperatures as low as 250°C, which is very low for borohydride-based systems. The low reaction temperature is possibly due to the catalytic nature of Ni in the $[NiH_4]^{4-}$ anion. Thus far, this system appears to be the only reversible destabilized system that reacts through a (new) direct reaction pathway. In contrast, the well-studied $LiBH_4/MgH_2$ system reacts sequentially with initial dehydrogenation of MgH_2 followed by reaction of Mg with $LiBH_4$. We believe that to achieve the full benefit of the mixed hydride system destabilization strategy, reaction through new pathways is essential. Thus, this system represents an important demonstration that such new pathways are possible.

As indicated, reaction of $LiBH_4$ with Mg_2NiH_4 leads to the formation of the ternary boride $MgNi_{2.5}B_2$. Formation of the ternary boride $MgNi_{2.5}B_2$ is significant because few boride phases reversibly hydrogenate under mild conditions (~100 bar H_2) [267]. In addition, ternary transition-metal-based boride phases have not previously been considered as hydrogen storage materials. Identification of this phase suggests that other ternary (or higher-order) transition-metal-based boride phases should be experimentally or computationally tested for reversible hydrogenation activity.

The good kinetics of the $LiBH_4/Mg_2NiH_4$ reaction allow its equilibrium to be characterized over the temperature range 270°C to 360°C, as shown by the van't Hoff plot in Figure 6.28. This range extends below the lowest temperatures measured for the $LiBH_4/MgH_2$ system (315°C) because of the improved kinetics, with the lowest temperature data point below the melting temperature for bulk $LiBH_4$ (T_m = 280°C). The pressure varies logarithmically with the inverse temperature characterized by a change in enthalpy ΔH of 15 kJ/mol H_2 and a change in entropy ΔS of 62 J/K-mol H_2 for the hydrogen release reaction.

This change in enthalpy is very low for a reversible system. A low enthalpy is advantageous for practical systems in which heat must be supplied to release hydrogen and dissipated during rehydrogenation. However, systems with low enthalpies (less than ~30 kJ/mol H_2) typically cannot be rehydrogenated because the equilibrium temperatures, given by $T_{eq} = \Delta H/\Delta S$, are too low. Remarkably, for this system ΔS is also very low, which raises the equilibrium temperature and enables reversibility. Vajo et al. [266] speculated that the low ΔS originates from the relatively high entropy of two complex hydride anion species, $[BH_4]^-$ and $[NiH_4]^{4-}$, in the hydrogenated phase. Overall, the capacity for the direct LT step shown is too low for practical use. However, Mg_2NiH_4 is a transition-metal-based complex hydride, of which there are numerous (>100) known examples. Therefore, the remarkable behavior of this system holds promise that other $LiBH_4$/transition-metal-based complex hydride systems could be found with higher hydrogen capacities.

Another example of destabilization concerns salts of the interesting intermediate $[B_{12}H_{12}]^{2-}$ anion. Ionic salts of these $[B_{12}H_{12}]^{2-}$ anions are typically undesired and thermodynamically stable intermediates in hydrogen release reactions involving borohydrides. The possibility of destabilizing these intermediates, allowing hydrogen release at lower temperatures, was predicted by Ozolins et al. [268] for a number of $B_{12}H_{12}$ systems. For

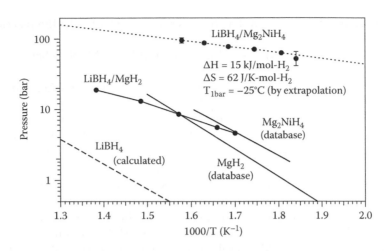

FIGURE 6.28

Van't Hoff plots for the LiBH$_4$/Mg$_2$NiH$_4$ destabilized system, pure LiBH$_4$, pure Mg$_2$NiH$_4$, the LiBH$_4$/MgH$_2$ system, and pure MgH$_2$. The (LiBH$_4$/Mg$_2$NiH$_4$) data shows equilibrium pressures at 0.67 wt% for the first reaction step shown in Reaction 6.64. This hydrogen content is at the midpoint of the reversible capacity for this step. The LiBH$_4$ behavior was calculated from tabulated thermodynamic data using HSC Chemistry for Windows software. The behavior for MgH$_2$ and Mg$_2$NiH$_4$ were obtained from the Sandia database. The data for LiBH$_4$/MgH$_2$ was obtained from measured isotherms. (Reproduced from J.J. Vajo, W. Li, P. Liu, *Chemical Communications*, 46 (2010) 6687–6689 with permission of the Royal Society of Chemistry.)

CaB$_{12}$H$_{12}$, it was predicted that destabilization with CaH$_2$ would lower ΔH considerably, thereby lowering the expected temperature for which 1 bar of hydrogen pressure would be attained (T_{1bar}):

$$CaB_{12}H_{12} \rightarrow CaB_6 + 6B + 6H_2 \qquad H\,(0K) = 65\ kJ/mol\ H_2 ;\ T(1bar) = 306°C$$

$$CaB_{12}H_{12} + CaH_2 \rightarrow 2CaB_6 + 7H_2 \qquad H = 38.6\ kJ/mol\ H_2 ;\ T(1bar) = 96°C$$

Stavila et al. [152] initiated experimental studies of the destabilization of Ca dodecahydro-*closo*-dodecaborate with CaH$_2$. It was found that introducing CaH$_2$ did indeed lower the thermodynamic requirements, although the material was severely kinetically limited. CaB$_{12}$H$_{12}$ by itself released only traces of hydrogen for temperatures below 650°C. However, when mixed with CaH$_2$, there was a dramatic reduction in the temperature required for H$_2$ release. As indicated in Figure 6.29, hydrogen release was observed at about 450°C for the destabilized system. These temperatures are of course too high for a practical hydrogen storage system and suggest a severe kinetic limitation to the material. Nevertheless, the results are a dramatic confirmation of destabilization theory [268] and indicate a strategy for liberating hydrogen from these stable [B$_{12}$H$_{12}$]$^{2-}$ intermediates.

As shown, the introduction of additives to a metal hydride can significantly destabilize the metal hydride, improving the thermodynamics for hydrogen release. This destabilization approach also complicates the solid-state system and opens up enormous possibilities for possible hydrogen storage systems. As shown, the dehydrogenated states of interesting hydrogen storage systems can be single metal atoms (e.g., Mg for MgH$_2$), two-component alloys (Mg$_2$Cu for the 3MgH$_2$ + MgCu$_2$ system), and even ternary three-component compositions (2MgNi$_{2.5}$B$_2$ as the dehydrogenated state for 4LiBH$_4$ + 5Mg$_2$NiH$_4$). Thus, from this dehydrogenated state perspective, one could ask how many possible three-component

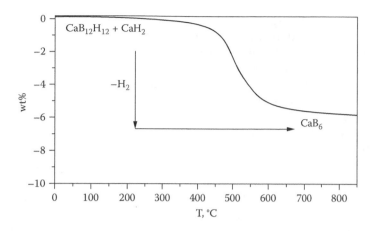

FIGURE 6.29
Destabilization of $CaB_{12}H_{12}$ with CaH_2.

materials are there of the nominal formula $A_xB_yC_z$ that could be explored as hydrogen storage systems? We can make a conservative estimate and assume $x = y = z = 1$ and further assume that any practical hydrogen storage system would have to consist of elements with mass less than that of iodine. With these assumptions, and further eliminating the rare gases and hydrogen itself, one finds that the number of theoretically possible combinations of the elements in a ternary system is 103,776. This is a big number, which is actually a gross underestimation because this does not take into account the possibility of additives beyond three elements and the enormous range of possible pressure and temperature conditions.

Theoretical Prediction of Complex Metal Hydride Materials

These hydrogen storage reactions are solid-state reactions with complexities concerning thermodynamics (ΔH, ΔS), issues of kinetics (reaction rates, activation energies), structural transitions even prior to hydrogen release, and potentialities of side reactions (NH_3, B_2H_6 release). The theoretical description of such reactions is a daunting theoretical challenge. Yet, our ability to theoretically predict such reactions has improved dramatically since 2003, and we review that progress now.

An example of the state of the art in theory from about 2004 is represented by the work of Wolverton et al. [269], which represents one of a relatively small group of studies [270–272] up to that time in which formation energies had been calculated from first principles. Wolverton et al. [269], in the course of their studies of the Al-H system using density functional theory (DFT), also calculated the energies of formation of 21 metal hydrides, including AlH_3, MgH_2, and $NaAlH_4$. The results showed that the DFT calculations in the generalized gradient approximation (GGA) provide an accurate and consistent picture of metal hydride structure and energetics. The formation energies agreed with experimental results (if available) to within about 20 kJ/mol H_2, and in many cases much better.

Although by 2004 DFT had been used to compute the thermodynamics of a number of individual metal hydride compounds, there was no way to predict phase diagrams/van't Hoff plots for metal hydrides, and theory could not account for the reaction complexities (e.g., metastable species and multistep reactions) that were being seen in solid-state H_2 desorption/absorption studies. In addition, there had been no theoretical work on amorphous metal hydride materials or robust attempts to account for the kinetics of these hydrogen storage systems. Furthermore, there had yet to be attempts to systematically predict solid-state reactions that might have desirable hydrogen storage properties. The years since 2004 have seen remarkable improvement in the ability of theorists to predict many (but not all) of these phenomena.

One approach to predicting new hydrogen storage materials is to use ab initio methods to screen the thermodynamic properties of a large number of novel materials that in many cases have not been experimentally examined. This was the approach of Alapati et al. [219, 273–278]. Although experiments are the final judge of a reaction's usefulness, it is not practical to systematically study every conceivable reaction mixture experimentally, as suggested for ternary compositions $A_xB_yC_z$. Therefore, these workers have performed systematic first-principles thermodynamic calculations in the search for desirable reaction mixtures [219, 273–278].

As stated, first-principles calculations have been shown to yield reaction-free energies that are accurate within about 20 kJ/mol H_2—a level of precision that is adequate for screening large numbers of potentially interesting reactions. We shall see that with more detailed consideration, the free energies of hydrogen storage reactions can be predicted with even greater accuracy.

Alapati et al. [219, 278] theoretically explored literally millions of possible reaction conditions consisting of different element spaces, compositions, and temperatures. The general process they used is as follows:

1. Compute DFT total energies for an extensive number of compounds and collect into a database. These total energies are for $T = 0$.

2. Specify a set of elements to screen (e.g., Li, B, Na, Mg; H is always selected); the program identifies all solid compounds having any of the elements.

3. Specify a pressure of H_2, typically $P_{H2} = 1$ bar.

4. Scan the composition of the solid phase (amount of H_2 in the system not specified).

5. Pick a starting and ending T and ramp the temperature in increments dT of 2 K.

6. At each T, minimize the grand potential Ω via linear programming to obtain equilibrium composition, using the DFT heats of formation and the chemical potential of H_2 at the specified T, P.

7. Chemical reactions are identified when the equilibrium composition changes between two successive temperatures.

8. Pick only reactions with hydrogen wt% greater than 6 and with hydrogen release enthalpy changes between 15 and 70 kJ/mol H_2. This requirement assumed the reaction entropy change ΔS was in the range 90–140 J/K mol H_2, and varied somewhat over the studies [219, 273–278]. The weight percentage requirement is driven by the expected hydrogen storage properties needed to satisfy the goals of the recent DOE hydrogen storage program targeting light-duty vehicle applications of hydrogen storage.

This method has been used to scan over 20 million different reaction compositions [219]. The vast majority of the reactions screened in this way were predicted to be "not useful,"

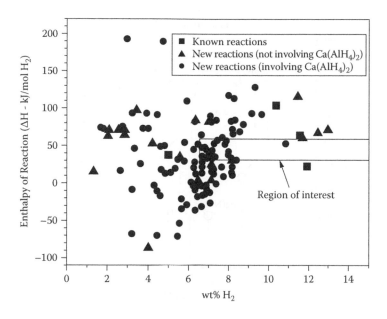

FIGURE 6.30

Theoretical predictions for reactions with weight percent hydrogen greater than 6 wt% and the desorption enthalpies in the range $30 \leq \Delta H \leq 60$ kJ/mol H_2. (Reproduced from S.V. Alapati, J.K. Johnson, D.S. Sholl, *Journal of Physical Chemistry B*, 110 (2006) 8769–8776 with permission of American Chemical Society.)

having either reaction enthalpies that are too high or too low or having inadequate hydrogen storage weight percentages. Yet, some of them are of interest and are also new reactions. The results are shown in Figure 6.30. For this plot, we are considering the ΔH for dehydrogenation. Results in the box labeled "region of interest" show a restricted range of enthalpies within the screening criteria above $30 \leq \Delta H \leq 60$ kJ/mol H_2.

Subsequent work [273, 278] used a somewhat wider ΔH screening criterion than that of Reference 219. Alapati et al. [273] screened literally millions (and millions) of potential reactions of the elements Al, B, Ca, N, H, Li, Mg, K, Na, Sc, Ti, V, C, and Si, thereby exceeding earlier efforts [219, 278] by orders of magnitude. The expanded enthalpic screening condition was ΔH between 15 and 75 kJ/mol H_2. They found 43 distinct reactions with weight percentage hydrogen greater than 6% and ΔH in the desired range. Many of these favorable reactions involved $LiBH_4$ and $Mg(BH_4)_2$. Those involving $LiBH_4$ are shown in Figure 6.31.

The reactions in Figure 6.31 are for "single-step" dehydrogenation reactions, proceeding from reactants to products in a single step, without intermediates. Note the prediction in Figure 6.31 that $LiBH_4/MgH_2$ would be an attractive destabilized system, as shown by the work of Vajo et al. [131] and presented in Figure 6.26.

This more recent work of Alapati included considerations of multistep reactions, the importance of which were emphasized by Siegel et al. [279] and Akbarzadeh et al. [220]. These workers emphasized that some reactions may be multistep in nature. If there are multiple steps in a reaction, although the overall value of ΔH may be attractive, if one of the steps has a very unattractive ΔH, then the reaction will not proceed to the end product due to the system being "trapped" by the step that is thermodynamically unfavorable. As a result, screening also needs to account for multistep character, which might present a highly undesirable ΔH.

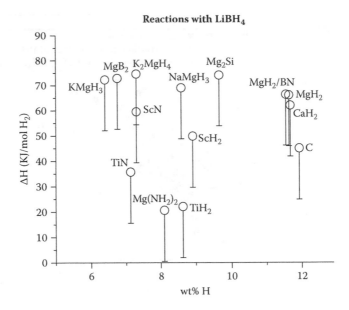

FIGURE 6.31
Reaction enthalpies computed at 0 K for single-step reactions involving $LiBH_4$. The reactants that combine with $LiBH_4$ are indicated in the figure. The error bars for each reaction span the predicted ΔG at 300 K. (Reproduced from S.V. Alapati, J.K. Johnson, D.S. Sholl, *Journal of Physical Chemistry C*, 112 (2008) 5258–5262 with permission of American Chemical Society.)

To account for the possibility that some of these single-step reactions may actually have multistep character, Alapati et al. [273] developed a method to allow the identification of metastable reactions that may lie near each other energetically. In other words, to account for possible intermediates that would introduce a multistep nature to a reaction path, a method was developed to see if nearly degenerate reaction pathways may exist in reactions previously thought to be single step. By this process, for a given single-step reaction, the space is scanned for possible nearby metastable reactions. The vast majority of reactions examined by the linear program have turned out to be solely single-step reactions in the sense discussed [273]. Therefore, the thermodynamic predictions are conceptually valid. However, a few reactions turned out to have this multistep character.

We described how the destabilization predictions made by Ozolins et al. [268] for a number of $[B_{12}H_{12}]^{2-}$ systems have been experimentally confirmed for the Ca case by Stavila et al. [152] in Figure 6.29. Likewise, a number of the predictions made by Alapati et al. [219, 273–278] have also been experimentally confirmed. Alapati et al. predicted that the system ($LiNH_2 + MgH_2$) would have attractive weight percentage and good thermodynamics. These predictions were experimentally confirmed by Lu et al. [218], as discussed previously for the amides. In addition, the prediction in Figure 6.31 that $LiBH_4/CaH_2$ would be an interesting system to examine received confirmation in the experimental work of Pinkerton and Meyer [280]. These workers found the system ($6LiBH_4 + CaH_2$) to reversibly release hydrogen with a ΔH value of 59 kJ/mol H_2, which agrees well with the prediction by Alapati et al. [273] of 62.1 kJ/mol H_2.

A few caveats are in order [273] for these theoretical results. These calculations were based on a catalog of known compounds. If a reaction involves unknown compounds, these theoretical methods cannot describe it. For example, the earlier work of Alapati et al.

[219] did not include the metal closoboranes ($M_nB_{12}H_{12}$, for M = Li, Mg, Ca, or K), which since that time have been observed or predicted as intermediates in the decomposition reactions of the metal borohydrides. The DFT calculations [219, 273–278] do not consider other gas-phase products that might be produced, for example, NH_3 or B_2H_6. Finally, the results, being based on thermodynamic analyses, do not contain information about chemical kinetics, which clearly can determine whether or not a reaction is observed.

As described, computational modeling of the thermodynamics of hydrogen storage materials was used to narrow the search in composition space and to guide experimental explorations. These theory results require a priori knowledge of the solid-phase crystal structures for the species involved in the reaction being considered. In some cases, the crystal structures have been experimentally measured. However, for many reactions that are new and have been predicted to have desirable hydrogen storage properties, the reactants may have unknown structure. Take, for example, the case of $Ca(BH_4)_2$ and its decomposition releasing hydrogen:

$$3Ca(BH_4)_2 \rightarrow CaB_6 + 2CaH_2 + 10H_2$$

In 2006, the geometric (crystal) structure of $Ca(BH_4)_2$ was unknown. To theoretically predict the thermodynamics of this material, there are two possible approaches to estimating the structure. In the Inorganic Crystal Structure Database (ICSD) method, a search is conducted of all materials with the same general formula as $Ca(BH_4)_2$, namely, AB_2X_8. Of the approximately 80,000 structures in the database, there are about 30 with the formula AB_2X_8.

At this point, a common but logically suspect assumption is made. The assumption is that the true structure type of $Ca(BH_4)_2$ is the same as one of the about 30 structures for the formula AB_2X_8. This assumption does not allow for possible new crystal structures that might exist. Regardless, for each of the 30 structures, a first-principles DFT calculation using the Vienna Ab initio Simulation Package (VASP) is performed for the total energy (at $T = 0$) for all 30 trial structures. The structure with the lowest calculated ground state is presumed to be the ground state of $Ca(BH_4)_2$.

The ICSD approach suffers from a central conceptual problem. As mentioned, it does not allow for the possibility that $Ca(BH_4)_2$ has a new structure not found in the database. In some cases, there may not be many trial structures for the repetitive VASP calculations, leading to greater uncertainty in the results. So, although database searching can yield the correct structure in many cases, there are also many exceptions.

Majzoub and Ozolins [281] sought a more rigorous method to predict crystal structures by incorporating the understanding that these complex anionic materials are dominated by electrostatic interactions. By treating the complex anionic units (such as BH_4^-) as a rigid unit and performing a global minimization of the electrostatic interactions with the counterion, one obtains a lowest-energy structure that can then be the basis of one VASP calculation of the total energy.

Describing this method in more detail, it is important to note that in metal-hydrogen compounds containing complex anions, the metal atoms are frequently alkali or alkaline earth and the complex anions one of $[NH]^{2-}$, $[NH_2]^-$, $[BH_4]^-$, $[AlH_4]^-$, and $[AlH_6]^{3-}$. Infrared and Raman vibrational spectra of existing alanates and borohydrides have established the nature of many of these compounds to be molecular ionic structures, with the bending and stretching modes of the anions distinctly separated from the crystal modes involving motion of the cations. A simplified model of the structure of complex metal hydrides consists of a rigid anion with appropriate charges on the center and vertex positions of, for example, the Al and H atoms, in $[AlH_4]^-$. These charges are conveniently provided by first-principles calculated Born effective charges from literature compounds. One may then optimize the total electrostatic energy of the crystal using a suitable optimization algorithm, such as simulated annealing metropolis Monte Carlo. Such a procedure has been developed for electrostatic interactions resulting in the Prototype Electrostatic Ground State (PEGS) method, the details of which can be found in the publication by Majzoub et al. [281].

This remarkably successful method predicts [281] ground-state structures in many of the complex hydrides, including $NaAlH_4$, $Mg(AlH_4)_2$, and even quite complicated structures such as the bialkali alanate K_2LiAlH_6, $Mg(BH_4)_2$ [121], and dodecahydro-*closo*-dodecaborate intermediates [268, 282]. The success of the PEGS method for structure determination of the complex hydrides indicates that the method captures the essential electrostatic interatomic interactions in these compounds.

All of the work discussed relied on the use of DFT calculations for predictions of the thermodynamics of new material systems. As discussed by these authors, the estimated error in predicted hydrogen desorption enthalpies ΔH is about 20 kJ/mol H_2. While this level of certainty is reasonable and appropriate for "screening" purposes, it does represent a roughly 30% error in the calculation of ΔH, and it is only natural to ask what the phenomena are, unaccounted for in the DFT work, that are leading to this error. Zarkevich and Johnson [283] showed that anharmonic vibrations at finite temperature need to be taken into account for a quantitative account of reaction enthalpies involving molecular solids at finite temperatures, as well as consideration of the physical phase from which hydrogen release occurs.

Molecular-based materials have significant contributions to the free energy arising from vibrations, of which there are two contributions: so-called harmonic vibrations (for which the potential energy varies with the square of the interatomic distance) and anharmonic vibrations (with a more complicated nonquadratic dependence of potential energy with bond distance). For $LiBH_4$, the negatively charged, tetrahedral $[BH_4]^-$ anions are charged balanced by Li^+ cations. The $[BH_4]^-$ anions are molecules that vibrate harmonically (e.g., typical molecular breathing and shear modes), whereas the Li^+ and $[BH_4]^-$ vibrate in response to each other's motion, and these are mostly anharmonic. Using ab initio molecular dynamics (MD), Zarkevich and Johnson showed [283] the importance of including both harmonic and anharmonic vibrations. Their theory provided the most accurate theoretical description to date of the $2LiBH_4 + MgH_2$ system.

Whereas DFT calculations [273] predict the $2LiBH_4 + MgH_2$ system to have a ΔH of 66.2 kJ/mol H_2 (at $T = 0$), the first-principles finite-temperature MD calculation of Zarkevich et al. [283], including the contribution of both harmonic and anharmonic vibrational modes, predicts the desorption enthalpy to be 41 kJ/mol H_2 at 600 K, in excellent agreement with the experimental value of 40.5 kJ/mol H_2 measured by Vajo et al. [131] over the temperature range 588–673 K. The estimated theoretical error for the ΔH value for desorption was ± 1 kJ/mol H_2. This is remarkable theoretical accuracy. Note that the theory permits proper consideration of the fact that hydrogen release occurs from the melted phase of $LiBH_4$, which exists for temperatures above 553 K [283]. The work of Zarkevich and Johnson [283] represents the current state-of-the-art in quantitatively accounting for reaction enthalpies in destabilized hydrogen storage reactions.

Summarizing these theoretical results for destabilized reactions, DFT methodology has significantly advanced so that DFT can now be routinely used to predict promising materials systems based on their predicted ΔH and ΔG. These promising material reactions are predicted by assessing a wide range of P, T, and composition, thereby focusing experimental efforts on promising compounds. Linear search methods have been implemented, allowing the scanning of literally millions of different reaction conditions (composition, T, P), and can also examine the influence of multistep reaction sequences. Theory can now predict the existence of important and surprising reaction intermediates that are being confirmed by experiment (e.g., $[B_{12}H_{12}]^{2-}$ intermediates). The PEGS method has been a breakthrough development for predicting crystal structures beyond the use of the ICSD, thereby increasing accuracy and enabling thermodynamic predictions for new structural phases of materials. With more refined ab initio finite-temperature MD calculations, the equilibrium thermodynamic processes and underlying physics of selected reactions can be more deeply understood with high accuracy.

Nanoscale Complex Metal Hydrides

One of the common themes from this review of complex metal hydride R&D is that the kinetics of hydrogen release and hydrogen absorption are almost always slower than desired. Often, this is generically attributed to the presumed slow rates of the diffusion of hydrogen and important intermediate species, which in turn originate from the covalent and ionic bonding characteristic of these light elements metal hydrides. Over the years, methods have been sought to improve the kinetics of these materials. The preceding discussion of alanates, borohydrides, and amides has shown that additives can in many cases improve the kinetic performance. This was shown for Ti promotion of $NaAlH_4$, and KH promotion of the kinetics of the $2LiNH_2/MgH_2$ system. Although kinetic promotion by additives is well documented, there is a lot to be learned about the mechanism by which these additives are improving the kinetics. It has also been found that by "nanosizing," by methods such as ball milling and nanoconfinement, the constituent materials can improve kinetics.

Ball milling has been shown to reduce the particle size and crystallite size of the reactants in these solid-state reactions, in addition to providing a source of energy to allow the creation of metastable phases of a material. A nice example of this is presented in the work of Huot et al. for MgH_2 [284]. The "β" phase of MgH_2 is commercially available, with an average particle size of about 20 μm. Starting with this initial material, Huot et al. found that energetic ball milling for 20 h converted about 18% of the β phase to the closely related γ phase and increased the surface area by about 8 times, from 1.2 m^2g^{-1} to 9.9 m^2g^{-1}. Milling also significantly reduced the average crystallite size from about 1 μm to about 15 nm. This is a greater than 50 times reduction in the average crystallite size.

Figure 6.32 shows that ball milling significantly enhanced the hydrogen desorption and absorption kinetics of the MgH_2 system [284].

As shown in Figure 6.32(a) desorption from the unmilled sample is quite slow, taking more than 2000 s at 623 K to desorb about 5 wt% hydrogen. However, for the milled sample, at 623 K, the sample is essentially completely dehydrogenated at 700 s. Similar kinetic enhancement via ball milling can be seen in Figure 6.32(b) for hydrogen absorption of the Mg metallic phase. It is interesting that with cycling, the initially formed γ phase of MgH_2 disappears, and one is left with β-MgH_2, yet the kinetic enhancements remain, indicating that the enhancement is related to the higher surface area or the much smaller crystallite size produced by ball milling [284]. Analyses showed that the activation energy for desorption was decreased from 156 kJ/mol for the unmilled sample, to 120 kJ/mol for the

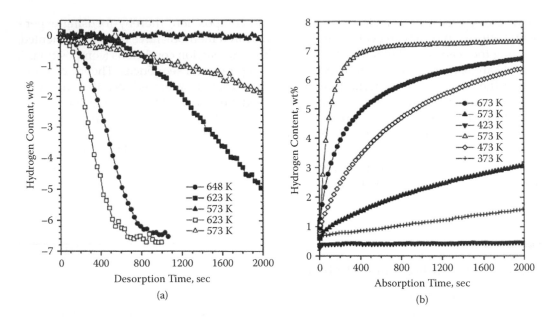

FIGURE 6.32

(a) Hydrogen desorption curves of unmilled MgH$_2$ (filled symbols) and ball-milled (open symbols) MgH$_2$ with desorption into a hydrogen pressure of 0.15 bar. (b) Hydrogen absorption curves of the unmilled (filled symbols) and ball-milled (open symbols) samples of Mg under a hydrogen pressure of 10 bar. (Reproduced from J. Huot, G. Liang, S. Boily, A. Van Neste, R. Schulz, *Journal of Alloys and Compounds*, 293–295 (1999) 495–500 with permission of Elsevier BV.)

milled sample. PCT measurements showed that the thermodynamics of the system were not changed by ball milling, indicating that the effects were purely kinetic [284].

It is important to ask the question why milling increases the kinetics. From an Arrhenius perspective, the rate constant for a process can be increased by decreasing the activation energy E_a or by increasing the preexponential factor A. It seems intuitive that a large increase in the surface area exposes a larger number of potential reaction sites at the surface, where H$_2$ can be desorbed into the gas phase, or alternatively, gas-phase hydrogen can find reactive Mg metal to hydrogenate. This effect would likely increase the preexponential factor. In addition, if a reaction requires the encounter of diffusing species (say H atoms) across a particle surface, then the smaller the particle, the shorter the diffusion distance, and the diffusion times can decrease markedly. In this way, a simple increase in surface area produced by a decrease in particle diameter can improve both the desorption and the absorption kinetics.

However, a reduction in the activation energy would involve a change in the relative energies of reactants and the transition-state species at the top of the activation barrier. It is not a priori obvious why milling would decrease the activation barrier. One possibility is that the reduction in the activation barrier is related to the formation during milling of a significant intercrystallite phase. From the results described, milling reduces the crystallite size more than the particle size, which necessitates formation of additional grain boundaries and the associated intercrystallite phase. This phase is likely amorphous or at least highly defective because it is not detected by diffraction. Within such a defective phase, the diffusion of hydrogen or new phase nucleation may be expected to occur with a lower activation barrier. This question relates to the deeper problem that there are really no comprehensive theories of the kinetics of solid-state hydrogen desorption and absorption reactions, and further work in this direction would be highly desirable. Clearly, the situation is complicated as it requires understanding not only the reaction mechanism but also other factors that affect solid-state kinetics, such as nucleation of nascent phases and diffusion of species along surfaces and across particle interfaces.

Although ball milling is a straightforward method to reduce particle dimensions (albeit with some risk of increased oxidation and premature degradation of reactants), in general nanoscale particles produced in this way do not survive multiple desorption/adsorption cycles. The problem is that the nanoscale particles tend to agglomerate together with repeated cycles of hydrogen absorption and desorption [285, 286], which reduces the initial kinetic benefit. Interestingly, in contrast, nanoscale crystallites within an intercrystallite phase appear to be relatively stable during cycling. This difference is probably related to the comparatively low solid-solid interfacial energy of the crystallites vs. the much higher free-surface energy of particles. An approach to circumvent particle agglomeration is to use nanoporous scaffolds to confine hydride materials at the nanometer length scale. If a hydride material can be incorporated into the pores of these nanoporous scaffolds, and if the scaffold can survive the temperatures and pressures of cycling, then it is possible to confine the particles to the nanoscale without agglomeration during cycling.

Gutowska and coworkers [287] were the first to demonstrate this approach, and showed that ammonia borane can have its kinetics increased through the formation of a nanoconfined "hybrid material." Ammonia borane (often abbreviated AB), NH_3BH_3, is a "chemical hydride" that releases hydrogen with high gravimetric density, favorable volumetric density, and favorable, although exothermic, thermodynamics [287]. However, the rate of hydrogen release at 85°C was low, and the purity of the hydrogen stream needed to be improved. In particular, borazine (the B-N analog of benzene) production during desorption needed to be suppressed. Gutowska et al. demonstrated the incorporation of ammonia borane into a high-surface-area mesoporous silica material called SBA-15. Mesoporous silica materials have extremely high surface areas and a highly ordered pore structure, which gives them the appearance of nanochanneled silica scaffolds [287]. Figure 6.33 shows the mesoporous silica SBA-15.

Figure 6.33(a) is a scanning electron microscopic (SEM) image of the mesoporous morphology, with Figure 6.33(b) displaying the remarkable channel structure with transmission electron microscopy (TEM). The pores are schematically rendered in Figure 6.33(c) and have a pore diameter of 7.5 nm. The relative sizes of a pore and the hydrogen-bonded ammonia borane network are shown in Figure 6.33(d).

Figure 6.34 shows the results of temperature-programmed desorption mass spectrometry measurements performed for both the neat AB material and AB confined to the mesoporous silica [287]. The first effect of nanoconfinement was the reduction in the peak temperature for hydrogen release by about 15°C, indicative of an increased rate of hydrogen release. Kinetic data indicated that the activation barrier for H_2 release from AB in the nanoscaffold was about 67 kJ/mol, much less than the value of about 184 kJ/mol obtained from neat AB. The second effect was a significant suppression of the volatile product borazine. Measurements indicated that borazine was not likely trapped within the scaffold and instead pointed to a change in the decomposition pathway such that less borazine was being produced. In this way, the scaffold was not merely acting to reduce particle dimensions but actually was affecting the reaction pathway.

The influence on reaction pathway could lead to a change in the thermodynamics of the system. In fact, this was observed. The enthalpy ΔH of desorption for neat AB was measured to be –21 kJ/mol, whereas that for the nanoconfined NH_3BH_3 was found to be –1 kJ/mol. It was found that desorption from neat AB led to more nonvolatile components

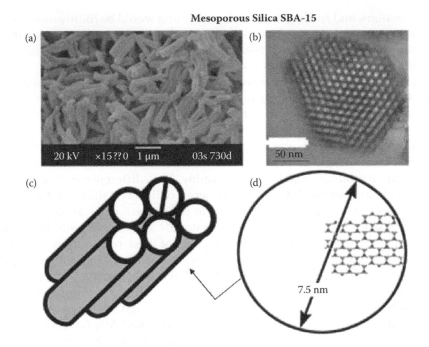

FIGURE 6.33
(a) SEM image of mesoporous silica; (b) TEM cross-sectional image of SBA-15 showing channels; (c) schematic representation of parallel channels in SBA-15; and (d) schematic representation of NH_3BH_3 hydrogen-bonded network inside 7.5-nm channel of SBA-15. (Reproduced from A. Gutowska, L. Li, Y. Shin, C.M. Wang, X.S. Li, J.C. Linehan, R.S. Smith, B.D. Kay, B. Schmid, W. Shaw, M. Gutowski, T. Autrey, *Angewandte Chemie International Edition*, 44 (2005) 3578–3582 with permission of Wiley-VCH.)

that increased the exothermicity of the desorption. These more exothermic reaction products were not formed in the AB confined to the mesoporous silica, leading to a less-exothermic (more favorable) overall reaction. Summarizing, this pioneering study showed that kinetics can be increased with nanoconfinement, that undesirable reaction products can be suppressed, and most importantly, the nanoscaffold leads to different reaction pathways that can alter the thermodynamics. In the case of AB confined in SBA-15, a favorable change in thermodynamics resulted.

About the same time as the work of Gutowska et al., theoretical interest in the effects of reduced dimensionality on hydrogen storage properties began to grow, mostly as a result of the ball-milling experiments referenced previously. Liang used DFT to predict that the ΔH for thin films of MgH_2 were lower than that of the bulk [288]. For a thin film of 2-unit cell thickness, the decrease in ΔH was predicted to be 5 kJ/mol H_2 [288]. Reducing the size along three dimensions, that is, producing clusters, in principle affects both the energy of the metallic particle and the energy of the hydrided form of the particle. As shown in Figure 6.35, the effect on hydrogen desorption energy is reflected in the difference between the energies of the hydrided and dehydrogenated materials, both of which are in principle different in clusters than they are in the bulk material.

Cheung et al. [289] performed reactive force field calculations for Mg_x and Mg_xH_y nano-clusters as a function of cluster size. The results showed a decreasing relative stability for hydrogenated magnesium clusters (as suggested by Figure 6.35) as particle size decreased below about 2 nm. This would translate to a decreased hydrogen desorption enthalpy with decreasing cluster size. The results of Cheung et al. are shown in Figure 6.36.

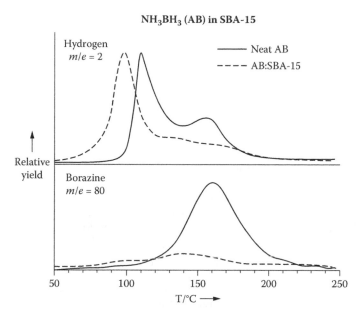

FIGURE 6.34

TPD/MS (1°C/min) of volatile products generated by heating neat ammonia borane (solid line) and AB confined in SBA-15 (dashed line). (Reproduced from A. Gutowska, L. Li, Y. Shin, C.M. Wang, X.S. Li, J.C. Linehan, R.S. Smith, B.D. Kay, B. Schmid, W. Shaw, M. Gutowski, T. Autrey, *Angewandte Chemie International Edition*, 44 (2005) 3578–3582 with permission of Wiley-VCH.)

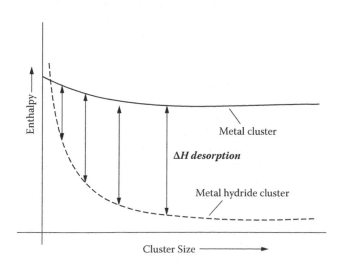

FIGURE 6.35

Schematic diagram showing that both metallic and metal hydride clusters have enthalpy affected by reduced cluster size, but the desorption enthalpy is controlled by the difference between the two. In this case, the desorption enthalpy is reduced with decreasing cluster size.

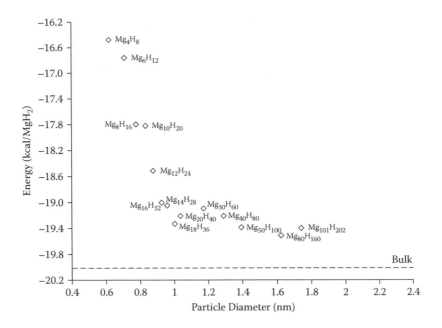

FIGURE 6.36

Variation of heat of formation of magnesium hydride particles as a function of particle size calculated using reactive field methods. The dashed line represents the heat of formation of bulk MgH_2 (–20 kcal/mol MgH_2 = 83.72 kJ/mol MgH_2). (Reproduced from S. Cheung, W.-Q. Deng, A.C.T. van Duin, W.A. Goddard, *Journal of Physical Chemistry A*, 109 (2005) 851–859 with permission of American Chemical Society.)

It is clear from Figure 6.36 that as the nanoparticle size decreased, the formation of Mg_xH_y became less exothermic. In particular, there was a steep increase below about 1 nm. Below about 1 nm, all Mg and H atoms were exposed to the surface [289]. It is in this region where deviations from the bulk really increased. It was also observed by Cheung et al. that for these cluster sizes, the surface hydrogen atoms were generally found to occupy less-stable surface sites (top and bridge sites) with fewer Mg-bonding partners [289]. Thus, these hydrogen atoms would desorb more readily. Reducing the particle size below 1 nm led to significant thermodynamic modification of the system. This is in general agreement with the ball-milling experiments [284] shown in Figure 6.32, in which the kinetic improvements induced by 2-nm particle sizes did not change the thermodynamics of the system. According to Cheung et al., further reductions in particle size would have been needed to alter the thermodynamics. The results of Cheung et al. were later confirmed by the Hartee-Fock/DFT calculations of Wagemans et al. [290] and by the quantum Monte Carlo simulations of magnesium hydride clusters by Wu et al. [291].

The example of Figure 6.36 suggests that a decrease in hydrogen desorption enthalpy arises when the hydrided form of the cluster becomes less stable more rapidly than the metallic form as the cluster size decreases. In a sense, this is a reduced dimensionality version of metal hydride destabilization discussed previously, although in that case the intrinsic nature of the bonding species was changing. However, one might suspect that, alternatively, reducing cluster size could destabilize the metallic form more than the hydrided form, and an increasing enthalpy of hydrogen desorption could result. The theoretical work of Kim et al. [292] showed that this can happen.

Kim et al. examined the effect of decreasing particle size on the hydrogen storage properties of a number of binary hydrides using a Wulff construction [292]. In this method, rather

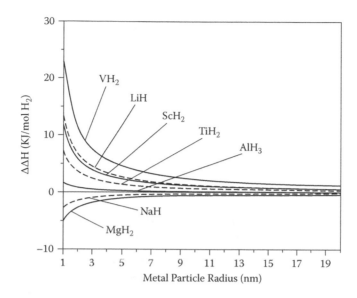

FIGURE 6.37
Change in the hydrogen desorption enthalpy as a function of metal particle radius for a number of binary metal hydrides. (Reproduced from K.C. Kim, B. Dai, J.K. Johnson, D.S. Sholl, *Nanotechnology*, 20 (2009) 204001. With permission.)

than calculate the energies of individual clusters, which can be prohibitively costly for a study of many systems, the authors of reference took a "top-down" approach in which the variations in energy of bulk crystal structures were modified in the nanoclusters as new surfaces appeared on the smaller structures. The net surface energy of the nanoparticle can be calculated from DFT, allowing an estimate of the reaction thermodynamics for a larger range of systems [292]. The results are shown in Figure 6.37 for the binary hydrides of V, Li, Sc, Ti, Al, Na, and Mg.

The results of Kim et al. [292] for Mg metal particle size agree qualitatively with the prior "bottom-up" cluster results [289–292] in that a reduction in the enthalpy of hydrogen desorption is predicted as the particle size shrinks. However, Figure 6.36 predicts that while this behavior is true for MgH_2 and NaH, it is not generally true, and that hydrogen desorption enthalpy can increase for other binary metal hydrides. Concurrent with this theoretical development, further efforts on investigating experimentally the effects of confining metal hydrides in nanoconfined structures were launched virtually simultaneously on $LiBH_4$ [293, 294], MgH_2 [295], and $NaAlH_4$ [13–16].

Gross et al. [294] conducted detailed studies of incorporating hydride materials into nanoporous scaffolds, such as carbon aerogels [296, 297]. Carbon aerogels [297] are mesoporous materials composed of three-dimensional networks of interconnected carbon particles measuring 5–10 nm in diameter. Two pore size regimes exist, as indicated in the schematic aerogel structure diagram of Figure 6.38 [297]. The larger mesopores of about 10 nm or more consist of the cavities between the carbon particle in the material. Smaller pores (<2 nm) exist within the constituent particles. The main features of carbon aerogels are high surface area (400–900 m^2/g) and high total pore volumes (1–3 cm^3/g).

Gross et al. measured isothermal dehydrogenation rates for $LiBH_4$ incorporated into carbon aerogels with 4- and 25-nm mesopore sizes [294]. Carbon aerogels were selected because they are relatively inert, and the mesopore size can be synthetically adjusted. $LiBH_4$ was introduced into the aerogel by heating together samples of $LiBH_4$ and aerogel

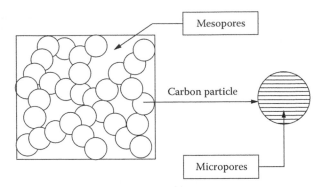

FIGURE 6.38
Carbon aerogel diagram. (From S. Gavalda, K.E. Gubbins, Y. Hanzawa, K. Kaneko, K.T. Thomson, *Langmuir*, 18 (2002) 2141–2151 with permission of American Chemical Society.)

$$LiBH_4 \rightarrow LiH + B + 1.5\ H_2$$

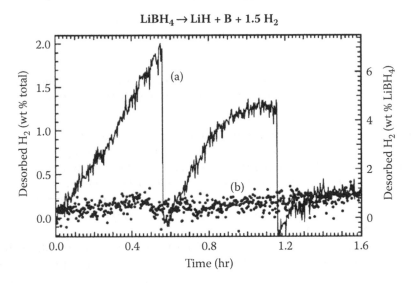

FIGURE 6.39
(a) Dehydrogenation of LiBH$_4$ at 300°C for LiBH$_4$ nanoconfined in 13-nm aerogel. (b) A control sample (dots) consisted of LiBH$_4$ mixed with nonporous graphite. (Reprinted with permission from A.F. Gross, J.J. Vajo, S.L. Van Atta, G.L. Olson, *Journal of Physical Chemistry C*, 112 (2008) 5651–5657. Copyright 2008, American Chemical Society.)

under inert atmosphere. At about 300°C, the LiBH$_4$ melts and infiltrates the nanoscaffold aerogel by capillary action [294]. These workers estimated that up to 90% of the available pore volume in the aerogel was filled with LiBH$_4$. A dramatic increase in the rate of hydrogen release from the nanoconfined LiBH$_4$ is shown in Figure 6.39.

For LiBH$_4$ incorporated into 13-nm aerogel (Figure 6.39(a)), isothermal dehydrogenation at 300°C proceeded at a nearly constant rate up to 6.4 wt% with respect to the LiBH$_4$. At this point, the hydrogen pressure was 0.045 bar. To ensure this hydrogen pressure did not limit the rate of desorption, the system was briefly evacuated, and desorption resumed (as shown in Figure 6.39(a)).

In contrast to the aerogel-confined sample, the analogous desorption experiments for LiBH$_4$ mixed with nonporous graphite showed very little dehydrogenation over the same time interval, as shown in Figure 6.39(b). Detailed assessments indicated that the rate of hydrogen release from LiBH$_4$ infiltrated in 13-nm aerogel was about 50 times faster than that of the control sample [294]. Furthermore, the activation energy for dehydrogenation decreased from 146 kJ/mol LiBH$_4$ (graphite sample) to 103 kJ/mol LiBH$_4$ for the 13-nm aerogel sample. This is analogous to the activation energy decrease observed in the AB meoporous silica studies of Reference 287. This reduction in activation energy again indicates that the effect of the nanoconfinement is not simply the reduction in diffusion distances. Rather, the pathway for the dehydrogenation is being altered.

The effect of nanoconfinement on the capacity retention during cycling was also studied [294]. In bulk LiBH$_4$, the capacity retention is poor as the material is cycled, as less than 30% of the capacity remains after three cycles. In contrast, when confined within a nanoporous scaffold the retained capacity can be as high as 70% after three cycles.

Additional studies of LiBH$_4$ confinement were performed by Liu et al. [293] in a variety of carbon templates that included highly ordered cylindrical nanoporous carbon with narrow channels of 4-nm diameter, along with carbon aerogels with 9- to 15-nm mesopore diameter. LiBH$_4$ was introduced via the melt. Improvements in desorption temperature and the conditions of rehydrogenation were seen with LiBH$_4$ infiltrating the carbon nanoscaffolds, consistent with the results of Gross et al. [294]. Interestingly, the authors also saw a reduction in diborane (B$_2$H$_6$) emission as the pore size decreased, again pointing to a change in the decomposition pathway relative to bulk LiBH$_4$ [293].

A significant synthetic breakthrough in the nanoconfinement studies came from the organometallic synthesis methods developed by Zhang et al. [295]. The original syntheses of nanoconfined LiBH$_4$ required infusing the carbon aerogel with liquid (i.e., melted) LiBH$_4$. However, this method is not very versatile and can damage some nanoframeworks since elevated temperatures are used. Zhang and coworkers developed a LT homogenous organometallic approach to incorporation of Al- and Mg-based hydrides into 13-nm carbon aerogels that has resulted in high loadings without degradation of nanoporous scaffold [295]. The strategy is shown in Figure 6.40. Figure 6.40 shows how precursor MgBu$_2$ can be infused in the aerogel scaffold using heptane as a solvent. After solvent removal, the nanoconfined MgBu$_2$ can be thermally decomposed using hydrogen, which simultaneously hydrogenates the Mg to give scaffold-incorporated MgH$_2$.

Studies of this approach have shown that high (9–16 wt%) MgH$_2$ loadings can be achieved in the 13-nm carbon aerogel without host degradation. XRD and TEM studies confirmed that the organometallic approach produced MgH$_2$ particles confined in the carbon mesopores. Figure 6.41 shows that the rate of dehydrogenation from MgH$_2$ incorporated in 13-nm pores in C aerogel was, at 252°C, five times faster than the initial rate found for bulk ball-milled MgH$_2$.

The rate remained the same over four cycles of dehyrogenation-rehydrogenation, indicating the Mg was well confined in the pores even when thermally cycled. One might

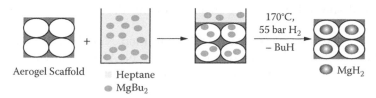

Aerogel Scaffold   Heptane   MgBu$_2$ 170°C, 55 bar H$_2$ – BuH   MgH$_2$

FIGURE 6.40

Organometallic approach to incorporation of metal hydrides into C aerogels. (Figure developed by C.M. Jensen based on S. Zhang, A.F. Gross, S.L. Van Atta, M. Lopez, P. Liu, C.C. Ahn, J.J. Vajo, C.M. Jensen, *Nanotechnology*, 20 (2009) 204027.)

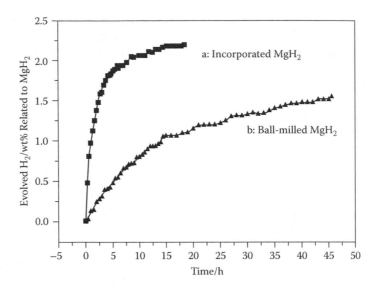

FIGURE 6.41

Isothermal dehydrogenation kinetic data at 252°C of (a) MgH$_2$ incorporated in carbon aerogel (13-nm pores) and (b) a control measurement of graphitic flakes ball milled with 16 wt% MgH$_2$. (Reproduced from S. Zhang, A.F. Gross, S.L. Van Atta, M. Lopez, P. Liu, C.C. Ahn, J.J. Vajo, C.M. Jensen, *Nanotechnology*, 20 (2009) 204027 with permission of IOP Publishing.)

wonder if the incorporation of MgH$_2$ into 13-nm pores altered the thermodynamics of the hydrogen release from MgH$_2$. It turns out that equilibrium studies of the plateau pressure of hydrogen at 250°C from MgH$_2$ confined to 13-nm pores agreed well with database plateau pressure for bulk MgH$_2$ to within 10% accuracy [298]. If there were a change in the thermodynamics, there would be a corresponding change in the equilibrium hydrogen pressure at a given temperature, which was not observed experimentally to within 10%. This lack of thermodynamic improvement was also confirmed by Nielsen et al. [299] in studies of MgH$_2$ confined to carbon aerogels of 7- and 22-nm pore sizes. Presumably, one needs to make the MgH$_2$ particle significantly smaller than 13-nm diameter to see a significant change in the thermodynamics of H$_2$ release, which is consistent with the theoretical predictions discussed.

In analogy with the work of Gutowska et al. on NH$_3$BH$_3$, Zheng and coworkers investigated the effect on hydrogen storage properties of infusing NaAlH$_4$ into ordered mesoporous silica (OMS) with about 10-nm pore diameter [300]. NaAlH$_4$, confined in the 10-nm channels, showed a somewhat lower desorption temperature (~175°C) compared to the desorption temperature (183°C) observed for a pristine sample of NaAlH$_4$. The nano-confined material showed a dramatic increase in the desorption kinetics, as shown in Figure 6.42.

The work of Stephens and coworkers [301] examined NaAlH$_4$ confinement in carbon aerogel. Stephens reported kinetic enhancement for NaAlH$_4$ confined to 13-nm carbon aerogel, as shown in Figure 6.43 [301]. From Figure 6.43, it is clear that the Ti-catalyzed sample of NaAlH$_4$ showed the best kinetic performance, as expected. However, it is also clear that the dehydrogenation kinetics of uncatalyzed NaAlH$_4$ confined in 13-nm aerogel was also dramatically superior to that of both the bulk and ball-milled (but uncatalyzed) samples of NaAlH$_4$. Although not as low in temperature as the Ti-catalyzed sample, the dehydrogenation onset of about 140°C for the nanoconfined NaAlH$_4$ clearly demonstrated

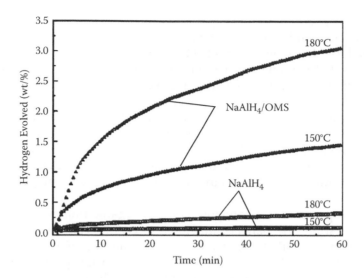

FIGURE 6.42
Dehydrogenation curves for pristine NaAlH$_4$ and NaAlH$_4$ confined in ordered mesoporous silica (OMS) of 10-nm pore diameter at 180°C and 150°C. (Reproduced from S.Y. Zheng, F. Fang, G.Y. Zhou, G.R. Chen, L.Z. Ouyang, M. Zhu, D.L. Sun, *Chemistry of Materials*, 20 (2008) 3954–3958.)

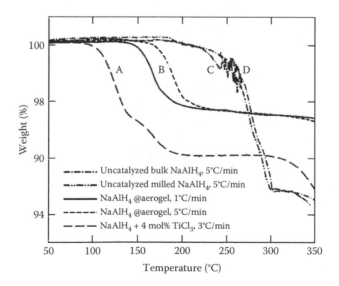

FIGURE 6.43
Thermogravimetric analysis (TGA) of four NaAlH$_4$ materials: (A) ball-milled NaAlH$_4$ catalyzed with 4 mol% TiCl$_3$; (B) uncatalyzed NaAlH$_4$ melt-infused into 13-nm aerogel; (C) uncatalyzed NaAlH$_4$ ball milled for 5 h; and (D) bulk uncatalyzed NaAlH$_4$. The heating rates are specified in the legend. (Reproduced from R.D. Stephens, A.F. Gross, S.L.V. Atta, J.J. Vajo, F.E. Pinkerton, *Nanotechnology*, 20 (2009) 204018 with permission of IOP Publishing.)

that incorporation of NaAlH$_4$ into nanoporous carbon aerogel substantially improved the desorption kinetics, even in the absence of a catalyst. Furthermore, the desorption temperature was reduced far below the onset of bulk melting of NaAlH$_4$ at about 250°C, which essentially eliminated foaming that would accompany hydrogen release from the bulk uncatalyzed samples. The results of Zheng and Stephens for nanoconfined NaAlH$_4$ were confirmed down to the 2- to 3-nm level by Gao et al. [302].

The effects of nanoconfinement have also been investigated for the LiBH$_4$/MgH$_2$ destabilized system [303–305]. Using direct melt infiltration of ball-milled LiBH$_4$/MgH$_2$, significant improvements in the kinetics were observed. Of particular note, dehydrogenation occurred in a single step, in contrast to the two-step dehydrogenation that occurs in bulk. Reviews of nanoconfinement can be found in the work of Jensen et al. [306], Fichtner [307], and Vajo [308].

Bhakta et al. proposed to use nanoporous metal organic frameworks (MOFs) as suitable templates to study the nanonconfinement effects of NaAlH$_4$ clusters and found remarkable changes in both kinetics and thermodynamics of hydrogen release compared to bulk [309, 310]. On infiltration of NaAlH$_4$ into the pores of Cu$_3$(btc)$_2$ (btc = 1,3,5–benzenetricarboxylate), also known as HKUST-1, the hydride undergoes a single-step dehydrogenation reaction in which the Na$_3$AlH$_6$ intermediate is not formed. Comparison of the thermodynamically controlled quasi-equilibrium reaction pathways showed that the nanoclusters were stabilized by confinement, having an H$_2$ desorption enthalpy that was 6 kJ/mol H$_2$ higher than the bulk material. The activation energy for desorption was only 53 kJ/mol H$_2$, more than 60 kJ/mol H$_2$ lower compared to bulk sodium alanate.

In addition to favorably altering the kinetics (and perhaps eventually the thermodynamics) of a confined hydride, use of a scaffold must be practical when considering the overall gravimetric and volumetric hydrogen capacities and the chemical and mechanical stabilities during cycling. The scaffold pore size, topology, and surface chemistry must be optimized for the hydride thermodynamics and kinetics; the specific pore volume must be optimized for the overall capacities; and the scaffold composition must be optimized for stability.

Ideally, the scaffold itself would also store hydrogen and contribute to the storage capacity. However, this is extremely challenging, and thus far, the scaffolds studied simply contributed extra weight to the system, thereby penalizing the system both gravimetrically and volumetrically. To be practical, these penalties must be reduced.

As analyzed by Vajo [308], the gravimetric capacity of a nanoconfined metal hydride including the mass of the scaffold relative to the bulk capacity [$C_{G,\text{scaffold}}/C_{G,\text{bulk}}$] depends on the hydride density [ρ_{hydride} (g/cm^3)] and the specific pore volume of the scaffold [PV (cm^3/g)]. This dependence, expressed as a percentage and denoted as the retained gravimetric capacity, is given by

$$C_{G,\text{scaffold}}/C_{G,\text{bulk}} = 100\% \cdot \rho_{\text{hydride}} \cdot PV/(\rho_{\text{hydride}} \cdot PV + 1)$$

A plot of the retained gravimetric capacity as a function of scaffold pore volume for different hydride materials is shown in Figure 6.44.

Overall, the retained capacity is increased for scaffolds with larger pore volume and denser hydride materials. For example, if a scaffold were used to confine LaNi$_5$H$_6$, which has a relatively high density of 6.4 g/cm^3, the retained capacity could be greater than 80% for a scaffold with a pore volume of about 1 cm^3/g. In contrast, for LiBH$_4$, which has one of the lowest densities of any hydride, a pore volume of about 4 cm^3/g is required for a retained capacity of 70%. The scaffolds used in the studies described had specific

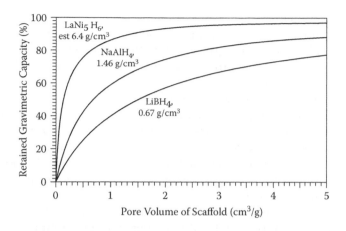

FIGURE 6.44
Retained gravimetric capacity for hydrides nanoconfined within scaffold hosts. Capacities are shown vs. scaffold pore volume for selected hydride densities. (Reproduced from J.J. Vajo, *Current Opinion in Solid State and Materials Science*, 15 (2011) 52–61 with permission of Elsevier.)

pore volumes of 0.5–1.5 cm^3/g. However, much larger specific pore volumes are possible [311]. For example, aerogels can be synthesized with pore volumes greater than 4 cm^3/g. The challenge will be combining a sufficient pore volume with the desired pore size in a structurally stable scaffold capable of withstanding multiple sorption cycles and infusing this material with a high-density metal hydride with high gravimetric hydrogen capacity.

Similarly, the retained volumetric capacity [$C_{V,\text{scaffold}}/C_{V,\text{bulk}}$] depends on the scaffold density [ρ_{scaffold} (g/cm^3)] and the pore volume as given by

$$C_{V,\text{scaffold}}/C_{V,\text{bulk}} = 100\% \cdot \rho_{\text{scaffold}} \cdot PV/(\rho_{\text{scaffold}} \cdot PV + 1).$$

A plot of the retained volumetric capacity for carbon-based scaffolds is given in Figure 6.45. From this plot, about 80% retained volumetric capacity can be achieved

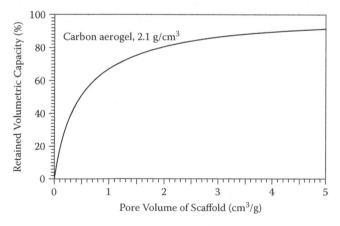

FIGURE 6.45
Retained volumetric capacity for hydrides nanoconfined within scaffold hosts. Capacities are shown vs. scaffold pore volume for a scaffold density of 2.1 g/cm^3, which is typical for porous carbon materials, including carbon aerogel. (Reproduced from J.J. Vajo, *Current Opinion in Solid State and Materials Science*, 15 (2011) 52–61 with permission of Elsevier.)

using carbon scaffolds with pore volumes of 1.5–2 cm^3/g. Silica-based scaffolds would be similar due to a similar density. Thus, for low-density hydride materials, retaining the volumetric capacity places a lower requirement on the scaffold pore volume than the gravimetric capacity. Overall, it appears that nanoconfined hydride materials with about 80% retained capacities are feasible.

R&D addressing the interactions with the confined metal hydride with the nanoporous host are only just beginning. Opalka et al. [312] experimentally investigated the reactive stability of $LiBH_4$ when mixed with inorganic aerogel materials ZrO_2, Al_2O_3, and SiO_2. These workers also used DFT calculations to model the metal hydride/nanopore thermodynamics to understand better the stability issues. ZrO_2 was found to be the least disruptive of the inorganic aerogels investigated [307] and would form the basis for future nanoconfinement experiments [313].

While the inorganic aerogels may be more chemically stable, the carbon aerogels are more widely available. For $LiBH_4$ confined in carbon aerogel, preliminary measurements [294] of CH_4 formation during dehydrogenation showed that CH_4 was detectable, but small (less than ~0.2 wt%). Further work is needed to understand the chemical stability of the scaffolds during repeated cycling.

Summarizing, nanoconfinement can improve metal hydride kinetics and reversibility and in some cases can suppress the release of undesired volatile species. However, the nanoscaffolds need to be made with larger pore volume if they are to form the basis of practical hydrogen storage materials. Thermodynamics can typically be modified for confinement in pores smaller than about 1-nm diameter.

Summary and Outlook

We have just described the significant experimental and theoretical advancements made in the area of alanates, borohydrides, amides, and the methods of destabilization and nanoconfinement that have been developed to improve these materials both thermodynamically and kinetically. Here, we summarize some thoughts about where future R&D needs to be directed to fully realize the hydrogen storage potential of these materials.

Destabilized Materials

The work discussed has shown that complex metal hydride systems can be destabilized by the addition of a second compound to provide lower thermodynamic thresholds for releasing hydrogen. It is clear from the studies that much additional fundamental work is needed on the mechanism of the destabilization and how the kinetics can be improved. The $LiBH_4/Mg_2NiH_4$ discussion showed that entirely new reaction pathways can emerge in these multicomponent metal hydride systems. The detailed mechanistic route by which destabilization is achieved needs to be examined in greater detail for existing promising destabilized systems and those yet to be discovered.

It is also clear from the studies that the destabilized systems can be kinetically limited, for example, in the destabilization of MgH_2 with Si. This kinetic limitation may be due to the fact that by introducing a second component into a metal hydride system, one is introducing a second material phase. In the "initial state" of the destabilized hydrogenated phase, two material phases exist, and these phase-phase interactions can determine how fast hydrogen is released. Certainly, improvements can be made on how intimately these phases are intermixed in the starting (hydrogenated) compound. On hydrogenation of the spent material, it seems logical that two separate material phases need to be generated to get back to the initial hydrogenated state. This phase separation must introduce kinetic barriers for the rates of hydrogenation. However, we do not understand these phase-induced kinetic limitations in any real way, and the thoughts given in this chapter are largely speculative.

Therefore, based on this situation, we recommend future fundamental studies be directed toward understanding the role of solid-solid interfaces on the kinetics of hydrogen desorption and absorption. In particular, more studies are needed to characterize the interface area and structure of multicomponent destabilized systems and how processing can affect them as well. In addition, experimental studies are needed that can correlate the structure and area of the interfaces between the two destabilized components with the kinetics that are actually observed. In other words, more study is needed to understand the role of the phase separation problem in the observed kinetics of hydrogen desorption and absorption.

Nanoconfinement

Our discussion has revealed that, in general, good things happen when one incorporates a metal hydride into a nanoporous material. For example, the kinetics of hydrogen release can be enhanced greatly ($LiBH_4$, MgH_2), the reversibility can be improved ($LiBH_4$), and in some cases precursor stability can be improved as well, as research on $NaTi(BH_4)_2 \cdot DME$ has shown [313]. In addition, foaming of materials can be eliminated ($LiBH_4$) if the kinetic improvement allows significant hydrogen release below a melting phase transition [131]. However, the kinetic improvement needs to be understood in detail.

The "going-in" assumption of why the kinetics might be improved is based on the general idea that diffusion times follow the relationship $t \sim l^2/D$, where t is the diffusion time, D is the diffusion coefficient, and l is the required diffusion distance. So, if nanoporous materials reduce the required diffusion distances, diffusion time should be reduced, and reaction rate should increase. This is only a very general concept, and in fact, the real atomistic mechanism for why the kinetics of the studied systems are being dramatically improved remains unknown.

If bulk diffusion is the rate-determining step for these hydrogen storage chemical reactions, then it follows that reducing materials to the nanoscale dimension would reduce bulk diffusion times as there is clearly less bulk material in existence. However, all sorts of other issues are created that can also affect kinetics, such as an increase in the number of particle-particle interfaces whose properties can affect kinetics, as well as increased contamination that can come from mechanically or chemically increasing the surface area. So, this situation is complicated. In a sense, the nanoscaffold materials have been developed as model systems to understand the prospects of nanoconfinement to enhance thermodynamics and kinetics. As has been pointed out, for these nanoconfined materials to serve as practical hydrogen storage materials, the gravimetric and volumetric penalties associated with the nanoscaffold need to be significantly reduced. Therefore, based on this situation, we recommend future fundamental studies be conducted on metal hydrides confined to nanoporous structures. In particular, (1) we recommend that fundamental studies be performed on understanding the nature and origin of fast-diffusing species presumed to exist in these nanoconfined systems. NMR will be a key analytical tool in these studies, as demonstrated recently for confined $LiBH_4$ [314]; (2) the nature of the scaffold-hydride interactions needs to be understood in much greater detail as these interactions are likely to drastically affect the stability and reversibility of the hydrogen storage system; (3) nanoscaffolds with about 5- to 10-nm pore dimensions, with greater than 3 cm^3/g pore volume, need to be developed to reduce the gravimetric penalty associated with using nanoconfinement. Additional fundamental synthetic and structural study needs to be performed on nanoscaffolds (both carbon-based and inorganic), with an emphasis on reducing the wall thicknesses between nanopockets.

Kinetics of Solid-State Reactions

Almost all of the complex anionic materials are severely kinetically limited. Our theoretical understanding of the kinetics of solid-state reactions in the hydrogen storage arena is almost nonexistent. Clearly, being able to address quantitatively the kinetics of these hydrogen desorption and absorption reactions requires detailed understanding of the mechanistic pathways, the intermediates involved, and understanding of the activation barriers that the reactions must overcome as well as the kinetic implications of the different morphologies (particle sizes) that these materials possess. The development of a robust theory of solid-state reaction kinetics is an extremely challenging problem and is worthy of the title "grand challenge."

Therefore, we recommend for the future that detailed theoretical methods be developed to predict and account for kinetics of model solid-phase hydrogen storage reactions. Furthermore, we recommend these theoretical studies be conducted in concert with experiments needed to validate the theory. In this way, some inroads (and therefore conceptual guidelines) can be developed to help guide our thinking about the kinetics of these materials and ultimately produce hydrogen storage materials with improved kinetics for hydrogen desorption and adsorption.

Effect of Additives on the Rates of Solid-State Reactions

It is well known that small mole percentages of additives can play a very important role in the kinetics of these reactions. The most famous of these is the discovery that about 2–4 mol% of Ti drastically improves the properties of $NaAlH_4$, particularly in hydrogenation of the spent material [5]. Over 10 years of effort have been invested by the community on trying to understand the role of this one elemental additive on the behavior of $NaAlH_4$, and the situation is still not completely understood.

In the discussion, we have seen examples of additives that are observed to play dramatic roles in the hydrogen storage reactions. For example, it was found that 4 mol% of KH increases the hydrogen absorption rate for the $2LiNH_2/MgH_2$ system by about two to three times. In general, we have little theoretical understanding of how these kinetic enhancements are occurring with such small (~2–4 mol%) levels of these additions. As a result, we have no guiding understanding to develop a strategy for choosing candidate catalysts for a particular hydrogen storage reaction. Given the importance of kinetics in these systems, it is very important to improve our understanding of how low-level addition of materials such as Ti, KH, and others can enhance the kinetics of both hydrogen desorption and absorption reactions. In particular, since 2–4 mol% additions do not dramatically cut the gravimetry storage capacity, understanding how kinetic enhancements can be achieved without an attendant gravimetric penalty is desirable.

Therefore, we recommend for the future that, although it will be a challenging task, detailed theoretical approaches need to be developed that will elucidate how specific catalytic agents are improving the kinetics of well-documented experimental hydrogen storage systems. This will require concerted theoretical and experimental work, which ultimately could then be used to predict new additives for enhancing the kinetic performance of the high-priority metal hydride storage systems.

Borohydrides

As discussed, many borohydride materials have been examined since 2005 as reversible hydrogen storage materials. One of their attractive features is the potential to offer high hydrogen gravimetric capacity, about 10–18 wt%. An interesting feature of the borohydrides is that their properties can vary widely. For example,

- $Mg(BH_4)_2$ releases H_2 at 295°C [315], yet $Na_2Zr(BH_4)$ starts to release H_2 at 40°C [13].
- $Li_2Zn(BH_4)_4$ releases H_2 with 30% B_2H_6 contamination [103], yet $Na_2Zr(BH_4)_6$ releases H_2 with no detectable B_2H_6 [13].
- $Ca(BH_4)_2 \bullet 2NH_3$ releases mainly NH_3 on heating [316], yet $Mg(BH_4)_2 \bullet 2NH_3$ only releases H_2 on heating [255].
- $Mg(BH_4)_2$ is about 90% reversible (with high T, P) [315], yet $Mn(BH_4)_2$ is not reversible at all [317].

We need to gain a fundamental understanding of these variations because if we can understand what controls these borohydride properties of reversibility, diborane release, ammonia release, and temperature for H_2 release, we should then have a reasonable chance for controlling these properties to produce a truly remarkable hydrogen storage material. Combined theoretical and experimental studies of borohydride stability and their tendency to release diborane are needed. Combined with these future borohydride studies would be more investigation of the role of $[B_{12}H_{12}]^{2-}$ species and how these enter into the reaction pathways for borohydride hydrogen desorption and absorption reactions.

References

1. G. Walker, *Solid-State Hydrogen Storage: Materials and Chemistry*, Woodhead Publishing Limited, Cambridge, England. 2008.
2. M. Hirscher, *Handbook of Hydrogen Storage*, 2010.
3. D.P. Broom, *Hydrogen Storage Materials: The Characterisation of Their Storage Properties*, 2011.
4. A.B. Züttel, L. Schlapbach, *Hydrogen as a Future Energy Carrier*, 2008.
5. B. Bogdanovic, M. Schwickardi, *Journal of Alloys and Compounds*, 253–254 (1997) 1–9.
6. M. Baricco, M. Palumbo, E. Pinatel, M. Corno, P. Ugliengo, *Advances in Science and Technology*, 72 (2010) 213–218.
7. L.H. Rude, T.K. Nielsen, D.B. Ravnsbaek, U. Bosenberg, M.B. Ley, B. Richter, L.M. Arnbjerg, M. Dornheim, Y. Filinchuk, F. Besenbacher, T.R. Jensen, *Physica Status Solidi a-Applications and Materials Science*, 208 (2011) 1754–1773.
8. C. Liang, Y.F. Liu, H.L. Fu, Y.F. Ding, M.X. Gao, H.G. Pan, *Journal of Alloys and Compounds*, 509 (2011) 7844–7853.
9. H.W. Li, Y.G. Yan, S. Orimo, A. Züttel, C.M. Jensen, *Energies*, 4 (2011) 185–214.
10. C. Li, P. Peng, D.W. Zhou, L. Wan, *International Journal of Hydrogen Energy*, 36 (2011) 14512–14526.
11. J. Graetz, *Chemical Society Reviews*, 38 (2009) 73–82.

12. B. Sakintuna, F. Lamari-Darkrim, M. Hirscher, *International Journal of Hydrogen Energy*, 32 (2007) 1121–1140.
13. S.I. Orimo, Y. Nakamori, J.R. Eliseo, A. Züttel, C.M. Jensen, *Chemical Reviews*, 107 (2007) 4111–4132.
14. W. Xu, Z.L. Tao, J. Chen, *Progress in Chemistry*, 18 (2006) 200–210.
15. A.J. Churchard, E. Banach, A. Borgschulte, R. Caputo, J.C. Chen, D. Clary, K.J. Fijalkowski, H. Geerlings, R.V. Genova, W. Grochala, T. Jaron, J.C. Juanes-Marcos, B. Kasemo, G.J. Kroes, I. Ljubic, N. Naujoks, J.K. Norskov, R.A. Olsen, F. Pendolino, A. Remhof, L. Romanszki, A. Tekin, T. Vegge, M. Zach, A. Züttel, *Physical Chemistry Chemical Physics*, 13 (2011) 16955–16972.
16. I.P. Jain, P. Jain, A. Jain, *Journal of Alloys and Compounds*, 503 (2010) 303–339.
17. G. Principi, F. Agresti, A. Maddalena, S. Lo Russo, *Energy*, 34 (2009) 2087–2091.
18. Q. Wang, Y.G. Chen, C.L. Wu, M.D. Tao, *Chinese Science Bulletin*, 53 (2008) 1784–1788.
19. J. Yang, A. Sudik, C. Wolverton, D.J. Siegel, *Chemical Society Reviews*, 39 (2010) 656–675.
20. S. McWhorter, C. Read, G. Ordaz, N. Stetson, *Current Opinion in Solid State and Materials Science*, 15 (2011) 29–38.
21. U. Eberle, *Oil Gas-European Magazine*, 37 (2011) 34–39.
22. W.I.F. David, *Faraday Discussions*, 151 (2011) 399–414.
23. T.K. Mandal, D.H. Gregory, *Proceedings of the Institution of Mechanical Engineers Part C—Journal of Mechanical Engineering Science*, 224 (2010) 539–558.
24. B. Bogdanovic, M. Felderhoff, G. Streukens, *Journal of the Serbian Chemical Society*, 74 (2009) 183–196.
25. P. Chen, M. Zhu, *Materials Today*, 11 (2008) 36–43.
26. J.J. Vajo, G.L. Olson, *Scripta Materialia*, 56 (2007) 829–834.
27. DOE targets for onboard hydrogen storage systems. http://www1.eere.energy.gov/hydrogenandfuelcells/storage/pdfs/targets_onboard_hydro_storage.pdf.
28. A.E. Finholt, A.C. Bond, H.I. Schlesinger, *Journal of the American Chemical Society*, 69 (1947) 1199–1203.
29. E.C. Ashby, G.J. Brendel, H.E. Redman, *Inorganic Chemistry*, 2 (1963) 499–504.
30. E. Wiberg, *Angewandte Chemie*, 65 (1953) 16–33.
31. M. Fichtner, O. Fuhr, O. Kircher, *Journal of Alloys and Compounds*, 356–357 (2003) 418–422.
32. H. Clasen, *Angewandte Chemie*, 73 (1961) 322–331.
33. T.N. Dymova, N.G. Eliseeva, S.I. Bakum, Y. Dergache, *Doklady Akademii Nauk Sssr*, 215 (1974) 1369–1372.
34. H. Morioka, K. Kakizaki, S.-C. Chung, A. Yamada, *Journal of Alloys and Compounds*, 353 (2003) 310–314.
35. T.N. Dymova, D.P. Aleksandrov, V.N. Konoplev, T.A. Silina, N.T. Kuznetsov, *Koordinatsionnaya Khimiya*, 19 (1993) 529–534.
36. P. Claudy, B. Bonnetot, J.-P. Bastide, L.T. Jean-Marie, *Materials Research Bulletin*, 17 (1982) 1499–1504.
37. J. Huot, S. Boily, V. Göether, R. Schulz, *Journal of Alloys and Compounds*, 283 (1999) 304–306.
38. E. Rönnebro, E.H. Majzoub, *Journal of Physical Chemistry B*, 110 (2006) 25686–25691.
39. B.M. Bulychev, K.N. Semenenko, K.B. Bitsoev, *Koordinatsionnaya Khimiya*, 4 (1978) 374–379.
40. M. Mamatha, C. Weidenthaler, A. Pommerin, M. Felderhoff, F. Schuth, *Journal of Alloys and Compounds*, 416 (2006) 303–314.
41. A. Mamatha, B. Bogdanovic, M. Felderhoff, A. Pommerin, W. Schmidt, F. Schuth, C. Weidenthaler, *Journal of Alloys and Compounds*, 407 (2006) 78–86.
42. N. Sklar, B. Post, *Inorganic Chemistry*, 6 (1967) 669–671.
43. B.C. Hauback, H.W. Brinks, H. Fjellvåg, *Journal of Alloys and Compounds*, 346 (2002) 184–189.
44. Z. Ma, M.Y. Chou, *Journal of Alloys and Compounds*, 479 (2009) 678–683.
45. B.C. Hauback, H.W. Brinks, C.M. Jensen, K. Murphy, A.J. Maeland, *Journal of Alloys and Compounds*, 358 (2003) 142–145.
46. A. Fossdal, H.W. Brinks, M. Fichtner, B.C. Hauback, *Journal of Alloys and Compounds*, 387 (2005) 47–51.

47. M. Fichtner, C. Frommen, O. Fuhr, *Inorganic Chemistry*, 44 (2005) 3479–3484.
48. C. Wolverton, V. Ozolins, *Physical Review B*, 75 (2007) 064101.
49. O.M. Løvvik, *Physical Review B*, 71 (2005) 144111.
50. H.W. Brinks, B.C. Hauback, C.M. Jensen, R. Zidan, *Journal of Alloys and Compounds*, 392 (2005) 27–30.
51. M.H. Sorby, H.W. Brinks, A. Fossdal, K. Thorshaug, B.C. Hauback, *Journal of Alloys and Compounds*, 415 (2006) 284–287.
52. O.M. Løvvik, O. Swang, *EPL (Europhysics Letters)*, 67 (2004) 607.
53. J. Graetz, Y. Lee, J.J. Reilly, S. Park, T. Vogt, *Physical Review B*, 71 (2005) 184115.
54. H. Grove, H.W. Brinks, R.H. Heyn, F.J. Wu, S.M. Opalka, X. Tang, B.L. Laube, B.C. Hauback, *Journal of Alloys and Compounds*, 455 (2008) 249–254.
55. W.D. Davis, L.S. Mason, G. Stegeman, *Journal of the American Chemical Society*, 71 (1949) 2775–2781.
56. O.M. Lovvik, S.M. Opalka, H.W. Brinks, B.C. Hauback, *Physical Review B*, 69 (2004).
57. T.N. Dymova, D.P. Aleksandrov, V.N. Konoplev, T.A. Silina, A.S. Sizareva, *Koordinatsionnaya Khimiya*, 20 (1994) 279–285.
58. W.E. Garner, E.W. Haycock, *Proceedings of the Royal Society of London Series A–Mathematical and Physical Sciences*, 211 (1952) 335–351.
59. R. Ehrlich, A.R. Young, G. Rice, J. Dvorak, P. Shapiro, H.F. Smith, *Journal of the American Chemical Society*, 88 (1966) 858–860.
60. V.I. Mikheeva, S.M. Arkhipov, *Russian Journal of Inorganic Chemistry*, 12 (1967) 1066–1071.
61. V.P. Balema, J.W. Wiench, K.W. Dennis, M. Pruski, V.K. Pecharsky, *Journal of Alloys and Compounds*, 329 (2001) 108–114.
62. V.P. Balema, V.K. Pecharsky, K.W. Dennis, *Journal of Alloys and Compounds*, 313 (2000) 69–74.
63. V.P. Balema, K.W. Dennis, V.K. Pecharsky, *Chemical Communications*, (2000) 1665–1666.
64. J.-W. Jang, J.-H. Shim, Y.W. Cho, B.-J. Lee, *Journal of Alloys and Compounds*, 420 (2006) 286–290.
65. R.A. Zidan, S. Takara, A.G. Hee, C.M. Jensen, *Journal of Alloys and Compounds*, 285 (1999) 119–122.
66. C.M. Jensen, R. Zidan, N. Mariels, A. Hee, C. Hagen, *International Journal of Hydrogen Energy*, 24 (1999) 461–465.
67. C.M. Jensen, K.J. Gross, *Applied Physics A. Materials Science and Processing*, 72 (2001) 213–219.
68. B. Bogdanović, R.A. Brand, A. Marjanović, M. Schwickardi, J. Tölle, *Journal of Alloys and Compounds*, 302 (2000) 36–58.
69. A. Zaluska, L. Zaluski, J.O. Ström-Olsen, *Journal of Alloys and Compounds*, 298 (2000) 125–134.
70. H.W. Brinks, C.M. Jensen, S.S. Srinivasan, B.C. Hauback, D. Blanchard, K. Murphy, *Journal of Alloys and Compounds*, 376 (2004) 215–221.
71. M.T. Kuba, S.S. Eaton, C. Morales, C.M. Jensen, *Journal of Materials Research*, 20 (2005) 3265–3269.
72. O. Palumbo, R. Cantelli, A. Paolone, C.M. Jensen, S.S. Srinivasan, *Journal of Physical Chemistry B*, 109 (2004) 1168–1173.
73. S. Gomes, G. Renaudin, H. Hagemann, K. Yvon, M.P. Sulic, C.M. Jensen, *Journal of Alloys and Compounds*, 390 (2005) 305–313.
74. J. Iniguez, T. Yildirim, *Journal of Physics-Condensed Matter*, 19 (2007).
75. J. Íñiguez, T. Yildirim, *Applied Physics Letters*, 86 (2005) 103109.
76. J. Íñiguez, T. Yildirim, T.J. Udovic, M. Sulic, C.M. Jensen, *Physical Review B*, 70 (2004) 060101.
77. S. Chaudhuri, J. Graetz, A. Ignatov, J.J. Reilly, J.T. Muckerman, *Journal of the American Chemical Society*, 128 (2006) 11404–11415.
78. S. Chaudhuri, J.T. Muckerman, *Journal of Physical Chemistry B*, 109 (2005) 6952–6957.
79. J.M. Bellosta von Colbe, W. Schmidt, M. Felderhoff, B. Bogdanović, F. Schüth, *Angewandte Chemie International Edition*, 45 (2006) 3663–3665.
80. J. Chen, N. Kuriyama, T. Kiyobayashi, H.T. Takeshita, H. Tanaka, N. Takeichi, T. Sakai, *Fundamentals of Advanced Materials for Energy Conversion*, 2002, 153–159.
81. J. Chen, N. Kuriyama, Q. Xu, H.T. Takeshita, T. Sakai, *Journal of Physical Chemistry B*, 105 (2001) 11214–11220.

82. V.P. Tarasov, S.I. Bakum, A.V. Novikov, *Russian Journal of Inorganic Chemistry*, 46 (2001) 409–414.
83. V.P. Tarasov, S.I. Bakum, A.V. Novikov, *Russian Journal of Inorganic Chemistry*, 45 (2000) 1890–1896.
84. M.E. Arroyo y de Dompablo, G. Ceder, *Journal of Alloys and Compounds*, 364 (2004).
85. J.R. Ares, K.F. Aguey-Zinsou, F. Leardini, I.J. Ferrer, J.F. Fernandez, Z.X. Guo, C. Sanchez, *Journal of Physical Chemistry C*, 113 (2009) 6845–6851.
86. M. Fichtner, O. Fuhr, *Journal of Alloys and Compounds*, 345 (2002) 286–296.
87. A. Fossdal, H.W. Brinks, M. Fichtner, B.C. Hauback, *Journal of Alloys and Compounds*, 404 (2005) 752–756.
88. R.A. Varin, C. Chiu, T. Czujko, Z. Wronski, *Journal of Alloys and Compounds*, 439 (2007) 302–311.
89. Y. Kim, E.K. Lee, J.H. Shim, Y.W. Cho, K.B. Yoon, *Journal of Alloys and Compounds*, 422 (2006) 283–287.
90. A. Klaveness, P. Vajeeston, P. Ravindran, H. Fjellvåg, A. Kjekshus, *Journal of Alloys and Compounds*, 433 (2007) 225–232.
91. V. Iosub, T. Matsunaga, K. Tange, M. Ishikiriyama, *International Journal of Hydrogen Energy*, 34 (2009) 906–912.
92. A. Züttel, P. Wenger, S. Rentsch, P. Sudan, P. Mauron, C. Emmenegger, *Journal of Power Sources*, 118 (2003) 1–7.
93. A. Züttel, S. Rentsch, P. Wenger, P. Sudan, P. Mauron, C. Emmenegger, Hydrogen storage materials: metals, carbon and complexes, 2003, *Processing and Fabricaton of Advanced Materials*, 407–412.
94. A. Züttel, S. Rentsch, P. Fischer, P. Wenger, P. Sudan, P. Mauron, C. Emmenegger, *Journal of Alloys and Compounds*, 356–357 (2003) 515–520.
95. H.I. Schlesinger, H.C. Brown, *Journal of the American Chemical Society*, 62 (1940) 3429–3435.
96. O. Friedrichs, F. Buchter, A. Borgschulte, A. Remhof, C.N. Zwicky, P. Mauron, M. Bielmann, A. Züttel, *Acta Materialia*, 56 (2008) 949–954.
97. D. Goerrig, in, Germany, 1958. Patent 10776644.
98. H.I. Schlesinger, H.C. Brown, B. Abraham, A.C. Bond, N. Davidson, A.E. Finholt, J.R. Gilbreath, H. Hoekstra, L. Horvitz, E.K. Hyde, J.J. Katz, J. Knight, R.A. Lad, D.L. Mayfield, L. Rapp, D.M. Ritter, A.M. Schwartz, I. Sheft, L.D. Tuck, A.O. Walker, *Journal of the American Chemical Society*, 75 (1953) 186–190.
99. P. Rittmeyer, U. Wietelmann, Hydrides, in *Ullmann's Encyclopedia of Industrial Chemistry*, Wiley-VCH, Berlin, 2000.
100. O. Friedrichs, A. Borgschulte, S. Kato, F. Buchter, R. Gremaud, A. Remhof, A. Zuettel, *Chemistry-a European Journal*, 15 (2009) 5531–5534.
101. N.N. Maltseva, N.S. Kedrova, V.I. Mikheeva, *Zhurnal Neorganicheskoi Khimii*, 18 (1973) 1989–1991.
102. S. Srinivasan, D. Escobar, M. Jurczyk, Y. Goswami, E. Stefanakos, *Journal of Alloys and Compounds*, 462 (2008) 294–302.
103. E. Jeon, Y. Cho, *Journal of Alloys and Compounds*, 422 (2006) 273–275.
104. P. Zanella, L. Crociani, N. Masciocchi, G. Giunchi, *Inorganic Chemistry*, 46 (2007) 9039–9041.
105. H. Hagemann, R. Cerny, *Dalton Transactions*, 39 (2010) 6006–6012.
106. P.M. Harris, E.P. Meibohm, *Journal of the American Chemical Society*, 69 (1947) 1231–1232.
107. J.P. Soulie, G. Renaudin, R. Cerny, K. Yvon, *Journal of Alloys and Compounds*, 346 (2002) 200–205.
108. A. Züttel, S. Rentsch, P. Fischer, P. Wenger, P. Sudan, P. Mauron, C. Emmenegger, *Journal of Alloys and Compounds*, 356 (2003) 515–520.
109. V. Dmitriev, Y. Filinchuk, D. Chernyshov, A.V. Talyzin, A. ilewski, O. Andersson, B. Sundqvist, A. Kurnosov, *Physical Review B*, 77 (2008) 174112.
110. K. Miwa, N. Ohba, S.-I. Towata, Y. Nakamori, S.-I. Orimo, *Physical Review B*, 69 (2004) 245120.
111. S.C. Abrahams, J. Kalnajs, *Journal of Chemical Physics*, 22 (1954) 434–436.
112. R.L. Davis, C.H.L. Kennard, *Journal of Solid State Chemistry*, 59 (1985) 393–396.
113. V.N. Konoplev, N.N. Maltseva, V.S. Khain, *Koordinatsionnaya Khimiya*, 18 (1992) 1143–1166.
114. P. Vajeeston, P. Ravindran, A. Kjekshus, H. Fjellvag, *Applied Physics Letters*, 89 (2006).
115. M.J.v. Setten, G.A.d. Wijs, M. Fichtner, G. Brocks, *Chemistry of Materials*, 20 (2008) 4952–4956.

116. B. Dai, D.S. Sholl, J.K. Johnson, *Journal of Physical Chemistry C*, 112 (2008) 4391–4395.
117. R. Cerny, Y. Filinchuk, H. Hagemann, K. Yvon, *Angewandte Chemie-International Edition*, 46 (2007) 5765–5767.
118. Y. Nakamori, K. Miwa, A. Ninomiya, H. Li, N. Ohba, S.I. Towata, A. Züttel, S.I. Orimo, *Physical Review B*, 74 (2006) 045126.
119. J.-H. Her, P.W. Stephens, Y. Gao, G.L. Soloveichik, J. Rijssenbeek, M. Andrus, J.-C. Zhao, *Acta Crystallographica Section B*, 63 (2007) 561–568.
120. Y. Filinchuk, R. Cerny, H. Hagemann, *Chemistry of Materials*, 21 (2009) 925–933.
121. V. Ozolins, E.H. Majzoub, C. Wolverton, *Physical Review Letters*, 100 (2008).
122. N.S. Kedrova, N.N. Maltseva, *Zhurnal Neorganicheskoi Khimii*, 22 (1977) 1791–1794.
123. M. Aoki, K. Miwa, T. Noritake, N. Ohba, M. Matsumoto, H.W. Li, Y. Nakamori, S. Towata, S. Orimo, *Applied Physics A. Materials Science and Processing*, 92 (2008) 601–605.
124. P. Vajeeston, P. Ravindran, H. Fjellvag, *Journal of Alloys and Compounds*, 446 (2007) 44–47.
125. E. Ronnebro, E.H. Majzoub, *Journal of Physical Chemistry B*, 111 (2007) 12045–12047.
126. K. Miwa, M. Aoki, T. Noritake, N. Ohba, Y. Nakamori, S. Towata, A. Züttel, S. Orimo, *Physical Review B*, 74 (2006).
127. M.D. Riktor, M.H. Sorby, K. Chlopek, M. Fichtner, F. Buchter, A. Zuettel, B.C. Hauback, *Journal of Materials Chemistry*, 17 (2007) 4939–4942.
128. E.M. Fedneva, V.L. Alpatova, V.I. Mikheeva, *Russian Journal of Inorganic Chemistry*, 9 (1964) 826–832.
129. D.S. Stasinevich, G.A. Egorenko, *Russian Journal of Inorganic Chemistry*, 13 (1968) 341–343.
130. S. Orima, Y. Nakamori, G. Kitahara, K. Miwa, N. Ohba, S. Towata, A. Züttel, *Journal of Alloys and Compounds*, 404 (2005) 427–430.
131. J.J. Vajo, S.L. Skeith, F. Mertens, *Journal of Physical Chemistry B*, 109 (2005) 3719–3722.
132. M. Au, A. Jurgensen, K. Zeigler, *Journal of Physical Chemistry B*, 110 (2006) 26482–26487.
133. M. Au, A. Jurgensen, Metal oxide modified lithium borohydrides for reversible hydrogen storage, in A. Dillion, C. Olk, C. Filiou, J. Ohi (Eds.), *Hydrogen Cycle-Generation, Storage and Fuel Cells*, Materials Research Society, Warrendale, PA, 2006, pp. 141–149.
134. M. Au, A. Jurgensen, *Journal of Physical Chemistry B*, 110 (2006) 7062–7067.
135. B.J. Zhang, B.H. Liu, Z.P. Li, *Journal of Alloys and Compounds*, 509 (2011) 751–757.
136. I. Saldan, *Central European Journal of Chemistry*, 9 (2011) 761–775.
137. F.C. Gennari, *International Journal of Hydrogen Energy*, 36 (2011) 15231–15238.
138. Y.H. Guo, X.B. Yu, L. Gao, G.L. Xia, Z.P. Guo, H.K. Liu, *Energy & Environmental Science*, 3 (2010) 465–470.
139. X.B. Yu, D.M. Grant, G.S. Walker, *Journal of Physical Chemistry C*, 113 (2009) 17945–17949.
140. D. Blanchard, Q. Shi, C.B. Boothroyd, T. Vegge, *Journal of Physical Chemistry C*, 113 (2009) 14059–14066.
141. N. Ohba, K. Miwa, M. Aoki, T. Noritake, S. Towata, Y. Nakamori, S. Orimo, A. Züttel, *Physical Review B*, 74 (2006).
142. S.I. Orimo, Y. Nakamori, N. Ohba, K. Miwa, M. Aoki, S. Towata, A. Züttel, *Applied Physics Letters*, 89 (2006).
143. S.-J. Hwang, R.C. Bowman, Jr., J.W. Reiter, J. Rijssenbeek, G.L. Soloveichik, J.-C. Zhao, H. Kabbour, C.C. Ahn, *Journal of Physical Chemistry C*, 112 (2008) 3164–3169.
144. V. Ozolins, E.H. Majzoub, C. Wolverton, *Journal of the American Chemical Society*, 131 (2009) 230–237.
145. H.W. Li, K. Miwa, N. Ohba, T. Fujita, T. Sato, Y. Yan, S. Towata, M.W. Chen, S. Orimo, *Nanotechnology*, 20 (2009) 204013.
146. L.L. Wang, D.D. Graham, I.M. Robertson, D.D. Johnson, *Journal of Physical Chemistry C*, 113 (2009) 20088–20096.
147. Z. Yongsheng, E. Majzoub, V. Ozolincedilscaron, C. Wolverton, *Physical Review B (Condensed Matter and Materials Physics)*, 82 (2010).

148. C.B. Minella, S. Garroni, D. Olid, F. Teixidor, C. Pistidda, I. Lindemann, O. Gutfleisch, M.D. Baro, R. Bormann, T. Klassen, M. Dornheim, *Journal of Physical Chemistry C*, 115 (2011) 18010–18014.

149. N. Verdal, W. Zhou, V. Stavila, J.-H. Her, M. Yousufuddin, T. Yildirim, T.J. Udovic, *Journal of Alloys and Compounds*, 509, Suppl. 2 (2011) S694–S697.

150. J.-H. Her, W. Zhou, V. Stavila, C.M. Brown, T.J. Udovic, *Journal of Physical Chemistry C*, 113 (2009) 11187–11189.

151. J.H. Her, M. Yousufuddin, W. Zhou, S.S. Jalisatgi, J.G. Kulleck, J.A. Zan, S.-J. Hwang, R.C. Bowman, T.J. Udovic, *Inorganic Chemistry*, 47 (2008) 9757–9759.

152. V. Stavila, J.H. Her, W. Zhou, S.J. Hwang, C. Kim, L.A.M. Ottley, T.J. Udovic, *Journal of Solid State Chemistry*, 183 (2010) 1133–1140.

153. P. Ngene, R. van den Berg, M.H.W. Verkuijlen, K.P. de Jong, P.E. de Jongh, *Energy & Environmental Science*, 4 (2011) 4108–4115.

154. V.N. Konoplev, *Zhurnal Neorganicheskoi Khimii*, 25 (1980) 1737–1740.

155. G.L. Soloveichik, Y. Gao, J. Rijssenbeek, M. Andrus, S. Kniajanski, R.C. Bowman, Jr., S.-J. Hwan, J.-C. Zhao, *International Journal of Hydrogen Energy*, 34 (2009) 916–928.

156. G.L. Soloveichik, M. Andrus, Y. Gao, J.C. Zhao, S. Kniajanski, *International Journal of Hydrogen Energy*, 34 (2009) 2144–2152.

157. H.W. Li, K. Kikuchi, Y. Nakamori, N. Ohba, K. Miwa, S. Towata, S. Orimo, *Acta Materialia*, 56 (2008) 1342–1347.

158. E. Roennebro, *Current Opinion in Solid State and Materials Science*, 15 (2011) 44–51.

159. G. Severa, E. Rönnebro, C.M. Jensen, *Chemical Communications*, 46 (2010) 421–423.

160. D.T. Shane, L.H. Rayhel, Z.G. Huang, J.C. Zhao, X. Tang, V. Stavila, M.S. Conradi, *Journal of Physical Chemistry C*, 115 (2011) 3172–3177.

161. R.J. Newhouse, V. Stavila, S.J. Hwang, L.E. Klebanoff, J.Z. Zhang, *Journal of Physical Chemistry C*, 114 (2010) 5224–5232.

162. J.H. Kim, S.A. Jin, J.H. Shim, Y.W. Cho, *Journal of Alloys and Compounds*, 461 (2008) L20-L22.

163. J.H. Kim, J.H. Shim, Y.W. Cho, in *1st Polish Forum on Fuel Cells and Hydrogen*, Zakopane, Poland, 2007, pp. 140–143.

164. J.H. Kim, J.H. Shim, Y.W. Cho, *Journal of Power Sources*, 181 (2008) 140–143.

165. I.B. Sivaev, V.I. Bregadze, S. Sjoberg, *Collection of Czechoslovak Chemical Communications*, 67 (2002) 679–727.

166. Y. Filinchuk, D. Chernyshov, V. Dmitriev, *Zeitschrift Fur Kristallographie*, 223 (2008) 649–659.

167. Y. Nakamori, H.W. Li, K. Kikuchi, M. Aoki, K. Miwa, S. Towata, S. Orimo, *Journal of Alloys and Compounds*, 446 (2007) 296–300.

168. Y. Nakamori, H.W. Li, M. Matsuo, K. Miwa, S. Towata, S. Orimo, *Journal of Physics and Chemistry of Solids*, 69 (2008) 2292–2296.

169. A. Züttel, A. Borgschulte, S.I. Orimo, *Scripta Materialia*, 56 (2007) 823–828.

170. S. Srinivasan, D. Escobar, Y. Goswami, E. Stefanakos, *International Journal of Hydrogen Energy*, 33 (2008) 2268–2272.

171. S.I. Orimo, Y. Nakamori, H.W. Li, M. Matsuo, T. Sato, N. Ohba, K. Miwa, S.I. Towata, Single- and double-cations borohydrides for hydrogen storage applications, 2009. *Materials Issues in a Hydrogen Economy*, 124–129.

172. Y. Nakamori, K. Miwa, A. Ninomiya, H.W. Li, N. Ohba, S.I. Towata, A. Züttel, S.I. Orimo, *Physical Review B*, 74 (2006).

173. Z.Z. Fang, X.D. Kang, J.H. Luo, P. Wang, H.W. Li, S. Orimo, *Journal of Physical Chemistry C*, 114 (2010) 22736–22741.

174. C. Kim, S.-J. Hwang, R.C. Bowman, Jr., J.W. Reiter, J.A. Zan, J.G. Kulleck, H. Kabbour, E.H. Majzoub, V. Ozolins, *Journal of Physical Chemistry C*, 113 (2009) 9956–9968.

175. H. Hagemann, M. Longhini, J.W. Kaminski, T.A. Wesolowski, R. Cerny, N. Penin, M.H. Sorby, B.C. Hauback, G. Severa, C.M. Jensen, *Journal of Physical Chemistry A*, 112 (2008) 7551–7555.

176. R. Cerny, G. Severa, D.B. Ravnsbaek, Y. Filinchuk, V. D'Anna, H. Hagemann, D. Haase, C.M. Jensen, T.R. Jensen, *Journal of Physical Chemistry C*, 114 (2010) 1357–1364.

177. R. Cerny, D.B. Ravnsbaek, G. Severa, Y. Filinchuk, V. D'Anna, H. Hagemann, D. Haase, J. Skibsted, C.M. Jensen, T.R. Jensen, *Journal of Physical Chemistry C*, 114 (2010) 19540–19549.
178. D.S. Aidhy, C. Wolverton, *Physical Review B*, 83 (2011).
179. R. Cerny, K.C. Kim, N. Penin, V. D'Anna, H. Hagemann, D.S. Sholl, *Journal of Physical Chemistry C*, 114 (2010) 19127–19133.
180. D.B. Ravnsbaek, C. Frommen, D. Reed, Y. Filinchuk, M. Sorby, B.C. Hauback, H.J. Jakobsen, D. Book, F. Besenbacher, J. Skibsted, T.R. Jensen, *Journal of Alloys and Compounds*, 509 (2011) S698–S704.
181. I. Lindemann, R. Domenech-Ferrer, L. Dunsch, Y. Filinchuk, R. Cerny, H. Hagemann, V. D'Anna, L.M.L. Daku, L. Schultz, O. Gutfleisch, *Chemistry-a European Journal*, 16 (2010) 8707–8712.
182. R. Cerny, N. Penin, V. D'Anna, H. Hagemann, E. Durand, J. Ruzicka, *Acta Materialia*, 59 (2011) 5171–5180.
183. D.B. Ravnsbaek, L.H. Sorensen, Y. Filinchuk, D. Reed, D. Book, H.J. Jakobsen, F. Besenbacher, J. Skibsted, T.R. Jensen, *European Journal of Inorganic Chemistry*, (2010) 1608–1612.
184. R. Ccaronernyacute, G. Severa, D.B. Ravnsbaeligk, Y. Filinchuk, V. D'Anna, H. Hagemann, D. Haase, C.M. Jensen, T.R. Jensen, *Journal of Physical Chemistry C*, 114 (2010).
185. T.J. Frankcombe, G.J. Kroes, A. Züttel, *Chemical Physics Letters*, 405 (2005) 73–78.
186. H.W. Li, S. Orimo, Y. Nakamori, K. Miwa, N. Ohba, S. Towata, A. Züttel, *Journal of Alloys and Compounds*, 446 (2007) 315–318.
187. F.W. Dafert, R. Miklauz, *Monatch. Chem*, 30 (1909) 649.
188. D.H. Gregory, P.M. O'Meara, A.G. Gordon, J.P. Hodges, S. Short, J.D. Jorgensen, *Chemistry of Materials*, 14 (2002) 2063–2070.
189. A. Rabenau, *Solid State Ionics*, 6 (1982) 277–293.
190. U.V. Alpen, A. Rabenau, G.H. Talat, *Applied Physics Letters*, 30 (1977) 621–623.
191. M. Matsuo, S. Orimo, *Advanced Energy Materials*, 1 (2011) 161–172.
192. O. Palumbo, A. Paolone, R. Cantelli, D. Chandra, *International Journal of Hydrogen Energy*, 33 (2008) 3107–3110.
193. Y. Nakamori, S. Orimo, *Journal of Alloys and Compounds*, 370 (2004) 271–275.
194. P. Chen, Z. Xiong, J. Luo, J. Lin, K.L. Tan, *Nature*, 420 (2002) 302–304.
195. S. Cui, W. Feng, H. Hu, Z. Feng, Y. Weng, *Solid State Communication*, 149 612 (2009).
196. C. Radley, Altergy Systems, private communication, June 10, 2012.
197. G.P. Meisner, F.E. Pinkerton, M.S. Meyer, M.P. Balogh, M.D. Kundrat, *Journal of Alloys and Compounds*, 404 (2005) 24–26.
198. T. Ichikawa, S. Isobe, N. Hanada, H. Fujii, *Journal of Alloys and Compounds*, 365 (2004) 271–276.
199. P. Chen, Z.T. Xiong, J.Z. Luo, J.Y. Lin, K.L. Tan, *Journal of Physical Chemistry B*, 107 (2003) 10967–10970.
200. J. Lamb, D. Chandra, W.M. Chien, D. Phanon, N. Penin, R. Cerny, K. Yvon, *Journal of Physical Chemistry C*, 115 (2011) 14386–14391.
201. W.F. Luo, S. Sickafoose, *Journal of Alloys and Compounds*, 407 (2006) 274–281.
202. H.Y. Leng, T. Ichikawa, S. Hino, T. Nakagawa, H. Fujii, *Journal of Physical Chemistry B*, 109 (2005) 10744–10748.
203. T. Ichikawa, N. Hanada, S. Isobe, H.Y. Leng, H. Fujii, *Journal of Alloys and Compounds*, 404 (2005) 435–438.
204. Z.T. Xiong, G.T. Wu, H.J. Hu, P. Chen, *Advanced Materials*, 16 (2004) 1522–1524.
205. Y. Nakamori, S. Orimo, *Materials Science and Engineering B*, 108 (2004) 48–50.
206. W.F. Luo, *Journal of Alloys and Compounds*, 381 (2004) 284–287.
207. T. Ichikawa, N. Hanada, S. Isobe, H.Y. Leng, H. Fujii, *Journal of Physical Chemistry B*, 108 (2004) 7887–7892.
208. Y. Nakamori, G. Kitahara, K. Miwa, N. Ohba, T. Noritake, S. Towata, S. Orimo, *Journal of Alloys and Compounds*, 404 (2005) 396–398.
209. S. Orimo, Y. Nakamori, G. Kitahara, K. Miwa, N. Ohba, T. Noritake, S. Towata, *Applied Physics A*, 79 (2004) 1765–1767.

210. W. Luo, V. Stavila, L.E. Klebanoff, *International Journal of Hydrogen Energy*, 37 (2012) 6646–6652.
211. J. Rijssenbeek, Y. Gao, J. Hanson, Q. Huang, C. Jones, B. Toby, *Journal of Alloys and Compounds*, 454 (2008) 233–244.
212. J. Hu, Y. Liu, G. Wu, Z. Xiong, P. Chen, *Journal of Physical Chemistry C*, 111 (2007) 18439–18443.
213. M. Aoki, T. Noritake, G. Kitahara, Y. Nakamori, S. Towata, S. Orimo, *Journal of Alloys and Compounds*, 428 (2007) 307–311.
214. L.E. Klebanoff, Metal Hydride Center of Excellence, presentation at the 2006 DOE Hydrogen Program Annual Merit Review, p. 17. 2006. http://www.hydrogen.energy.gov/annual_review06_storage.html#metal.
215. Z.T. Xiong, J.J. Hu, G.T. Wu, P. Chen, W.F. Luo, K. Gross, J. Wang, *Journal of Alloys and Compounds*, 398 (2005) 235–239.
216. Private communication from W. Luo to L.E. Klebanoff, June 10, 2006.
217. J. Wang, T. Liu, G. Wu, W. Li, Y. Liu, C.M. Araujo, R.H. Scheicher, A. Blomqvist, R. Ahuja, Z. Xiong, P. Yang, M. Gao, H. Pan, P. Chen, *Angewandte Chemie-International Edition*, 48 (2009) 5828–5832.
218. J. Lu, Z.Z.G. Fang, Y.J. Choi, H.Y. Sohn, *Journal of Physical Chemistry C*, 111 (2007) 12129–12134.
219. S.V. Alapati, J.K. Johnson, D.S. Sholl, *Journal of Physical Chemistry B*, 110 (2006) 8769–8776.
220. A.R. Akbarzadeh, V. Ozolins, C. Wolverton, *Advanced Materials*, 19 (2007) 3233–3239.
221. J. Lu, Y.J. Choi, Z.Z. Fang, H.Y. Sohn, *Journal of Power Sources*, 195 (2010) 1992–1997.
222. J.J. Hu, E. Rohm, M. Fichtner, *Acta Materialia*, 59 (2011) 5821–5831.
223. D.L. Anton, C.J. Price, J. Gray, *Energies*, 4 (2011) 826–844.
224. Y. Liu, K. Zhong, M. Gao, J. Wang, H. Pan, Q. Wang, *Chemistry of Materials*, 20 (2008) 3521–3527.
225. W. Osborn, T. Markmaitree, L.L. Shaw, *Journal of Power Sources*, 172 (2007) 376–378.
226. M. Aoki, K. Miwa, T. Noritake, G. Kitahara, Y. Nakamori, S. Orimo, S. Towata, *Applied Physics A: Materials Science & Processing*, 80 (2005) 1409–1412.
227. F.E. Pinkerton, G.P. Meisner, M.S. Meyer, M.P. Balogh, M.D. Kundrat, *Journal of Physical Chemistry B*, 109 (2005) 6–8.
228. P.A. Chater, W.I.F. David, S.R. Johnson, P.P. Edwards, P.A. Anderson, *Chemical Communications*, (2006) 2439–2441.
229. Y.E. Filinchuk, K. Yvon, G.P. Meisner, F.E. Pinkerton, M.P. Balogh, *Inorganic Chemistry*, 45 (2006) 1433–1435.
230. H. Wu, W. Zhou, T.J. Udovic, J.J. Rush, T. Yildirim, *Chemistry of Materials*, 20 (2008) 1245–1247.
231. P.A. Chater, W.I.F. David, P.A. Anderson, *Chemical Communications*, (2007) 4770–4772.
232. G.P. Meisner, M.L. Scullin, M.P. Balogh, F.E. Pinkerton, M.S. Meyer, *Journal of Physical Chemistry B*, 110 (2006) 4186–4192.
233. J. Yang, A. Sudik, D.J. Siegel, D. Halliday, A. Drews, R.O. Carter, III, C. Wolverton, G.J. Lewis, J.W.A. Sachtler, J.J. Low, S.A. Faheem, D.A. Lesch, V. Ozolins, *Angewandte Chemie-International Edition*, 47 (2008) 882–887.
234. S.S. Srinivasan, M.U. Niemann, J.R. Hattrick-Simpers, K. McGrath, P.C. Sharma, D.Y. Goswami, E.K. Stefanakos, *International Journal of Hydrogen Energy*, 35 (2010) 9646–9652.
235. M.U. Niemann, S.S. Srinivasan, A. Kumar, E.K. Stefanakos, D.Y. Goswami, K. McGrath, Processing analysis of the ternary LiNH(2)-MgH(2)-LiBH(4) system for hydrogen storage, 2010 Imece, Vol. 6, p. 35–39.
236. J.R. Hattrick-Simpers, J.E. Maslar, M.U. Niemann, C. Chiu, S.S. Srinivasan, E.K. Stefanakos, L.A. Bendersky, *International Journal of Hydrogen Energy*, 35 (2010) 6323–6331.
237. M.U. Niemann, S.S. Srinivasan, A. Kumar, E.K. Stefanakos, D.Y. Goswami, K. McGrath, *International Journal of Hydrogen Energy*, 34 (2009) 8086–8093.
238. G.J. Lewis, J.W.A. Sachtler, J.J. Low, D.A. Lesch, S.A. Faheem, P.M. Dosek, L.M. Knight, L. Halloran, C.M. Jensen, J. Yang, A. Sudik, D.J. Siegel, C. Wolverton, V. Ozolins, S. Zhang, *Journal of Alloys and Compounds*, 446 (2007) 355–359.
239. A. Sudik, J. Yang, D. Halliday, C. Wolverton, *Journal of Physical Chemistry C*, 112 (2008) 4384–4390.

240. J. Yang, A. Sudik, D.J. Siegel, D. Halliday, A. Drews, R.O. Carter, III, C. Wolverton, G.J. Lewis, J.W.A. Sachtler, J.J. Low, S.A. Faheem, D.A. Lesch, V. Ozolins, *Journal of Alloys and Compounds*, 446 (2007) 345–349.
241. Y. Zhang, Q.F. Tian, *International Journal of Hydrogen Energy*, 36 (2011) 9733–9742.
242. X.Y. Chen, Y.H. Guo, X.B. Yu, *Journal of Physical Chemistry C*, 114 (2010) 17947–17953.
243. H.L. Chu, Z.T. Xiong, G.T. Wu, J.P. Guo, X.L. Zheng, T. He, C.Z. Wu, P. Chen, *Chemistry—An Asian Journal*, 5 (2010) 1594–1599.
244. H.L. Chu, Z.T. Xiong, G.T. Wu, J.P. Guo, T. He, P. Chen, *Dalton Transactions*, 39 (2010) 10585–10587.
245. H. Chu, G. Wu, Y. Zhang, Z. Xiong, J. Guo, T. He, P. Chen, *Journal of Physical Chemistry C*, 115 (2011) 18035–18041.
246. X. Zheng, Z. Xiong, Y. Lim, G. Wu, P. Chen, H. Chen, *Journal of Physical Chemistry C*, 115 (2011) 8840–8844.
247. J. Lu, Z.Z. Fang, *Journal of Physical Chemistry B*, 109 (2005) 20830–20834.
248. M.U.D. Naik, S.U. Rather, C.S. So, S.W. Hwang, A.R. Kim, K.S. Nahm, *International Journal of Hydrogen Energy*, 34 (2009) 8937–8943.
249. Y. Nakamori, A. Ninomiya, G. Kitahara, M. Aoki, T. Noritake, K. Miwa, Y. Kojima, S. Orimo, *Journal of Power Sources*, 155 (2006) 447–455.
250. Z.T. Xiong, G.T. Wu, J.J. Hu, P. Chen, *Journal of Power Sources*, 159 (2006) 167–170.
251. Y. Kojima, M. Matsumoto, Y. Kawai, T. Haga, N. Ohba, K. Miwa, S.I. Towata, Y. Nakamori, S. Orimo, *Journal of Physical Chemistry B*, 110 (2006) 9632–9636.
252. J.F. Mao, Z.P. Guo, H.K. Liu, X.B. Yu, *Journal of Alloys and Compounds*, 487 (2009) 434–438.
253. S.A. Jin, J.H. Shim, Y.W. Cho, K.W. Yi, O. Zabara, M. Fichtner, *Scripta Materialia*, 58 (2008) 963–965.
254. D.B. Ravnsbaek, T.R. Jensen, *Journal of Physics and Chemistry of Solids*, 71 (2010) 1144–1149.
255. G. Soloveichik, J.H. Her, P.W. Stephens, Y. Gao, J. Rijssenbeek, M. Andrus, J.C. Zhao, *Inorganic Chemistry*, 47 (2008) 4290–4298.
256. S.R. Johnson, W.I.F. David, D.M. Royse, M. Sommariva, C.Y. Tang, F.P.A. Fabbiani, M.O. Jones, P.P. Edwards, *Chemistry—An Asian Journal*, 4 (2009) 849–854.
257. H. Chu, G. Wu, Z. Xiong, J. Guo, T. He, P. Chen, *Chemistry of Materials*, 22 (2010) 6021–6028.
258. X.L. Zheng, G.T. Wu, W. Li, Z.T. Xiong, T. He, J.P. Guo, H. Chen, P. Chen, *Energy & Environmental Science*, 4 (2011) 3593–3600.
259. J.J. Reilly, R.H. Wiswall, *Inorganic Chemistry*, 6 (1967) 2220.
260. J.J. Reilly, R.H. Wiswall, *Inorganic Chemistry*, 7 (1968) 2254.
261. A. Zaluska, L. Zaluski, J.O. Strom-Olsen, *Applied Physics A. Materials Science and Processing*, 72 (2001) 157–165.
262. J.J. Vajo, T.T. Salguero, A.E. Gross, S.L. Skeith, G.L. Olson, *Journal of Alloys and Compounds*, 446 (2007) 409–414.
263. J.J. Vajo, F. Mertens, C.C. Ahn, R.C. Bowman, Jr., B. Fultz, *Journal of Physical Chemistry B*, 108 (2004) 13977–13983.
264. P. Mauron, F. Buchter, O. Friedrichs, A. Remhof, M. Bielmann, C.N. Zwicky, A. Züttel, *Journal of Physical Chemistry B*, 112 (2008) 906–910.
265. S. Orimo, H. Fujii, *Applied Physics A. Materials Science and Processing*, 72 (2001) 167–186.
266. J.J. Vajo, W. Li, P. Liu, *Chemical Communications*, 46 (2010) 6687–6689.
267. W. Li, J.J. Vajo, R.W. Cumberland, P. Liu, S.J. Hwang, C. Kim, R.C. Bowman, Jr., *Journal of Physical Chemistry Letters*, 1 (2010) 69–72.
268. V. Ozolins, E.H. Majzoub, C. Wolverton, *Journal of the American Chemical Society*, 131 (2009) 230–237.
269. C. Wolverton, V. Ozolins, M. Asta, *Physical Review B*, 69 (2004).
270. L.G. Hector, J.F. Herbst, T.W. Capehart, *Journal of Alloys and Compounds*, 353 (2003) 74–85.
271. K. Miwa, A. Fukumoto, *Physical Review B*, 65 (2002).
272. K.M. Ho, H.J. Tao, X.Y. Zhu, *Physical Review Letters*, 53 (1984) 1586–1589.

273. S.V. Alapati, J.K. Johnson, D.S. Sholl, *Journal of Physical Chemistry C*, 112 (2008) 5258–5262.

274. S.V. Alapati, J.K. Johnson, D.S. Sholl, *Journal of Alloys and Compounds*, 446 (2007) 23–27.

275. S.V. Alapati, J.K. Johnson, D.S. Sholl, *Physical Review B*, 76 (2007) 104108.

276. S.V. Alapati, J.K. Johnson, D.S. Sholl, *Journal of Physical Chemistry C*, 111 (2007) 1584–1591.

277. K.C. Kim, A.D. Kulkami, J.K. Johnson, D.S. Sholl, *Physical Chemistry Chemical Physics*, 13 (2011) 7218–7229.

278. S.V. Alapati, J.K. Johnson, D.S. Sholl, *Physical Chemistry Chemical Physics*, 9 (2007) 1438–1452.

279. D.J. Siegel, C. Wolverton, V. Ozolins, *Physical Review B*, 76 (2007) 134102.

280. F.E. Pinkerton, M.S. Meyer, *Journal of Alloys and Compounds*, 464 (2008) L1–L4.

281. E.H. Majzoub, V. Ozolins, *Physical Review B*, 77 (2008).

282. V. Ozolins, A.R. Akbarzadeh, H. Gunaydin, K. Michel, C. Wolverton, E.H. Majzoub, First-principles computational discovery of materials for hydrogen storage, in H. Simon (Ed.) *Scidac 2009: Scientific Discovery through Advanced Computing*, 2009. Scidac, Vol. 180, p. 76.

283. N.A. Zarkevich, D.D. Johnson, *Physical Review Letters*, 100 (2008) 040602.

284. J. Huot, G. Liang, S. Boily, A. Van Neste, R. Schulz, *Journal of Alloys and Compounds*, 293–295 (1999) 495–500.

285. P.A. Huhn, M. Dornheim, T. Klassen, R. Bormann, *Journal of Alloys and Compounds*, 404–406 (2005) 499–502.

286. Z. Dehouche, T. Klassen, W. Oelerich, J. Goyette, T.K. Bose, R. Schulz, *Journal of Alloys and Compounds*, 347 (2002) 319–323.

287. A. Gutowska, L. Li, Y. Shin, C.M. Wang, X.S. Li, J.C. Linehan, R.S. Smith, B.D. Kay, B. Schmid, W. Shaw, M. Gutowski, T. Autrey, *Angewandte Chemie International Edition*, 44 (2005) 3578–3582.

288. J.J. Liang, *Applied Physics A: Materials Science & Processing*, 80 (2005) 173–178.

289. S. Cheung, W.-Q. Deng, A.C.T. van Duin, W.A. Goddard, *Journal of Physical Chemistry A*, 109 (2005) 851–859.

290. R.W.P. Wagemans, J.H. van Lenthe, P.E. de Jongh, A.J. van Dillen, K.P. de Jong, *Journal of the American Chemical Society*, 127 (2005) 16675–16680.

291. Z. Wu, M.D. Allendorf, J.C. Grossman, *Journal of the American Chemical Society*, 131 (2009) 13918–13919.

292. K.C. Kim, B. Dai, J.K. Johnson, D.S. Sholl, *Nanotechnology*, 20 (2009) 204001.

293. X.F. Liu, D. Peaslee, C.Z. Jost, T.F. Baumann, E.H. Majzoub, *Chemistry of Materials*, 23 (2011) 1331–1336.

294. A.F. Gross, J.J. Vajo, S.L. Van Atta, G.L. Olson, *Journal of Physical Chemistry C*, 112 (2008) 5651–5657.

295. S. Zhang, A.F. Gross, S.L. Van Atta, M. Lopez, P. Liu, C.C. Ahn, J.J. Vajo, C.M. Jensen, *Nanotechnology*, 20 (2009) 204027.

296. N. Hüsing, U. Schubert, *Angewandte Chemie International Edition*, 37 (1998) 22–45.

297. S. Gavalda, K.E. Gubbins, Y. Hanzawa, K. Kaneko, K.T. Thomson, *Langmuir*, 18 (2002) 2141–2151.

298. C. Jensen, private communication to L.E. Klebanoff May 28, 2007.

299. T.K. Nielsen, K. Manickam, M. Hirscher, F. Besenbacher, T.R. Jensen, *ACS Nano*, 3 (2009) 3521–3528.

300. S.Y. Zheng, F. Fang, G.Y. Zhou, G.R. Chen, L.Z. Ouyang, M. Zhu, D.L. Sun, *Chemistry of Materials*, 20 (2008) 3954–3958.

301. R.D. Stephens, A.F. Gross, S.L.V. Atta, J.J. Vajo, F.E. Pinkerton, *Nanotechnology*, 20 (2009) 204018.

302. J.B. Gao, P. Adelhelm, M.H.W. Verkuijlen, C. Rongeat, M. Herrich, P.J.M. van Bentum, O. Gutfleisch, A.P.M. Kentgens, K.P. de Jong, P.E. de Jongh, *Journal of Physical Chemistry C*, 114 (2010) 4675–4682.

303. R. Gosalawit-Utke, T.K. Niesen, K. Pranzas, I. Saldan, C. Pistidda, F. Karimi, D. Laipple, J. Skibsted, T.R. Jensen, T. Klassen, M. Dornheim, *Journal of Physical Chemistry C*, 116 (2012) 1526–1534.

304. R. Gosalawit-Utke, T.K. Nielsen, I. Saldan, D. Laipple, Y. Cerenius, T.R. Jensen, T. Klassen, M. Dornheim, *Journal of Physical Chemistry C*, 115 (2011) 10903–10910.

305. T.K. Nielsen, U. Bösenberg, R. Gosalawit, M. Dornheim, Y. Cerenius, F. Besenbacher, T.R. Jensen, *ACS Nano*, 4 (2010) 3903–3908.
306. T.K. Nielsen, F. Besenbacher, T.R. Jensen, *Nanoscale*, 3 (2011) 2086–2098.
307. M. Fichtner, *Physical Chemistry Chemical Physics*, 13 (2011) 21186–21195.
308. J.J. Vajo, *Current Opinion in Solid State and Materials Science*, 15 (2011) 52–61.
309. R.Bhakta, S. Maharrey, V. Stavila, A. Highley, T. Alam, E. Majzoub, M. Allendorf, *Physical Chemistry Chemical Physics*, 14 (2012) 8160–8169.
310. R.K. Bhakta, J.L. Herberg, B. Jacobs, A. Highley, R. Behrens, N.W. Ockwig, J.A. Greathouse, M.D. Allendorf, *Journal of the American Chemical Society*, 131 (2009) 13198–13199.
311. T.F. Baumann, private communication to John Vajo on June 1, 2008.
312. S.M. Opalka, X. Tang, B.L. Laube, T.H. Vanderspurt, *Nanotechnology*, 20 (2009) 204024.
313. X. Tang, private communication to L.E. Klebanoff on December 15, 2011.
314. D.T. Shane, R.L. Corey, C. McIntosh, L.H. Rayhel, R.C. Bowman, J.J. Vajo, A.F. Gross, M.S. Conradi, *Journal of Physical Chemistry C*, 114 (2010) 4008–4014.
315. G.L. Soloveichik, Y. Gao, J. Rijssenbeek, M. Andrus, S. Kniajanski, R.C. Bowman, Jr., S.-J. Hwang, J.-C. Zhao, *International Journal of Hydrogen Energy*, 34 (2009) 916–928.
316. H.I. Chu, G.T. Wu, Z.T. Xiong, J.P. Guo, T. He, P. Chen, *Chemistry of Materials*, 22 (2010) 6021–6028.
317. R. Cerny, N. Penin, H. Hagemann, Y. Filinchuk, *Journal of Physical Chemistry C*, 113 (2009) 9003–9007.

7

Storage Materials Based on Hydrogen Physisorption

Channing Ahn and Justin Purewal

CONTENTS

Introduction 213
Definitions 214
The Need to Consider Sorbents 215
Heats of Adsorption... 215
Absolute Uptake and the Langmuir Model 217
The Gibbs Surface Excess.. 220
The Mechanism of Physisorption 222
Electrostatic Interactions 222
Orbital Interactions.. 222
Size of Molecular Hydrogen... 223
Adsorbents 223
    Graphite... 224
    Fullerenes 226
    Activated Carbons... 227
    Zeolites.. 228
    Metal Organic Frameworks 229
    Intercalation Compounds 230
Future Directions.. 231
Summary 233
References... 234

Introduction

We have seen how hydrogen atoms can be bound within the interstitial sites of intermetallic metal hydrides (Chapter 5) and in the covalent chemical bonds of complex metal hydrides (Chapter 6). Here, we discuss the phenomenon whereby hydrogen gas is introduced in molecular form, adsorbs onto a surface as a molecule, and is desorbed as a molecule. The generally low activation energy associated with this phenomenon results in a conceptually and technologically simple means for storage that is not limited by cycle life, as the adsorbed molecule and the adsorbent remain relatively unchanged during the adsorption/desorption cycle. This weak interaction, however, results in the requirement of low temperatures if reasonably large uptake values are to be expected.

In adsorption, gas molecules can generally be found in greater concentration at the surface of a substrate than in the free gas volume as determined by real (vs. ideal) gas law behavior. This physical adsorption (or physisorption) can provide the rationale for the use of adsorbents for hydrogen storage systems, provided that the so-called surface excess [1] value that is measured typically using a volumetric apparatus under isothermal conditions reaches a value consistent with the desired properties of the complete storage system. As the present means of storing hydrogen is through the use of liquefaction or compressed gas technology, both of which are energy intensive, alternatives that can make use of physisorption may help to obviate the use of low temperatures or high pressures.

Given the weak intermolecular forces [2] in hydrogen gas, any discussion of hydrogen adsorption requires an almost simultaneous consideration of the sorbent morphology (which includes both surface area and micropore volume) as well as the hydrogen/adsorbent enthalpy. In this chapter, we discuss the behavior of hydrogen gas in the presence of adsorbents and consider the limitations of these systems as governed by the thermodynamics of adsorption.

We start with a brief introduction to the terminology of physisorption and then proceed to develop a more detailed description of the physical parameters that are required for a high-uptake sorbent. Due to the weak interactions between molecular hydrogen and typical substrates, requiring low temperatures, we also discuss more recent work that suggested that sorbents might be useful for hydrogen storage at ambient temperatures.

While the literature on physisorption of molecular hydrogen is extensive, we do not attempt a comprehensive accounting of all results, but limit our discussion to results that we feel are best illustrative of the principles needed to understand the motivation behind hydrogen sorbent development. A number of excellent overviews and reviews are already available in the literature [3] that cover specific material classes for hydrogen storage, and we do not attempt to duplicate those materials here.

Definitions

The recent literature on hydrogen adsorption has resulted in some ambiguity over the description of the quantities that are measured. This is due in part to the use of technical targets established by the U.S. Department of Energy (DOE) [4] that are expressed in percentages that are different from the values that are measured in the traditional literature on gas adsorption [5].

In this work, we generally use mass fractions (with typical units of milligrams/gram or moles/gram), although we defer to units used in references if appropriate. Ambiguity has arisen due to the use of "weight percentage" as a figure of merit. The traditional literature on adsorption generally expresses the quantity of adsorbed gas as

$$\frac{m_{gas}}{m_{sorbent}} \tag{7.1}$$

where m_{gas} is the surface excess value of the mass of adsorbed gas per mass of sorbent $m_{sorbent}$. We denote this quantity as the mass fraction. We understand weight percentage (wt%) as

$$\frac{m_{gas}}{m_{gas} + m_{sorbent}} \tag{7.2}$$

As noted, the quantity of adsorbed gas is expressed as the "surface excess" value as originally denoted by Gibbs [1], and that is the only quantity that is determined empirically either gravimetrically or volumetrically. Simply put, the surface excess is the quantity of gas that the surface adsorbs beyond the quantity that the real gas law would normally indicate exists at the surface. This quantity is distinguished from "absolute" adsorption as initially envisioned by Langmuir [6] (and which is the quantity that is generally expressed in the computational/model-based literature) and is the net amount of gas adsorbed onto a surface. We elaborate on the difference between the two in a later section. We note also that the expression *total uptake* has also become part of the recent literature on hydrogen adsorption. Presumably, this term is meant to apply to the sum of the surface excess and the gas law contribution to an interior unit cell volume for an adsorbent that has a large free interior volume. From a technological standpoint, however, the total uptake has no meaning unless the storage volume consists of a single crystal or unless this term is used in conjunction with an adsorbent packing density within a storage volume. Beyond these two cases, the use of total uptake is discouraged.

A single quantity is often used to describe the adsorption enthalpy. However, this single quantity or Henry's law [7] value typically only applies to the initial molecule adsorbed onto a surface and is formally referred to as the *differential enthalpy of adsorption at zero coverage*. This is to be distinguished from the change of enthalpy as a function of coverage or pressure, known as the *isosteric enthalpy of adsorption*. This nomenclature provides an important distinction as the sorption enthalpy will typically decrease due to surface site heterogeneities and sorbate-sorbate interactions.

The Need to Consider Sorbents

The fundamental difficulty of storing hydrogen as a liquid or compressed gas is evident in the hydrogen p-T phase diagram shown in Figure 7.1 [8]. At the triple point ($T = 13.803$ K, $p = 0.0704$ bar), the solid density is $\rho_s = 86.48$ kg m^{-3} and is slightly higher than the liquid density of $\rho_l = 77.03$ kg m^{-3}, and the vapor pressure is a modest 0.07 bar. In the narrow range between the triple point and the critical point, hydrogen is a liquid with a boiling point of 20.39 K. If liquid hydrogen is stored in a closed vessel, continuous boil-off can lead to pressures of 10,000 bar. The critical point ($T_c = 32.98$ K, $p_c = 13.25$ bar) of H_2 occurs at a temperature that is low compared to other gases. Above the critical point, hydrogen cannot be liquefied by increasing the pressure but can instead take the form of a supercritical fluid. Therefore, if a storage system is to operate at higher temperatures, the hydrogen will exist in the gas phase. From the constant density contours in Figure 7.1, we can see that at room temperature a pressure of over 1000 bar is required to achieve densities on the order of the liquid or solid phases. This may be possible using carbon-fiber-reinforced high-pressure cylinders but is undesirable for on-board vehicle storage and other applications.

Heats of Adsorption

Part of the goal of the Sorption and the Metal Hydride Hydrogen Storage Centers of Excellence [9] was to develop "on-board" reversible hydrogen storage materials that would

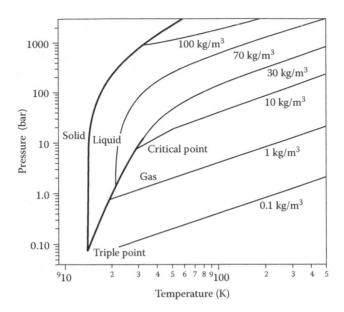

FIGURE 7.1
Phase diagram for H_2 calculated using the equation of state from Reference [8]. The melting line and the liquid-vapor line are drawn as bold lines. The triple point and the critical point are both labeled. Constant density contours are drawn as thin lines.

operate in a tank capable of being rehydrogenated (refueled) with times that are consistent with the present-day experience of refueling with gasoline. The challenge that this poses in the presence of a sorbent can be appreciated by noting that the ab(d)sorption process for a reversible system will be exothermic. The heat released during an exothermic phase change is given by $-\Delta H$, where ΔH the sorption enthalpy. For physisorption systems, ΔH can range from −4 to −10 kJ/mol H_2 [10]. For metal hydrides, the enthalpy of rehydrogenation can range from about −20 to −75 kJ/mol H_2 [11]. The heat dissipation rate requirements due to the presence of a sorbent can be summarized as in Figure 7.2, where this dissipation rate is plotted as a function of hydrogenation time (provided the reaction proceeds homogeneously). The supplying and removal of heat for engineered solid-state hydrogen storage systems are considered in greater detail in Chapters 10 and 11.

Physical adsorption is a process by which gas admolecules bind weakly onto the adsorbent surface by van der Waals forces. Chemical bonds are not formed. The equilibrium adsorption amount $n(T,p)$ is determined by the effective surface area of the adsorbent and the strength of the surface interaction. Following the analysis of Bhatia and Myers [12], the adsorbed layer and the bulk gas are in equilibrium, so the Gibbs free energies must be equal: $G_{gas} = G_{ads}$. Substituting $G = H - TS$ yields

$$H_{ads} - H_{gas} = T(S_{ads} - S_{gas}) \tag{7.3}$$

Adsorption involves a reduction in the degrees of freedom of the gas molecules so a tentative assumption $S_{gas} \gg S_{ads}$ can be made. If $S_{gas} \gg S_{ads}$, then $H_{ads} < H_{gas}$, and adsorption is thermodynamically favored. For many adsorbents, a change in entropy of −8R can be estimated. A simple estimate of the required enthalpy for room temperature storage is

$$\Delta H = H_{ads} - H_{gas} \approx -8RT_{rt} = -20 \text{ kJ mol} \tag{7.4}$$

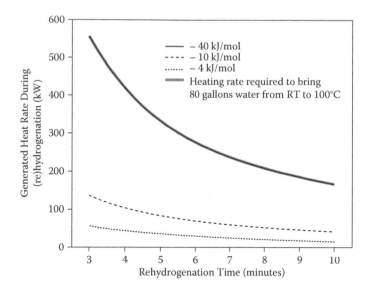

FIGURE 7.2

Plots of the rate of heat generated (−ΔH) during rehydrogenation as a function of time for −4, −10 and −40 kJ/mol H$_2$ adsorption enthalpies.

Absolute Uptake and the Langmuir Model

While the "absolute" uptake offers a simpler physical picture and has formed the basis for the preponderance of modeling and analytical work over the past century, the ability to use these models has posed difficulties in the interpretation of high-pressure measurement data. While of limited value for the rigorous analysis of real sorbents under technologically relevant operating conditions, the original work of Langmuir [6] does offer a simple conceptual foundation from which to consider sorbent requirements for gas storage technology. The assumption behind this model is that each lattice site has, and maintains, the same site potential and entropy that is independent of the fractional coverage of adsorbed molecules. Figure 7.3 depicts hydrogen molecules distributed across a surface lattice.

The Langmuir adsorption isotherm for noncompetitive, nondissociative adsorption can be written: $\theta = K \cdot p / (1 + K \cdot p)$, where θ is the fractional coverage, K is the equilibrium constant, and p is the pressure. The Langmuir isotherm can be derived either kinetically [13] or, because it represents an equilibrium state, through statistical methods [14].

Such a model gives rise to the type 1 isotherm as shown Figure 7.4; the adsorption reaches a plateau, and the general shape is similar to that obtained in high-pressure hydrogen uptake measurements. The asymptotic limit reached in this case has been described as micropore filling [15] of the sorbent by the molecular gas, as an analogue of the model of monolayer adsorption onto a surface, described originally by Langmuir.

A somewhat more realistic treatment of adsorption onto a surface that takes into account the interaction between gas molecules at higher fractional coverage was considered by Fowler and Guggenheim [16]. Still, their analysis, which considered pairs of near-neighbor molecules, falls short of a description that requires, in the case of engineering materials, an ultimate accounting of surface heterogeneities that yield differences in site potentials.

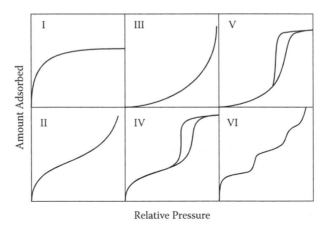

+ Lattice sites   ◯ Adsorbed molecule

FIGURE 7.3
Depiction of lattice sites and adsorbed molecules.

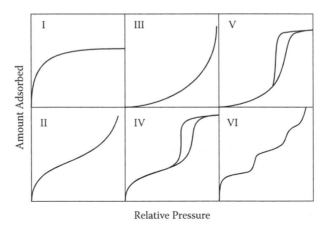

Relative Pressure

FIGURE 7.4
Classical adsorption types as defined by the International Union of Pure and Applied Chemistry (IUPAC). (Reproduced from K. S. W. Sing, D. H. Everett, R. A. W. Haul, L. Moscou, R. A. Pierotti, J. Rouquerol, and T. Siemieniewska, Reporting physisorption data for gas solid systems with special reference to the determination of surface-area and porosity (recommendations 1984), *Pure Appl Chem* **57** (4), 603–619 (1985). With permission.)

Unfortunately, any treatment that considers surface heterogeneities will need to rely on a fairly comprehensive understanding of surface morphology. This can be especially difficult as in the case of high-surface-area activated carbons used as sorbents [17].

The importance of adsorption thermodynamics can be best illustrated by rearranging the Langmuir model [18] to better show the pressure dependence. For an equilibrium pressure p, the amount of adsorbed gas $n(p)$ is given by

$$n(p) = n_{\max}\left(\frac{K \cdot p}{1 + K \cdot p}\right) \tag{7.5}$$

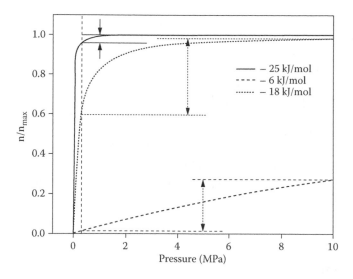

FIGURE 7.5
Langmuir isotherms for three different adsorption enthalpies ΔH. Hydrogen delivery Δn for each isotherm is indicated by the vertical distance between the adsorption amounts at 0.3 MPa (3 bar) and 10 MPa (100 bar) in this example.

where n_{max} is the maximum adsorption capacity of the material, and K is the equilibrium constant (which depends on both the temperature and the change in Gibbs free energy). Because of the requirements of proton exchange membrane (PEM) fuel cells, hydrogen should be delivered with a minimum pressure of $p_{min} = 3$ bar. The storage system cycles between storage tank operational pressure limits, and the total delivered hydrogen is the difference between adsorbed amounts $n(p_{max})$ and $n(p_{min})$. For convenience, we use 100 bar as the value for p_{max} as the ideal gas law and equation of state values are similar at this pressure.

We can now determine the enthalpy that optimizes the hydrogen delivery between these pressure limits [12]. For an operating temperature of about 295 K, realistic estimates of the optimum adsorption enthalpy typically give values between –15 and –20 kJ/mol H_2 [12]. In Figure 7.5, Langmuir adsorption isotherms are plotted for various enthalpies of adsorption as determined from Equation 7.5, where

$$K = \frac{1}{p_0} \exp\left(\frac{S}{R}\right) \exp\left(\frac{-H}{RT}\right)$$

and $p_0 = 0.1$ MPa, $T = 295$ K, ΔS is the molar entropy of adsorption, and ΔH is the enthalpy of adsorption ($\Delta H = -6, -18, -25$ kJ/mol). The adsorption entropy was estimated as $\Delta S = -8R$.

It can be seen that the hydrogen delivery depends strongly on the enthalpy. Because the isotherm for $\Delta H = -25$ kJ mol^{-1} is steep at low pressures, for example, most hydrogen remains adsorbed when the pressure cycles down to 3 bar (0.3 MPa) from 100 bar (10 MPa). On the other hand, while the low-value –6 kJ/mol isotherm requires high pressure to maximize the ultimate quantity of adsorbed hydrogen (which would be as close to $n/n_{max} = 1$ as possible), a much larger quantity Δn of hydrogen is available under isothermal conditions. An optimal adsorption enthalpy would occur between these two values, providing the largest amount of deliverable hydrogen capacity Δn.

This example demonstrates the importance of *adsorption enthalpy* in determining the properties of a physisorption-based storage system. Because heterogeneous adsorption sites and hydrogen-hydrogen interactions are omitted from the model, the adsorption enthalpy remains constant as a function of n. If the constant ΔH is replaced by one that decreases as a function of n (as is the case in virtually all adsorbents that have been studied to date), then we can expect hydrogen delivery to be reduced under isothermal conditions. While true that a sorbent with a nonvarying isosteric enthalpy of adsorption simplifies storage tank optimization over that of a decreasing isosteric enthalpy, there are few examples of an adsorbent with such an attribute. The ideal adsorbent for an isothermal storage system would ideally have a constant ΔH of –15 to –20 kJ mol H_2, although the specific values will depend on the operational pressure and temperature ranges.

The Gibbs Surface Excess

The true empirically determined quantity that is measured in any experiment is the Gibbs surface excess, which differs from textbook treatments that consider absolute absorption onto the surface in that a gas law component is subtracted. This quantity is defined at the so-called Gibbs dividing surface [19]. Several simplified treatments that reduce the original general multicomponent treatment of Gibbs to a single-component gas can be found in the literature [20]. The surface excess can be depicted schematically as shown in Figure 7.6. The dotted line roughly corresponds to the distance at which the local concentration of the adsorbed hydrogen reaches the equilibrium bulk density ρ of the gas phase. Here, the sorbent is represented by the shaded area on the left. The total quantity of surface excess gas n_{ex} is depicted as solid circles. The absolute amount of adsorbed gas is represented by open and solid circles to the left of the dotted line contained within a volume v_a. The volume to the right of the dotted line is v. The total amount of gas n is represented by the sum of solid and open molecules within a total void volume $v_o = v_a + v$. We can then express the surface excess as $n_{ex} = n - v_o\rho$. To account for the volume occupied by the sorbent, it is customary to measure the skeletal density using He gas.

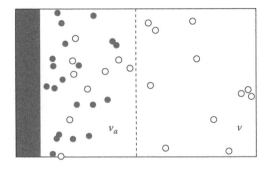

FIGURE 7.6
Schematic of the analysis for determination of the surface excess for adsorbed molecules in volume v_a. The bulk gas density (open circles) is subtracted from the absolute amount (open + filled circles) to yield the "surface excess."

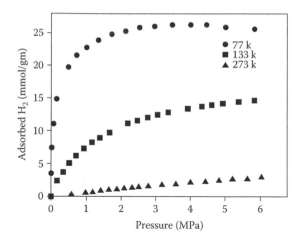

FIGURE 7.7

Several isotherm traces from an AX-21 high-surface-area carbon. The 77 K data show the maximum of the surface excess. We can also see how the higher-temperature isotherm data binds hydrogen more weakly with increasing temperature. Note 6 MPa corresponds to 60 bar pressure. (Reproduced from Y. A. Zhou and L. Zhou, Experimental study on high-pressure adsorption of hydrogen on activated carbon, *Sci China (Ser. B)* **39** (6), 10 (1996) with permission.)

Because the surface excess is a difference, there are conditions for which this value can attain a maximum and decrease, unlike that typical of the type 1 isotherm. Figure 7.7 shows [21] perhaps the first example of hydrogen surface excess isotherms for an AX-21-type activated carbon where this maximum can be seen. Once isotherms are obtained as a function of pressure for a number of temperature ranges of interest, it is possible to calculate two types of enthalpies. As noted, the enthalpy associated with the first molecule adsorbing onto an otherwise-pristine surface is the differential enthalpy of adsorption at zero coverage, which can be determined by the variation of the Henry's law constant as a function of temperature. This single-value enthalpy is often misleading as the enthalpy will generally change as a function of uptake due to the adsorbate-adsorbate interactions that eventually occur with higher gas loading. We refer to this continuum as the *isosteric enthalpy of adsorption*, a quantity that does decrease as more gas is adsorbed onto a surface.

The isosteric enthalpy of adsorption is calculated through an analogous form of the Clausius-Clapeyron equation [18],

$$H_{iso} = \frac{RT_1 T_2}{T_2 - T_1} \ln \frac{p_2}{p_1} \tag{7.6}$$

where ΔH_{iso} is assumed to be constant versus temperature, and $p_1(T_1)$ and $p_2(T_2)$ represent two distinct points at constant hydrogen loading. We do note that in a strict sense, the value ΔH_{iso} that we have defined should be referred to as the "isoexcess" [20] enthalpy of adsorption. Recent work that models the absolute adsorption [22–25] showed that when modeled absolute adsorption isotherms are used, a slight decrease over the range of enthalpy values occurs. This type of treatment can presumably address the apparent "retrograde" trend in the isosteric enthalpy of adsorption shown in some recent literature [26]. We can expect that all literature values are isoexcess enthalpies unless explicitly stated otherwise.

The Mechanism of Physisorption

In the previous section, it was illustrated that the enthalpy of adsorption plays a vital role in the operational characteristics of a physisorption-based hydrogen storage system. To determine the thermodynamic limits of such a storage system, a detailed understanding of this interaction energy is critical. Unfortunately, ab initio modeling of physisorption is much less developed than it is for chemisorption. Moreover, the complexity of surfaces in systems based on high-surface-area materials complicates the unit cells required as the starting point for computational work. Although density functional theory (DFT) works well for chemically bound systems, the inaccuracy of treating long-range interactions of weakly bound systems serves as a further hindrance to tractable computation. As dispersion forces [1, 15, 27–29] play a large role in physisorption, first-principles methods are required that are capable of treating the electron correlation, such as second order Moller-Plesset perturbation theory (MP2) or the coupled cluster (CCSD(T)) method. However, these methods are perhaps too computationally intensive to apply to realistic adsorbent systems.

Electrostatic Interactions

The hydrogen molecule does not have a permanent dipole moment. Due to the prolate, nonspherical shape of the H_2 molecule, the first nonzero multipole moment is the quadrupole moment. Consider a hydrogen molecule interacting with a point charge at a distance r. This situation might arise, for example, when an H_2 molecule is adsorbed on a metal-organic-framework (MOF) material containing coordinatively unsaturated metal centers. The hydrogen molecule is polarizable in the presence of external fields. For a hydrogen molecule interacting with a unit charge at a distance of 3 Å, the ion-quadrupole interaction energy is about 3.5 kJ/mol with an associated ion-induced-dipole interaction energy of about 6.8 kJ/mol [30]. Therefore, hydrogen binding affinity can be enhanced by these electrostatic interactions.

Orbital Interactions

Molecular hydrogen contains a ground-state bonding orbital σ_g with an energy level of about –11.7 eV. There is a relatively large energy gap between the bonding orbital σ_g and the unoccupied antibonding orbital σ_u^* with a magnitude of about 2 eV [30]. Interactions between filled molecular orbitals are primarily repulsive. However, interactions between filled and unfilled orbitals can result in charge transfer, donor-acceptor bonding, and overall stabilization. Orbital interactions have shorter bond lengths and larger binding energies than dispersion interactions. Charge transfer causes an elongation of the H_2 bond length, which is directly measurable by a softening of the intramolecular vibrational modes. Transition metals (TMs) are known to form H_2 coordination complexes in which H_2-σ to TM-d electron transfer is coupled with TM-σ to H_2-σ^* electron backdonation [31].

In reality, however, hydrogen does not easily donate or accept charge due to the large separation between its σ_g and σ_u^* orbitals. In fact, the adsorption mechanism in many metal-organic frameworks with exposed TM sites can actually be explained in terms of electrostatic interactions, without the need for orbital interactions [32].

Size of Molecular Hydrogen

We now discuss the geometric factors that should be considered with hydrogen adsorbents in relation to the size of molecular hydrogen. Relative to the sub-Ångstrom dimensions of hydrogen in atomic form, molecular hydrogen is large, with a kinetic diameter measured at about 2.9 Å [33] but with a size as determined by molecular sieve experiments of about 3.6 Å (an actual spheroid geometry of 3.44 to 3.85 Å) [34]. We can compare this value to the closest near-neighbor distance as measured by neutron scattering of molecular hydrogen adsorbed onto graphite, with a 3.51-Å distance [35]. We note also that solid hydrogen ($P6_3/mmc$ structure) has a lattice parameter $a = 3.76$ Å and $c = 6.14$ Å [36].

At the conditions typical for hydrogen sorption studies, hydrogen does not liquefy and multilayer adsorption should not occur. We can therefore apply simple geometric means in probing the upper limits of uptake that can be expected based on simple packing density models [37].

Adsorbents

Having considered elements of the appropriate thermodynamics and interactions of relevance, we can now discuss specific categories of materials of interest for adsorption. Carbon adsorbents are attractive for physisorption storage systems due to their simplicity, light weight, and generally low synthesis cost. Considerable empirical work has been performed in this field. Carbons can have a large range of specific surface area (SSA) that spans 2 m^2 g^{-1} to 3800 m^2 g^{-1}, depending on the material processing used.

A general description of carbons of relevance envisions a porous material that contains narrow internal cavities or channels of microporous dimensions. Pore sizes are classified by the International Union of Pure and Applied Chemistry (IUPAC) as micropores (pore width < 2 nm), mesopores (pore width 2–50 nm), and macropores (pore width > 50 nm) [5]. Microporous carbons are of particular importance for hydrogen storage since they contain large surface areas that should maximize the volumetric density of stored gas. Due to heterogeneities in the types of surface features for carbons, especially for high-surface-area ones, different types of binding sites are possible. However, certain trends in adsorption behavior of hydrogen in carbons have been noted.

The earliest trends relating both micropore volume and surface area were observed by Chahine and Bénard [38] in noting the trends at 77 K hydrogen uptake at 35 bar pressure for a number of different activated carbons. These uptake measurements of the "surface excess" are reproduced in Figure 7.8. The 35 bar pressure that was used in this work is not arbitrary and corresponds to a value close to or at the maximum of the surface excess for

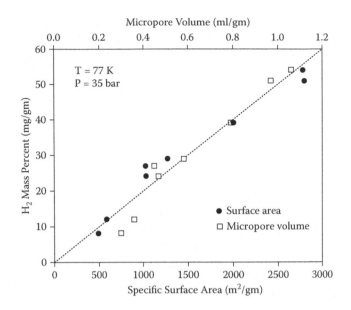

FIGURE 7.8
Mass percentage uptake as a function of surface area (bottom axis) and micropore volume (upper axis). (Data reproduced from R. Chahine and T. K. Bose, Characterization and optimization of adsorbents for hydrogen storage, *Hydrogen Energy Prog* Xi, **1–3**, 1259–1263 (1996) with permission.)

graphitic carbons. The line that is drawn through this plot corresponds to 1 mass percent uptake per 500 m² g⁻¹ surface area. Similarly, we might argue that we can expect 1 mass percent for every 0.2 mL/g of micropore volume.

Graphite

Graphite is one of the more familiar forms of carbon and consists of alternating layers of sp^2-bonded trigonal planer sheets. These sheets are also referred to as *graphene* sheets. Neighboring planes interact by overlapping bonds between the unhybridized carbon $2p$ orbitals. The stacking sequence of the hexagonal graphite planes along the c-axis follows a staggered –abab– pattern, so that half of the carbon atoms in a given plane sit between the hexagon centers of the layers above and below it. The carbon-carbon bond length in the basal plane is $a = 1.421$ Å, and the interlayer spacing is 3.354 Å. Graphitic carbons are generally nonporous, with surface areas typically less than 20 m² g⁻¹ and negligibly small hydrogen uptake at low temperature. Given the dimensions of the interplanar spacing and the size of molecular hydrogen, we would expect adsorption only on the surface of graphitic carbons. The measured adsorption enthalpy of hydrogen on graphite [40] is –3.8 kJ/mol, well below the targeted values of engineering interest.

One strategy for increasing the binding energy for carbon adsorbents is to open up space between the layer planes to accommodate guest molecules. The spaces in-between the graphene sheets are called *slit-pores*. In fact, this hypothetical graphene slit-pore structure has been the subject of several computational studies [41–44]. Due to the overlapping potential fields from opposing slit-pore walls, the heat of adsorption is enhanced. The optimal interlayer spacing should be large enough to accommodate two hydrogen monolayers (i.e., one monolayer per slit pore wall). Any additional interlayer expansion would not be useful

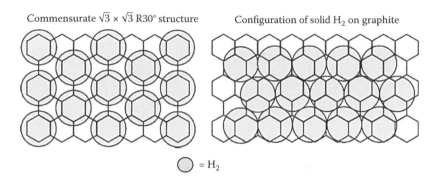

Commensurate $\sqrt{3} \times \sqrt{3}$ R30° structure   Configuration of solid H_2 on graphite

$\bigcirc$ = H_2

FIGURE 7.9

Two possible structures for a hydrogen monolayer on a graphene sheet. The diameters as depicted here for hydrogen molecules (shaded circles) are 3.51 Å and, for the incommensurate depiction on the right, 3.74 Å.

because hydrogen only adsorbs in monolayers at supercritical temperatures. Although the graphene slit pore structure is well suited for optimizing the carbon hydrogen-binding interaction, the general consensus is that the gravimetric density is intrinsically low due to the geometry.

This can be illustrated by considering H_2 monolayers on an ideal graphene sheet as was done by Brown and Dresselhaus [37]. An estimate of the graphene SSA is obtained from the fact that a single hexagon contains a net total of two carbon atoms with an area of $1.5 \sqrt{3a^2}$. This gives an SSA of 2630 m² g⁻¹ for both sides of graphene. Two possible configurations for a hydrogen monolayer on a graphene surface are illustrated in Figure 7.9.

Commensurate structures are presumably energetically favorable since the hydrogen molecules sit in the hexagon centers [35], but this configuration results in a lower gravimetric density than a close-packed structure. Provided both sides of a graphene surface are occupied, then the gravimetric density is about 5 wt% for the commensurate $\sqrt{3} \times \sqrt{3}$R30° structure. The gravimetric density of a close-packed H_2 monolayer can be obtained by using the standard method of estimating the Brunauer, Emmett, Teller (BET) theoretical cross-sectional area (although we should note that standard BET analysis that uses N_2 with an effective diameter of 4.5 Å may underestimate slightly the surface area accessible to H_2).

For a hexagonal close-packed structure, the cross-sectional area of a hydrogen molecule is given by

$$\sigma = f \left(\frac{M}{\rho N_a} \right)^{2/3} \tag{7.7}$$

where the hexagonal-close-packed (hcp) packing factor f is 1.091, ρ is the density of liquid hydrogen, M is the molar mass, and N_a is Avogadro's number. Taking the liquid H_2 density as $\rho = 77.03$ kg/m³, the cross-sectional area is $\sigma = 0.124$ nm². If the carbon SSA is a(SSA), then the hydrogen monolayer adsorption in weight percentage is

$$wt\% = \left(\frac{a(SSA)}{\sigma} \right) \left(\frac{M}{N_a} \right) \times 100 \tag{7.8}$$

Double layer of H_2, x-z cross-section

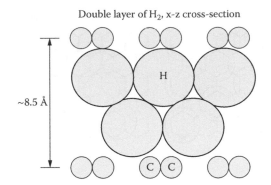

FIGURE 7.10
Schematic of hydrogen stacking required in graphite as an analog to adsorption on both surfaces of a graphene sheet. No consideration of van der Waals distances is taken into account in this depiction.

This gives 1.34 wt% per 500 m^2 g^{-1} carbon surface area. Double-sided graphene has a surface area of a(SSA) = 2630 m^2 g^{-1}, which translates to a maximum hydrogen adsorption of about 7 wt%. This represents the theoretical limit for H_2 density on graphene where both surfaces are accessible. Empirical work in the area of resolving the structure of hydrogen on graphite was done by Nielsen et al. [35] using neutron surface studies to identify the actual molecular position of hydrogen with respect to the graphitic surface. In that work, the nearest-neighbor distance for molecular hydrogen was 3.51 Å.

Of relevance for consideration of volumetric densities in an analysis of this type is the stacking of molecular hydrogen in the *c* direction of a graphitic lattice. As noted, while the *a-b* stacking sequence is separated by about 3.35 Å in graphite, and molecular hydrogen is known not to intercalate into graphite [45], a schematic of the layer spacing required for accommodating H_2 is depicted in Figure 7.10.

Given a bulk density of graphite of 2.1 g/cm^3, the idealized slit-pore structure shown in Figure 7.10 would correspond to the removal of two layers of graphite, giving an optimal slit-pore density of 0.7 g/cm^3. Provided that incommensurate loading at LH$_2$ densities is possible at higher than 20 K temperatures, this would correspond to a volumetric density of about 44 g/L. The synthesis of a graphite with a slit-pore geometry as depicted would require so-called pillars to separate the graphitic layers that are normally weakly held together via van der Waals attraction [46]. While some work has progressed in this area, the structures that have been synthesized still fall short of the idealized slit-pore configuration.

Fullerenes

Another strategy to enhance hydrogen binding in carbon adsorbents is to use a curved carbon surface. Fullerenes, such as single-walled carbon nanotubes [47] (SWCN) and C$_{60}$ buckyballs [48], are well-known examples. They are formed from graphene-like sheets composed of five- or six-member rings. The presence of pentagonal rings results in the curvature of the carbon planes, allowing the formation of C$_{60}$ spheres. Similarly, the hollow cylindrical structure of single- and multiwall carbon nanotubes is obtained by rolling up graphene sheets along different directions. Unfortunately, a large amount of variation and irreproducibility has plagued both experimental and theoretical work on hydrogen adsorption in fullerenes. Recent studies have indicated that carbon nanotubes have the same adsorption properties as activated carbons and other amorphous carbons [49, 50].

Nevertheless, as the only ordered allotropes of carbon that adsorb hydrogen, fullerenes provide a unique platform for rigorously studying the hydrogen-carbon interaction from both an experimental and a theoretical level. For example, the diameter of the SWCNs is a tunable parameter that can be used to optimize the adsorption behavior.

Activated Carbons

Activated carbons are predominantly amorphous structures that can be processed to have large surface areas and micropore pore volumes, with BET areas in excess of 3800 m^2 g^{-1}. They are best described as "a twisted network of defective carbon layer planes cross-linked by aliphatic bridging groups" [51]. The pore structure of an activated carbon is complex and ill-defined, making it challenging to study. An example of this complex cross-linked structure is illustrated in Figure 7.11 [52].

Unlike graphene, the surface areas of these materials can attain higher SSAs over that of graphite due to the zigzag and armchair terminations that are now part of the structure of material in this form, and these terminations are themselves part of the available surface onto which gases can adsorb. Structures of this type are not expected to retain dangling bonds under equilibrium conditions but to have oxygen or hydrogen terminations [53] that lower the adsorption enthalpy at those sites.

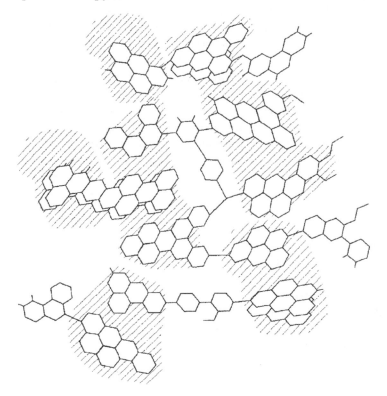

FIGURE 7.11
Schematic representation of an activated carbon. Note the edge terminations that make up part of the surface area, allowing for values in materials of this type well beyond the 2630 m^2/g of a single graphene sheet. (Reproduced from M. Winter, J. O. Besenhard, M. E. Spahr, and P. Novak, Insertion electrode materials for rechargeable lithium batteries, *Adv Mater* **10** (10), 725–763 (1998). With permission.)

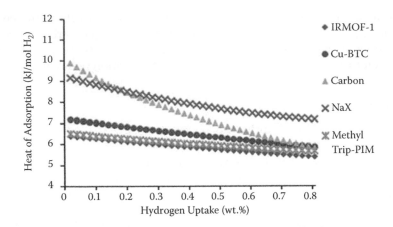

FIGURE 7.12

Isosteric heat of adsorption versus hydrogen uptake for a range of porous materials: activated carbon, Cu-BTC MOF, IRMOF-1 (i.e., MOF-5), zeolite NaX, and triptycene-based polymer of intrinsic microporosity (PIM). Reproduced with permission from Ref. [54].

We can develop an appreciation of the impact of site heterogeneity and adsorbate-adsorbate interaction in activated carbons by considering the "isoexcess" isosteric heat of adsorption for a highly microporous activated carbon in comparison to other microporous sorbents, as shown in Figure 7.12 [54]. Here, we observe the monotonic heat decrease as a function of weight percentage uptake, which can be as high as about 10 kJ/mol but can decrease to about 5 kJ/mol. While microporous carbons have relatively high Henry's law enthalpies, the continuous decrease in enthalpy with adsorption results in engineering complications due to the need to account for hydrogen release under differing binding conditions.

That having been said, a huge advantage in the use of activated carbons is the ability to produce these materials in industrial quantities, typically from a carbon-rich precursor by a physical or chemical activation process. Physically activated carbons commonly use bituminous coal or coconut shells as a starting material. The two-stage activation process consists of carbonization, by which oxygen and hydrogen are burned off, and gasification, by which the char is heated in a steam or carbon dioxide atmosphere to create a highly porous structure from carbon burn-off.

Carbon aerogels are a separate class of amorphous carbons that are similar to activated carbons. They can be prepared via a sol-gel polymerization process and can be activated using the standard methods [55, 56]. Because of the more open structure of aerogels, the densities are somewhat lower than in activated carbons, and 0.3 g/cc and below are typical values. In both activated carbons and carbon aerogels [57], gravimetric hydrogen uptakes on the order of 5.5 wt% at 77 K are common, and this value typically scales with BET surface area as noted previously [38].

Zeolites

Zeolites are crystalline materials composed of SiO_4 or AlO_4 building blocks. They contain an intracrystalline system of channels and cages that can trap guest H_2 molecules. The adsorption capacity of zeolites at 77 K is typically below 2 wt%. A theoretical capacity of 2.86 wt% has been suggested as an intrinsic geometric constraint of zeolites [49, 58]. Due to this low gravimetric density, zeolites are not typically considered feasible hydrogen storage materials. Isosteric heats on the order of 6–7 kJ mol^{-1} are typical for hydrogen-zeolite systems [59]. Zeolite structures such as Zeolite NaA can have intracrystalline cavities on the order of the H_2 diameter itself. They function as molecular sieves, blocking adsorption of larger gas molecules by steric barriers. They also exhibit quantum-sieving effects on hydrogen isotopes. When confined inside a molecular-size cavity, the heavier D_2 molecule is adsorbed preferentially over the lighter H_2 molecule due to its smaller zero-point motion [60].

Metal Organic Frameworks

Metal organic frameworks (MOFs) are synthetic crystalline materials that consist of organic linker molecules coordinatively bound to inorganic clusters to form a porous framework structure. The MOF nomenclature is a rebranding of what has traditionally been referred to as "coordination polymers" [61]. However, we continue to use the term MOF to identify these structures in connection with gas sorption. One of the early examples is the use of MOF-5 [62] for hydrogen storage, which has a structure (also referred to as IRMOF-1) based on Zn_4O units connected by 1,4 benzene-dicarboxylate (BDC) linkers in a simple cubic fashion.

The feature of interest, in this class of structures that has developed rapidly, is the dependence on gravimetric gas uptake with surface area [63], as is the case with high-surface-area carbons. The internal volume of these materials will fall somewhere in-between micro- and mesoporous, so unlike the case of microporous carbons, for which we noted a dependence on both surface area and micropore volume, in MOFs we can expect hydrogen uptake to depend predominantly on surface area. The surface areas that can be obtained in these materials is due presumably to the edge-dominant nature of the organic linkers. Consequently, this class of materials has surface areas that are linker length dependent. From the standpoint of adsorption enthalpy, however, instead of having a linker structure that has a substantial graphitic hexagonal component, the dominance of edge component in the linkers that have hydrogen or oxygen terminations will yield adsorption enthalpies that are typically smaller than seen in graphitic carbons [64].

While straightforward structures of MOFs have been discerned, initial work that was attempted to evaluate the hydrogen storage properties of these materials suffered inconsistencies. Given the processing required during the synthesis of this material, which typically requires extensive use of solvents and specific temperatures, differences in SSA could arise due to subtle variations in these synthesis conditions. As the maximum of the surface excess of hydrogen uptake in MOFs also has a strong surface area dependence, these differences in surface area could yield surface excess maxima anywhere from 4 to 7 wt% [65]. The difference in surface area responsible for the difference in uptake measurements can be attributed to trapped solvent within the large internal volume. Processes that can mitigate the presence of this solvent produce materials with surface areas approaching theoretical values for a given structure [66].

As with all prospective sorbents, an engineered system that can make use of higher sorption enthalpies would be of interest if higher-temperature storage containment is to be realized. In the case of MOFs that have so-called coordinatively unsaturated metal centers (CUMCs), Henry's law behavior showing a higher local chemical potential [67] has been observed in Mn^{2+}-based material and in a MOF-74 compound [68]. In the latter, the Zn metal centers have been shown to enhance the initial hydrogen binding [69] in the vicinity of the metal atoms as determined by neutron diffraction. As might be expected, once the metal center sites in this class of materials adsorb hydrogen, the rest of the structure will

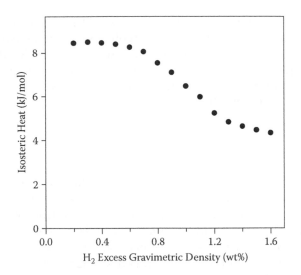

FIGURE 7.13

Change in isosteric heat ($-\Delta H$) in a MOF-74 CUMC showing that the initially high heat release associated with binding to the Zn sites drops off as hydrogen adsorbs onto these sites. (Reproduced from Y. Liu, H. Kabbour, C. M. Brown, D. A. Neumann, and C. C. Ahn, Increasing the density of adsorbed hydrogen with coordinatively unsaturated metal centers in metal-organic frameworks, *Langmuir* **24** (9), 4772–4777 (2008). With permission.)

have enthalpies reflective of the remaining surface potentials. This can be seen readily as a drop-off in isosteric enthalpy as shown in Figure 7.13 [69]. The nature of electrostatic interactions in CUMC materials yields unanticipated implications for hydrogen loading over that discussed previously using a hard sphere model on a graphitic carbon.

The density of molecular hydrogen in the presence of strong electrostatic interactions is increased considerably in several other systems with CUMCs as shown in Figure 7.14. Here, the surface packing density of the *y*-axis is defined as the maximum of the surface excess over the BET surface area. At 77 K, hydrogen molecules are typically adsorbed as a monolayer on the material surface. The dotted trace is the 500 m²/g gravimetric uptake noted previously for high-surface-area carbons.

While a high surface packing density that exceeds that which might be expected for carbon based on enhanced electrostatic interactions is desirable, this effect in MOF-type structures will probably always be limited at the local site of the unsaturated metal center unless substantial charge transfer can be effected in such materials in a way that would yield a more homogeneous charge distribution of the adsorbent surface. The difficulty that the localization of charge poses for an engineering system that works under isothermal conditions is that such a system still requires conditions that can accommodate weakly bound hydrogen. Hydrogen that is bound to metal centers will ultimately require higher temperature to deliver the equilibrium pressures of interest for most storage systems designed for fuel cell applications.

Intercalation Compounds

Systems that do in fact exhibit a fairly constant sorption enthalpy are stage 2 type alkali-intercalated graphites [70, 71]. While these materials generally have low gravimetric

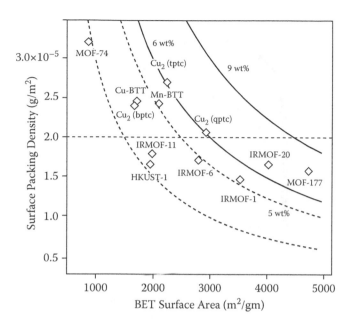

FIGURE 7.14

The H_2 surface packing density (SPD) as a function of N_2 BET surface area. The MOFs with CUMCs, in general, show larger SPD compared to those without CUMCs. The horizontal dashed line shows the SPD for typical carbon material, where every 500 m^2/g of N_2 BET surface area can adsorb 1 wt% of H_2. Solid curves show the lines for gravimetric uptake of 6 and 9 wt%. (Reproduced from Y. Liu, H. Kabbour, C. M. Brown, D. A. Neumann, and C. C. Ahn, Increasing the density of adsorbed hydrogen with coordinatively unsaturated metal centers in metal-organic frameworks, *Langmuir* **24** (9), 4772–4777 (2008). With permission.)

densities of limited value to the transportation sector, the electron back-donation from the alkali metal to the graphitic host ensures that the local chemical potential is maintained while molecular hydrogen is accommodated within the alkali-metal containing layers. So-called stage 1 compounds will generally undergo a chemisorption reaction, but the stage 2 compounds that have a stoichiometry of MC_{24} with alkali metal atoms M distributed along every second layer of graphite can adsorb molecular hydrogen [72]. In an intercalation system of this type wherein only a single layer of molecular hydrogen can be accommodated in the alkali-metal-containing layer, this isotherm will be type 1 (see Figure 7.4) as no gas law component need be taken into account as in surface excess measurements and analyses.

In Figure 7.15, we show an example of isotherms for an RbC_{24} intercalation compound measured at several temperatures [73]. Note that the maximum value can be reached in a system of this type, with high and reasonably constant isosteric heat, as shown in Figure 7.16, at higher temperatures, given a high enough pressure as a driving force.

Future Directions

The challenge of finding a hydrogen sorbent with both high gravimetric and volumetric density is still outstanding. Given the infrastructure that is presently being deployed for high-pressure storage solutions and that will operate at 350 to 700 bar, sorbents that might

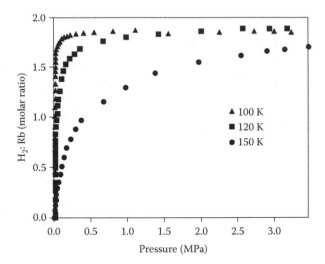

FIGURE 7.15
Hydrogen sorption in the RbC$_{24}$ intercalation compound showing that the near-maximum value (H$_2$:Rb = 2) can be reached at temperatures higher than 77 K given a high enough pressure.

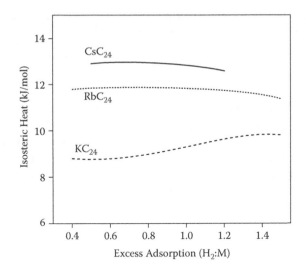

FIGURE 7.16
Isosteric heats measured from three stage 2 compounds showing nearly constant values as a function of increased hydrogen loading.

otherwise offer advantages over purely high-pressure systems can serve to provide near-term engineering possibilities. Because hydrogen fuel delivery is typically at –50°C, any near-term solution will effectively be required to operate at ambient temperatures and not at 77 K. Few measurements are taken for sorbents at 350 to 700 bar under ambient conditions. However, some data are beginning to emerge that offer some possibilities that a near-term physisorbent-based system might be a possibility [74].

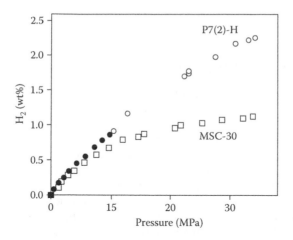

FIGURE 7.17

A comparison of 30°C isotherm data for an AX-21-type activated carbon (MSC-30) and a highly microporous zeolite carbon labeled as P7(2)-H. (Reproduced from H. Nishihara, P. X. Hou, L. X. Li, M. Ito, M. Uchiyama, T. Kaburagi, A. Ikura, J. Katamura, T. Kawarada, K. Mizuuchi, and T. Kyotani, High-pressure hydrogen storage in zeolite-templated carbon, *J Phys Chem C* **113** (8), 3189–3196 (2009) with permission.)

Zeolite-templated carbons are a class of pure carbons that are synthesized using zeolites as a starting matrix. Carbon is essentially deposited onto the surface of the zeolite, and the zeolite is dissolved away, leaving only a microporous carbon in its place [74]. Such a system might represent the ultimate microporous adsorbent, and the 30°C isotherm data are presented in Figure 7.17. Ultimately, the sorption enthalpies of a system like this will be similar to typical carbons. Chemical modifications that might increase the enthalpy over a range of hydrogen loading will be the requisite goal to minimize the engineering challenges for hydrogen storage systems based on hydrogen physisorption.

Summary

In this chapter, we have described the physicochemical factors that should be considered in the development and analysis of physisorbents for hydrogen storage systems. These factors include (a) the low enthalpy associated with the van der Waals interactions between gas and adsorbent, which results in modest thermal heat transfer requirements during rehydrogenation but at the cost of low temperature storage; (b) the role of surface area and the size of molecular hydrogen, which imposes limits on the volumetric density that can be attained if high-surface-area materials are used and; (c) the difficulties in the use of schemes to promote high Henry's law enthalpies but that show decreasing heats as a function of hydrogen loading. Because we can generally employ gas law and geometric considerations in our analyses, we can, given a reasonably comprehensive description of an adsorbent structure, draw reasonable conclusions of sorbent properties that can be used subsequently for storage tank design. Chapter 10 will describe hydrogen storage tank design based on physisorption.

References

1. J. W. Gibbs, *The Scientific Papers of J. Willard Gibbs: In Two Volumes* (Ox Bow Press, Woodbridge, CT, 1993).
2. V. A. Parsegian, *Van der Waals Forces: A Handbook for Biologists, Chemists, Engineers, and Physicists* (Cambridge University Press, New York, 2006).
3. F. L. Darkin, P. Malbrunot, and G. P. Tartaglia, Review of hydrogen storage by absorption on carbon nanotubes, *Int J Hydrogen Energy* **27**, 193 (2002).
4. S. Satyapal, J. Petrovic, C. Read, G. Thomas, and G. Ordaz, The U.S. Department of Energy's National Hydrogen Storage Project: progress towards meeting hydrogen-powered vehicle requirements, *Catal Today* **120** (3–4), 246–256 (2007).
5. K. S. W. Sing, D. H. Everett, R. A. W. Haul, L. Moscou, R. A. Pierotti, J. Rouquerol, and T. Siemieniewska, Reporting physisorption data for gas solid systems with special reference to the determination of surface-area and porosity (recommendations 1984), *Pure Appl Chem* **57** (4), 603–619 (1985).
6. I. Langmuir, The adsorption of gases on plane surfaces of glass, mica and platinum, *J Am Chem Soc* **40**, 1361–1403 (1918).
7. F. Rouquerol, J. Rouquerol, and K. S. W. Sing, *Adsorption by Powders and Porous Solids: Principles, Methodology, and Applications* (Academic Press, San Diego, CA, 1999).
8. J. Leachman, R. Jacobsen, S. Penoncello, and E. Lemmon, Fundamental equations of state for parahydrogen, normal hydrogen and orthohydrogen, *J. Phys. Chem. Ref. Data*, **38** (3), 721–748 (2009).
9. C. Read, J. Petrovic, G. Ordaz and S. Satyapal, The DOE National Hydrogen Storage Project: recent progress in on-board vehicular hydrogen storage, *Hydrogen Cycle-Generation Storage Fuel Cells* **885**, 125–134 (2006).
10. M. Dinca, S. S. Kaye, H. J. Choi, A. Demessence, S. Horike, L. J. Murray, and J. R. Long, PHYS 54-hydrogen storage in microporous metal-organic frameworks with exposed metal sites, *Abstr Pap Am Chem Soc* **237** (2009).
11. R. C. Bowman and B. Fultz, Metallic hydrides. 1. I: Hydrogen storage and other gas-phase applications, *MRS Bull* **27** (9), 688–693 (2002).
12. S. K. Bhatia and A. L. Myers, Optimum conditions for adsorptive storage, *Langmuir* **22** (4), 1688–1700 (2006).
13. D. H. Everett, The thermodynamics of adsorption. 2. Thermodynamics of monolayers on solids, *Trans Faraday Soc* **46** (11), 942–957 (1950).
14. R. H. Fowler, A statistical derivation of Langmuir's adsorption isotherm, *Proc Camb Philol Soc* **31**, 260–264 (1935).
15. S. J. Gregg and K. S. W. Sing, *Adsorption, Surface Area, and Porosity*, 2nd ed. (Academic Press, London, 1982).
16. R. H. Fowler and E. A. Guggenheim, *Statistical Thermodynamics; A Version of Statistical Mechanics for Students of Physics and Chemistry* (Macmillan, University Press, New York, 1939).
17. M. J. Sparnaay, Physisorption on heterogeneous surfaces, *Surf Sci* **9** (1), 100–118 (1968).
18. K. G. Denbigh, *The Principles of Chemical Equilibrium, with Applications in Chemistry and Chemical Engineering*, 3rd ed. (Cambridge University Press, Cambridge, UK, 1971).
19. O. Talu and A. L. Myers, Molecular simulation of adsorption: Gibbs dividing surface and comparison with experiment, *AIChE J* **47** (5), 1160–1168 (2001).
20. S. Sircar, Gibbsian surface excess for gas adsorption—revisited, *Ind Eng Chem Res* **38** (10), 3670–3682 (1999).
21. Y. A. Zhou and L. Zhou, Experimental study on high-pressure adsorption of hydrogen on activated carbon, *Sci China (Ser. B)* **39** (6), 10 (1996).
22. R. Staudt, G. Saller, M. Tomalla, and J. U. Keller, A note on gravimetric measurements of gas-adsorption equilibria, *Ber Bunsen Phys Chem* **97** (1), 98–105 (1993).

23. L. Zhou and Y. P. Zhou, Linearization of adsorption isotherms for high-pressure applications, *Chem Eng Sci* **53** (14), 2531–2536 (1998).

24. F. O. Mertens, Determination of absolute adsorption in highly ordered porous media, *Surf Sci* **603** (10–12), 1979–1984 (2009).

25. W. Zhou, H. Wu, M. R. Hartman, and T. Yildirim, Hydrogen and methane adsorption in metal-organic frameworks: a high-pressure volumetric study, *J Phys Chem C* **111** (44), 16131–16137 (2007).

26. S. S. Kaye and J. R. Long, Hydrogen storage in the dehydrated Prussian blue analogues M-3[Co(CN)(6)](2) (M = Mn, Fe, Co, Ni, Cu, Zn), *J Am Chem Soc* **127** (18), 6506–6507 (2005).

27. F. London, The general theory of molecular forces, *Trans Faraday Soc* **33** (1937).

28. J. E. Lennard-Jones, Processes of adsorption and diffusion on solid surfaces, *Trans Faraday Soc* **28**, 0333–0358 (1932).

29. D. M. Young and A. D. Crowell, *Physical Adsorption of Gases* (Butterworths, London, 1962).

30. R. C. Lochan and M. Head-Gordon, Computational studies of molecular hydrogen binding affinities: the role of dispersion forces, electrostatics, and orbital interactions, *Phys Chem Chem Phys* **8** (12), 1357–1370 (2006).

31. G. J. Kubas, Fundamentals of H(2) binding and reactivity on transition metals underlying hydrogenase function and H(2) production and storage, *Chem Rev* **107** (10), 4152–4205 (2007).

32. W. Zhou and T. Yildirim, Nature and tunability of enhanced hydrogen binding in metal-organic frameworks with exposed transition metal sites, *J Phys Chem C* **112** (22), 8132–8135 (2008).

33. A. C. Dillon and M. J. Heben, Hydrogen storage using carbon adsorbents: past, present and future, *Appl Phys A-Mater* **72** (2), 133–142 (2001).

34. J. Koresh and A. Soffer, Study of molecular-sieve carbons. 2. Estimation of cross-sectional diameters of non-spherical molecules, *J Chem Soc Farad Trans 1* **76**, 2472–2485 (1980).

35. M. Nielsen, J. P. McTague, and W. Ellenson, Adsorbed layers of D2, H2, O2, and 3He on graphite studied by neutron scattering, *J Phys Coll C4* **38** (suppl 10), 9 (1977).

36. H. Landolt, R. Börnstein, and K. H. Hellwege, *Numerical Data and Functional Relationships in Science and Technology, New Series III/14a* (Springer-Verlag, Berlin, 1988).

37. S. D. M. Brown, G. Dresselhaus, and M. S. Dresselhaus, Reversible hydrogen uptake in carbon-based materials, *Mater Res Soc Symp P* **497**, 157–163 (1998).

38. R. Chahine and P. Bénard, Adsorption storage of gaseous hydrogen at cryogenic temperatures, *Adv Cryogr Eng* **43**, 1257–1264 (1998).

39. R. Chahine and T. K. Bose, Characterization and optimization of adsorbents for hydrogen storage, *Hydrogen Energy Prog* Xi, **1–3**, 1259–1263 (1996).

40. E. L. Pace and A. R. Siebert, Heat of adsorption of parahydrogen and orthodeuterium on Graphon, *J Phys Chem-US* **63** (9), 1398–1400 (1959).

41. Q. Y. Wang and J. K. Johnson, Molecular simulation of hydrogen adsorption in single-walled carbon nanotubes and idealized carbon slit pores, *J Chem Phys* **110** (1), 577–586 (1999).

42. M. Rzepka, P. Lamp, and M. A. De La Casa-Lillo, Physisorption of hydrogen on microporous carbon and carbon nanotubes, *J Phys Chem B* **102** (52), 10894–10898 (1998).

43. D. D. Do and H. D. Do, Adsorption of argon from sub- to supercritical conditions on graphitized thermal carbon black and in graphitic slit pores: a grand canonical Monte Carlo simulation study, *J Chem Phys* **123** (8), 084701 (2005).

44. S. Patchkovskii and T. Heine, Evaluation of the adsorption free energy of light guest molecules in nanoporous host structures, *Phys Chem Chem Phys* **9** (21), 2697–2705 (2007).

45. M. Watanabe, M. Tachikawa, and T. Osaka, On the possibility of hydrogen intercalation of graphite-like carbon materials—electrochemical and molecular orbital studies, *Electrochim Acta* **42** (17), 2707–2717 (1997).

46. Z. Jin, W. Lu, K. J. O'Neill, P. A. Parilla, L. J. Simpson, C. Kittrell, and J. M. Tour, Nano-engineered spacing in graphene sheets for hydrogen storage, *Chem Mater* **23** (4), 923–925 (2011).

47. S. Iijima, Helical microtubules of graphitic carbon, *Nature* **354** (6348), 56–58 (1991).

48. R. F. Curl and R. E. Smalley, Probing C-60, *Science* **242** (4881), 1017–1022 (1988).

49. M. G. Nijkamp, J. E. M. J. Raaymakers, A. J. Van Dillen, and K. P. De Jong, Hydrogen storage using physisorption—materials demands, *Appl Phys A-Mater* **72** (5), 619–623 (2001).

50. B. Panella, M. Hirscher, and S. Roth, Hydrogen adsorption in different carbon nanostructures, *Carbon* **43** (10), 2209–2214 (2005).

51. F. S. Baker, C. E. Miller, A. J. Repik, and E. D. Tolles, in *Kirk-Othmer Encyclopedia of Chemical Technology*, Vol. 4 (Wiley, New York, 2003).

52. M. Winter, J. O. Besenhard, M. E. Spahr, and P. Novak, Insertion electrode materials for rechargeable lithium batteries, *Adv Mater* **10** (10), 725–763 (1998).

53. J. R. Dacey, in *The Solid-Gas Interface*, Vol. 2, edited by E. A. Flood (Dekker, New York, 1967) pp. 995–1022.

54. S. Tedds, A. Walton, D. Broom, and D. Book, Characterisation of porous hydrogen storage materials: carbons, zeolites, MOFs and PIMs, *Faraday Discussions*, **151**, 75–94 (2011).

55. J. Biener, M. Stadermann, M. Suss, M. Worsely, M. Biener, K. Rose, T.F. Baumann, Advanced carbon aerogels for energy applications, *Energy Environ. Sci.*, **4**, 656–667 (2011).

56. R. W. Fu, T. F. Baumann, S. Cronin, G. Dresselhaus, M. S. Dresselhaus, and J. H. Satcher, Formation of graphitic structures in cobalt- and nickel-doped carbon aerogels, *Langmuir* **21** (7), 2647–2651 (2005).

57. H. Kabbour, T. F. Baumann, J. H. Satcher, A. Saulnier, and C. C. Ahn, Toward new candidates for hydrogen storage: high-surface-area carbon aerogels, *Chem Mater* **18** (26), 6085–6087 (2006).

58. M. Felderhoff, C. Weidenthaler, R. Von Helmolt, and U. Eberle, Hydrogen storage: the remaining scientific and technological challenges, *Phys Chem Chem Phys* **9** (21), 2643–2653 (2007).

59. S. H. Jhung, J. W. Yoon, S. Lee, and J. S. Chang, Low-temperature adsorption/storage of hydrogen on FAU, MFI, and MOR zeolites with various Si/Al ratios: effect of electrostatic fields and pore structures, *Chem-Eur J* **13** (22), 6502–6507 (2007).

60. A. V. A. Kumar, H. Jobic, and S. K. Bhatia, Quantum effects on adsorption and diffusion of hydrogen and deuterium in microporous materials, *J Phys Chem B* **110** (33), 16666–16671 (2006).

61. R. Robson, Design and its limitations in the construction of bi- and poly-nuclear coordination complexes and coordination polymers (aka MOFs): a personal view, *Dalton Trans* (38), 5113–5131 (2008).

62. N. L. Rosi, J. Eckert, M. Eddaoudi, D. T. Vodak, J. Kim, M. O'Keeffe, and O. M. Yaghi, Hydrogen storage in microporous metal-organic frameworks, *Science* **300** (5622), 1127–1129 (2003).

63. B. Panella, M. Hirscher, H. Putter, and U. Muller, Hydrogen adsorption in metal-organic frameworks: Cu-MOFs and Zn-MOFs compared, *Adv Funct Mater* **16** (4), 520–524 (2006).

64. M. Schlichtenmayer, B. Streppel, and M. Hirscher, Hydrogen physisorption in high SSA microporous materials—a comparison between AX-21_33 and MOF-177 at cryogenic conditions, *Int J Hydrogen Energ* **36** (1), 586–591 (2011).

65. S. S. Kaye, A. Dailly, O. M. Yaghi, and J. R. Long, Impact of preparation and handling on the hydrogen storage properties of Zn4O(1,4-benzenedicarboxylate)(3) (MOF-5), *J Am Chem Soc* **129** (46), 14176–14177 (2007).

66. M. Latroche, S. Surble, C. Serre, C. Mellot-Draznieks, P. L. Llewellyn, J. H. Lee, J. S. Chang, S. H. Jhung, and G. Ferey, Hydrogen storage in the giant-pore metal-organic frameworks MIL-100 and MIL-101, *Angew Chem Int Ed* **45** (48), 8227–8231 (2006).

67. M. Dinca, A. Dailly, Y. Liu, C. M. Brown, D. A. Neumann, and J. R. Long, Hydrogen storage in a microporous metal-organic framework with exposed Mn^{2+} coordination sites, *J Am Chem Soc* **128** (51), 16876–16883 (2006).

68. J. L. C. Rowsell and O. M. Yaghi, Effects of functionalization, catenation, and variation of the metal oxide and organic linking units on the low-pressure hydrogen adsorption properties of metal-organic frameworks, *J Am Chem Soc* **128** (4), 1304–1315 (2006).

69. Y. Liu, H. Kabbour, C. M. Brown, D. A. Neumann, and C. C. Ahn, Increasing the density of adsorbed hydrogen with coordinatively unsaturated metal centers in metal-organic frameworks, *Langmuir* **24** (9), 4772–4777 (2008).

70. M. Colin and A. Herold, Systems graphite-alkali metal-hydrogen. 2. Hydrogenation of phases of type KC_8, RbC_8, CsC_8—reactions of hydrogen with phases MC_{24} systems graphite-potassium-deuterium, *B Soc Chim Fr* (6), 1982 (1971).

71. P. Lagrange, D. Guerard, J. F. Mareche, and A. Herold, Hydrogen storage and isotopic protium deuterium-exchange in graphite potassium intercalation compounds, *J Less-Common Metals* **131**, 371–378 (1987).

72. K. Watanabe, T. Kondow, M. Soma, T. Onishi, and K. Tamaru, Molecular-sieve type sorption on alkali graphite intercalation compounds, *Proc R Soc London Ser A–Math Phys Eng Sci* **333** (1592), 51–67 (1973).

73. J. Purewal, previously unpublished results.

74. H. Nishihara, P. X. Hou, L. X. Li, M. Ito, M. Uchiyama, T. Kaburagi, A. Ikura, J. Katamura, T. Kawarada, K. Mizuuchi, and T. Kyotani, High-pressure hydrogen storage in zeolite-templated carbon, *J Phys Chem C* **113** (8), 3189–3196 (2009).

8

Development of Off-Board Reversible Hydrogen Storage Materials

Jason Graetz, David Wolstenholme, Guido Pez, Lennie Klebanoff,
Sean McGrady, and Alan Cooper

CONTENTS

Introduction 240
Aluminum Hydride...241
  Synthesis of AlH$_3$..242
    Structure and Thermodynamics of AlH$_3$...244
    Kinetics of H$_2$ Release 247
    Rehydrogenation of Al to Form AlH$_3$...250
  Organometallic Approach to Regeneration of AlH$_3$.......................................252
  Electrochemical Reversible Formation of AlH$_3$...255
  LiAlH$_4$ 256
Ammonia Borane 258
  Solid-State Chemistry 258
    Structure..258
    Thermal Decomposition 260
  Solution-Phase Decomposition 267
  Computational Studies...269
  Tailoring the Hydrogen Release Properties..273
    Metal-Mediated Dehydrogenation 273
    Acid-Catalyzed Dehydrogenation 275
    Thermal Decomposition in Ionic Liquids 277
    The Release of Hydrogen Using Scaffolds..278
    Changes in the Chemical Composition 280
  Regeneration of AB from Spent Dehydrogenated Products 284
  Outlook for Ammonia Borane..286
Liquid Organic Hydrogen Carriers (LOHCs)...287
  Catalytic Hydrogenation/Dehydrogenation Fundamentals...............................287
  The Cyclohexane-Benzene-Hydrogen Cycle..289
  Methylcyclohexane-Toluene-Hydrogen Cycle...290
    Innovative Catalytic Reactor Designs..290
    Vehicular Applications of the MTH Cycle 291
    Electrical Energy Storage..292
  Decalin-Naphthalene-Hydrogen Cycle..292
    DEC Dehydrogenation Catalysis 292
    Innovative Reactor/Reaction Systems 293
    Membrane Reactors..293

Catalytic Dehydrogenation under Nonequilibrium Conditions 293
Reactor Development—Toward Technology Implementation 296
Polycyclic Aromatic Carriers: Toward Lower Dehydrogenation Enthalpy Systems ... 297
Polyaromatic Hydrocarbons 297
Polyaromatic Hydrocarbons with Nitrogen and Oxygen Heterotoms 299
The N-Ethyldodecahydrocarbazole/N-Ethylcarbazole (NEDC/NEC) Cycle 301
Hydrogenation of N-Ethylcarbazole 301
Catalytic Dehydrogenation of N-Ethyldodecahydrocarbazole 305
Dehydrogenation Reactor Development 309
Potential N- and O-Heterocyclic Molecule Carriers 309
The Perhydro-4,7-Phenanthrolene/4,7-Phenanthrolene System 309
Piperidine and Octahydroindole Systems: Effects of Ring Substitution 310
The Perhydrodibenzofuran-Dibenzofuran System 311
Boron-Nitrogen Heterocyclics 312
Autothermal Hydrogen Storage 312
Coupled Dehydrogenation Reactions 312
Dehydrogenation Coupled with Selective Partial Oxidation 314
Miscellaneous Potential Liquid Carriers 315
Electrochemical "Virtual" Hydrogen Storage with Organic Liquid Carriers 316
Outlook for LOHC 318
Acknowledgments 320
References 321

Introduction

The hydrogen storage materials described in Chapter 5 (interstitial metal hydrides), Chapter 6 (on-board reversible complex metal hydrides), and Chapter 7 (physisorption materials) all aim to develop a material that can deliver hydrogen to a fuel cell or a hydrogen internal combustion engine (ICE) and then be recharged "on-board" the particular application with molecular hydrogen. The application could be a hydrogen-powered light-duty vehicle or a piece of fuel-cell-powered equipment, such as the Fuel Cell Mobile Light described in Chapter 2. When the hydrogen storage material is "spent," the tankage and hydrogen storage material within the tankage stay on-board the application, and hydrogen, typically at an elevated pressure, is applied to recharge the material. This situation mimics current refueling of vehicles with gasoline; the vehicle with empty tank arrives at the filling station, fills up, and then leaves.

A different approach to recharging the material is to conduct "off-board" recharging or rehydrogenation. In one version of this approach, when the hydrogen storage material is spent, the tank is removed from the application, and a new tank with fully hydrogenated storage material is installed on the vehicle. The "empty" tank is then sent to a central facility where the material can be recharged "off-board" the application. In another version of this approach, the tank may remain on-board the application, but the spent hydrogen storage material itself is somehow removed from the tank at the refueling station and replaced with new hydrogenated material within the same tank. Perhaps the hydrogen storage material is in a "slurry" form in both hydrogenated and dehydrogenated states, allowing it to be removed from the tank by pumping, followed by refueling of the tank with new hydrogenated slurry material. The dehydrogenated material, now removed from the tank, can be sent to a rehydrogenation facility, where it can be recharged (or reversed) off-board the application.

This off-board reversing of the material requires many of the same material properties as the on-board reversible materials, namely, a high level of reversibility, good thermodynamics, and rapid kinetics. However, in off-board rehydrogenation, there would be more latitude about the kinds of chemical approaches that could be used to rehydrogenate the material. Clearly, off-board regeneration would be a major philosophical change from the current refueling of hydrocarbon-fueled vehicles.

This chapter considers a number of hydrogen storage media that have been investigated with off-board rehydrogenation in mind. These include the metal hydride AlH_3; the complex metal hydride $LiAlH_4$; ammonia borane (NH_3BH_3; AB), which stores hydrogen in covalent N-H and B-H bonds; and liquid organic hydrogen carriers (LOHCs) that store hydrogen in covalent C-H bonds. We begin with AlH_3, often called *alane*.

Aluminum Hydride

In many ways, aluminum hydride (AlH_3) represents a nearly ideal hydrogen storage material from a desorption point of view. AlH_3 has a gravimetric capacity of 10 wt% and volumetric capacity of 149 g H_2/L, along with a hydrogen desorption temperature of about 60°C to 175°C (depending on particle size, surface coatings, and catalyst concentration). Due to aluminum hydride's low temperature of decomposition and its ability to store 10% hydrogen by weight, this material has been the subject of study for decades, as reviewed by Sandrock et al. [1, 2] and more recently by Graetz et al. [3].

Over the past 50 years, aluminum hydride has been used as an explosive, a reducing agent, a solid rocket propellant, a hydrogen source for portable power systems, and for the deposition of Al films. Interest in aluminum hydride dates to the 1950s, when independent military programs in the United States and the former Soviet Union (FSU) first synthesized the crystalline form of α-AlH_3. In the United States, much of the early work on aluminum hydride was conducted by the Dow Chemical Company in collaboration with the U.S. Air Force. Their work was focused on developing a stable form of alane as a high-energy rocket propellant. The program ended around 1970 and resulted in a continuous method for preparing alpha alane [4] capable of producing 100 lb/day [5].

A slightly different process used in the FSU resulted in a higher-purity material with better crystallinity and morphology. Although the exact procedure remains unknown, an effort to reproduce the FSU material was initiated by ATK Thiokol Incorporated in 2001. This effort led to a procedure for preparing high-purity, highly crystalline α-AlH_3 (similar to FSU material) in batches up to 25 g. Currently, there are no chemical suppliers preparing large quantities of aluminum hydride, and despite attempts by various companies in recent years, it is still not commercially available. Scanning electron microscopic (SEM) pictures of these two types (from the United States and FSU) of α-AlH_3 are shown in Figure 8.1.

The more recent renaissance in hydrogen storage research for automotive and portable electronic applications has generated renewed interest in AlH_3 due to its light weight and low decomposition temperature. However, a number of challenges remain. When used as an explosive or a propellant, the material releases its hydrogen (and energy) nearly

(a)   (b)

FIGURE 8.1
α-AlH$_3$ prepared by (a) Dow Chemical Company (United States) showing slightly irregular 100-μm cuboids with some porosity and (b) the more regular and less-porous material prepared by the former Soviet Union with a crystallite size of about 20 μm. (Reproduced from Graetz, J.; Reilly, J.J.; Yartys, V.A.; Maehlen, J.P.; Bulychev, B.M.; Antonov, V.E.; Tarasov, B.P.; Gabis, I.E.; *J. Alloys Compd.* 2011, *509S*, S517, with permission from Elsevier.)

instantaneously. If alane is to be used as a hydrogen source for fuel cell or H$_2$ ICE applications, the rate of hydrogen release must be controllable. Since aluminum hydride is thermodynamically unstable at room temperature, the hydrogen release must be controlled by a kinetic barrier (kinetically stabilized), and considerable effort has been recently devoted to understanding how the structure, morphology, and composition affect the decomposition rate (hydrogen release rate). The other challenge with aluminum hydride is that it is not easily formed by direct hydrogenation (Al + 3/2H$_2$ → AlH$_3$) at reasonable pressures; therefore, the cost and complexity of preparing the hydride remain an issue even for off-board regeneration schemes. A number of ongoing efforts are focused on developing low-cost regeneration routes for the formation of AlH$_3$ from the spent fuel (e.g., Al powder).

Synthesis of AlH$_3$

Aluminum hydride was first prepared in 1942 as an alane amine complex AlH$_3$·2N(CH$_3$)$_3$ by Stecher and Wiberg [6]. An ethereal solution of aluminum hydride was prepared in 1947 using a simple reaction between LiH (or LiAlH$_4$) and AlCl$_3$ in diethyl ether (Et$_2$O) [7].

$$3\text{LiH} + \text{AlCl}_3 + n\text{Et}_2\text{O} \rightarrow \text{AlH}_3 \cdot n\text{Et}_2\text{O} + 3\text{LiCl} \tag{8.1}$$

$$3\text{LiAlH}_4 + \text{AlCl}_3 + n\text{Et}_2\text{O} \rightarrow 4\text{AlH}_3 \cdot n\text{Et}_2\text{O} + 3\text{LiCl} \tag{8.2}$$

The first nonsolvated form of AlH$_3$ was prepared by Chizinsky [8] in 1955 and was later identified as γ-AlH$_3$ from x-ray diffraction (XRD) data [4]. In this process, the etherated alane solution was filtered into an inert liquid (e.g., pentane) followed by drying under high vacuum for at least 12 h. Over the next 20 years, alane was developed as a rocket propellant by the United States and the FSU. In 1976, Brower and coworkers (Dow Chemical Company) published a summary of their work on nonsolvated AlH$_3$, during which time they identified a total of seven different pure and impure polymorphs [4]: α, α', β, γ, δ, ε, ζ. Similar to Chizinsky and Finholt, their synthesis method involved an ethereal reaction between LiAlH$_4$ and AlCl$_3$ to form etherated AlH$_3$ (Reaction 8.3), which was followed up with a desolvation step at 60–70°C (Reaction 8.4). The Dow team improved on the desolvation step by using excess $LiAlH_4$ and, in some cases, excess $LiBH_4$ in the alane-ether solution, which reduced the time and temperature of desolvation and increased the yield of pure AlH_3. It was later speculated that the successful desolvation by Chizinsky was somewhat serendipitous due to the inadvertent use of excess $LiAlH_4$ [4].

The Dow microcrystallization reaction, with excess $LiAlH_4$ and $LiBH_4$, remains the simplest and most reliable synthesis procedure for the preparation of α-AlH_3 (100- to 200-nm particle diameter):

$$4LiAlH_4 + AlCl_3 + Et_2O \rightarrow 4AlH_3 \cdot nEt_2O + 3LiCl\downarrow + LiAlH_4 + Et_2O \qquad (8.3)$$

$$4AlH_3 \cdot nEt_2O + LiAlH_4 + xLiBH_4 + Et_2O \rightarrow 4AlH_3\downarrow + Et_2O\uparrow + LiAlH_4\downarrow + xLiBH_4\downarrow \quad (8.4)$$

In the final step, the excess $LiAlH_4$ and $LiBH_4$ is removed by washing with diethyl ether (AlH_3 is insoluble in Et_2O).

The other polymorphs of AlH_3 are prepared by slight modifications of these reactions. For example, if no $LiBH_4$ is used in Reaction 8.4 ($x = 0$), then the γ polymorph will be the first phase to form during the desolvation. On further heating, the γ phase transforms into the α phase, but if the reaction temperature is kept low (~60°C) and is stopped before the transformation, then a pure phase of γ-AlH_3 can be isolated. γ-AlH_3 can also be prepared by carefully grinding a dry mixture of alane etherate ($AlH_3 \cdot nEt_2O$) with $LiAlH_4$ (in a 4:1 mixture) and heating to 70–80°C under dynamic vacuum. β-AlH_3 is typically formed using the standard synthesis procedure (Reactions 8.3 and 8.4) with $x \sim 1$ and typically forms within the first 1–2 h along with γ-AlH_3 and rarely (never) appears in pure form. α'-AlH_3 has been prepared by a few different methods; Brower et al. prepared it by performing the desolvation step in a sealed pressure reactor [4], while Brinks et al. prepared an impure form of α' by reacting $LiAlH_4$ with $AlCl_3$ in a cryomill [9]. The other polymorphs (δ, ε, and ζ) identified by Brower et al. were formed either during the continuous reaction or in reactions using other solvents [4], but little information is known about these phases since no reproducible synthesis method has been established.

Pure or etherated AlH_3 can also be synthesized using any of the binary hydrides or tetrahydroaluminates (alanates) from the first and second groups through a reaction with a Brønsted-Lewis acid (e.g., $AlCl_3$) [10–17]. The crystallization process can be accomplished through a process that avoids the formation of alane etherate as an independent step. As with the other synthesis routes, the use of the excess of lithium alanate, introduced in solution or as a solid mixture, is necessary to help with the desolvation. Other synthesis methods have been identified in recent years, including the "direct" synthesis of nonsolvated alane by a reaction between an alanate and aluminum bromide in an aromatic hydrocarbon in the absence of ether [18]. Another example is a high-temperature synthesis of α-AlH_3 similar to Reaction 8.3, but performed by mixing the solutions of alanate and aluminum chloride in an ether-toluene solution at high temperature (>90°C) [19]. Although all of these alternative routes are effective, the yield and purity are typically lower than what is achieved with the standard synthesis method (Reactions 8.3 and 8.4).

Larger quantities of α-AlH_3 are typically prepared using a continuous reaction [4]. In this process, the ethereal solution is prepared as in Reaction 8.3, but the desolvation step occurs in a crystallization flask equipped with a fractionation column, as shown in Figure 8.2. The ethereal AlH_3 solution (with excess $LiAlH_4$ and $LiBH_4$) is added continuously to benzene (or toluene), and the solution is heated to about 77°C to distill off the Et_2O and precipitate AlH_3. A number of different solvents can be used in the continuous reaction, provided one of the liquid components is nonsolvating with a higher boiling point than diethyl ether

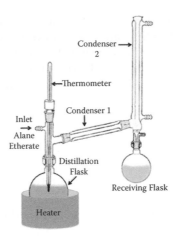

FIGURE 8.2
Example of a continuous reactor in which alane etherate is fed into a distillation flask, toluene is refluxed, and AlH_3 precipitates as ether is removed.

(normally belonging to the group of aromatic hydrocarbons). This procedure allows for more control over the rate of precipitation and therefore the size of the crystallites, which are typically 10–100 μm. The continuous synthesis method was used by Dow in the 1970s to prepare many kilograms of α-AlH_3. This process was improved slightly with the application of a postsynthesis wash in dilute acid, which results in an oxide/hydroxide surface coating, reduces the number of smaller crystallites, and helps to remove impurities. In pure water, hydrolysis of AlH_3 causes the pH to increase rapidly, accelerating the reaction and leading to complete loss of hydrogen. The acid helps to maintain a level pH (~7.0) causing the reaction to stop after the formation of a thin passivation layer.

Structure and Thermodynamics of AlH₃

Seven polymorphs of AlH_3 (α, α′, β, γ, ε, δ, and ζ) were originally reported by Brower et al. [4], but only four of those phases (α, α′, β, and γ) have reliably been prepared in subsequent studies. Considerable effort has been devoted to characterizing the thermodynamics and crystal structures of the most stable phases to better understand the hydrogen storage properties of the material. The structure of the first and most stable aluminum hydride phase, α-AlH_3, was solved in 1969 by Turley and Rinn. α-AlH_3 has a hexagonal unit cell in the trigonal space group $R\bar{3}c$, with lattice parameters $a = 4.449$ Å and $c = 11.804$ Å [20, 21]. In this structure, the AlH_6 octahedra are corner connected, with each H bridging two AlH_6 units, and the shortest bond distances are 1.715 (Al-H), 2.418 (H-H), and 3.236 (Al-Al) Å, as shown in Figure 8.3. The Al layer in AlH_3 exhibits face-centered cubic packing and is very similar to the packing in Al metal; however, the density of AlH_3 is only half that of Al metal due to the larger interatomic distances in AlH_3.

During thermal decomposition, α-AlH_3 decomposes into Al metal and H_2 gas in a single endothermic step: α-$AlH_3 \rightarrow Al + 3/2H_2$, as shown in Figure 8.4b. Thermodynamic values determined from calorimetric studies of the hydrogen release give a α-AlH_3 formation entropy of –129 J/K mol H_2 [22], an enthalpy of formation ranging from –6.0 to –7.6 kJ/mol H_2 [22–24], and a Gibbs free energy of approximately 31.1–32.7 kJ/mol H_2 at room temperature [22, 23]. It is important to note that α-AlH_3 is thermodynamically unstable and has an equilibrium hydrogen pressure of 0.7 GPa (fugacity of 50 GPa) at 300 K [25].

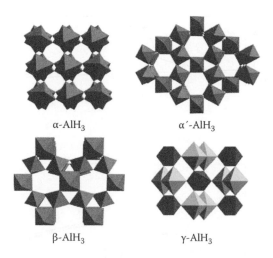

α-AlH₃   α′-AlH₃

β-AlH₃   γ-AlH₃

FIGURE 8.3
Crystal structures of the AlH_3 polymorphs showing α-AlH_3, α'-AlH_3, β-AlH_3, and γ-AlH_3. (Reproduced from Graetz, J.; Reilly, J.J.; Yartys, V.A.; Maehlen, J.P.; Bulychev, B.M.; Antonov, V.E.; Tarasov, B.P.; Gabis, I.E.; *J. Alloys Compd.* 2011, *509S*, S517, with permission from Elsevier.)

Little was known about the structure and thermodynamics of the other, less-stable polymorphs of AlH_3 until 2005, when Ke et al. used density functional theory (DFT) to propose two stable structures of AlH_3, orthorhombic *Cmcm* and cubic $Fd\bar{3}m$ [26]. The following year, the crystal structure of the α' phase was experimentally determined by Brinks et al. using a combination of XRD and neutron diffraction [9]. In these experiments, the hydrogen atoms were replaced with deuterium, which is chemically identical but has a much lower incoherent neutron scattering cross section, making it more suitable for neutron diffraction studies. α'-AlD_3 is orthorhombic, *Cmcm*, with unit cell dimensions a = 6.470 Å, b = 11.117 Å, and c = 6.562(2) Å [9]. It has two distinct Al sites, and the shortest bond distances are 1.68 (Al-D), 2.31 (D-D), and 3.22 (Al-Al) Å. Similar to the structure of the α phase, the α' phase consists of corner-connected octahedra (each H [or D] shared with two octahedra), but the α' structure is more open, consisting of a three-dimensional network of octahedra with large 3.9-Å channels (see Figure 8.3).

α'-AlH_3 is less stable than the α phase, and during thermal decomposition it can decompose directly to the elements (Al + H_2) or undergo an exothermic transition to α-AlH_3 (releasing 1.1 kJ/mol H_2 of heat) [27], followed by decomposition and H_2 release (Figure 8.4a). Previous studies have shown that the direct decomposition is favored at lower temperatures (<100°C). The total formation enthalpy for α'-AlH_3 is –4.9 to –6.5 kJ/mol H_2, giving a Gibbs free energy of 32.2–33.8 kJ/mol H_2 at room temperature [3, 27].

The structure of β-AlD_3 was determined in 2007 by Brinks et al., also using a combination of synchrotron XRD and neutron diffraction studies. β-AlD_3 has a cubic crystal structure ($Fd\bar{3}m$) with a unit cell dimension of a = 9.004 [28]. Similar to the α' structure, the β structure consists of corner-connected octahedra forming a three-dimensional network with 3.9-Å channels (Figure 8.3). The structure has one Al site, and the shortest bond distances are 1.712 (Al-D), 2.358 (D-D), and 3.183 (Al-Al) Å.

β-AlH_3 is also less stable than α-AlH_3, and during thermal decomposition it can decompose directly to the elements (Al + H_2) or undergo an exothermic transition to α-AlH_3 (releasing 1.5 kJ/mol H_2 of heat), followed by decomposition and H_2 release (Figure 8.4b, middle trace) [24]. Previous studies have shown that the $\beta \rightarrow \alpha$ transition is favored above

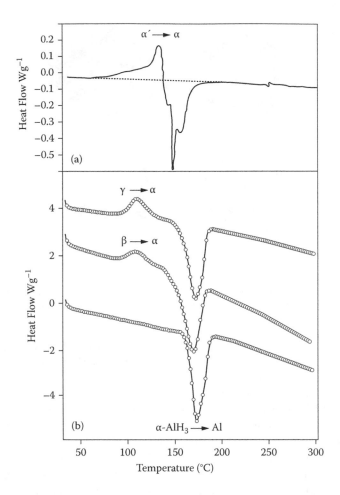

FIGURE 8.4

(a) Differential scanning calorimetry (DSC) trace from α'-AlH$_3$ (5°C/min) showing the exothermic $\alpha' \rightarrow \alpha$ transition at about 120°C. (b) DSC traces from α-AlH$_3$, β-AlH$_3$, and γ-AlH$_3$ (10°C/min) showing the decomposition of the α phase at about 170°C along with the exothermic polymorphic transitions $\beta \rightarrow \alpha$ and $\gamma \rightarrow \alpha$ at about 110°C. (Reproduced from Graetz, J.; Reilly, J.J.; *Scripta Materialia* 2007, *56*, 835, with permission.)

100°C, but at lower temperatures, the direct decomposition reaction becomes more favorable [3, 23, 29]. The total formation enthalpy for β-AlH$_3$ is −4.5 to −6.1 kJ/mol H$_2$, giving a Gibbs free energy of 32.6–34.2 kJ/mol H$_2$ at room temperature [23].

The structure of γ-AlH$_3$ was simultaneously determined by Yartys et al. [30] and Brinks et al. [31] in 2007 using combined synchrotron and neutron diffraction. γ-AlH$_3$ crystallizes with an orthorhombic unit cell (*Pnnm* space group) with unit cell dimensions $a = 5.3806(1)$ Å, $b = 7.3555(2)$ Å, $c = 5.77509(5)$ Å, and $V = 228.561(7)$ Å. The crystal structure of γ-AlH$_3$, shown in Figure 8.3, contains two types of AlH$_6$ octahedral units with edge-sharing double-bridge bonds, Al-2H-Al bonds, and corner-connected single-bridge bonds, Al-H-Al, found in all other alane structures. The double-bridge bond has short Al-Al (2.61-Å) and Al-H (1.68–1.70-Å) bonds, while the Al-H distances for Al atoms in the single-bridge bonds are 1.769–1.784 Å. The crystal structure of γ-AlH$_3$ contains large cavities between the AlH$_6$ octahedra; therefore, the density is 11% less than for α-AlH$_3$.

γ-AlH$_3$ is the least stable of the four primary AlH$_3$ polymorphs, and similar to α′-AlH$_3$ and β-AlH$_3$, thermal decomposition occurs by two possible routes: direct decomposition to the elements (Al + H$_2$) or an exothermic transition to α-AlH$_3$ (releasing 2.8 kJ/mol H$_2$ of heat), followed by decomposition and H$_2$ release (Figure 8.4b, top trace). Previous studies have shown that the γ → α transition is favored above 100°C, but the direct decomposition reaction becomes more favorable at lower temperatures [3, 23, 29]. The total formation enthalpy for γ-AlH$_3$ is −3.2 to −4.8 kJ/mol H$_2$, giving a Gibbs free energy of 33.9–35.5 kJ/mol H$_2$ at room temperature [23].

Kinetics of H$_2$ Release

The rate of H$_2$ desorption from aluminum hydride varies considerably with the properties of the material (e.g., structure, morphology/size, surface composition, and catalyst concentration). In all cases, the thermal decomposition curve has a similar sigmoidal shape and consists of three distinct regions as shown in Figure 8.5: (I) an induction period at the onset of the reaction, corresponding to a breakup of the surface layer and nucleation of aluminum; (II) an acceleratory period, attributed to growth of the aluminum phase in two and three dimensions; and (III) a decay period, where the concentration of the hydride phase is limited and the growing aluminum particles begin to overlap [3].

Early research efforts on aluminum hydride were primarily focused on stabilization, preparing an air-stable hydride (solid rocket fuel) that was stable against decomposition from years to decades. Since even the most stable form of alane, α-AlH$_3$, is unstable at room temperature, a variety of stabilization methods were developed over the years. In one example, magnesium-based compounds, LiAlH$_4$·Mg(AlH$_4$)$_2$ and LiCl·Mg(AlH$_4$)$_2$ [32], were used to significantly increase the energy barrier to decomposition. Other groups

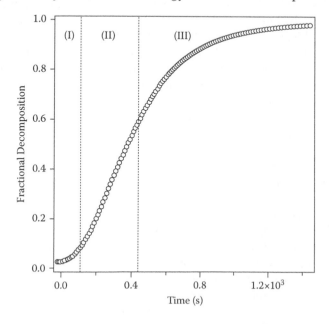

FIGURE 8.5

Isothermal decomposition of α-AlH$_3$ (Dow) at 180°C showing (I) induction period, (II) acceleratory period, and (III) decay period. (Reproduced from Graetz, J.; Reilly, J.J.; Yartys, V.A.; Maehlen, J.P.; Bulychev, B.M.; Antonov, V.E.; Tarasov, B.P.; Gabis, I.E.; *J. Alloys Compd.* 2011, *509S*, S517, with permission from Elsevier.)

FIGURE 8.6
Large single crystals of α-AlH₃ (a) before and (b) after H₂ desorption. (Reproduced from Graetz, J.; Reilly, J.J.; Yartys, V.A.; Maehlen, J.P.; Bulychev, B.M.; Antonov, V.E.; Tarasov, B.P.; Gabis, I.E.; *J. Alloys Compd.* 2011, *509S*, S517, with permission from Elsevier.)

demonstrated that the alane could also be stabilized by altering the surface chemistry of the crystallites through the application of a surface oxide or hydroxide layer [33–36]. Typically, this passivation layer is introduced through a controlled hydrolysis reaction, such as by washing in a dilute acid solution or an alcohol plus 2% water solution (or both).

The rate of hydrogen transport through a surface layer of Al_2O_3 decreases rapidly with oxide thickness (1–3 nm) [37]. Rapid desorption of H_2 from the surface of AlH_3 requires the breakup of the oxide layer, the first step in the decomposition [38]. Once cracks or channels have formed in the surface layer, small nucleation sites of metallic Al form on the hydride surface and initiate for further decomposition. This leads to the formation of a porous aluminum structure, as shown in Figure 8.6, which provides short diffusion paths for hydrogen traveling from the metal-hydride boundary to the surface. The greatest level of stability was achieved by the Dow Chemical Company through the application of a thin organic layer. Although their procedure remains unknown, the α-AlH₃ prepared by the Dow is extremely stable, with a demonstrated shelf life of over 30 years in air [1].

In the late 1970s and early 1980s, Herley et al. investigated the thermal and photolytic decomposition of α-AlH₃ prepared by Dow Chemical Company, a material consisting of large cuboid crystallites (50–100 μm) coated in a thin organic layer [39–43]. The Dow material had an activation energy for the acceleratory and decay period of approximately 157 kJ/mol. Irradiation of the material prior to thermal decomposition resulted in a reduction of the induction period likely due to an increase in the number of nucleation sites.

The hydrogen release rates from the thermal decomposition of small crystallites (100–150 nm) of α-, β-, and γ-AlH₃ (no coatings) were investigated to determine the reaction pathways and activation energies [44]. At high temperature (≥100°C), the reaction rates for all three polymorphs are similar, as shown in Figure 8.7. At these temperatures, the less-stable β and γ phases quickly transform in α-AlH₃ prior to decomposition, rendering the rates from all three samples similar. At lower temperatures (<100°C), the rate curves begin to diverge, and the H_2 evolution occurs more rapidly from the β and γ phases. In this temperature range, two decomposition pathways are observed, direct decomposition (e.g., γ-AlH₃ → Al + 3/2H₂) along with indirect decomposition (e.g., γ-AlH₃ → α-AlH₃ → Al + 3/2H₂), as previously discussed [23, 29]. In situ synchrotron XRD studies of the thermal decomposition of γ-AlH₃ heated at 2°C/min under vacuum showed that around 60% of the hydrogen is released in the direct decomposition process (γ-AlH₃ → Al + 3/2H₂), while the remaining 40% of the γ-AlH₃ first transforms into the α-AlH₃ and then decomposes [45]. The divergence of the rate curves in Figure 8.7 indicates that the fraction of material that undergoes direct decomposition increases with decreasing temperature.

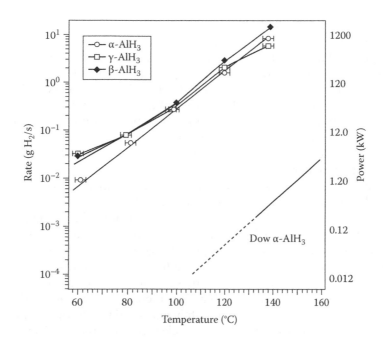

FIGURE 8.7

Rate of H_2 release as a function of temperature during thermal decomposition of α-, β-, and γ-AlH$_3$ and α-AlH$_3$ prepared by Dow [23, 29, 41]. The right axis shows the equivalent power based on the lower heating value of H_2 (120 kJ/g). (Reproduced from Graetz, J.; Reilly, J.J.; Yartys, V.A.; Maehlen, J.P.; Bulychev, B.M.; Antonov, V.E.; Tarasov, B.P.; Gabis, I.E.; *J. Alloys Compd.* 2011, *509S*, S517, with permission from Elsevier.)

The measured activation energies for the three AlH$_3$ polymorphs were 102 kJ/mol (α-AlH$_3$), 92 kJ/mol (β-AlH$_3$), and 79 kJ/mol (γ-AlH$_3$) over the temperature range of 60–140°C [44]. It is important to note that the energies for β- and γ-AlH$_3$ are averaged over both possible decomposition pathways (direct and indirect). A comparison of decomposition rates measured from larger crystallites (20 μm) of uncoated α-AlH$_3$ reveals an activation energy of 104 kJ/mol, which is similar to the value measured from the smaller 100- to 200-nm particles (102 kJ/mol) [46]. On the other hand, a much greater activation energy (~157 kJ/mol) [41] was measured in α-AlH$_3$ (~50 μm) coated with a thin organic layer (prepared by Dow Chemical). This seems to suggest that the enhanced stabilization observed in the Dow material is primarily attributed to the surface composition rather than the crystallite size, which has a smaller effect.

Recent interest in aluminum hydride for low-temperature fuel cell applications has shifted the focus toward destabilizing (rather than stabilizing) the hydride and enhancing desorption rates. This has been motivated by the need for lightweight and compact hydrogen storage media for fuel cells that require rapid hydrogen evolution rates at low temperature (≤100°C). Based on the previously reported kinetic values, we can estimate the hydrogen supply rate for all three AlH$_3$ polymorphs (α, β, and γ) is about 2 g H_2/s at 120°C, based on 100 kg of material. This H_2 rate is more than three orders of magnitude greater than that of NaAlH$_4$ (undoped) and exceeds the Department of Energy (DOE) full-flow target of 1 g H_2/s for a 50-kW fuel cell (although the temperature of 120°C is slightly higher than the delivery target of 85°C).

The activation energy and rate of decomposition are both highly dependent on the synthesis conditions, sample purity, morphology, and surface composition. The application of a postsynthesis thermal treatment can be used to eliminate the induction period but has no effect on the activation energy. Aluminum hydride can also be destabilized by exposure to light [39, 42, 43], which has been shown to decrease the activation energy by 15–20% compared to nonirradiated "fresh" samples [47]. Ball milling AlH_3 results in an enhancement of the H_2 desorption rates likely due to a decrease in crystallite size and an increase in clean (oxide-free) surfaces [1, 48]. The enhanced desorption rates achieved with each of these activation methods (thermal, irradiation, ball milling) can also be explained by the small amount of desorption that occurs during the treatment. The metallic Al particles formed on the hydride surface act as nucleation sites and reduce, or in some cases eliminate, the incubation period.

The most significant destabilization effect, resulting in an enhancement of hydrogen evolution rates, has been achieved with the use of additives, such as alkali hydrides [1, 2] and transition metals, [48, 49] which catalyze the desorption of H_2. The addition of titanium has resulted in a reduction of the activation energy by at least 50%, making the storage and even synthesis of catalyzed alane quite dangerous [50]. Although the addition of a transition metal catalyst (e.g., Ti) is extremely effective, the catalyst is most active when it is added in solution during the AlH_3 synthesis. The addition of only a few parts per million of $TiCl_3$ into Reaction 8.3 results in a clear increase in the rate of H_2 evolution [3]. Isothermal decomposition curves from Ti-catalyzed α-AlH_3 (~10 ppm) and uncatalyzed α-AlH_3 are shown in Figure 8.8 [3, 51]. There is a clear reduction in the induction period at the onset of decomposition (most evident at higher temperatures) along with an increase in the slope or rate of H_2 release (most evident at lower temperatures). Higher Ti concentrations were found to be completely unstable, and in many cases the hydride decomposed at room temperature before any measurements could be made. These results clearly demonstrated that Ti has a dramatic destabilizing effect on α-AlH_3, especially when it is finely distributed throughout the material.

Rehydrogenation of Al to Form AlH_3

After the initial research phase characterizing the structure and hydrogen release properties of AlH_3, attention turned quickly to the problem of regeneration. The primary barrier to using AlH_3 as a hydrogen source for mobile fuel cell or hydrogen ICE applications is the energy cost required to regenerate the hydride from the dehydrogenated state, namely, Al metal. A number of high-pressure [25, 52–55] and thermodynamic studies [22, 23] on the Al-H system have been performed to establish a pressure-temperature phased diagram, shown in Figure 8.9 [3]. From this plot, it is easy to see the problem—the direct hydrogenation of aluminum at room temperature requires pressures greater than 7 kbar, much too high to be practical on a large scale, making on-board rehydrogenation with molecular H_2 virtually impossible. Although AlH_3 is easily prepared by reacting $LiAlH_4$ with $AlCl_3$ in ether (Reactions 8.3 and 8.4), this method produces stable salt by-products (e.g., LiCl) that are energetically costly to recycle, making this route too expensive for most applications.

Due to alane's nearly ideal hydrogen desorption characteristics, and the fact that onboard reversibility seems to be impossible, the possibility for off-board regeneration was investigated. One off-board method of alane regeneration considered the formation of AlH_3 from Al and H_2 under supercritical fluid conditions. This method, however, proved to be ineffective. This left two other methods, both of which proved successful. The first, performed by Brookhaven National Laboratory (BNL), is the regeneration of AlH_3 using a two-step organometallic process; the second is the regeneration of AlH_3 via electrochemical

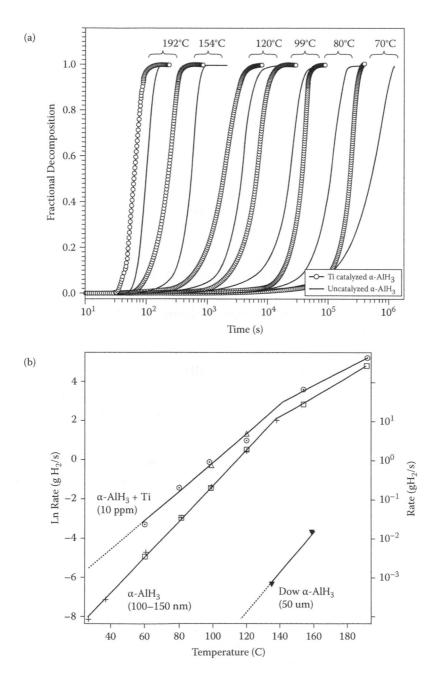

FIGURE 8.8
(a) Isothermal decomposition curves for catalyzed (circles) and uncatalyzed (thin line) α-AlH₃ along with (b) the corresponding hydrogen evolution rate plots compared with rates obtained from α-AlH₃ prepared by Dow Chemical. (Reproduced from Graetz, J.; Reilly, J.J.; Yartys, V.A.; Maehlen, J.P.; Bulychev, B.M.; Antonov, V.E.; Tarasov, B.P.; Gabis, I.E.; *J. Alloys Compd.* 2011, *509S*, S517, with permission from Elsevier.)

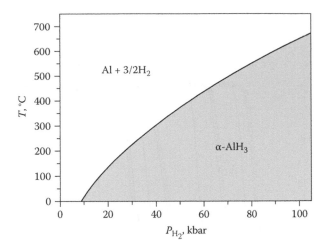

FIGURE 8.9

Pressure-temperature phase diagram of the Al-H system showing a solid line representing the calculated equilibrium for $AlH_3 \leftrightarrow Al + 3/2H_2$. (Reproduced from Graetz, J.; Reilly, J.J.; Yartys, V.A.; Maehlen, J.P.; Bulychev, B.M.; Antonov, V.E.; Tarasov, B.P.; Gabis, I.E.; *J. Alloys Compd.* 2011, *509S*, S517, with permission from Elsevier.)

means (conducted by Savannah River National Laboratory, SRNL). These two approaches are summarized next.

Organometallic Approach to Regeneration of AlH$_3$

The BNL approach to regeneration made use of a two-step process involving hydride stabilization followed by separation. Although the direct hydrogenation of Al to form AlH_3 requires extremely high pressures, this barrier can be substantially reduced by the addition of a stabilizing agent that forms an adduct with alane (e.g., an ether or an amine). The stabilizing influence of ethers was seen in the original syntheses of alane, shown in Equations 8.1 and 8.2. Once the hydride is formed, the second step involves the removal of the stabilizing agent and the recovery of the pure AlH_3. Since alane is a Lewis acid, an appropriate stabilizing agent is a Lewis base, such as an amine (NR_3), which will allow the hydrogenation to occur at low pressure. A generic example of the regeneration procedure is as follows:

$$Al + NR_3 + 3/2H_2 \rightarrow AlH_3\text{-}NR_3 \rightarrow AlH_3 + NR_3\uparrow \qquad (8.5)$$

In early 2008, BNL demonstrated that this adduct-assisted formation of AlH_3 was possible at low temperatures and low pressures using triethylenediamine (TEDA) and catalyzed aluminum powder. In this reversible reaction, the catalyzed Al powder is stirred into a solution of TEDA in tetrahydrofuran (THF). The application of just a few bars of hydrogen pressure (at 50°C) results in the formation of AlH_3-TEDA, which is insoluble and precipitates out of solution as a white solid [56].

In this first step, catalyzed aluminum is hydrogenated along with an amine (NR_3) in a liquid medium to form an amine alane (AlH_3-NR_3) [3]. Figure 8.10a shows the drop in hydrogen pressure associated with the hydrogenation of Al and dimethylethylamine (DMEA) to form AlH_3-DMEA. A number of other tertiary amines are also effective at stabilizing

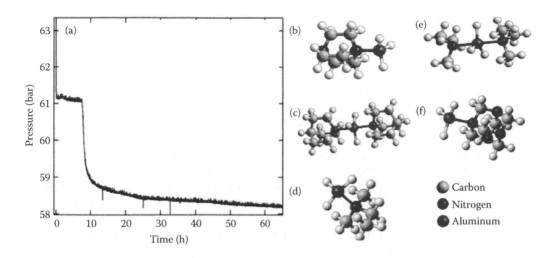

FIGURE 8.10
(a) Pressure drop due to hydrogenation of Al + DMEA to form AlH_3-DMEA and molecular structures of various hydrogenation products showing (b) triethylenediamine (TEDA) alane [56], (c) trimethylamine (TMA) alane [57], (d) quinuclidine alane [58], (e) hexamine alane, and (f) dimethylethylamine (DMEA) alane. [59]

AlH_3, including TEDA [56], trimethylamine (TMA) [57], quinuclidine [58], hexamine, and dimethylethylamine (DMEA) [59], shown in Figures 8.10(b)–(f), respectively.

The chief difficulty with all of the amine alane adducts shown in Figure 8.10 is that they are all too stable. The Al-N bond is typically strong, making the direct separation of AlH_3-NR_3 more difficult than the alane etherates. The recovery of intact AlH_3 from most amine-alane adducts is difficult since the temperatures required for adduct separation are typically greater than the AlH_3 decomposition temperature (~100°C). If one attempted to recover pure AlH_3 by heating the AlH_3-TEDA adduct (e.g., to desorb the amine), the H_2 is also lost from AlH_3. This led to a search for a less-stable amine alane adduct that could be separated at less than 100°C.

The only known alane-amine complex that can be separated under a partial vacuum at 70°C is triethylamine (TEA) alane, AlH_3-TEA. However, despite numerous attempts, AlH_3-TEA is not easily formed by direct hydrogenation. To circumvent this problem, BNL showed that one could successfully stabilize AlH_3 formation using one amine adduct and then convert that adduct to a AlH_3-TEA adduct whose TEA moiety could be more easily removed by thermal processing. That conversion of one amine adduct to another amine adduct is called "transamination." The BNL team demonstrated that DMEA and TMA in AlH_3-DMEA and AlH_3-TMA could be exchanged with TEA to form AlH_3-TEA under mild conditions [57, 59].

The final step in the regeneration process involves the separation of AlH_3-TEA into TEA and AlH_3 by heating liquid AlH_3-TEA to 75°C under a nitrogen sweep or partial vacuum for about 4 h. The TEA is removed as a gas (recovered in a trap) and the remaining solid is crystalline AlH_3. An XRD pattern of the material recovered from AlH_3-TEA shows primarily AlH_3, with a small amount of Al (~10%). The full regeneration procedure for AlH_3 is as follows:

Step 1: Formation of AlH_3-DMEA by direct hydrogenation:

$$Al + DMEA + 3/2H_2 \rightarrow AlH_3\text{-DMEA}$$

Step 2: Transamination, exchange DMEA with TEA:

$$AlH_3\text{-}DMEA + TEA \rightarrow AlH_3\text{-}TEA + DMEA\uparrow$$

Step 3: Separation of AlH_3-TEA:

$$AlH_3\text{-}TEA \rightarrow AlH_3 + TEA\uparrow$$

BNL demonstrated each of the steps necessary for alane regeneration using the spent fuel (catalyzed aluminum) and hydrogen gas and different stabilizing amines, DMEA, and TMA. In both cases, the amine alanes (AlH_3-DMEA and AlH_3-TMA) were formed by direct hydrogenation and transaminated to form AlH_3-TEA. The recovery of AlH_3 from AlH_3-TEA was also demonstrated, thereby demonstrating a new low-energy method to regenerate Ti-catalyzed AlH_3 from catalyzed aluminum and hydrogen gas. The results are summarized in Figure 8.11.

A major concern with such an off-board regeneration method is the energy required to execute these chemical steps. In 2011, an independent analysis of AlH_3 regeneration using the TMA approach of Figure 8.11 was conducted by R. K. Ahluwalia of Argonne National Laboratory [60]. Assuming 75% yields in the transamination step and 75% in the TEA decomposition and recovery step, this study found that to make 1 kg of H_2 in alane in a slurry (combustion energy = 120.9 MJ, lower heating value [LHV]), it takes about 335 MJ of primary energy (natural gas and electricity) for the regeneration process. This energy consumption excludes the energy it takes to make the 1 kg of hydrogen in the first place from natural gas. Thus, the ratio of (hydrogen energy produced)/(regeneration energy) is 0.36. Since the temperatures for all regeneration steps in Figure 8.11 are typically low (<80°C), improvements in efficiency can be achieved by using industrial waste heat rather than natural gas for the heating requirements. Using waste heat would effectively lower the needed regeneration energy.

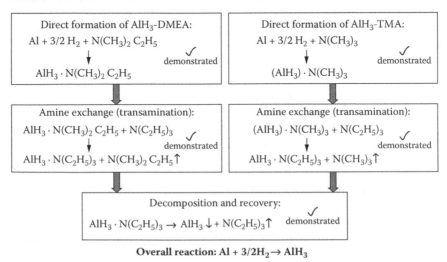

Overall reaction: Al + 3/2H$_2$→ AlH$_3$

FIGURE 8.11
Regeneration of AlH_3 via demonstrated chemical routes. Although not indicated, catalytic levels of Ti were present in the original Al being processed.

Further work on this organometallic approach should emphasize increasing the yield of the different chemical steps and the purity of the products obtained. In general, the yields for forming AlH_3-DMEA in the pressurized reactor were good. Difficulties occurred in the vacuum distillation transamination step, where trace amounts of metallic aluminum were observed. It is known that trace amounts of aluminum promote the decomposition of AlH_3. This fact restricts the window for cleanly separating AlH_3 from AlH_3-TEA. Hence, after repeated attempts to recover pure AlH_3 from AlH_3-TEA, the final product always consisted of both AlH_3 and aluminum. This problem needs to be resolved experimentally.

Electrochemical Reversible Formation of AlH$_3$

Another possible regeneration method for aluminum hydride at low pressures involves the electrochemical hydrogenation of aluminum. The electrical potential E required to drive hydrogen into a metal varies as the logarithm of the pressure: $E \propto -\ln[P]$. Hence, the application of even a low electrochemical potential results in high hydrogen fugacity. Investigations of electrochemical hydrogenation of aluminum date back to the early 1960s, when Clasen et al. used an aluminum anode and an iron/mercury cathode with an electrolyte of $NaAlH_4$ in THF [61, 62] to form AlH_3. In the late 1990s, Birnbaum et al. used aqueous solutions of H_2SO_4 and HCl along with a recombination poison ($NaAsO_4$) to achieve high hydrogen fugacities for cathodic charging of aluminum [63]. They demonstrated that hydrogen enters vacancies in the Al structure and achieved concentrations of greater than 1000 ppm, but no transformation to α-AlH_3 was observed.

Zidan et al. [64] at SRNL investigated the possibility of regenerating AlH_3 electrochemically. The relatively low potential required to achieve high hydrogen pressure is illustrated in Faraday's equation as

$$E = -\frac{RT}{2F} \ln P_{H_2} \qquad (8.6)$$

where E is the electrical potential, F is Faraday's constant, R is the gas constant, and T is the temperature.

From Equation 8.6, it is evident that the driving potential is proportional to the logarithm of the hydrogen pressure, resulting in modest driving potential requirements. However, the use of electrochemistry has to take into account that Al and AlH_3 oxidize in aqueous environment, thereby prohibiting the use of all protic solvents as electrolytes. For this reason, SRNL developed a novel route using a nonaqueous solvent system. A polar aprotic solvent such as THF is used in the electrolytic cells. These cells are operated at ambient pressure and temperature. Following cell design and fabrication, a search for useful electrodes and operating conditions was conducted.

Details of the electrochemical regeneration route are beyond the scope of this summary and can be found in the publication by Zidan and coworkers [64]. However, the general idea can be shown in Figure 8.12. The anode in the cell is a pure aluminum electrode. The counterelectrode is platinum foil or platinum coiled wire. The electrolyte in the cell is made of alanates such as $NaAlH_4$ or $LiAlH_4$ dissolved in anhydrous THF or diethyl ether. The electrolysis is carried out in an electrochemically stable, aprotic, and polar solvent such as THF or ether. $MAlH_4$ (M = Li, Na) is dissolved in this solvent, forming the ionic solution that is used as the electrolyte.

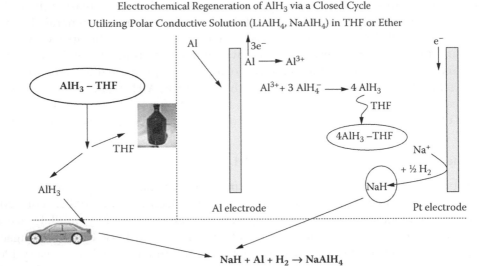

FIGURE 8.12
Electrochemical regeneration of AlH_3 from Al metal.

Starting with spent Al from the desorption of H_2 from AlH_3, the Al is removed from the particular application (say, a light-duty hydrogen-powered vehicle), and the off-board processing begins. The Al is reacted with NaH and H_2 to form $NaAlH_4$. This $NaAlH_4$ is dissolved in THF and placed in the cell. Alternatively, Al from the vehicle can enter the electrochemical environment as the Al anode. When the cell is polarized, Al metal can be converted to Al^{3+}, which reacts with dissolved AlH_4^- ions to produce AlH_3 complexed with THF. The AlH_3-THF precipitate is then filtered out from the system.

Once the AlH_3-THF adduct is formed as a white solid, the THF must be removed from the AlH_3. This proved to be problematic as the temperatures required to remove the THF also led to some dehydrogenation of AlH_3, leading to a mixture of both Al and AlH_3 in the final product. Things improved dramatically when another proprietary adduct besides THF was introduced, which led to facile removal of the adduct, forming pure AlH_3 in gram quantities. XRD confirmation of AlH_3 as the product is shown in Figure 8.13.

Improvements in yield and efficiency were achieved by the use of $LiAlH_4$ in THF and the introduction of LiCl, which acts as an electrocatalytic additive (ECA). The use of LiCl greatly enhanced the electrochemical process, yielding higher cell efficiency and more alane adduct produced.

$LiAlH_4$

As discussed in Chapter 6, lithium aluminum hydride ($LiAlH_4$) is a promising compound for hydrogen storage with a high gravimetric and volumetric hydrogen density and a low decomposition temperature. Its hydrogen release is described by the following two-step reactions:

$$3LiAlH_4 \rightarrow Li_3AlH_6 + 2Al + 3H_2 \qquad 5.3 \text{ wt\% H} \qquad \Delta H = -10 \text{ kJ/mol} \qquad (8.7)$$

$$LiAlH_6 \rightarrow 3LiH + Al + 1.5H_2 \qquad 2.6 \text{ wt\% H} \qquad \Delta H = +25 \text{ kJ/mol} \qquad (8.8)$$

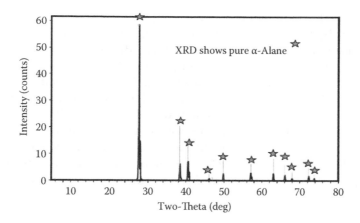

FIGURE 8.13

XRD confirmation that the electrochemically synthesized material is AlH₃. (Reproduced from Zidan, R.; Garcia-Diaz, B.L.; Fenwox, C.S.; Stowe, A.C.; Gray, J.R.; Harter, A.G.; *Chem. Commun.* 2009, *25*, 3717, with permission.)

Similar to other metastable hydrides, LiAlH₄ does not form by direct hydrogenation of LiH and Al at reasonable hydrogen pressures, and as such, LiAlH₄ is disqualified as an on-board reversible material. However, it desorbs hydrogen at a low temperature, making it worthwhile to explore off-board approaches to regenerate the material from the dehydrogenated products LiH and Al.

BNL demonstrated a low-energy route to regenerate LiAlH₄ from LiH and Ti-catalyzed Al. The BNL team adopted a regeneration process similar to that used for AlH₃ (Reaction 8.5), involving hydride stabilization followed by separation. However, in this case THF was used to stabilize the hydride (rather than an amine) as follows:

$$LiH + Al + n THF + 3/2H_2 \rightarrow LiAlH_4 - n THF \tag{8.9}$$

The thermodynamics of this reversible reaction were investigated by measuring pressure-composition isotherms, and the free energy was found to be small and slightly negative ($\Delta G = -1.1$ kJ/mol H₂), suggesting an equilibrium hydrogen pressure of just under 1 bar at 300 K. BNL also demonstrate that the adduct LiAlH₄•4THF can be desolvated at low temperature to yield crystalline LiAlH₄ [65]. More recently, this same approach has been used by Jensen and McGrady et al. using dimethyl ether (DME), rather than THF, as the stabilizing ligand [66].

$$LiH + Al[Ti] \xrightarrow[\text{r.t.; 24 h}]{Me_2O/H_2 \ (100 \ \text{bar})} LiAlH_4[Ti] \tag{8.10}$$

In this process, the use of DME over THF may be more convenient since the DME is a gas at room temperature, and it is easily removed after hydrogenation by venting. In the full regeneration procedure, the dehydrogenated product LiH + Al + Ti catalyst is dissolved in Me₂O at a hydrogen pressure of 100 bar (Me₂O is a liquid at this pressure). After 24 h at room temperature, the gases H₂ and Me₂O are released, and one is left with a quantitative conversion to catalyzed LiAlH₄. Studies of the hydrogen storage properties of the LiAlH₄ (Ti catalyzed) produced by this method (Equation 8.10) showed a hydrogen release of 7 wt% (80–180°C) over five regeneration cycles.

The prospects for this process to be a low-energy route to LiAlH$_4$ regeneration are considerable. A study by Argonne indicated that as the amount of DME required for this process is decreased, the thermal efficiency increases dramatically.

Summarizing to this point, a great deal of progress has been made in developing off-board regeneration methods for AlH$_3$ and LiAlH$_4$. An organometallic approach [3] was developed in which AlH$_3$ could be generated from H$_2$ in the presence of a stabilizing agent, with that stabilizing agent eventually removed to yield pure AlH$_3$. An electrochemical route was also established [64] in which spent aluminum could be converted to AlH$_3$ with high purity and in good yield. In addition, a remarkably facile method to regenerate LiAlH$_4$ was developed that nearly satisfies the energy efficiency requirements [66]. All of these methods provide for viable off-board regeneration of these metal hydrides.

Ammonia Borane

Ammonia borane (NH$_3$BH$_3$; AB) has been the subject of intensive research owing to its promise as an effective synthetic reagent, polymer precursor, and potential reversible off-board hydrogen storage material. AB is a colorless solid at ambient temperatures and pressures (melting point = 110–114°C), whereas its isoelectronic counterpart ethane (melting point = –181°C) exists in its gaseous form [67]. This remarkable difference results from the ability of AB to form intricate hydrogen-bonding networks involving acidic N-H and basic B-H moieties that stabilize its solid-state structure (see the following section on structure), whereas the relatively nonpolar nature of the C-H bonds in ethane precludes the formation of analogous supramolecular interactions.

The existence of hydrogen-rich AB as a solid at ambient temperatures endows this chemical hydride with favorable volumetric and gravimetric hydrogen storage capacities, a prerequisite for any potential hydrogen storage candidate. Indeed, AB possesses a theoretical hydrogen storage capacity of 19.6 wt%, which surpasses many other chemical hydrides that have been proposed for hydrogen storage (e.g., 10.1 wt% AlH$_3$; 7.9 wt% LiAlH$_4$) [66, 68]. However, the complete dehydrogenation of AB yields boron nitride as the end product, which is a highly stable compound ($\Delta H°_{f,298K} = -251 \pm 1.5$ kJ/mol), and hence a considerable energy investment is required to regenerate AB [67]. This effectively restricts the system to about 13.1 wt% accessible hydrogen (two-thirds of its theoretical hydrogen content), and the majority of efforts to develop AB for hydrogen storage have focused on controlling the kinetics and thermodynamics associated with the release of these first two equivalents of hydrogen, along with developing methods of regenerating AB from spent fuel [67].

The following sections provide a detailed account of the chemistry of AB, with particular emphasis on attempts that have been made to tailor its hydrogen release properties and to develop methods of rehydrogenating the materials formed through the thermolysis of AB.

Solid-State Chemistry

Structure

The solid-state chemistry of AB is determined in large part by the presence of both protic and hydridic hydrogen atoms capable of participating in the evolution of hydrogen. In fact, a better understanding of the solid-state structure of AB at various temperatures and

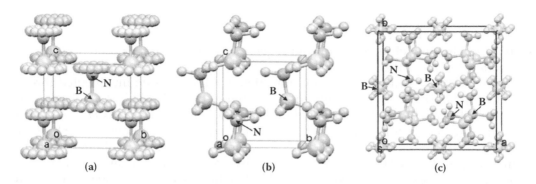

(a)   (b)   (c)

FIGURE 8.14

Solid-state structure of (a) tetragonal phase of NH_3BH_3 (*I4mm*, #107 > 225 K); (b) orthorhombic phase of NH_3BH_3 (*Pmn2_1*, #31 < 225 K); and (c) tetragonal $[NH_3BH_2NH_3]^+[BH_4]^-$ (*I4/mcm*, #140). The small letters o, a, b, and c indicate crystallographic axes o-a, o-b, and o-c, with o at the origin. (Figures 8.14a and 8.14b reproduced from Bowden, M.; Autrey, T.; *Curr. Opin. Solid State Mater. Sci.* 2011, 15, 73–79, with permission of Elsevier. Copyright 2011 Elsevier. Figure 8.14c reproduced from Bowden, M.; Heldebrant, D.J.; Karkamkar, A.; Proffen, T.; Schenter, G.K.; Autrey, T.; *Chem. Commun.* 2010, 46, 8564–8566, with permission of RSC Publishing. Copyright 2010 RSC Publishing.)

pressures has assisted in the rational design of methods to improve the thermal behavior and reversibility of this system [67]. The original x-ray powder diffraction study of AB indicated a tetragonal (*I4mm*, #107) structure [69]. However, this study was unable to establish hydrogen atom positions since these either were involved in significant rotation about the N-B bond or else were statistically disordered. This latter shortcoming is due to the crystallographic symmetry associated with the tetragonal cell, in which the B and N atoms are positioned on a fourfold rotation axis, in contrast with the threefold symmetry exhibited by the molecular motif. Hence, the larger number of symmetry-degenerate sites in the tetragonal cell leads to a model with partial occupancies, with the hydrogen atoms able to occupy more than one location. This feature is highlighted in Figure 8.14a, which shows how a recent model of disorder in AB described the positions of the hydrogen atoms using four different locations for each atom (i.e., 12 overall hydride and proton positions) [70, 71].

A phase transition to lower symmetry occurs at about –48°C, below which AB adopts an orthorhombic cell (*Pmn2_1*, #31; Figure 8.14b) [72]. This allows the hydrogen atoms to orient in fixed positions and affords an estimate of the $N-H^{\delta+}\cdots^{\delta-}H-B$ contacts ($d_{H\cdots H}$ = 2.02 Å), a value considerably shorter than the sum of the van der Waals radii for two interacting neutral hydrogen atoms (2.4 Å) [73]. It is noteworthy that a favorable orientation of the B-H and N-H moieties in the tetragonal structure can also produce similar proton-hydride distances ($d_{H\cdots H}$ = 1.91 Å), although the nature of the disorder prevents a precise characterization of these stabilizing supramolecular interactions. However, a recent neutron and molecular dynamics study of this disordered phase suggested that the $N-H^{\delta+}\cdots^{\delta-}H-B$ proton-hydride network in the tetragonal structure is weaker than those in its orthorhombic counterpart [71].

This apparent loosening of the proton-hydride bonding at higher temperatures may play a fundamental role in the decomposition pathway of AB since recent evidence suggests that breaking of these interactions is followed by isomerization to form the diammoniate of diborane ($[NH_3BH_2NH_3]^+[BH_4]^-$; DADB), with the subsequent release of hydrogen. This ionic tautomer is believed to play a crucial role in the release of hydrogen from AB (see the next section) [74]. The crystal structure of DADB was determined by Bowden et al. through a combined XRD and neutron diffraction study [75]. This showed DADB to adopt a tetragonal structure (*I4/mcm*, #140), in which the borohydride (BH_4^-) anions are disordered over two orientations related by a 90° rotation (Figure 8.14c). The shortest N–H$^{\delta+}$···$^{\delta-}$H-B proton-hydride contacts range from 1.8 to 2.1 Å, indicating a strengthening of these interactions with respect to those observed in AB and suggesting a reason for the differences in stability and reactivity of AB and DADB.

The effects of pressure on the phase transition in AB have been investigated through combined powder XRD experiments and DFT studies [76]. Filinchuk et al. observed the formation of a new phase of AB at 1.1 and 1.4 GPa (14,000 bar), for which a $Cmc2_1$ (#36) structure was assigned. This high-pressure phase displayed the same number of N–H$^{\delta+}$···$^{\delta-}$H-B proton-hydride interactions as the low-temperature $Pmn2_1$ structure, although different configurations of these H···H interactions were observed. This new arrangement may play a role in the hydrogen release properties of AB at higher pressures, although this remains speculative at this stage. To provide a complete phase diagram for AB, a calculated intermediate $P2_1$ (#4) structure was postulated to connect the high-pressure phase with the standard $Pmn2_1$ conformation.

Finally, the structures of the solid-state decomposition products have yet to be fully characterized, but these are believed to contain a combination of linear, branched, cross-linked, and coiled substructures based on experimental solid-state nuclear magnetic resonance (NMR) results and DFT calculations [77–79]. In a similar vein, the majority of products and intermediates formed through the solution-based thermal decomposition of AB have remained elusive. However, structural characterization of the products resulting from the dehydrogenation of N,N-dimethylamine-borane (Me$_2$AB) has been reported, and these provide a degree of guidance regarding the outcome of the AB decomposition products [80]. However, a detailed discussion is beyond the scope of this review.

Thermal Decomposition

The thermolysis of AB is often described as a stepwise process (Equations 8.11–8.13) in which the exact temperature range for each stage is highly dependent on the experimental conditions (heating rate, pressure, etc.) [67]. The first detailed description of this process was reported by Hu et al., who employed thermogravimetric analysis (TGA) and differential thermal analysis (DTA) techniques to monitor the progression of the reaction [81]. These authors observed initial onset of hydrogen release at about 120°C (5°C/min heating rate), which resulted in a mass loss of 31.6 wt%, considerably higher than the expected theoretical yield. This discrepancy was later shown to arise from a loss of heavier volatile by-products, such as borazine [67]. The thermodynamics of this process were revealed from the DTA measurements, which exhibited slightly different decomposition temperatures than the corresponding TGA results (10°C/min heating rate). The DTA trace in Figure 8.15 shows that an endothermic melt transition occurs at about 112°C, followed by a large exothermic event at 117°C. This latter exothermic peak corresponds to the initial loss of volatile gases (hydrogen). In addition, these authors reported foaming of the sample at the temperatures described, highlighting a common problem in the thermal decomposition of AB. The release of the second equivalent of hydrogen was then observed as a broad exothermic peak between 150°C and 200°C.

$$NH_3BH_3 \rightarrow {}^1/_n[NH_2BH_2]_n + H_2 \tag{8.11}$$

$$[NH_2BH_2]_n \rightarrow [NHBH]_n + H_2 \tag{8.12}$$

$$[NHBH]_n \rightarrow [NB]_n + H_2 \tag{8.13}$$

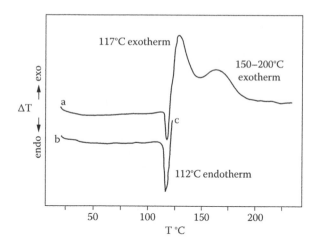

FIGURE 8.15

DTA curves for NH_3BH_3 with a 10°C/min heating rate: (a) open sample holder; (b) sealed glass capillary; and (c) sample tube explodes. (Reproduced from Hu, M.G.; Geanangel, R.A.; Wendlandt, W.W.; *Thermochim. Acta* 1978, *23*, 249–255, with permission of Elsevier. Copyright 1978 Elsevier.)

In an attempt to resolve the apparent disagreement between the TGA and DTA results, Sit and coworkers performed a similar experiment with a reduced heating rate of 2°C/min [82]. In this instance, the endothermic melt appeared at 114°C, preceded by the slow evolution of hydrogen gas (~30°C lower than the earlier experiments). The first exothermic event in this process was also shifted to a higher temperature (~125°C); this was attributed to the formation of hydrogen and $NH_2 = BH_2$ as evidenced by mass spectrometry (MS). However, this putative intermediate is highly reactive and polymerizes rapidly to form polyaminoboranes (PABs: $[NH_2BH_2]_n$; Equation 8.11) [83]. These PABs where found to be unstable at about 155°C, giving rise to the release of a second equivalent of hydrogen and the formation of polyiminoborane (PIBs: $[NHBH]_n$; Equation 8.12). The final stage in the decomposition of AB (Equation 8.13) occurs at much higher temperatures (>1200°C), which greatly exceeds the operational limit of modern-day proton exchange membrane (PEM) fuel cells and makes this final step inaccessible for practical on-board vehicular applications [67].

In a related study, Wolf et al. analyzed the temperature dependence of differential scanning calorimetry (DSC) experiments on AB [84]. These authors found that heating a sample of AB at 1°C/min resulted in an exothermic peak at 95°C, which was not completely separated from the previously observed endothermic melt. Further heating of the sample gave rise to two additional exothermic events at 113°C and 125°C. In marked contrast, lowering the heating rate to 0.05°C/min resulted in the appearance of the first exotherm at 82°C, and no melt was observed. This study emphasized that given enough time AB will decompose below its melting point. Hence, these authors suggested that the previously observed endothermic melt only exists due to the presence of residual AB.

Figure 8.16 displays hydrogen release from the isothermal heating of AB at various temperatures and shows that decomposition can occur as low as 75°C, but this requires extended periods of time (>1 day) to reach completion [77]. However, increasing the temperature to 85°C significantly accelerates this process (<8 h). This type of study was later extended to even lower temperatures; the decomposition of AB at 50°C took several months, while increasing the temperature by 10°C only reduced the reaction time to weeks. Taken together, these studies demonstrated that the rate and mechanism of hydrogen release by

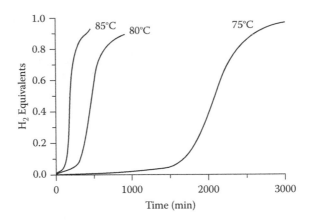

FIGURE 8.16
Isothermal hydrogen release from NH_3BH_3 at various temperatures (75°C, 80°C, 85°C). (Reproduced from Bowden, M.; Autrey, T.; *Curr. Opin. Sol. State Mater. Sci.*, 2011, 15, 73–79, with permission from RSC Publishing.)

AB is highly dependent on the reaction conditions, and that reducing the heating rate leads to the onset of hydrogen evolution at lower temperatures. Moreover, prolonged isothermal heating provides a means of releasing the first equivalent of hydrogen (Equation 8.1) at temperatures well below the melting point of AB. In conclusion, the solid-state decomposition of this chemical hydride is complex, and a greater understanding of this process will be required for the degree of control necessary to satisfy the hydrogen storage targets laid out for the automotive industry.

The thermal decomposition of AB under high pressures (up to 600 bar) of hydrogen has also been investigated in considerable detail [85]. It is noteworthy that this medium was chosen to monitor whether the reaction was reversible under these conditions. Unfortunately, no reversible process was evident in these experiments. The resulting DSC trace from this study, shown in Figure 8.17, displayed the first two exothermic events at 87°C and 132°C, respectively. The slow heating rate used throughout this reaction (0.05°C/min) resulted in no endothermic melt as decomposition of AB occurred well below its melting point. Moving to lower pressures (250 bar) resulted in no changes in the decomposition behavior of the material. However, these experiments were conducted at pressures below those required thermodynamically for the crystallographic phase transition (see the previous section on structure).

Raman spectroscopy was used to investigate the thermolysis of AB at pressures up to 9 GPa (90,000 bar) using a 500-μm culet diamond press, and this revealed dramatic changes in the thermal behavior [86]. These experiments revealed two distinct decomposition steps; the first produced PABs, and the second gave rise to a hexagonal $(BN)_x$ species. Increasing the pressure resulted in higher decomposition temperatures: At 10,000 bar, the first exothermic peak was observed around 127°C, with the subsequent release of hydrogen at temperatures above 200°C at an average heating rate of 1°C/min. This process was deduced from the disappearance of the corresponding N-H and B-H stretching frequencies in the Raman spectrum. However, when a pressure of 5.5 GPa (55,000 bar) was applied using a diamond anvil cell, the spectrum began to change only at 140°C (phase transition). In this instance, hydrogen was evolved in two events between 192–200°C and 230–237°C, but no rehydrogenation to AB was observed under these conditions.

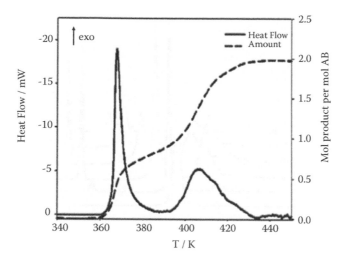

FIGURE 8.17

Temperature dependence of the heat flow and the amount of volatile products released in the thermal decomposition of NH_3BH_3 at 600 bar and 0.05°C/min. (Reprinted from Baitalow, F.; Wolf, G.; Grolier, J.P.E.; Dan, F.; Randzio, S.L.; *Thermochim. Acta* 2006, *445*, 121–125, with permission of Elsevier. Copyright 2006 Elsevier.)

It is important to note here that as the reactions described in Equations 8.11 and 8.12 are exothermic, and as the production of hydrogen in these reactions will result in a positive entropy change, the direct hydrogenation of PAB and PIB to regenerate AB is nonspontaneous under all achievable conditions of temperature and pressure, and alternative chemical methods have to be invoked (see the section on regeneration of AB from spent dehydrogenated products). Operation at external pressures lower than ambient has also been explored: When a sample of AB was held at 90°C under 0.05 bar of argon, hydrogen was released within 3.5 h, whereas nearly twice as long was required at ambient pressure [87]. Unfortunately, this method resulted in an increase of toxic volatile by-products (i.e., borazine), making this an unattractive route to obtain pure hydrogen.

Characterization of the intermediates formed in the decomposition of AB has proven rather difficult as these species are generally amorphous powders that are highly insoluble. Nevertheless, Autrey and coworkers have observed a significant induction period for the initiation of decomposition in the solid-state, during which a number of changes occur that facilitate the release of hydrogen [74, 88, 89]. Initially, an increase in libration within the crystal lattice results in a weakening of the $N-H^{\delta+}\cdots^{\delta-}H-B$ proton-hydride bonding network. This feature is revealed in Figure 8.18, which shows a sharpening and shift of the resonances corresponding to the BH_3 moiety in the ^{11}B solid-state NMR spectrum. This relaxation facilitates the isomerization of AB to DADB as shown in Scheme 8.1, with new BH_4 (–38.2 ppm) and BH_2 (–13.2 ppm) signals appearing in the ^{11}B NMR spectrum concomitant with the aforementioned changes in the BH_3 resonance (–24 ppm for AB vs. –22.5 ppm for new-phase AB). The N-H moieties of the nascent DADB then react with their B-H counterparts in AB, with the elimination of H_2 and the formation of longer-chain cations with a PAB backbone. The decomposition of AB is thus an intermolecular process, in accord with isotopic labeling experiments in which heating of a mixture of ND_3BD_3 and NH_3BH_3 resulted in production of H_2, HD, and D_2 [88].

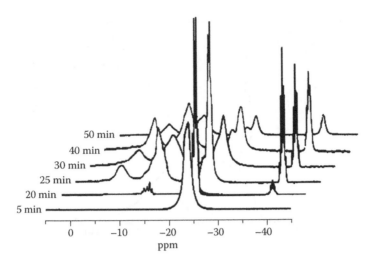

FIGURE 8.18
Solid-state ^{11}B NMR spectra for the thermal decomposition of NH_3BH_3. (Reprinted from Stowe, A.C.; Shaw, W.J.; Linehan, J.C.; Schmid, B.; Autrey, T.; *Phys, Chem. Chem. Phys.* 2007, 9, 1831–1836, with permission of RSC Publishing. Copyright 2007 RSC Publishing.)

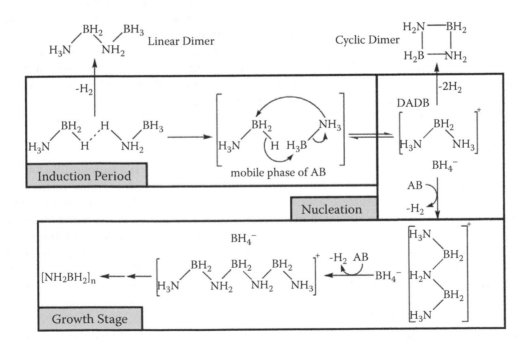

SCHEME 8.1
Proposed induction-nucleation-growth mechanism for the solid-state thermal decomposition of NH_3BH_3. (Reprinted from Stowe, A.C.; Shaw, W.J.; Linehan, J.C.; Schmid, Autrey, T.; *Phys, Chem. Chem. Phys.* 2007, 9, 1831–1836, with permission of RSC Publishing. Copyright 2007 RSC Publishing. With permission.)

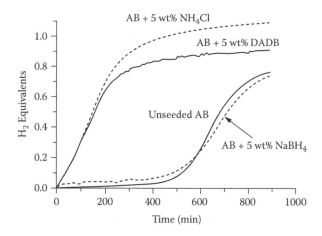

FIGURE 8.19

Isothermal decomposition of NH_3BH_3 with 5 wt% additives $NaBH_4$, $[NH_3BH_2NH_3]^+[BH_4]^-$ (DADB), and NH_4Cl. (Reprinted from Bowden, M.; Autrey, T.; *Curr. Opin. Solid State Mater. Sci.* 2011, *15*, 73–79, with permission of Elsevier. Copyright 2011 Elsevier.)

This proposed mechanism gained support through the direct reaction of DADB (5 wt% addition) and AB, which eliminated the induction period and accelerated the reaction. Figure 8.19 shows that the addition of ammonium chloride (NH_4Cl) to a sample of AB resulted in comparable kinetics as the AB/DADB reaction, whereas sodium borohydride ($NaBH_4$) did not alter the hydrogen release properties of this reaction. This deviation in the hydrogen release properties of AB on the addition of different additives was attributed to the formation of an $[(NH_3)_2BH_2]^+[Cl]^-$ intermediate for the NH_4Cl reaction, which is similar to DADB allowing for improved kinetics, whereas the $NaBH_4$ reaction does not produce a related ion pair, resulting in no change in the evolution of hydrogen from neat AB. This elegant study by Autrey et al. has provided fundamental insight into the mechanism and kinetics that characterize the early stages in the decomposition of AB, at the same time demonstrating the complexity of the process.

The production of three different isotopomers of hydrogen from mixtures of ND_3BD_3 and NH_3BH_3 could potentially indicate that more than one hydrogen release pathway is involved in the thermal decomposition of AB. A recent study by Wolstenholme et al. investigated the solid-state decomposition of selectively labeled isotopomers of AB (ND_3BH_3 and NH_3BD_3) [90]. The hydrogen evolved was detected by $^1H/^2H$ NMR spectroscopy, and the mass loss was monitored by TGA. In these experiments, a hydrogen release pathway mediated solely by $N-H^{\delta+}\cdots^{\delta-}H-B$ interactions would result in only HD being evolved, whereas any contribution from an alternative $B-H^{\delta-}\cdots^{\delta-}H-B$ pathway would produce additional amounts of H_2 and D_2, respectively. In this regard, it is worth mentioning that hydrogen release by the related alkali metal borohydrides (MBH_4) can only occur through such a hydride-hydride pathway. As shown in Figure 8.20, the presence of $H_2(D_2)$ along with HD in the ^1H and ^2H NMR spectra demonstrates unambiguously the involvement of both proton-hydride and hydride-hydride pathways in this process, notwithstanding the absence of short hydride-hydride contacts in the crystal structure of AB ($d_{H\cdots H} = 3.05$ Å).

These results could also be plausibly explained by hydrogen scrambling, with hydrogen atoms migrating between the NH_3 and BH_3 moieties (e.g., ND_2HBH_2D). However, Raman spectra collected at temperatures up to 110°C showed no evidence of N-H or N-D stretching modes in ND_3BH_3 and NH_3BD_3, respectively [90]. A notable feature of this study

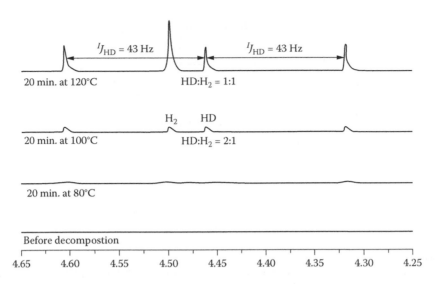

FIGURE 8.20

^{1}H NMR spectra (in toluene) of the gaseous products evolved from the decomposition of ND$_3$BH$_3$. The corresponding ^{2}H NMR spectra can be found in the original manuscript. (Reprinted from Wolstenholme, D.J.; Traboulsee, K.T.; Hua, Y.; Calhoun, L.A.; McGrady, G.S.; *Chem. Commun.* 2012, 48, 2597–2599, with permission from RSC Publishing. Copyright 2012 RSC Publishing.)

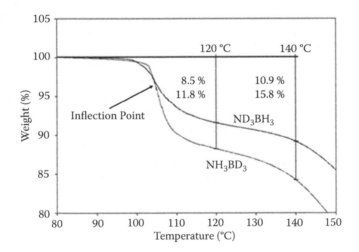

FIGURE 8.21

TGA plot of the thermal decomposition of ND$_3$BH$_3$ and NH$_3$BD$_3$, in which the difference in the mass loss ratio between these two species are 1.38 at 120°C and 1.44 at 140°C, while the HD:H$_2$(D$_2$) ratios for both isotopomers are 1:1 at 120°C and 2:3 at 140°C. (Reprinted from Wolstenholme, D.J.; Traboulsee, K.T.; Hua, Y.; Calhoun, L.A.; McGrady, G.S.; *Chem. Commun.* 2012, 2012, 48, 2597–2599. With permission from RSC Publishing. Copyright 2012 RSC Publishing.)

was the ratio of HD to H$_2$(D$_2$) formed on decomposition of these two isotopomers of AB. Figure 8.21 displays the TGA traces for the ND$_3$BH$_3$ and NH$_3$BD$_3$, which demonstrate that almost-equal contributions from the proton-hydride and hydride-hydride pathways occurring up to about 120°C, with the latter becoming dominant above the melting point of AB. The discovery and central involvement of a counterintuitive hydride-hydride pathway in the formation of hydrogen warrants further investigation and may provide new and important insight into the complex decomposition chemistry of AB.

Solution-Phase Decomposition

The dehydrogenation of AB in solution was first investigated in 1988 by Wang and Geanangel to ascertain the nature of the elusive $NH_2 = BH_2$ intermediate postulated from analogous solid-state reactions [91]. In these experiments, the authors heated (i.e., refluxed) several 0.15 M samples of AB in a wide variety of polar solvents (acetonitrile, diglyme, glyme, 2-methyl-tetrahydrofuran, pyridine, and THF). The progression of these reactions was monitored by ^{11}B NMR spectroscopy, which allowed subtle variations in chemical environment of the BH_3 moiety to be detected. The reactions carried out in ether solvents showed mainly decomposition products, while acetonitrile and pyridine appeared to react with the AB substrate. The corresponding spectra for these reactions each displayed a signal at about –11 ppm, which was tentatively assigned to a cyclic $[NH_2BH_2]_3$ product, cyclotriborazane, based on comparison with an authentic sample. However, the authors noted that the formation of small cyclic oligomers of different sizes may also explain these results. In addition, these spectra displayed peaks corresponding to borazine (30.2 ppm) and µ-aminoborane (–28.9 ppm). These findings led the authors to propose a two-step pathway for the process (Equations 8.14 and 8.15).

$$NH_3BH_3 \rightarrow {}^1/_n[NH_2BH_2]_n + H_2 \tag{8.14}$$

$$3/n[NH_2BH_2]_n \rightarrow [NHBH]_3 + H_2 \tag{8.15}$$

The current interest in AB as a hydrogen storage candidate prompted Shaw et al. [92] to revisit this class of reactions recently, with in situ variable-temperature ^{11}B NMR experiments permitting considerably higher resolution than in the earlier study of Wang and Geanangel [91]. These authors found that the decomposition of AB followed second-order kinetics in glyme, with a strong dependence on the concentration of the AB solutions. This study provided persuasive evidence that the unimolecular formation of $NH_2 = BH_2$ and subsequent [2 + 2] cycloaddition reactions of this monomer were an unlikely process, in contrast to the mechanism postulated for the decomposition of the related *N,N*-dimethylamine-borane (NHMe$_2$BH$_3$) [93]. Figure 8.22 shows results from the time-dependent (3.5-h) decomposition of a 1 M solution of AB at 80°C monitored by ^{11}B NMR spectroscopy, which provided considerable insight into the pathway for this reaction. Initial heating gave rise to a peak at –10.8 ppm, which was assigned to cyclodiborazane (CDB). Prolonged heating then resulted in the appearance of three new peaks, which grew at similar rates (BH = –5.0 ppm; BH$_2$ = –10.9 ppm; and BH$_3$ = –23.4 ppm), assigned to *B*-(cyclodiborazanyl)amino-borohydride (BCDB). Finally, the appearance of a peak at –10.6 ppm corresponded to the cyclotriborazane (CTB) observed in the earlier study of Wang and Geanangel.

In light of these observations, Shaw et al. proposed a novel mechanism for the dehydrogenation of AB in solution, as shown in Scheme 8.2. The initial stage of this reaction occurs through a bimolecular process, analogous to the solid-state decomposition of AB, in which isomerization occurs to form DADB prior to the evolution of significant quantities of hydrogen. The reactive DADB can then release hydrogen in three different ways: (1) loss of 1 equivalent of hydrogen to produce a linear dimer; (2) loss of 2 equivalents of hydrogen, resulting in the elusive $NH_2 = BH_2$ species; or (3) loss of 2 equivalents of hydrogen followed by ring closure to give cyclodiborazane. It is noteworthy that this cyclic

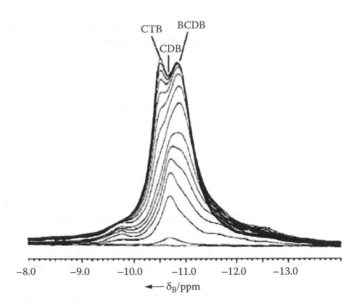

FIGURE 8.22
Time-dependent $^{11}B(^1H)$ NMR spectra for the first 3.5 h in the isothermal decomposition of a 1 M solution of AB. CTB = cyclotriborazane, CDB = cyclodiborazane and BCDB = B-(cyclodiborazanyl)aminoborohydride. (Reprinted from Shaw, W.J.; Linehan, J.C.; Szymczak, N.K.; Heldebrant, D.J.; Yonker, C.; Camaioni, D.M.; Baker, R.T.; Autrey, T.; *Angew. Chem. Int. Ed.* 2008, 47, 7493–7496, with permission of Wiley VCH. Copyright 2008 Wiley VCH.)

intermediate was the only product observed in these experiments. Subsequent addition of AB resulted in the loss of hydrogen to produce *B*-(cyclodiborazanyl)aminoborohydride, which can readily convert to cyclotriborazane. It is clear that decomposition of AB in the solid state (Scheme 8.1) and in solution (Scheme 8.2) occurs through different reaction pathways, and this aspect of AB chemistry requires further investigation and elucidation.

In an attempt to understand better the initial stages in the solution phase decomposition of AB, Chen et al. carefully monitored the reaction between NH_3 and BH_3-THF and proposed a three-stage mechanism for the formation of AB and DADB, as shown in Scheme 8.3 [94]. The first stage of this reaction involved displacement of the coordinating solvent molecule in BH_3-THF through an S_N2-type reaction. This process was evident in the ^{11}B NMR spectra for the early stages of this reaction, shown in Figure 8.23, whereby a decrease in the BH_3-THF signal (–0.9 ppm) was accompanied by the appearance of a new peak corresponding to AB (–22.5 ppm). Subsequent reaction between AB and remaining BH_3-THF produced two additional peaks at –13.8 and –25.5 ppm, which the authors assigned to ammonia diborane, $NH_3BH_2(\mu\text{-}H)BH_3$; AaDB. This intermediate can then react with NH_3 to provide either DADB or AB depending on the approach of the nucleophile (i.e., in the direction of the NH_3BH_2 or the BH_3 moieties in AaDB, respectively). Remarkably, the AB reaction pathway is favored over its DADB counterpart (59.9 kJ/mol for AB vs. 128.1 kJ/mol for DADB), despite the clear presence of significant quantities of DADB observed in the ^{11}B NMR spectra. However, theoretical studies showed that N-H$^{\delta+}\cdots^{\delta-}$H-B proton-hydride interactions between AaDB and AB lower the energy barrier of this second pathway, providing compelling evidence that these H$\cdots$H interactions are essential for the formation of DADB.

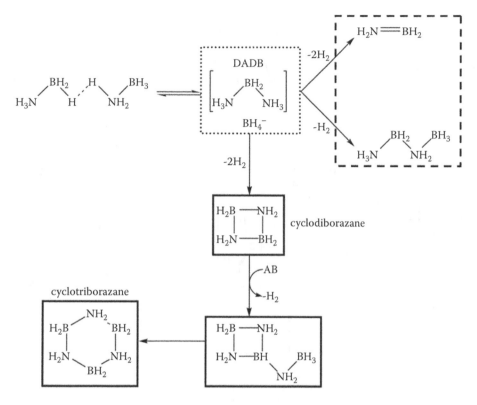

SCHEME 8.2
Proposed solution-based thermal decomposition of NH_3BH_3 (solid boxes = observed, dotted boxes = observed in low concentration, and dashed box = not observed). (Reprinted from Shaw, W.J.; Linehan, J.C.; Szymczak, N.K.; Heldebrant, D.J.; Yonker, C.; Camaioni, D.M.; Baker, R.T.; Autrey, T.; *Angew. Chem. Int. Ed.* 2008, *47*, 7493–7496, with permission of Wiley VCH. Copyright 2008 Wiley VCH.)

Computational Studies

The experimental results obtained from solid-state and solution-phase decomposition studies of AB have provided a wealth of knowledge concerning its hydrogen release characteristics and the mechanisms associated with these reactions. Elucidation of this complex process has also benefited from high-level theoretical calculations, which have the potential to reveal subtle details that cannot readily be obtained through experiments. For example, initial mechanistic studies concerning the thermal decomposition of AB (gas phase) predicted a large kinetic barrier (134–138 kJ/mol) for the loss of the first equivalent of hydrogen [95]. However, a reexamination of this reaction revealed that the dissociation energy for AB was lower than this transition state [96], leading the authors to speculate that free BH_3 (liberated through this dissociation process) could act as a Lewis acidic catalyst via an interaction with AB.

Figure 8.24a shows how these calculations predict the initial formation of a strong complex between AB and BH_3 (ba-com1), which is 74.9 kJ/mol (17.9 kcal/mol) more stable than the AB + BH_3 asymptote. This complex then releases hydrogen via three possible transition states. The highest energy route (TSba-BB) involves the formation of a $B-H^{\delta-}\cdots^{\delta-}H-B$ hydride-hydride interaction between the AB and BH_3 moieties, along with a weakening of

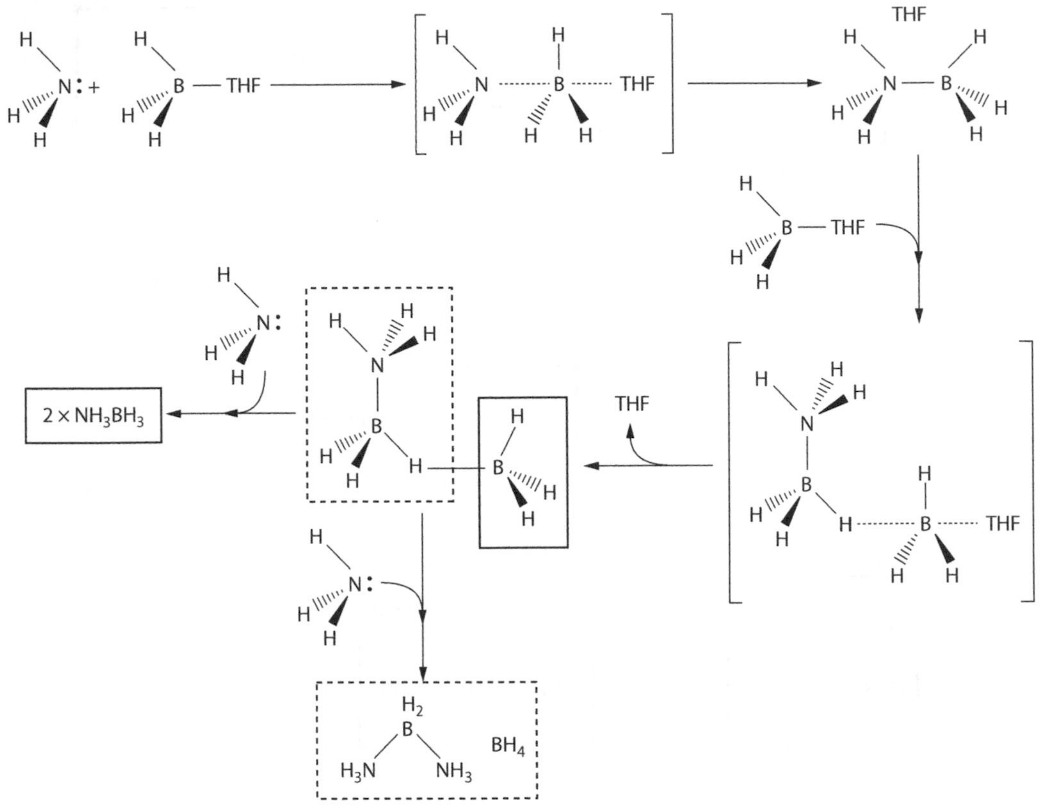

SCHEME 8.3
Proposed mechanism in the formation of NH_3BH_3 and $[NH_3BH_2NH_3]^+[BH_4]^-$ via a reaction of NH_3 and BH_3-THF. The solid and dashed boxes correspond to the different methods in which NH_3 can interact with AaDB: NH_3 + $BH_3(\mu\text{-H})BH_2NH_3 \rightarrow 2NH_3BH_3$ or NH_3 + $BH_2NH_3(\mu\text{-H})BH_3 \rightarrow [NH_3BH_2NH_3]^+[BH_4]^-$, respectively. (Reprinted from Chen, X.; Bao, X.; Zhao, J.-C.; Shore, S.G.; *J. Am. Chem. Soc.* 2011, *133*, 14172–14175, with permission of the American Chemical Society. Copyright 2011 American Chemical Society.)

an N-H bond as the proton is transferred to the BH_3 group. The second pathway (81.7 kJ/mol lower in energy) corresponds to a process in which the BH_3 group interacts directly with the NH_3 portion of AB. This results in a weakening of the N-B bond and facilitates an interaction between the BH_3 moiety (behaving as a classical Lewis acid) and the lone pair of the N atom, which stabilizes the framework for the release of hydrogen. Interestingly, this pathway gives rise to a three-membered ring (ba-ring) product in preference to the expected $NH_2 = BH_2$ species. The lowest energy transition state (TSba-BN) resembles the TSba-BB route, in which a N-H$^{\delta+}\cdots^{\delta-}$H-B proton-hydride interaction now mediates the evolution of hydrogen (176.3 kJ/mol lower than TSba-BB and 94.6 kJ/mol lower than TSba-lew). The identification of these three different transition states provides fundamental insight into this process, and the TSba-BB route corresponds to the hydride-hydride pathway for hydrogen formation deduced experimentally by Wolstenholme et al. [90]. In a related study, Weismiller et al. employed reactive force field calculations and concluded that the free BH_3 molecule favors recombination with AB [97], with unimolecular elimination of hydrogen being faster than the catalytic route at higher temperatures. These computational studies provided early evidence to support the experimental findings by Shore et al., in which BH_3 helps mediate the release of hydrogen from AB [94].

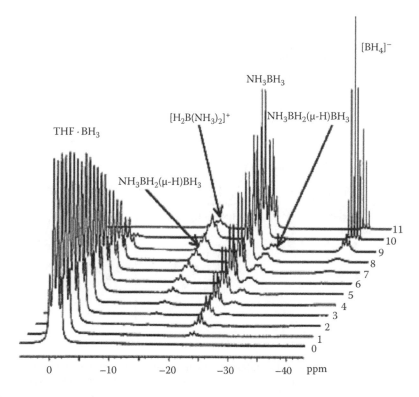

FIGURE 8.23

^{11}B NMR spectra for the reaction of NH$_3$ and BH$_3$-THF at −78°C highlighting the formation of NH$_3$BH$_3$, NH$_3$BH$_2$(μ-H)BH$_3$, and [NH$_3$BH$_2$NH$_3$]$^+$[BH$_4$]$^-$ products. (Reprinted from Chen, X.; Bao, X.; Zhao, J.-C.; Shore, S.G.; *J. Am. Chem. Soc.* 2011, *133*, 14172–14175, with permission of the American Chemical Society. Copyright 2011 American Chemical Society.)

The discovery of DADB and its role in the decomposition of AB has prompted considerable efforts to model the process computationally [98]. Coupled-Cluster with Single and Double and perturbative Triple excitations CCSD(T) calculations predicted DADB to be about 44.4 kJ/mol higher in energy than the most stable head-to-tail arrangement of AB dimers. It was also found that DADB could be generated through three different methods: (1) from two monomers; (2) from two NH$_3$ molecules successively reacting with diborane; and (3) from the successive addition of BH$_3$ and NH$_3$ to AB. These potential reactions each involve different pathways, but all share the same rate-determining barrier. The crucial step in the release of hydrogen from DADB can occur through two different reaction pathways, as displayed in Figure 8.24b. The highest-energy route involves a transition state that consists of a N-H$^{\delta+}$···$^{\delta-}$H-B proton-hydride interaction between neighboring NH$_3$ and BH$_2$ components of the cation (DADB-ts2). Alternatively, the BH$_4$ anion can interact with a terminal NH$_3$ group of the cation to release hydrogen and form a linear dimer (DADB-ts1). This last transition state was predicted to be 128.1 kJ/mol lower in energy than the former one and represents a more rational route to the experimentally observed PAB and cyclodiborazane decomposition products.

Although these calculations provided important information concerning the dehydrogenation of AB, the details of this process deviate significantly depending on the conditions of the reaction (i.e., solid state or solution). Miranda and Ceder attempted to model this process in the solid state using high-level periodicity calculations, defining the reactions

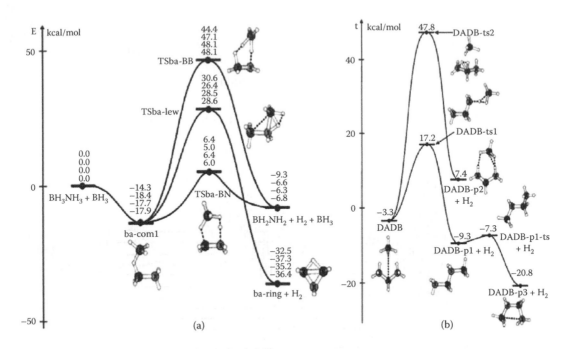

FIGURE 8.24
(a) Computational prediction of the BH_3-catalyzed dehydrogenation of NH_3BH_3 with relative energies in kilocalories per mole. (b) Calculated mechanism for the release of hydrogen from $[NH_3BH_2NH_3]^+[BH_4]^-$. (Figure 8.24a reprinted from Nguyen, M.T.; Nguyen, V.S.; Matus, M.H.; Gopakumar, G.; Dixon, D.A.; *J. Phys. Chem. A* 2007, *111*, 679–690, with permission of the American Chemical Society. Copyright 2007 American Chemical Society. Figure 8.24b reprinted from Nguyen, V.S.; Matus, M.H.; Grant, D.J.; Nguyen, M.T.; Dixon, D.A.; *J. Phys. Chem. A* 2007, *111*, 8844–8856, with permission of the American Chemical Society. Copyright 2007 American Chemical Society.)

based on previous experimental TGA and DSC results [99]. These authors used the structure of AB reported by Klooster et al. [72] as a starting point for their calculations and employed polyethylene to model PAB, taking into account the errors associated with the different conformations of the two structures. This study showed that the formation of the decomposition products PAB and PIB was always exothermic, and that this exothermicity increased at elevated temperatures. In addition, the solid-state and gas-phase calculations showed relatively good agreement for the first stage in the decomposition of AB. However, the energetics associated with the loss of 2 equivalents of hydrogen to form PIB resulted in considerable deviations between the two methods ($\Delta H \cong 31.0$ kcal/mol in the gas phase vs. −9.57 kcal/mol in the solid state), with the former producing better agreement with the experimental results. Solid-state periodicity calculations may thus afford more accurate models of this process, although the lack of experimental structures for the intermediate products limits the efficacy of this approach at present.

These studies demonstrated the utility of computational methods for gaining useful insight into the processes surrounding the evolution of hydrogen from AB. This feature is exemplified in the dissociation of the adduct into NH_3 and BH_3, with the latter acting as a catalyst in the efficient release of hydrogen from subsequent AB molecules. It is noteworthy that Dixon et al. predicted this pathway several years prior to its experimental validation [96]. This prescient study also foretold the B-H$^{\delta-}$···$^{\delta-}$H-B hydride-hydride pathway recently proposed by Wolstenholme et al. [90], highlighting the utility of computational methods in helping to unravel the mysteries of AB and its complicated decomposition reactions.

Tailoring the Hydrogen Release Properties

The solid-state and solution-based decomposition of AB occurs over a broad temperature range (ca. 100–200°C), and this will need careful control to release sufficient quantities of hydrogen and satisfy the requirements of modern-day PEM fuel cells (which operate optimally at 60–120°C) [67]. As hydrogen release from AB is exothermic, and as the production of hydrogen gas ensures that the corresponding entropy change is positive, the process is spontaneous under all achievable conditions of temperature and pressure. Hence, AB is a kinetically, rather than thermodynamically, stable material, and its release of hydrogen cannot be controlled through establishment of a temperature-determined equilibrium plateau pressure. Rather, an AB-based hydrogen fuel tank will need to be configured to ensure that the amount of material heated at any time is carefully controlled to deliver hydrogen to the fuel cell or other hydrogen conversion device, at the precise rate needed for optimal operation. The existence of an induction period, and the rather sluggish evolution of hydrogen from AB at temperatures below 100°C, compounded by the production of significant levels of impurities deleterious to fuel cell performance, highlight the advantages of using a catalyst to accelerate and control this process at temperatures close to ambient. The following sections outline several of the leading methods currently under investigation to tailor the hydrogen release properties of AB.

Metal-Mediated Dehydrogenation

Initial reports concerning the catalytic dehydrocoupling of AB involved the addition of 1.5 mol% of $[Rh(1,5\text{-}cod)(\mu\text{-}Cl)]_2$ in diglyme or tetraglyme at 45°C for 3 days [100]. This reaction resulted in the elimination of 2 equivalents of hydrogen, giving rise to borazine in low yields (~10%), in which the major products were believed to be nonvolatile oligomeric species [101]. In a similar vein, Denney and coworkers demonstrated that Brookhart's iridium pincer complex, $[(POCOP)Ir(H)_2]$; $POCOP = [\mu^3\text{-}1,3\text{-}(OP t Bu_2 C_6 H_3)]$ acts as an effective catalyst (0.5 mol%) for the dehydrogenation of AB in dilute THF within 14 min [102]. The relative insolubility of the resulting products made their characterization rather difficult. Nonetheless, these authors used infrared (IR), powder XRD, and solid-state ^{11}B NMR techniques to propose a cyclopentaborazane product, noting at the same time that linear, oligomeric, or polymeric products could not be discounted [103].

These findings prompted Paul and Musgrave to propose a potential mechanism, in which they used a slightly modified model [104]. They analyzed two potential catalyst pathways: (1) via a 14-electron iridium species and (2) via a 16-electron system. They concluded that the 14-electron pathway could not occur at room temperature, while the 16-electron route seemed more probable. The first step in this latter process involved the substrate binding to the catalyst through σ B-H bonds, as illustrated in Scheme 8.4. This was then followed by a transition state that transferred the protic hydrogen atom of the NH_3 moiety to the metal center, followed by subsequent dissociation of the dehydrogenated AB product. This mechanism is consistent with the presence of the experimentally observed tetrahydride intermediate. The second transition state involved the formation and release of hydrogen from this intermediate, with the barrier associated with this process comparable to the previous transition state, indicating that two rate-determining steps may exist for this reaction.

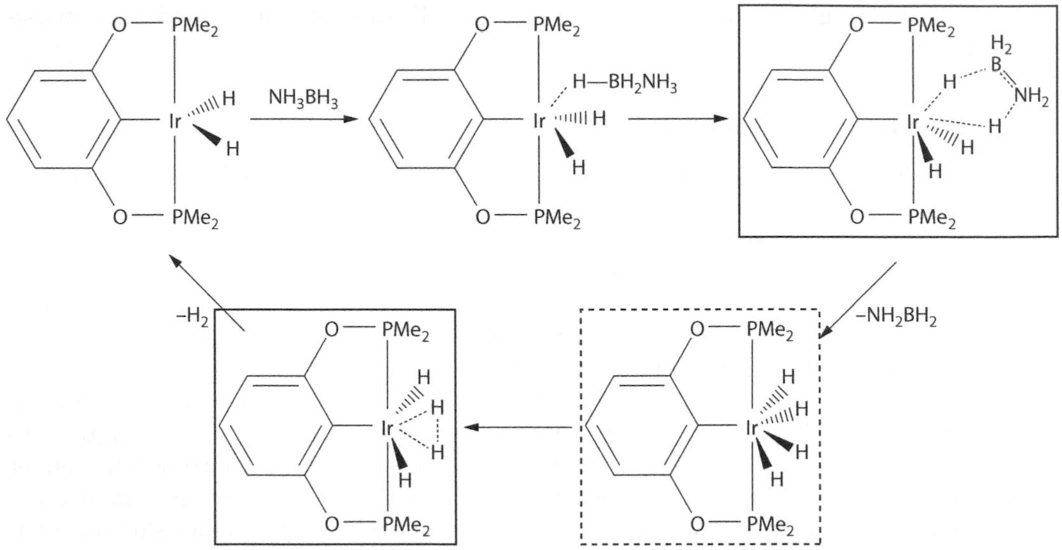

SCHEME 8.4
Proposed pathway for the dehydrogenation of NH_3BH_3 through the use of an Ir pincer catalyst. Solid and dashed boxes correspond to transition states and experimentally observable intermediates, respectively. (From Paul, A.; Musgrave, C.B.; *Angew. Chem. Int. Ed.* 2007, *46*, 8153–8156. With permission.)

Although late transition metal complexes exhibit impressive catalytic activity, the high costs of the metals involved preclude their widespread use. Accordingly, catalysts based on nonprecious metals have been the target of recent research efforts. Keaton et al. investigated Ni-based catalysts for the dehydrogenation of AB [105]. These authors mixed a $Ni(cod)_2$ solution in C_6D_6 with various *N*-heterocyclic carbene ligands, then added a solution of AB in diglyme (~1:10 ratio at 60°C). This resulted in the immediate evolution of gas, with 2.8 equivalents of hydrogen being released within 4 h. Characterization of the products was achieved through ^{11}B NMR spectroscopy; as shown in Figure 8.25, this revealed a doublet between 18 and 40 ppm, which was attributed to the formation of cross-linked borazine-type products. In addition, a peak at about –36 ppm was tentatively assigned to a by-product consisting of the carbene ligand interacting with a BH_3 group (2% of the observed [11]B signals).

Measurement of the kinetic isotope effect for this reaction using partially and fully deuterated forms of AB provided an insight into the mechanism of this metal-mediated process. Fully deuterated AB exhibited the largest kinetic isotope effect, consistent with cleavage of both the N-H and B-H bonds as the rate-determining steps since these two steps have comparable rates. Computational studies exploring the reaction mechanism concurred that the carbene ligand participates in the proton abstraction process, but there is less agreement over the subsequent stages, with most studies favoring an interlinked catalytic cycle for this reaction [106].

These studies into the metal-mediated dehydrogenation of AB have revealed the relative ease of removing hydrogen under mild conditions through the use of a catalyst. In general, these reactions are believed to proceed via a reactive $NH_2 = BH_2$ intermediate, which has proven extremely difficult to isolate and characterize [107, 108]. However, an elegant recent study of the stoichiometric dehydrogenation of AB by Alcaraz and coworkers using a bis(dihydrogen) Ru complex, $[RuH_2(\eta^2-H_2)_2(PCy_3)_2]$; (Cy = C_6H_{12}), permitted these authors to trap the ephemeral $NH_2 = BH_2$ intermediate via the

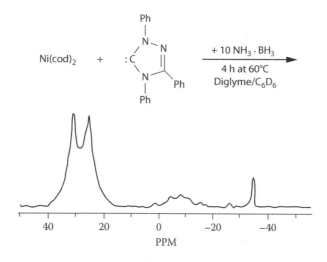

FIGURE 8.25

^{11}B NMR spectrum in C_6D_6/diglyme from the catalytic reaction of NH_3BH_3 with a mixture of $Ni(cod)_2$ and *N*-heterocyclic carbene after 4 h at 60°C. (Reprinted from Keaton, R.J.; Blacquiere, J.M.; Baker, R.T.; *J. Am. Chem. Soc.* 2007, *129*, 1844–1845, with permission of the American Chemical Society. Copyright 2007 American Chemical Society.)

formation of an η^2-coordinated $[RuH_2(\eta^2{:}\eta^2\text{-}H_2BNH_2)(PCy_3)_2]$ complex [109, 110]. The ^{1}H NMR spectrum for this system exhibits a characteristic broad singlet at −6.80 ppm for the coordinating B-H bonds, while the more shielded Ru-H signal appears at −11.85 ppm with equal intensity. The ^{11}B NMR spectrum revealed a broad signal at about 46 ppm, attributed to the $NH_2 = BH_2$ ligand.

The most interesting outcome of this study was the structural characterization of $[RuH_2(\eta^2{:}\eta^2\text{-}H_2BNH_2)(PCy_3)_2]$ displayed in Figure 8.26. In this complex, the Ru center adopts a pseudo-octahedral geometry with the phosphine ligands in the axial positions. The $NH_2 = BH_2$ group coordinates to the metal through two σ B-H bonds in an η^2-fashion *trans* to the Ru-H bonds in the equatorial plane. The isolation and characterization of this intermediate has permitted a comparison of its geometry with the structural features of AB. The trapped $NH_2 = BH_2$ moiety exhibits a short N-B bond [1.396(3) vs. 1.58(2) Å for AB], confirming the double-bond character of this species. An elongation of the (coordinated) B-H bonds [1.22(3)-1.25(2) vs. 1.15(3)-1.18(3) Å for AB] also demonstrates the activation of these moieties. DFT calculations on a PMe_3-substituted version of the experimental $[RuH_2(\eta^2{:}\eta^2\text{-}H_2BNH_2)(PCy_3)_2]$ complex were conducted to determine the ideal coordination of the $NH_2 = BH_2$ ligand. Figure 8.27 reveals four different isomers identified on the potential energy surface, with the σ-complex displaying the lowest energy. The related π adducts were considerably less stable (332.6 and 250.6 kJ/mol), in marked contrast to the metal coordination chemistry of the isoelectronic $CH_2 = CH_2$. This study has provided key information concerning the nature of the evanescent $NH_2 = BH_2$ intermediate and the structural changes that occur on the loss of 1 equivalent of hydrogen from AB.

Acid-Catalyzed Dehydrogenation

The acid-catalyzed hydrolysis of amine-borane adducts represents an efficient means of dehydrogenating this class of molecular hydrides, in particular AB (reaction rate = 600 × 10^2 k_2/L mol^{-1} s^{-1}). Kelly and Marriott proposed that this reaction could proceed via either

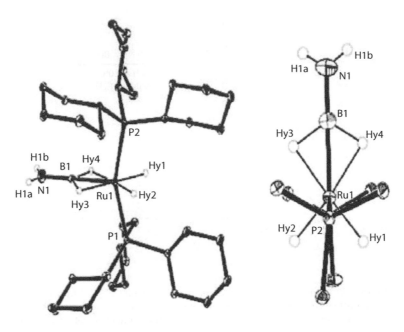

FIGURE 8.26
Solid-state structure of $[RuH_2(\eta^2:\eta^2-H_2BNH_2)(PCy_3)_2]$. (Reprinted from Alcaraz, G.; Sabo-Etienne, S.; *Angew. Chem. Int. Ed.* 2010, *49*, 7170–7179, with permission of Wiley VCH. Copyright 2010 Wiley VCH.)

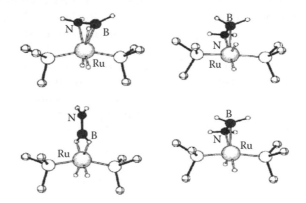

FIGURE 8.27
Optimized geometries (B3PW91) for the four potential coordination modes of $NH_2 = BH_2$ and a substituted RuH_2 complex. (Reprinted from Alcaraz, G.; Vendier, L.; Clot, E.; Sabo-Etienne, S.; *Angew. Chem. Int. Ed.* 2010, *49*, 918–920, with permission of Wiley VCH. Copyright 2010 Wiley VCH.)

(1) protonation of the NH_3 group, resulting in the dissociation of BH_3 for hydrolysis; or (2) a direct reaction with the BH_3 moiety via a five-coordinate boron intermediate [111]. These authors concluded that the former mechanism was more likely based on the analysis of isotope effects and reaction rates.

In spite the facile nature of acid-catalyzed hydrolysis of AB, the first example of such a process under nonhydrolytic conditions was only recently reported by Stephens et al. [112]. In this instance, hydrogen was evolved at about 20°C when an appropriate Lewis $[B(C_6F_5)_3]$ or Brønsted $(HOSO_2CF_3)$ acid was added to an ethereal solution of AB. The reaction was found to proceed rapidly at 60°C. The authors postulated that a strong Lewis acid

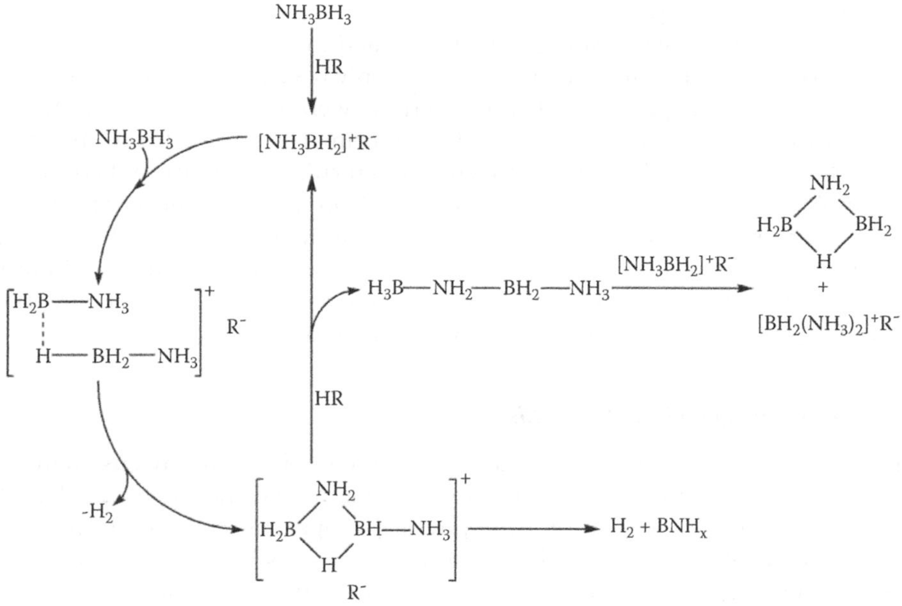

SCHEME 8.5
Proposed mechanism for the acid-catalyzed dehydrogenation of NH_3BH_3. HR indicates a Lewis acid or Bronsted acid. (From Stephens, F.H.; Baker, R.T.; Matus, M.H.; Grant, D.J.; Dixon, D.A.; *Angew. Chem. Int. Ed.* 2007, 46, 746–749. With permission.)

would afford a boronium salt $[NH_3BH_2]^+$ through a hydride abstraction reaction, as shown in Scheme 8.5. This cationic species would then react with an additional AB molecule to release hydrogen and form an AaDB-type intermediate, in a process similar to that proposed by Chen et al. for the production of DADB from AB [94]. The subsequent dehydrogenation of this AaDB-type intermediate can lead to the formation of linear and branched acyclic aminoborane oligomers, which will cyclize at low temperatures to produce the desired products. However, high acid concentrations resulted in a competing side reaction, in which a hydride moiety is transferred from AB to an ever-growing cationic chain, forming oligomers that can react with the boronium initiator to produce μ-aminodiborane via NH_3 transfer. This side reaction can be avoided by limiting the amount of acid present in solution (≤1 mol%). Although this approach offers advantages over the uncatalyzed desorption of hydrogen from AB, further improvements in the rates are necessary if it is to achieve practical applications in transportation.

Thermal Decomposition in Ionic Liquids

In attempts to lower the decomposition temperature of AB, Bluhm et al. moved from a conventional solvent to an ionic liquid medium ([BMIM][Cl] = 1-butyl-3-methylimidazolium chloride) [113]. An ionic liquid can be considered as a salt that exists in the liquid state at an arbitrary temperature (including room temperature) and consists of discrete ions and short-lived ion pairs [114]. The respective protic and hydridic N-H and B-H moieties in AB may be expected to experience significant activation in such a medium, encouraging the formation of H_2 at lower temperatures. In this regard, the isothermal decomposition of AB at 85°C produces negligible amounts of hydrogen within 3 h. However, the analogous reaction in [BMIM][Cl] evolved about 1 equivalent of hydrogen over the same period. Raising the temperature to 95°C resulted in the release of an additional 0.5 equivalent (1.5 equivalent total), as compared with only 0.8 equivalent at this temperature in the ionic liquid-free experiment. These reactions were monitored by ^{11}B NMR spectroscopy, and the products were determined to be primarily acyclic linear and branched oligomers. Moreover, the early stages of this reaction resulted in significant concentrations of BH_4^- and BH_2^+ consistent with the formation of DADB. These results confirmed that an ionic liquid medium does indeed facilitate evolution of hydrogen from AB. Although the mechanistic role of the ionic liquids in this process has yet to be determined, Bluhm et al. speculated that the medium may reduce the energy associated with the polar transition states, allowing the reaction to proceed at lower temperatures.

The Release of Hydrogen Using Scaffolds

The nanoconfinement of AB was introduced in Chapter 6. Here, we discuss nanoscaffold encapsulation of AB in more detail. The use of a macromolecule to encapsulate AB has been shown by several groups to alter the physical properties of the molecule [115–120]. Gutowska et al. developed a system based on mesoporous silica (7.5-nm pore diameter), into which AB was intercalated [115]. This allowed for hydrogen release to occur at much faster rates and at lower temperatures. DTA measurements using a 1°C/min heating rate showed that the AB-silica peak dehydrogenation temperature was shifted by about 12°C (98°C vs. 110°C for AB). This change in the thermodynamics of this system was believed to be caused by the high internal surface area available to AB within the pores of the silica. This type of structural variation is known to lower the temperature at which AB undergoes its phase transition, potentially accounting for the changes in the hydrogen release properties of this system.

To explore the role of the mesoporous silica in more detail, Wang et al. performed a study of the AB-silica system through hyperpolarized ^{129}Xe NMR spectroscopy [116]. This method affords a means of determining the nature of the size, shape, and connectivity of the pores, along with the behavior of the adsorbents within these cavities, since these features are highly correlated with the changes in the ^{129}Xe chemical shift. In these experiments, a sample of AB-free mesoporous silica displayed two signals at 0 ppm and between about 50 and 100 ppm, as shown in Figure 8.28. These correspond to free ^{129}Xe and to the gas adsorbed into the cavities of the silica, respectively. The addition of about 33 wt% AB resulted in the signal from the encapsulated ^{129}Xe shifting upfield due to interaction with the isolated AB molecules. Higher concentrations of AB gave rise to a new signal in the vicinity of free ^{129}Xe gas, which the authors attributed to interparticle spacing resulting from excessive AB. This study showed that a larger saturation loading of AB (33 wt%) was obtained than predicted, emphasizing that AB is present in quantities greater than a monolayer. The authors speculated that the incorporation of AB into the mesoporous silica may disrupt the proton-hydride bonding that dominates the solid-state structure of the material, facilitating the formation of DADB at lower temperatures.

In a related study, the AB-silica system was investigated by highly sensitive anelastic spectroscopy and DSC measurements [117]. The former method (at 0.5 GPa) clearly demonstrated that a change in the Young's modulus (used to measure the stiffness of an elastic material) for AB occurs when 1 equivalent of AB is infused into the mesoporous silica scaffold, whereas the addition of only 0.5 equivalents resulted in no deviation. The Young's modulus data are shown in Figure 8.29. Analogously, the corresponding DSC traces for

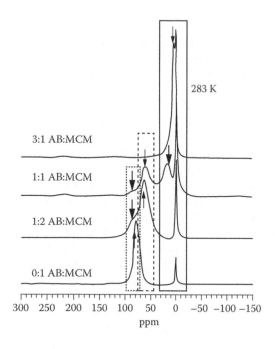

FIGURE 8.28
Hyperpolarized ^{129}Xe spectra of mesoporous silica (MCM—mobile crystalline materials) nanocomposites with NH_3BH_3 in various ratios. The dotted, dashed, and solid boxes correspond to cavities without NH_3BH_3, cavities coated with NH_3BH_3, and excess NH_3BH_3 aggregated outside the cavities, respectively. (Reprinted from Wang, L.-Q.; Karkamkar, A.; Autrey, T.; Exarhos, G.J.; *J. Phys. Chem. C* 2009, *113*, 6485–6490, with permission of the American Chemical Society. Copyright 2009 American Chemical Society.)

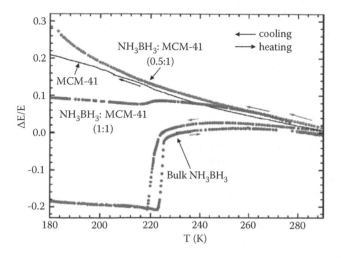

FIGURE 8.29
Temperature dependence of the variation of the Young modulus relative to its room temperature value for MCM-41 (MCM = mobil crystalline materials), bulk NH_3BH_3, MCM-41:NH_3BH_3 (0.5:1), and MCM-41:NH_3BH_3 (1:1). (Reprinted from Paolone, A.; Palumbo, O.; Rispoli, P.; Autrey, T.; Karkamkar, A.; *J. Phys. Chem. C* 2009, *113*, 10319–10321, with permission of the American Chemical Society. Copyright 2009 American Chemical Society.)

these two systems revealed similar features, in which the phase transition (orthorhombic ↔ tetragonal) was dramatically reduced for the 1:1 mixture and disappeared completely for the 0.5:1 sample. These findings confirm the absence of a phase transition for AB in a silica scaffold and underline the radically different thermal behavior of the material in this mesoporous host.

The substitution of mesoporous silica by a 24 wt% carbon cryogel (~10-nm pore diameter), and AB nanocomposite produced an even lower decomposition threshold of 90°C, with no further reaction being observed at higher temperatures [118]. In addition, use of the carbon cryogel produced no unwanted volatile byproducts (i.e., borazine) as determined by MS. Surprisingly, this host gave rise to a more exothermic reaction than neat AB (−28.7 kcal/mol), while the mesoporous silica decreased the exothermicity of this process (−0.2 kcal/mol) [119]. This observation shows that the two scaffold materials engage in different interactions with AB. The ^{11}B NMR spectrum for the carbon cryogel/AB mixture at 85°C showed the gradual formation of a weak peak at about −22.1 ppm (possibly a mobile phase of AB), along with signals corresponding to DADB. In addition, the presence of a signal for polyborazylene-like species (28 ppm) was observed, which the authors interpreted as polymerization of borazine and the clean release of hydrogen. The spectrum also revealed a peak at 16 ppm consistent with related borate species. Fourier transform infrared (FTIR) studies of this composite supported the presence of B-O interactions [120], indicating how the reactive cavities in the material can perturb and activate the AB guest molecules toward hydrogen release in an analogous manner to that discussed for ionic liquids in the section on thermal decomposition in ionic liquids.

Changes in the Chemical Composition

The hydrogen release properties of AB can be altered dramatically through the use of an additive. This phenomenon is exemplified in the thermal decomposition of a 2:1 mixture of lithium amide ($LiNH_2$) and AB in THF, in which heating the sample to 45°C results in a two-step reaction [122]. This process involves the initial loss of NH_3 followed by the release of about 10 wt% of hydrogen, as confirmed by MS and manometric measurements. ^{11}B NMR analysis showed that $LiNH_2$ initially interacts with AB to form lithium amidoborane ($LiNH_2BH_3$), which is then responsible for the release of hydrogen. The reduced decomposition temperature for this complex metal hydride has sparked considerable interest in the hydrogen storage properties of a range of alkali or alkaline-earth metal amidoborane complexes MNH_2BH_3 and $M(NH_2BH_3)_2$, respectively, for which a complete description is beyond the scope of this review. The following section focuses exclusively on the MNH_2BH_3 derivatives; information concerning the related $M(NH_2BH_3)_2$ complexes can be obtained from several excellent recent reviews [67, 122].

The substitution of a protic hydrogen atom of the NH_3 moiety in AB for an alkali metal significantly alters the structure and reactivity of the resulting MNH_2BH_3 complexes [122]. The shift from a molecular to ionic material has been proposed to enhance the dehydrogenation properties of these systems, along with suppressing the formation of deleterious volatile by-products that are observed in the thermal decomposition of AB. Xiong et al. first proposed these amidoboranes as off-board hydrogen storage materials based on a study of $LiNH_2BH_3$ and $NaNH_2BH_3$ [123, 124]. These complexes possess theoretical gravimetric storage capacities of about 11.0 and 7.4 wt% hydrogen, respectively. The onset of hydrogen release from these materials occurs at about 85°C, and this is considerably less exothermic than for the parent AB (−3 kJ/mol for $LiNH_2BH_3$ and −5 kJ/mol for $NaNH_2BH_3$), pointing to a greater likelihood of engineering a reversible process for these derivatives.

The $NaNH_2BH_3$ reaction was later reexamined by Fijalkowski and Grochala, and considerably different thermal behavior was reported by these authors [125]. In this instance, $NaNH_2BH_3$ released slightly less hydrogen (6.6 wt%), and this was accompanied by the evolution of NH_3 as monitored by IR and MS. This outcome represents a major hurdle for the practical application of these systems, as volatile NH_3 can react irreversibly with the noble metal catalysts and the proton exchange membrane in PEM fuel cells, degrading their performance. The DSC trace reported by Fijalkowski and Grochala for $NaNH_2BH_3$ exhibited a broad endotherm between 45°C and 80°C, followed by a large exothermic event. These authors suggested that the broad endothermic feature was due to head-to-tail dimerization of $NaNH_2BH_3$ (Equation 8.16), which could account for the observed NH_3 formed throughout this process (Equation 8.17). The resulting Na^+ salt would then undergo decomposition to release hydrogen (Equation 8.18). In marked contrast, Chen et al. observed an endothermic event at 57°C, which they attributed to a melt transition, despite the lack of visible evidence for this process.

$$2NaNH_2BH_3 \rightarrow [NH_3Na]^+[BH_3(NHNa)BH_3]^- \tag{8.16}$$

$$[NH_3Na]^+[BH_3(NHNa)BH_3]^- \rightarrow Na^+[BH_3(NHNa)BH_3]^- + NH_3 \tag{8.17}$$

$$Na^+[BH_3(NHNa)BH_3]^- \rightarrow BH_3(NHNa)BH_2 + NaH \tag{8.18}$$

More recently, the thermal decomposition of KNH_2BH_3 has been shown to result in similar hydrogen release properties to those reported by Chen et al. for its lithium and sodium congeners, with no borazine or NH_3 being observed [123, 124, 126].

The mechanism associated with the formation and decomposition of $LiNH_2BH_3$ was initially explored through two different computational approaches [127, 128]. Lee and McKee modeled this reaction with a $(LiH)_4$ cluster and predicted initial adsorption of AB onto this cluster, giving rise to $(LiH)_3 \cdot LiNH_2BH_3$ and the release of hydrogen. This process was characterized by a relatively low activation barrier, in contrast to the study of Kim et al., who observed no barrier for their analogous model involving only one LiH unit. Lee and McKee proposed that this low barrier could potentially account for the lack of borazine in the experimental studies since the corresponding activation barrier for the formation of $NH_2 = BH_2$ from AB is much higher.

These authors also observed an alternative pathway for the reaction, but they quickly discounted this mechanism due to its higher activation barrier. They also explored the decomposition pathway for $LiNH_2BH_3$ (dimers) and found a one-step reaction energetically unfavorable ($7 \rightarrow TS7/9 \cdot H_2 \rightarrow 9 \cdot H_2$; $\Delta G = 223.2$ kJ/mol). In contrast, a two-step process ($7 \rightarrow TS7/8 \rightarrow 8 \rightarrow TS8/9 \cdot H_2 \rightarrow 9 \cdot H_2$; $\Delta G = 151.6$ kJ/mol), involving transition states with a bridging Li···H···Li moiety bound to the dehydrogenated $NH_2 = BH_2$ fragment, was much more likely, as shown in Figure 8.30. In this instance, the release of hydrogen occurs through an interaction between the bridging hydride and a protic hydrogen atom of the $NH_2 = BH_2$ groups ($TS8/9 \cdot H_2$), resulting in the formation of a $LiNHBH_2 \cdot LiNH_2BH_3$ complex (9), which can further decompose cleanly to $LiNHBH_2$. The subsequent release of hydrogen from $LiNHBH_2$ was predicted to be unfavorable, although the authors speculated that this process could occur if lattice energy stabilization accompanies the liberation of hydrogen. This study revealed the crucial role played by the Li^+ ion in bringing the hydridic and protic components into close enough proximity to facilitate the release of hydrogen.

In an attempt to shed more light on the mechanism of dehydrogenation of MNH_2BH_3 complexes, Luedtke and Autrey investigated the decomposition of $LiNH_2BH_3$ and its

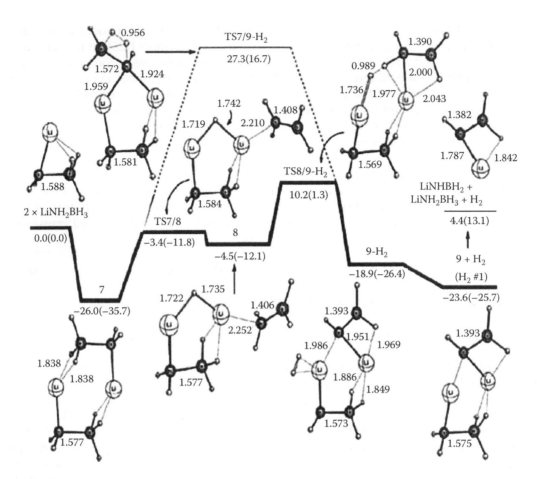

FIGURE 8.30
Free energy surface for the dehydrogenation of LiNH₂BH₃ dimers. (Reprinted from Lee, T.B.; McKee, M.L.; *Inorg. Chem.* 2009, *48*, 7564–7575, with permission of the American Chemical Society. Copyright 2009 American Chemical Society.)

partially deuterated isotopomers LiND₂BH₃ and LiNH₂BD₃ [129]. They observed that 0.5 equivalent of hydrogen was released at similar rates for the LiNH₂BH₃ and LiND₂BH₃, while the LiNH₂BD₃ isotopomer required twice as long to release the same amount, indicating a kinetic isotope effect for the first stage of the reaction and showing that the stronger B-D bonds are activated more slowly than their B-H counterparts. The lack of any rate difference between LiNH₂BH₃ and LiND₂BH₃ then indicates that breaking of the B-H(D) moiety represents the rate-determining step in this stage of the process. The products obtained from the loss of 1 equivalent of hydrogen were subsequently characterized by in situ magic angle spinning (MAS) ¹¹B NMR spectroscopy. Heating a sample of LiNH₂BH₃ at 90(5)°C resulted in a gradual broadening and eventual loss of intensity for the peak corresponding to the starting material. This was replaced by a new signal at 29 ppm, which is inconsistent with a tetrahedral BH₂ moiety (5 and –15 ppm) [130]. However, the absence of these signals does not preclude the involvement of a species containing this structural motif; it may play a role but be consumed rapidly under the reaction conditions. Thus, the major product in the thermolysis of LiNH₂BH₃ will contain BH and BH₃ groups. The authors tentatively assigned the reaction products as LiNH = BHLiNHBH₃ (1a) or LiNH₂BH

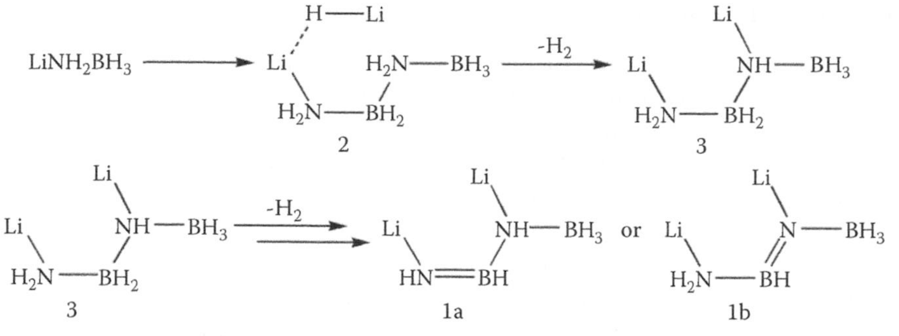

SCHEME 8.6
Proposed mechanism for the solid-state decomposition of $LiNH_2BH_3$. (From Luedtke, A.T.; Autrey, T.; *Inorg. Chem.* 2010, 49, 3905–3910. With permission.)

$= NLiBH_3$ (1b) depending on the location of the $N = B$ bond. However, they noted that this material contains reactive BH_3 and LiNH groups, whereas the actual products formed from the solid-state decomposition of this system are likely a mixture of compounds with varying chain lengths containing sp^2 BH and terminal BH_3 groups.

The kinetic studies and ^{11}B NMR results led these authors to propose a mechanism similar to that deduced by the previous computational studies [127, 128]. The presence of a kinetic isotope effect for $LiNH_2BD_3$ and the absence of this phenomenon for its $LiND_2BH_3$ counterpart provides persuasive evidence that the N-H bond is broken after the rate-determining step, consistent with the loss of LiH and the formation of $LiNH_2BH_2NHBH_3$ (2), as shown in Scheme 8.6. This stage in the reaction is then followed by rapid deprotonation of 2 by the metal hydride (i.e., metal-assisted hydride transfer), resulting in the liberation of hydrogen to form $LiNH_2BH_2LiNHBH_3$ (3). However, this species was not detected in the MAS ^{11}B NMR spectra, indicating that it quickly loses hydrogen to form 1a or 1b. This mechanism also helps to explain the experimental trends in the decomposition of these MNH_2BH_3 complexes: An increase in the electropositive character of the metal (Li < Na < K) decreases the strength of the M-H and M-N bonds, consistent with the reaction rates of these systems (K > Na > Li) [129].

The different thermal behaviors of the partially deuterated forms of $LiNH_2BH_3$ prompted Wolstenholme et al. to investigate the atomic provenance of the hydrogen evolved from $LiND_2BH_3$ [131]. In this endeavor, they employed similar ^1H NMR experiments as the previous AB study [131], in which proton-hydride and hydride-hydride pathways should appear as HD or H_2, respectively. An observed HD/H_2 ratio of about 3:1 indicates that the counterintuitive hydride-hydride pathway plays a significant role in the release of hydrogen from this system. However, the authors noted that HD may also potentially arise from a similar scenario since the highest barrier transition state in Lee and McKee's theoretical model (TS8/9·H_2 in Figure 8.30) involved an Li-H···H-N(Li) interaction, in which the protic hydrogen atom formed a weak interaction with Li^+ [127]. Therefore, this transition state may also involve a hydride-hydride interaction that would result in the appearance of HD in the corresponding ^1H NMR spectrum. However, based solely on the relative intensity of the HD/H_2 resonances from $LiND_2BH_3$, it appears that the hydride-hydride pathway for hydrogen production is less significant than in the case of the parent AB, in spite of a short $B-H^{\delta-}$···$^{\delta-}H-B$ contact (2.10 Å) in the crystal structure of $LiNH_2BH_3$. This study provided further evidence of the important but hitherto invisible role played by counterintuitive hydride-hydride interactions in the thermal desorption of hydrogen from amine-borane complexes.

Regeneration of AB from Spent Dehydrogenated Products

To this point, the focus has been entirely on the manner in which AB releases hydrogen, which is more complicated than one might have suspected! What about the reverse process, regeneration of AB from the spent dehydrogenated material? The release of hydrogen from the solid-state and solution-phase decomposition of AB remains an exothermic process, in spite of various approaches that have been tried to alter the thermal behavior of this system. This feature implies that regeneration of the decomposition products obtained from these reactions will require off-board chemical manipulation. To date, only a handful of such methods have been reported successfully for AB [132–137]. In 2007, Ramachandran et al. demonstrated that the product obtained from the metal-mediated solvolysis of AB (i.e., $[NH_4]^+[B(OMe)_4]$) can be readily converted back to the starting material at ambient temperature through a reaction with ammonium chloride (NH_4Cl) and lithium aluminum hydride ($LiAlH_4$) [132]. However, this method only recovers about 81% of the starting material, making the approach unattractive for large-scale technological application. Moreover, the formation of a product containing a strong B-O bond limits the efficiency of this process as this will need to be reduced to reform AB.

In recent years, Hausdorf et al. developed a method of recycling these undesirable products by first digesting these materials in HCl as a superacid mixture with $AlCl_3$ to form BCl_3 and NH_4Cl [133]. A similar approach was also adopted by Sneddon, using an HBr/$AlBr_3$ mixture [134]. The challenge associated with this route lies in the hydrodechlorination of the BCl_3 intermediate, which requires heating to high temperatures (>600°C) in the presence of hydrogen gas to engineer partial reduction to $BHCl_2$, which can then disproportionate into BCl_3 and B_2H_6 on separation from the HCl by-product. In an earlier study, Taylor and Dewing noted that addition of NMe_3 to BCl_3 drastically reduced the temperature threshold of this reaction to about 200°C under 2000 atm pressure [135], at the same time avoiding the formation of hazardous B_2H_6. Unfortunately, this process requires a considerable energy investment and results in low yields (25%). In a similar vein, Sneddon et al. demonstrated that BBr_3 coordinated to *N,N'*-diethylaniline resulted in an acid-base adduct that can easily be reduced by triethylsilane under moderate conditions [134]. This finding prompted Hausdorf et al. [133] to use MgH_2 and triethylsilane to reduce their NR_3BCl_3 adducts. In these instances, the coordination of a weak base (i.e., NPh_3) pushes the reaction to completion within 20 min at 80°C, whereas stronger bases (i.e., NEt_3) hinder the reduction of the BCl_3 moiety. The resulting NR_3BH_3 species can then be converted to AB through a displacement reaction with NH_3. Although this approach avoids detrimental B-O products, the efficiency of this reaction is less than ideal (>60% yield), and a large number of steps are required to regenerate AB.

In an alternative approach, Davis et al. initially performed DFT calculations on the conversion of borazine (a model for the expected polyborazylene products) to AB using 1,2-benzenedithiol as a reactant [136]. These predicted an exothermic reaction with an enthalpy of –20.4 kcal/mol in the condensed phase. Indeed, the reaction of this reagent with polyborazylene in THF under reflux conditions resulted in the formation of $(C_6H_4S_2)BHNH_3$ and $[NH_4][B(C_6H_4S_2)_2]$, as confirmed by ^{11}B NMR spectroscopy. The latter salt could then be converted to the desired $(C_6H_4S_2)BHNH_3$ product through the addition of Bu_3SnH at 60°C; this was subsequently converted to AB and $C_6H_4S_2Sn(nBu)_2$ using nBu_2SnH_2. However, this process resulted in poor yields (67%) and required toxic tin hydrides; hence, it is unlikely to represent a viable industrial process. Sutton et al. modified this procedure by introducing a Lewis base (NMe_3) and using $nBuSnCl$, which avoids the complications associated

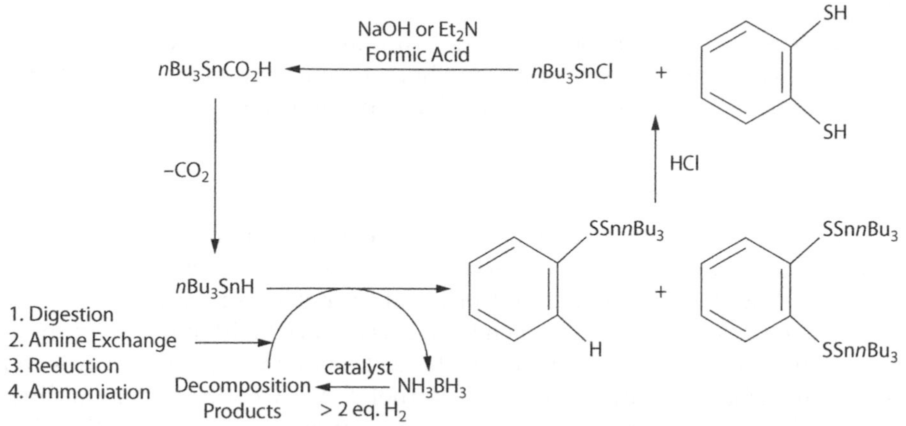

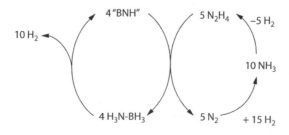

SCHEME 8.7
Proposed mechanism for converting polyborazylene to AB via Sn complexes. (From Sutton, A.D.; Davis, B.L.; Bhattacharyya, K.X.; Ellis, B.D.; Gordon, J.C.; Power, P.P.; *Chem. Commun.* 2010, 46, 148–149. With permission.)

with nBu_3SnH [137]. This approach results in slightly different Sn by-products, as shown in Scheme 8.7, which could be converted to the starting reagents through treatment with HCl.

The approaches described here to recover AB from the dehydrogenated spent fuels highlight the feasibility of this goal. Nevertheless, the disadvantages (low yields and forcing conditions) generally outweigh the advantages of these methods. In light of these issues, Sutton et al. explored the potential of hydrazine (N_2H_4; Hz) as a reductant to regenerate AB from its decomposition products [138]. The solubility of the spent fuels in polar solvents prompted these researchers to carry out initial experiments in THF at room temperature. Indeed, this approach resulted in upgrading of the spent material to systems that contained only $-BH_3$ moieties. However, this procedure resulted in strong coordination of Hz to the resulting BH_3 groups, with consequent minimal recovery of AB, as evidenced by ^{11}B NMR spectroscopy. This finding indicated that the conversion of this Hz-BH_3 adduct to AB would be the main challenge in finding a viable route to regenerating the starting material. Addition of the Hz-BH_3 species to a sealed vessel containing liquid NH_3, followed by mild heating to 60°C, resulted in about 85% conversion to AB within 24 h (Scheme 8.8). It is noteworthy that the analogous room temperature reaction resulted in no AB formation, highlighting the thermal activation needed to break the N-B bond of the Hz-BH_3 intermediate. Removal of

SCHEME 8.8
Proposed mechanism for the regeneration of AB decomposition products using N_2H_4 in NH_3. (Reprinted from Sutton, A.D.; Burrell, A.K.; Dixon, D.A.; Garner, E.B., III; Gordon, J.C.; Nakagawa, T.; Ott, K.C.; Robinson, J.P.; Vasiliu, M.; *Science* 2011, 331, 1426–1429, with permission of the American Association for the Advancement of Science. Copyright 2011 American Association for the Advancement of Science. With permission.)

volatile by-products in the workup, including residual Hz, improved the yield of AB to 95%. The authors observed no conversion to AB when similar reactions were performed (14 days) in the absence of Hz. The discovery of this method currently provides the most promising route to developing AB as an off-board-reversible hydrogen storage candidate.

Outlook for Ammonia Borane

The hydrogen-rich nature and tunable hydrogen release properties associated with AB have attracted considerable interest in this chemical hydride as a vehicular hydrogen storage material. Although the structure and physical properties of AB are now well established, the thermal desorption of hydrogen from this material will require manipulation of its kinetic and thermodynamic behavior (see the section on tailoring the hydrogen release properties) to interface with modern-day PEM fuel cells. A better understanding of the mechanism and pathways that characterize the formation of hydrogen will assist in this regard, as well as helping to minimize the formation of deleterious volatile by-products, such as ammonia, diborane, and borazine. The following challenges remain to be solved by researchers in the quest for a viable AB-based hydrogen storage system.

The dehydrogenation of AB as a melt (solid-state decomposition) requires further investigation since this represents the physical state in which hydrogen is typically evolved from this material. What type of interactions mediate this process N-H$^{\delta+}$···$^{\delta-}$H-B or B-H$^{\delta-}$···$^{\delta-}$H-B, and to what extent? Answers to these questions will help to develop strategies for improving the hydrogen release properties of AB and developing efficient, high-yield processes for the regeneration of this material from its spent fuels. The isothermal decomposition of AB below its melting point is also poorly characterized. To what extent does the rigidity of the crystalline material hinder the isomerization to DADB, which has been shown to be so important in the release of hydrogen from the molten phase? For example, can a catalyst be cocrystallized with AB that will allow the liberation of hydrogen under mild conditions (about 60–80°C)?

The solid-state decomposition products of AB (spent fuel) are surprisingly poorly defined. For example, PAB and polyiminoborane (PIB), as described in the section on thermal decomposition and Equations 8.1 and 8.2, are believed to consist of a combination of linear, branched, cross-linked, and coiled substructures, but the exact nature of these systems is unknown. These materials are notoriously difficult to characterize on account of their complex amorphous nature and poor solubility. Multinuclear solid-state NMR appears to offer the most viable means of teasing out structural information from these recalcitrant materials at the local level.

The unfavorable thermodynamics associated with the regeneration of spent fuel represents a major obstacle to the widespread adoption of AB as a hydrogen storage material. Significant progress has been made in the development of viable approaches to regeneration, as exemplified by the elegant and efficient hydrazine cycle of Sutton et al. [138]. However, hydrazine is itself a highly endoergic material ($\Delta H°_{f,298K} = +50.6$ kJ/mol) [139], and its preparation as a reagent for this process will require investment of considerable amounts of energy. Other unconventional methods of rehydrogenating the decomposition products of AB should be explored, including unconventional solvents and phases, whose advantages have been demonstrated, for example, in circumventing the unfavorable thermodynamics associated with the rehydrogenation of lithium alanate (LiAlH$_4$) [140–142].

Attention should be focused on exerting greater control over the connectivity of the dehydrogenated material. For example, what factors favor the formation of cyclic oligomers rather than linear polymers? What types of interaction are involved in this process? The ability to produce more homogeneous and well-defined, ideally crystalline, and soluble, forms of PAB and PIB would represent a significant advantage to researchers tasked with the development of a simple and efficient route back to AB.

The alkali metal and alkaline earth metal amidoborane derivatives of AB (see the section on changes in chemical composition) remain underdeveloped; they offer several potential advantages (cleaner, less-exothermic H_2 release, lower decomposition temperature, better solubility, etc.) over their parent amine-borane adduct. Efforts should focus on characterization of these materials as possible alternatives to, or coadditives with, AB as second-generation amine-borane hydrogen storage material [143].

In summary, the development of a sustainable and efficient hydrogen-based automotive industry using AB as a hydrogen storage material remains a possible but challenging outcome, and creative strides have been made in this direction over the past decade, as advocated in 2003 by the U.S. DOE [144]. However, the major obstacles to success remain the complexity of the hydrogen release by AB, the intractability of the resulting spent material and the unfavorable thermodynamics of its conversion back to AB.

Liquid Organic Hydrogen Carriers (LOHCs)

Unsaturated organic molecules, typically olefins and aromatics, can be induced to react with molecular hydrogen in the presence of a metal catalyst to provide the corresponding alkane. This process of catalytic hydrogenation is a well-established unit operation in the chemical and petroleum-refining industries. When rendered reversible by an appropriate off-board process combining liquid substrates and hydrogenation/dehydrogenation catalysis, organic molecules provide the means for potential storage, transportation, and delivery of hydrogen. The substrates, which we refer to as liquid organic hydrogen carriers (LOHC), are occasionally referred to as "chemical hydrides" or "liquid organic hydrides" in the literature. This is unfortunate since the compositions are all organic molecules with essentially covalent C-H bonds and certainly have no hydridic (i.e., H^- anion-like) character.

Here we give a critical account of progress by investigators in this field in view of the end goal of providing a viable liquid carrier technology for a hydrogen energy infrastructure.

Catalytic Hydrogenation/Dehydrogenation Fundamentals

Recall that a hydrogenation reaction can be fully quantified in terms of the basic thermodynamic parameters, enthalpy ΔH, entropy ΔS, Gibbs free energy ΔG, and the equilibrium constant K, for the reversible addition of hydrogen to substrate A, forming A-2H by the familiar thermodynamic relationship $\Delta G = -RT \ln K = \Delta H - T\Delta S$, where R is the ideal gas constant and T is the reaction temperature (in degrees kelvin). The equilibrium can be written,

$$A + H_2 \xleftrightarrow{\quad K \quad} A - 2H$$

Typically, the LOHC hydrogenation reaction (e.g., converting benzene to cyclohexane) is conducted in the presence of a transition metal catalyst at modest temperatures and under hydrogen pressure. With aromatic substrates, the addition of hydrogen is almost always spontaneous and exothermic ($\Delta H < 0$) with a rate that is determined by the efficacy of the catalyst and the specific reaction conditions. The reverse dehydrogenation reaction, however, is thermodynamically limited and only occurs because of an increase in entropy for the system that is largely associated with the liberation of hydrogen gas. Therefore, dehydrogenation is generally conducted at elevated temperatures to generate hydrogen at useful equilibrium pressures.

For example, the dehydrogenation of cyclohexane to benzene, a system that has been investigated for hydrogen storage, has an experimental enthalpy change at standard conditions (1 atm pressure, 25°C), $\Delta H°$, of 68.7 kJ/mol H_2. The ΔG at 80°C (a typical operating temperature for a PEM fuel cell) is 25.8 kJ/mol H_2, with a corresponding equilibrium constant of 3.43×10^{-12} atm. The ΔG for this reaction approaches zero only at about 280°C, and the temperature required for about a 95.5% conversion of cyclohexane to benzene at equilibrium under 1 atm H_2 is 319°C. It is important to note that the dehydrogenation enthalpy, the heat required for liberating the hydrogen from cyclohexane (68.7 kJ/mol H_2), is almost one-third of the LHV for H_2 (242 kJ/mol).

The development of a more easily reversible hydrogenation/dehydrogenation system, one that is thermodynamically "well balanced," is the most fundamental challenge toward the goal of realizing a practical organic liquid carrier. With organic substrates, the determining equilibrium constant K is largely dictated by the enthalpy change ΔH since the corresponding entropy contribution representing the change from bound to free (gas-phase) hydrogen is approximately the value of S^0 for H_2 (130.7 J/mol). For an idealized carrier system ($AH_6 \leftrightarrow A + 3H_2$) with 98% dehydrogenation conversion at 80°C under hydrogen pressure of 1 atm, and with the capability of being rehydrogenated under reasonable conditions (98% conversion under 100 atm H_2 pressure at 150°C), the dehydrogenation enthalpy, ΔH_D, would be about 40 kJ/mol H_2, as also discussed in Chapter 5.

This idealized enthalpy value is the minimal energy cost for "containing" the hydrogen, which has to be supplied for its release from the carrier. It is significantly lower than the corresponding dehydrogenation enthalpy for cyclohexane/benzene + $3H_2$ (68.7 kJ/mol H_2) and the enthalpy for *trans*-decalin (DEC)/naphthalene + $5H_2$ (66.6 kJ/mol H_2) and also significantly less than that for any experimentally evaluated organic liquid carriers. A low desorption enthalpy is also desirable for minimizing the quantity of heat that has to be transferred in the respective (exothermic) hydrogenation and (endothermic) dehydrogenation steps of the hydrogen storage and delivery cycle. It is interesting to note that while LOHC dehydrogenation suffers from "too tight" binding of hydrogen (as covalent C-H bonds), the microporous adsorption material approach to hydrogen storage, reviewed in Chapter 7, has the opposite problem, namely, an insufficient binding of molecular hydrogen for an adequate loading of the gas at near-ambient temperatures.

Hydrogenation of organic substrates is commonly conducted using catalysts from the platinum group metals (usually Pt, Rh, and Ru) on metal oxide or carbon supports or as finely divided forms of nickel and cobalt. While similar catalyst compositions may also function for the reverse reaction, finding adequately performing dehydrogenation reaction catalysts is often a more demanding task. An endothermic (dehydrogenation) reaction may be expected to have a higher energy of activation barrier than the reverse (hydrogenation) process. Also, there is relatively far less prior art on dehydrogenation catalysis particularly at lower temperatures (<200°C).

Other requirements for organic liquid carriers, from physical to environment-related properties, are addressed in the following discussion of specific systems.

The Cyclohexane-Benzene-Hydrogen Cycle

The use of conventional hydrogenation and dehydrogenation technology, coupled with relatively inexpensive aromatic molecules such as benzene and toluene, formed the basis of the experimental investigations of liquid organic compounds for efficient hydrogen storage starting in the late 1970s. Bélanger [145] was one of the principal investigators of what became known as the cyclohexane-benzene-hydrogen (CBH) cycle:

$$C_6H_{12} \leftrightarrow C_6H_6 + 3H_2 \qquad (8.19)$$

The gravimetric capacity of this cycle is 7.1 wt% with a volumetric storage density of 55.5 kg H_2/m^3. The standard enthalpy change ΔH^0_{gas} for dehydrogenation for cyclohexane is $\Delta H^0 = 68.7$ kJ/mol H_2; the standard free energy change for hydrogen release is $\Delta G^0_{gas} = 32.6$ kJ/mol H_2.

In initial scoping experiments, vaporized cyclohexane was passed through a fixed bed of Pd, Pt, and Ni on alumina catalysts at 400–450°C, which resulted in modest-to-high conversions to benzene and hydrogen. The dehydrogenation of methylcyclohexane (MCH) was also demonstrated at somewhat lower conversions. Touzani et al. [146] also provided a schematic of an organic liquid-carrier-based hydrogen fuel infrastructure, as shown in Figure 8.31.

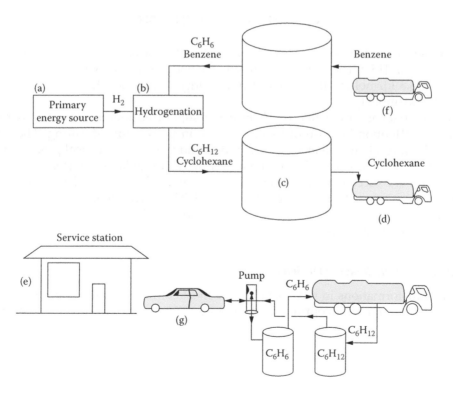

FIGURE 8.31
Schematic of an organic liquid-carrier-based hydrogen fuel infrastructure, as illustrated by the cyclohexane-benzene-hydrogen (CBH) cycle. (Reproduced from Touzani, A.; Klvana, D.; Bélanger, G.; *Int. J. Hydrogen Energy* 1984, *9*, 929, with permission from the International Association for Hydrogen Energy.)

In this infrastructure, hydrogen generated from a primary energy source is used to catalytically hydrogenate benzene to cyclohexane. The stored cyclohexane carrier is transported by tanker truck to a service station for distribution in vehicles. On-board the vehicle, the liquid cyclohexane is vaporized and passed through a catalytic reactor, where it is reconverted to benzene and hydrogen. The hydrogen would then be burned in a hydrogen internal combustion engine (H_2ICE) or used to power a hydrogen fuel cell, as described in Chapter 2. Waste heat from the hydrogen conversion device could be employed for vaporizing the cyclohexane and maintaining the reactor at temperature. Upon refueling, benzene is removed from the vehicle and replaced with fresh cyclohexane.

These early concepts were quantified in the engineering/economics evaluation by Cacciola et al. [147] on the storage and transmission of hydrogen via a CBH cycle and in the design and simulation study of a dehydrogenation reactor [148]. Despite the high capacity (one H per C atom) and apparently good reversibility of the CBH system, owing to the toxicity of benzene, most studies were performed using the MCH carrier.

Methylcyclohexane-Toluene-Hydrogen Cycle

The methylcyclohexane-toluene-hydrogen (MTH) cycle concept for hydrogen storage (first proposed by Sultan and Shaw [149] in 1975) is based on a reversible dehydrogenation of MCH indicated in Equation 8.20 and Scheme 8.9:

$$C_6H_{11}CH_3 \quad \leftrightarrow \quad C_6H_5CH_3 + 3H_2 \tag{8.20}$$

The gravimetric capacity of the MTH cycle is 6.1 wt%, with a volumetric storage density of 47.0 kg H_2/m^3 [150]. The standard enthalpy change ΔH^0 for dehydrogenation for MCH is $\Delta H^0{}_{gas} = 68.3$ kJ/mol H_2; the standard free energy change ΔG^0 for hydrogen release is $\Delta G^0{}_{gas} = 31.5$ kJ/mol H_2.

The dehydrogenation reaction of MCH using supported metal catalysts has received considerable attention because of its importance in petroleum reforming. The research comprises both fundamental mechanistic studies and the more empirical process improvement work, which has been critically reviewed by Alhumaidan et al. [150]. In summary, Pt/Al$_2$O$_3$ and its successor Pt-Re/Al$_2$O$_3$ are the most effective catalysts. In the literature, there is some controversy regarding the MCH dehydrogenation kinetics, poisoning mechanisms (e.g., sulfur-containing compounds), and the lifetime and deactivation (coking, etc.) of the catalysts.

Innovative Catalytic Reactor Designs

Catalytic transformations in the gas phase are most commonly performed using a fixed-bed catalytic reactor in a continuous flow system. The conversion of a substrate for a

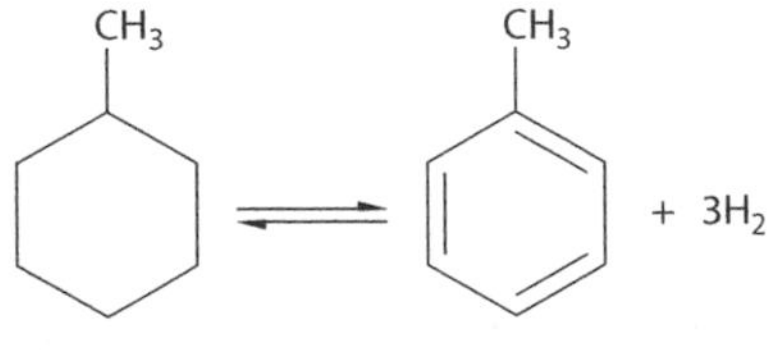

SCHEME 8.9
Dehydrogenation of methylcyclohexane to toluene, releasing hydrogen.

reversible reaction is a function of the inherent activity of the catalyst, the liquid hourly space velocity (LHSV), and, ultimately, by the reaction thermodynamics at the operating temperature. For the dehydrogenation of MCH, innovative reactor designs have been employed for mitigating the thermodynamic limitations and high heat input requirements of the reaction.

A pulsed-spray reactor, wherein an atomized spray of MCH is directed at a heated Pt/LaYNi perovskite surface, is reported to provide hydrogen with high catalyst activity and complete selectivity for the dehydrogenation reaction [151]. The use of reactors that operate under similar nonequilibrium conditions for the dehydrogenation of MCH and benzene is discussed further in the context of our review of the DEC-naphthalene carrier system.

Membrane reactors that provide an in situ removal of hydrogen from the dehydrogenation reaction zone have been used for obtaining a greater-than-equilibrium production of hydrogen from MCH. Ali et al. employed a membrane reactor in a tube-and-shell configuration with the catalyst in the annular space (the reaction zone) and a thin-wall Pd-Ag alloy inner tube that functioned as the selectively H_2-permeable membrane and conduit for the product hydrogen [152]. At 300–400°C, MCH conversions were up to four times greater than the estimated equilibrium values. A pilot-scale model reactor was constructed, and the performance for H_2 production from MCH was modeled in terms of the dehydrogenation kinetics and H_2 permeation rate [153]. With a reactor constructed with an amorphous microporous silica membrane, Oda and coworkers were able to obtain remarkably high-purity hydrogen from MCH, attesting to the membrane's high selectivity for permeation of the gas [154].

For providing the MCH dehydrogenation heat requirements, Kerleau, Pitault, and coworkers [155] devised an ingenious autothermal reactor device wherein the reaction heat is supplied by the combustion of less than 10% of the by-product toluene (6% for the chemical reaction endotherm and 4% for thermal losses). The heat exchanger/reactor consisted of a stack of plates with channeled surfaces, containing a powdered dehydrogenation catalyst on one side and a catalyst wash coat for the combustion reaction on the other side. Various configurations of the apparatus were tested with the objective of realizing the best heat transfer between the two zones while maintaining ideal plug flow conditions for each reaction. With a second-generation reactor using a cocurrent flow of MCH and toluene/air, the desired autothermal operating conditions were achieved with total combustion of the toluene feed.

Vehicular Applications of the MTH Cycle

Taube and coworkers have provided an engineering and economics model for H_2-powered heavy vehicles in the context of a liquid carrier infrastructure [156, 157]. Hydrogen generated by electrolysis of water using energy from a nuclear power station is employed for the hydrogenation of toluene, and the hydrogenation reaction exotherm can be used for local heating. In a demonstration, a 17-ton truck was outfitted with a 150-kW ICE and a 710-L tank for MCH. Heat extracted from the engine exhaust was used to drive the MCH dehydrogenation reaction, but combustion of some stored toluene was needed to provide a cold start. A photograph of the truck shows the combination of reactor, fuel tanks, heat exchanger, and associated instrumentation occupying about 80% of the bed of the vehicle—"a movable laboratory" [158]. In a second-generation prototype truck, the fuel storage and hydrogen generation plant, which supplies 3 g of H_2/s, was much more compact, making for a seemingly more realistic vehicle [159].

Electrical Energy Storage

A number of publications [145, 146, 149, 156, 159] have noted the potential of exploiting the CBH and MTH cycles for short-term (electrical load leveling) as well as seasonal electrical energy storage (capturing hydroelectric power that is generated in summer, especially at nighttime, and using it to augment existing supplies during winter). The basic concept is that the LOHC is "loaded" with electrolysis-generated hydrogen at times or locations of relatively low-cost electric power. On demand, the liquid is catalytically dehydrogenated with appropriate heat input for a subsequent "reelectrification" process of the generated hydrogen. Scherer [160] studied the economics of three possible options for this conversion using molten carbonate fuel cells (MCFCs), solid oxide fuel cells (SOFCs), and hydrogen gas combustion turbines. PEM and phosphoric acid fuel cells (PAFCs) operate at temperatures of about 80°C and about 180°C, respectively, far too low for satisfying the dehydrogenation enthalpy for the MTH cycle. The best case of a MTH-SOFC combination could not compete pricewise with electricity generated in the winter from fossil fuel sources. Possible mitigating factors were cited such as low CO_2 emissions and the strategic energy independence that an MTH hydrogen liquid-carrier-based technology could provide.

Decalin-Naphthalene-Hydrogen Cycle

The decalin-naphthalene-hydrogen (DNH) cycle is based on a reversible dehydrogenation of decalin (DEC; liquid with boiling point of 195.6°C [*cis*], 187.2°C [*trans*]) to naphthalene (solid, with melting point of 80.5°C) with tetralin, $C_{10}H_{12}$, as a reaction intermediate. The reaction is shown in Equation 8.21 and Scheme 8.10.

$$C_{10}H_{18} \leftrightarrow [C_{10}H_{12} + 3H_2] \leftrightarrow C_{10}H_8 + 2H_2 \qquad (8.21)$$

The gravimetric capacity of the DNH cycle is 7.2 wt%, with a volumetric storage density of 64.9 kg H_2/m³ [150]. The ΔH^0_{gas} and ΔG^0_{gas} for dehydrogenation from the *cis* isomer of DEC are 63.9 kJ/mol H_2 and 27.6 kJ/mol H_2, respectively. The ΔH^0_{gas} and ΔG^0_{gas} values for hydrogen release from the *trans* decalin isomer are 66.6 kJ/mol H_2 and 30.1 kJ/mol H_2, respectively.

DEC Dehydrogenation Catalysis

Decalin or decahydronaphthalene is produced industrially via a catalytic hydrogenation of naphthalene [161, 162] and is commercially available as a mixture of the *cis* and *trans* stereoisomers, with the latter being thermodynamically more stable.

In continuing work directed at seasonal energy storage, Newson et al. [163] were able to substitute DEC for MCH as the carrier molecule, citing significant engineering advances. In addition to exploiting the higher capacity of DEC, it was possible to conduct both the hydrogenation and dehydrogenation steps with the same catalyst

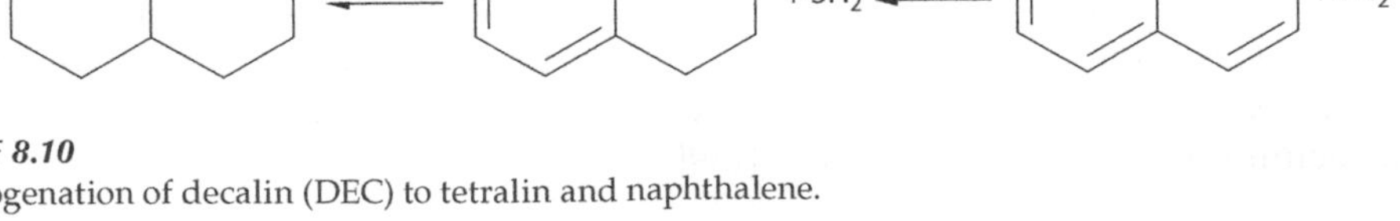

SCHEME 8.10
Dehydrogenation of decalin (DEC) to tetralin and naphthalene.

and reactor system, thus saving costs for capital and equipment. A 98% conversion of DEC at 440°C under 11-bar hydrogen pressure and rehydrogenation of the resulting naphthalene at 220°C and 16-bar hydrogen pressure was reported using an unspecified commercial catalyst.

A kinetic model was derived from a detailed study of the dehydrogenation of DEC in the range of 250–350°C over supported Pt catalysts using a fixed bed flow reactor [164]. The study was performed with the objective of providing a process for the production of pure hydrogen for a fuel cell. At 350°C and 1 atm pressure, DEC conversions of greater than 98% to naphthalene were realized. The DEC feed material consisted of a mixture of 24% *cis* and 76% *trans* isomers. The *cis* isomer proved to be the more reactive than the *trans* isomer, with conversions of 98% and 66.8%, respectively, at 300°C. The kinetic modeling, which was performed for the entire reaction system, accounted for the *cis*-to-*trans* isomerization reaction of DEC, the formation of the tetralin intermediate (only one ring dehydrogenated), and formation of the end product, naphthalene.

A new catalyst type consisting of 1% Pt supported on a stacked cone carbon nanotube composition was evaluated for the dehydrogenation of DEC in a continuous flow system employing a fixed-bed plug flow tubular reactor [165]. This catalyst was found to be significantly more active for this reaction than Pt, Pd, or Rh on a γ-alumina support. At 240°C, a 13% conversion of DEC isomers to tetralin and naphthalene was demonstrated, and the *cis* isomer was the more active isomer for dehydrogenation. A homogeneous (soluble) catalyst, the so-called iridium pincer complex, $IrH_2[C_6H_3-2,6-(CH_2PBu^t_2)_2]$, was reported to catalyze the transfer dehydrogenation of cyclic alkanes, including DEC, to the corresponding aromatic molecules in the presence of *tert*-butylethylene as the hydrogen acceptor [166].

Innovative Reactor/Reaction Systems

The desire of effecting a dehydrogenation of DEC at the lowest possible temperature, to enable a more facile heat integration and longer catalyst lifetimes without the need for a hydrogen cofeed, led to a considerable body of research on nontraditional ways of performing this reaction.

Membrane Reactors

Loutfy [167] conducted a dehydrogenation reaction with DEC in a membrane reactor in the presence of a Pt/C catalyst in the temperature range of 250–320°C. Removing some of the hydrogen through a permselective membrane resulted in a shift of the equilibrium toward the hydrogen and naphthalene products. In a comparison between the calculated and experimental equilibrium conversion, it was evident that the temperature of DEC dehydrogenation was significantly lowered by using the membrane reactor. Dehydrogenation of MCH and DEC was conducted using a membrane reactor where the reaction was thermally coupled to a Fischer Tropsch (CO/H_2 to hydrocarbons) process, thus providing both the necessary heat and a separation of the product hydrogen from the reaction zone [168].

Catalytic Dehydrogenation under Nonequilibrium Conditions

As conducted in conventional plug flow reactors, stirred tank reactor systems, and even membrane reactors (allowing for the desired selective hydrogen permeation), the extent of dehydrogenation of cycloalkanes to the corresponding aromatic molecules is usually limited by unfavorable thermodynamics. Research groups principally at the Tokyo University of Science [169–176] and at Hokkaido University [177–180] have developed experimental techniques for performing catalytic reactions under nonequilibrium conditions with resultant higher conversions at milder temperatures. The catalytic dehydrogenation methods have been referred to in the literature as "liquid-film-type" dehydrogenation and reaction under "wet-dry-multiphase conditions."

As described by Hodoshima et al. [174, 175] the experimental apparatus for the liquid film technique consists of a reaction flask attached to a reflux condenser, which is in turn connected to a gas buret for measuring hydrogen flow. Batchwise DEC dehydrogenation reactions were conducted in the presence of a fixed charge of catalyst (0.75 g of 5 wt% Pt on C) and varying amounts of added DEC. With heating at 210°C and cooling (with the condenser) at 5°C, the liquid was refluxing at a significantly higher temperature, about 22°C above the boiling point of DEC. As shown in Figure 8.32, the conversion of DEC after 2.5 h and the hydrogen evolution rate showed a remarkable dependence on the amount of added DEC. With 5–10 mL of DEC, the system functioned as a suspension in a refluxing liquid (referred to as "the suspended state"), and the conversion was only marginally higher than the estimated equilibrium value. With a 3-mL charge of substrate (3 mL DEC/0.75 g of Pt/C), there was about a sevenfold increase in conversion and H_2 evolution rate.

Here, the catalyst appeared wetted throughout and was covered with a thin film of liquid substrate, hence its designation as the *liquid film state* or *superheated liquid film state*. With lesser amounts of added DEC, the catalyst surface slowly dried out (the *sand bath state*), and with a deficiency of substrate, the conversion and reaction rate decreased. A significant enhancement in the dehydrogenation of MCH to toluene at liquid film catalysis conditions was also demonstrated.

These phenomena were quantified using a Langmuir-like model for the dehydrogenation reaction kinetics, $\vartheta = k/[1 + K \text{ (aromatic by-product)}]$, where ϑ is the reaction rate, k is the reaction rate constant, and K is the retardation rate constant for dehydrogenation by-product. Passing from the liquid-rich suspended state to the liquid film state, there was about a fivefold increase in k and a decrease of similar amount in K, indicative of a weaker adsorption of by-product at these optimum conditions for the catalysis. Sebastián et al.

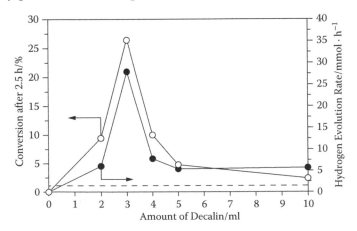

FIGURE 8.32
Relationship between catalytic dehydrogenation activity and the amount of added liquid decalin under boiling and refluxing conditions. Full circles represent the initial rate of H_2 evolution and open circles the percentage conversion of decalin after 2.5 h of reaction. The dotted line is the calculated equilibrium conversion (at 210°C, 1-bar pressure). (Reproduced from Hodoshima, S.; Arai, H.; Saito, Y.; *Int. J. Hydrogen Energy* 2003, *28*, 197, with permission from the International Association of Hydrogen Energy.)

[181] later conducted similar experiments for the dehydrogenation of DEC at liquid film reactive distillation conditions. At 260°C with an optimized Pt/C catalyst, a clear but less-pronounced dependence of reaction rate on the feed-to-catalyst ratio (2.6 mL DEC/0.3 g Pt/C) was noted [181a].

Similar experimental approaches had been employed by Saito et al. [182] for investigating the catalytic dehydrogenation of 2-propanol to acetone, which was significantly enhanced by operating at liquid film vis-à-vis suspended-state conditions. The dehydrogenation of DEC and MCH at such relatively mild conditions [171, 172, 175] when the system is operating in the liquid film state is accounted for by the following operative phenomena:

1. A higher temperature at the catalyst/liquid interface than the boiling point of the liquid. When liquid ≫ catalyst, the suspension will be refluxing at close to the boiling point of the substrate regardless of the heating temperature. However, when there is only a thin layer of liquid covering the catalyst surface, significant thermal gradients and a higher temperature on the catalyst surface liquid may be expected (the *superheated liquid film* state).

2. Vigorous bubble formation, which is an irreversible (nonequilibrium) process, should assist the dehydrogenation by stimulating the removal of hydrogen and aromatic by-product from the catalyst surface.

3. A selective evaporation of more volatile catalyst-inhibiting by-products (naphthalene in the case of DEC dehydrogenation) at the reactive distillation conditions of operation should be advantageous.

In continuing research by this group, it was possible to achieve nearly 100% conversion of DEC with external heating at 280°C. A nearly complete conversion of MCH to toluene at 240°C was also realized [175]. The work was performed in laboratory systems in both batch mode and continuous operation [169, 173]. A concept for an industrial prototype reactor was offered [175, 176].

The group of Ichikawa et al. [177–180] at Hokkaido University conducted similar studies, using essentially the same laboratory equipment, on a catalytic dehydrogenation of cyclohexane, MCH, and DEC under what was called "wet-dry multiphase" conditions. The effects of temperature, dependence on the liquid substrate-to-catalyst ratio, and catalyst composition on the reaction rate were surveyed. A Langmuir-type kinetic model was used for deriving the kinetic constant k and by-product retardation constant K. For the conversion of cyclohexane and MCH at 300°C over Pt/petroleum coke, there was a strong dependence of reaction rate on the amount of substrate. Using DEC, this dependence was quite modest, in apparent contrast to the data of Hodoshima et al. [169, 170] (whose experiments were performed at a much lower temperature, 210°C).

The required liquid-to-catalyst ratio for optimum dehydrogenation catalysis was found to vary substantially depending on the surface area, pore size distribution, and particle size of the carbon support. It was suggested that the wetting of the support at reaction conditions may be an important factor. Use of γ-alumina resulted in low reaction rates and no dependence on the amount of liquid substrate relative to catalyst. The retardation constant K was lower than that for the carbon supports, consistent with the generally greater affinity of carbons for organic adsorbents. It is evident that the conditions of optimum catalysis are highly dependent on the physical structure and adsorptive properties of the catalyst support, which is not surprising considering that it is an interfacial phenomenon.

Kariya et al. [177] admitted that the dehydrogenation reaction conditions at the liquid film state and wet-dry multiphase conditions are basically the same. They suggested that the liquid film state description implies a static system, while their terminology better describes the dynamic phase changes (from liquid to gas) and rapid desorption (to partial dryness) of the liquid reactants and products from the catalyst surface as depicted in Figure 8.33.

Reactor Development—Toward Technology Implementation

The concept of alternating wet-dry conditions most likely inspired later development of a pulsed-spray mode reactor for dehydrogenation [178, 179] (Kobayashi et al. [183] had proposed such a system for the catalytic hydrogenation of 2-propanol to acetone). The reactor consisted of a cylindrical glass vessel equipped with a condenser for collecting the products and containing at its base a catalyst surface on a circular plate heater. Liquid reactant was introduced intermittently from a nozzle above the catalyst. An anodized aluminum plate impregnated with Pt provided the best catalysis at 375°C. In the initial studies with this apparatus, a 25–35% conversion of cyclohexane, MCH, tetralin, and DEC was obtained. The dehydrogenation kinetics were shown to be a function of the liquid feed conditions and could be optimized by varying the dosing intervals. With limited dosing, the catalyst surface is mostly dry with minimal reaction taking place, while high doses lead to a fully wet and cooler catalyst surface [180]. The in-between state is where the substrate is spread as a thin film, the dry-wet multiphase conditions shown in Figure 8.33, and provide the optimum catalysis.

This technology was further developed by industry consortia and scaled up to commercial prototype equipment used for providing hydrogen for vehicle fueling stations [179]. An on-board reactor was devised for a dual-fuel vehicle; the hot exhaust from the H_2 internal combustion engine supplied the heat of dehydrogenation. MCH and an "alkyl decalin" (for liquidity) were used as hydrogen carriers. In the review article, Ichikawa [179]

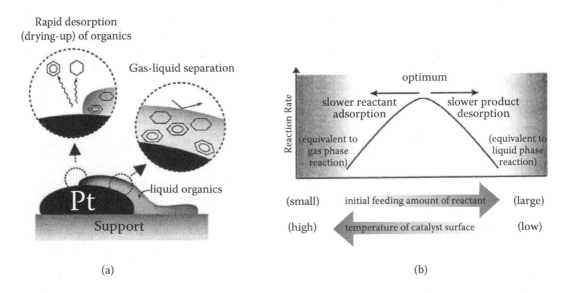

(a)   (b)

FIGURE 8.33
(a) Illustration of processes at catalyst-liquid-gas interface at "wet-dry multiphase conditions" as illustrated for a dehydrogenation of a liquid cycloalkane. (b) Schematic view of the effect of substrate feed amount and temperature on dehydration rate at these conditions. (Figures reproduced with permission from Reference [177].)

discussed a futuristic hydrogen delivery network (the "hydrogen highway") that interconnects large-scale hydrogen sources (petroleum refineries, solar/wind power plants, etc.) with local fuel cell electricity generation sites. In this scenario, organic liquid carriers replace conventional high-voltage power lines as the energy transmission medium. Among the technology needs for improved dehydrogenation catalysts and reactors, providing the necessary heat for the dehydrogenation reaction is clearly cited as a technical barrier toward the fulfillment of this vision.

Polycyclic Aromatic Carriers: Toward Lower Dehydrogenation Enthalpy Systems

Computational chemistry, as in the use of algorithms based on quantum mechanics (QM), has proved to be of great value for predicting the thermodynamic properties of perhydrogenated aromatic/aromatic "molecule pairs." As practiced by several research groups [183–187, 190–193], computational chemistry has provided valuable guidance for identifying systems that have favorable dehydrogenation enthalpies and thus the potential for lower hydrogen desorption temperatures.

Polyaromatic Hydrocarbons

In the early 2000's, Pez and coworkers at Air Products and Chemicals, Inc. [184, 185], conducted a systematic search for molecule pairs that would offer lower dehydrogenation enthalpies ΔH_D, ideally, about 40 kJ/mol H_2 (9.5 kcal/mol H_2) as described in the section on catalytic hydrogenation/dehydrogenation fundamentals. Calculations were used to identify trends in ΔH_D for series of molecules of varying structural type and atom content. The first set of calculations was performed on series of perhydrogenated polyaromatic/ polyaromatic hydrocarbon pairs. Necessitated by the large size of some of the molecules, a semi-empirical molecular orbital method was initially employed to provide heats of formation, ΔH_f, from which ΔH_D was calculated using the value of $\Delta H_f = 0$ for hydrogen. While the estimated energies shown in Figure 8.34 are somewhat lower than the experimental energies, the observed trends were considered to be meaningful. The plot in the lower section of Figure 8.34 (Curve I) represents the variation in ΔH_D for the series from cyclohexane/benzene to decalin/naphthalene to the 14-ring chain perhydropolyene/polyene. Note that the ΔH_D values decrease going up the ordinate axis in Figure 8.34.

The higher ΔH_D for cyclohexane vs. decalin is as experimentally observed, and the remaining trend to now increasing values of ΔH_D in Curve I was ascribed to a diminishing aromaticity (lower proportion of aromatic sextets) for the dehdrogenated molecules. Fusing the rings in a staggered arrangement (as in perhydrophenanthrene/phenanthrene, n = 3 for Curve II) led to a smaller ΔH_D that now becomes even more favorable with an increasing number of fused rings. The most favorable situation (toward the ideal 9.5-kcal/ mol H_2) is found for Curve III showing the most "circular" hydrogenated/dehydrogenated molecule pairs, exemplified by pyrene ($C_{16}H_{10}$), coronene ($C_{24}H_{12}$) and hexabenzocoronene ($C_{42}H_{18}$). These molecules and the others on Curve III have the largest proportion of aromatic sextets, implying a greater aromaticity.

For experimentally testing these concepts, the catalytic hydrogenation and dehydrogenation reactions of pyrene ($C_{16}H_{10}$), coronene ($C_{24}H_{12}$), and hexabenzocoronene ($C_{42}H_{18}$) were investigated using a stirred tank reactor that was capable of high hydrogen pressures and included a means for an in situ grinding of the solid or liquid reaction components (carrier and catalyst). As an example, a partially hydrogenated pyrene liquid ($C_{16}H_{20}$– $C_{16}H_{26}$) and an admixed 5% Rh/C catalyst under H_2 (1.7 bar) on heating at 150°C, with

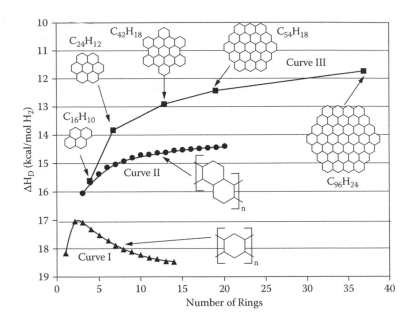

FIGURE 8.34

Predicted variation in ΔH_D with the number of rings for the three series (Curves I-III) of perhydrogenated aromatic/aromatic molecule pairs. Note the diminishing values of ΔH_D along the vertical axis. (Adapted from Pez, G.P.; Scott, A.R.; Cooper, A.C.; Cheng, H.; U.S. Patent 7,101,530.)

moderate mechanical grinding, resulted in a liquid mixture ($C_{16}H_{10}$–$C_{16}H_{26}$) with release of 25% of the stored hydrogen. Significantly, under the same reaction conditions there was no detectable catalytic dehydrogenation of DEC (a reaction that is usually performed at > 210°C; Sotoodeh et al. [188] used DEC as a reaction solvent for the dehydrogenation of perhydro-N-ethylcarbazole, and it was verified that no dehydrogenation of DEC took place below 180°C in the presence of supported Ru and Pd catalysts). The more facile dehydrogenation of the larger polycyclic hydrocarbons at 150°C was thus viewed as consistent with their lower (calculated) dehydrogenation enthalpies.

It is interesting to note that a sample of coronene impregnated with Pd nanoparticles was found to undergo substantial hydrogenation (at 150°C, 68 bar H_2) and dehydrogenation (at 150–200°C, 1.4 bar H_2) in a static system with no mechanical grinding or agitation [185]. It was proposed that the solid-state hydrogenation and dehydrogenation (up to a reversible 4.6 wt% H_2 uptake) occurred via a poorly understood "spillover" of H atoms to and from the palladium catalyst [189].

Additional computational studies [186, 187] were performed using the B3LYP hybrid functional and a 6–311G** basis set for the calculation of electronic energies, vibrational frequencies, thermodynamic properties, and an estimated temperature for 95% hydrogen release. Data were collected for polycyclic hydrocarbons, including systems for which results could be compared with experimental data. The survey comprised molecule pairs that contained five- and six-membered rings; surprisingly, it was found that an inclusion of five-membered ring structures was advantageous. For example, by replacing the center six-membered ring of anthracene with a five-membered ring (as in fluorene), there is a decrease in ΔH_D for the corresponding perhydrogenated molecules from 66 to 60 kJ/mol H_2. A discussion of experimental results on the perhydrofluorene/fluorene system as a potential hydrogen carrier is provided in this review in the section on autothermal hydrogen storage.

Polyaromatic Hydrocarbons with Nitrogen and Oxygen Heterotoms

Calculations (by the Air Products group) of ΔH_D were performed for the same series of perhydrogenated polyaromatic/polyaromatic hydrocarbon pairs, including substitution of carbon atoms with N heteroatoms [184, 185]. It is clear from the comparisons provided in Figure 8.35, Curves I, II, III, and IV, that the introduction of even one atom of N per ring results in a significant drop in ΔH_D. The trend of diminishing ΔH_D with an increasing number of rings is shared with the heteroatom-free molecules.

These studies provided thermodynamic data and hydrogen desorption temperatures for a modest set of polycyclic molecules comprising nitrogen and oxygen heteroatoms. The beneficial effect of having nitrogen (and to a lesser extent, oxygen) heteroatoms was confirmed by the high-level calculations. Experimental carriers identified as a result of these calculations, principally N-ethylcarbazole (NEC) and dibenzofluorene-based systems, are discussed in this review in the sections on N-ethyldodecahydrocarbazole (NEDC)/NEC cycle and potential N- and O-heterocyclic molecule carriers.

In subsequent and independent computational and experimental studies, Moores and Crabtree et al. [190] likewise proposed that nitrogen and oxygen heteroatom substitution in carbocycles favors dehydrogenation to the corresponding aromatic molecules. A comparison of calculated ΔH_D for cyclohexane and several N-containing heterocycles was provided, which clearly indicated that, thermodynamically, hydrogen release was facilitated by the replacement of one CH_2/C-H fragment with NH/N. A 1,3 arrangement of two nitrogen atoms was found to be more effective, with the lowest estimated ΔH_D value found for triazinane, from which one would expect the molecule to dissociate spontaneously

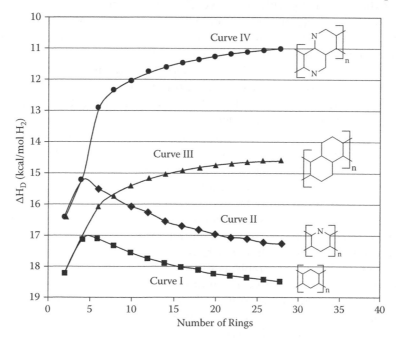

FIGURE 8.35
Predicted variation in ΔH_D with the number of rings for series of perhydrogenated polyaromatic/aromatic pairs (Curves I and III) vs. the same molecule pairs containing nitrogen heteroatoms (Curves II and IV). Note the diminishing values of ΔH_D along the vertical axis. (Adapted from Pez, G.P.; Scott, A.R.; Cooper, A.C.; Cheng, H.; U.S. Patent Appl. Publ. 2004/0223907.)

into triazene and hydrogen. This beneficial effect of nitrogen was ascribed to the slightly weaker N-H vs. C-H bonds and the well-established effect of a weakening in aliphatic amines of C-H bonds that are adjacent to nitrogen. The radical remaining on carbon following a homolytic dissociation of a C-H linkage is stabilized by an interaction with the nitrogen lone pair.

Several model compounds were used to determine their relative tendency toward dehydrogenation in refluxing toluene under an argon sweep. The most significant result was a total dehydrogenation of indoline to indole in the presence of a Pd/C catalyst, conditions for which there was no conversion of indane (the corresponding hydrocarbon substrate). Of six-membered heterocycles, only 1,2,3,4-tetrahydroquinoline was fully dehydrogenated to quinoline.

In a following study with Clot et al. [191], a relatively high-level DFT (B3PW91/aug-cc-pVDZ) computational method was used to provide thermodynamic data and predicted H_2-release temperatures, T_D (where $\Delta G = 0$), for approximately 30 molecule pairs. Included were several cyclic hydrocarbons, but most of the molecule pairs were comprised of five- and six-membered ring structures with 1–3 N atoms in ring and substituent positions. Calculated values of ΔH_D were said to be in excellent agreement with available experimental data (which were provided in the text). This survey provided valuable molecular structure-dehydrogenation potential relationships and several main conclusions. First, the H_2 desorption temperature was increasingly reduced by having from 1 to 3 N atoms in nonadjacent positions (no N-N linkages) in a six-membered ring. Second, a nitrogen in a substituent group can be more effective than a ring nitrogen (e.g., the T_D for aniline is slightly lower than that for pyridine). However, from a practical viewpoint, it is evident that the introduction of aliphatic substituent groups on a ring will necessarily result in a lower gravimetric hydrogen capacity of the carrier. Last, molecules with five- rather than six-membered rings are preferred. A provided explanation is that the five-ring molecules have fewer C-H bonds to break to attain aromaticity. Thus, the indicated overall most effective strategy is in the use of multiply ring-substituted five-membered N-heterocycles.

In a more recent study [192], the effects on ΔH_D and T_D of substituting up to four atoms of nitrogen for carbon in the DEC-naphthalene system was evaluated using the G3 high-level computational method, which is known to provide accurate thermodynamic data. As now expected, substitution with one or more nitrogen atoms had a favorable effect depending on their relative positions. With four nitrogen atoms distributed in a 1,3 arrangement in each ring, a H_2 desorption temperature as low as 73°C was predicted.

The predictions from these computational modeling studies of very low H_2 release temperatures are highly enticing to experimentalists. However, quite apart from the serious catalyst activity issues, there are more potential thermodynamic challenges. Catalytic hydrogenation and dehydrogenation reactions usually proceed in multiple steps, and each step has to be at least as favorable as the overall change in Gibbs free energy for the entire sequence. Otherwise, the reaction will be limited to an intermediate stage product. Cui et al. [193] commented that in the dehydrogenation of two-ring systems, the ΔH_D for each ring has to be "roughly equal," and if one is much greater than the other, only partial dehydrogenation will be favored. Particularly when working with substrates of multiple functionality, an *a priori* calculation of the energetics (ΔH_D) of each step of conceivable reaction pathways can be instructive (additional discussion in the section on the perhydrodibenzofuran-dibenzofuran system).

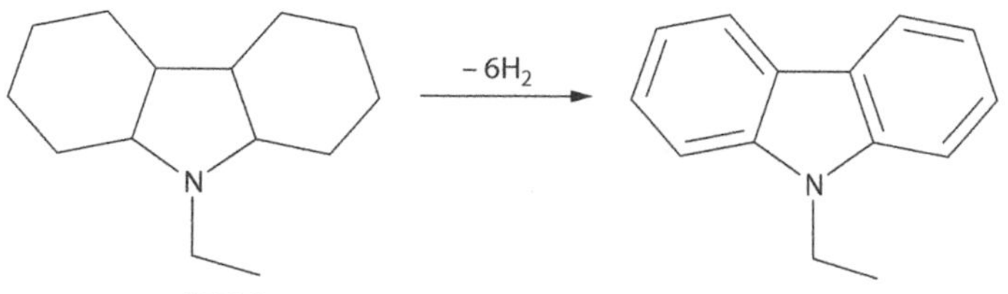

SCHEME 8.11
Catalytic dehydrogenation of N-ethyldodecahydrocarbazole (NEDC) to N-ethylcarbazole (NEC).

The N-Ethyldodecahydrocarbazole/N-Ethylcarbazole (NEDC/NEC) Cycle

The NEDC/NEC cycle is based on a reversible catalytic dehydrogenation of NEDC (liquid, boiling point 148–150°C at 21 torr [194], 124–125°C at 8 torr [195]) to NEC (solid, melting point 68–70°C), depicted in Scheme 8.11.

The gravimetric capacity of the NEDC/NEC cycle is 5.8 wt% with a volumetric storage density of 47.4 kg H_2/m^3 (the latter value calculated assuming that NEDC has the same density as DEC, 0.896 g/mL). The enthalpy of dehydrogenation ΔH_D = 49 kJ/mol H_2 (as measured by reaction calorimetry for the corresponding hydrogenation reaction at 150°C [186]).

In recent years, this system has received considerable attention in view of the good reversibility and catalytic dehydrogenation that can be conducted with the NEDC substrate in a liquid state at milder temperatures than "traditional" carriers. The last feature is consistent with a significantly lower ΔH_D for NEDC than those measured for MCH and DEC. As for the DEC-naphthalene cycle, however, the fully unloaded carrier, NEC, is a low-melting solid.

Hydrogenation of N-Ethylcarbazole

The catalytic hydrogenation of carbazole, N-methylcarbazole, and NEC over Raney nickel and copper chromite catalysts under forcing conditions (200–250°C, 250–300 bar) provides the fully hydrogenated compounds [195]. However, the use of a 5% Pd/C catalyst allows the hydrogenation of N-methylcarbazole in DEC solvent to take place under milder reaction conditions (150–200°C, 34 bar) [196]. In our laboratory, the hydrogenation of NEC, as a neat substrate, was typically conducted using a 5% Ru/lithium aluminate catalyst at 160°C for 2.5 h under 60 bar of H_2 pressure [186]. The NEDC was obtained as a free-flowing liquid at a greater than 99% selectivity (by gas chromatography mass spectroscopy [GC/MS]) with dodecahydrocarbazole and dicyclohexyl as the only discernible trace-level impurities. The GC/MS analysis also revealed the NEDC reaction product consisted of several isomers (configurational isomers or stereoisomers), at least three of which could be resolved in a GC column. The isomerism can arise from the possible different relative disposition of H atoms in the central five-member ring. In retrospect, this isomerism was not surprising since, as reported by Adkins et al. [195], dodecahydrocarbazole prepared by the hydrogenation of carbazole over nickel at 260°C was found to consist of two isomers, one liquid and the other a solid with a melting point of 73–74°C.

In view of its considered importance for hydrogen storage, the catalytic hydrogenation of NEC was investigated at some depth by three other research groups. Sotoodeh et al.

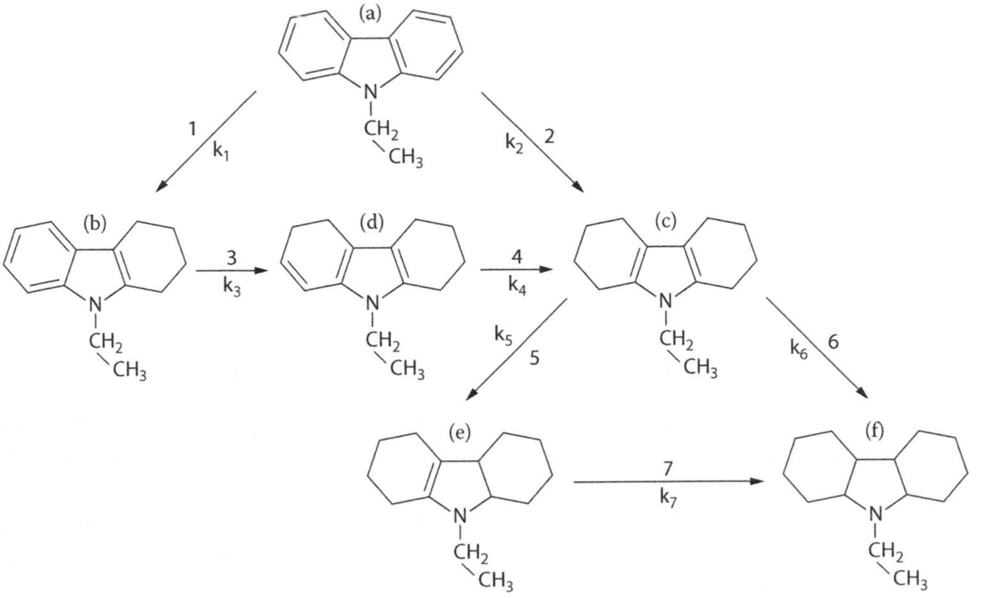

FIGURE 8.36
Hydrogenation reaction path of N-ethylcarbazole as a solution in decalin over a 5% Ru/Al$_2$O$_3$ catalyst: (a) N-ethylcarbazole; (b) N-ethyltetrahydrocarbazole (H-4); (c) N-ethyloctahydrocarbazole (H-8); (d) N-ethylhexahydrocarbazole (H-6); (e) N-ethyldecahydrocarbazole (H-10); and (f) the product, N-ethyldodecahydrocarbazole (H-12). (Reproduced from Sotoodeh, F.; Smith, K.J.; *Ind. Eng. Chem. Res.* 2010, *49*, 1018, with permission from the American Chemical Society.)

[188, 197] used a continuous stirred tank reactor (CSTR) containing a 3.5–6 wt% solution of NEC in DEC. The hydrogenation was carried out over a H$_2$-reduced 5% Ru/Al$_2$O$_3$ catalyst at 130–150°C and 70-bar H$_2$ pressure. Small samples of liquid were periodically removed and analyzed by GC. From the distribution of product NEDC (as a single isomer) over time, a hydrogenation pathway was suggested as shown in Figure 8.36.

Data from three different temperatures were used to model the reaction kinetics, providing values for the kinetic constants k_1 to k_6 and apparent activation energies. It was proposed that the reaction proceeds with a stepwise saturation of the double bonds with parallel pathways to N-ethyltetrahydrocarbazole (H-4) and N-ethyloctahydrocarbazole (H-8) as the initial products. The conversion of H-4 to H-8 proceeds via a N-ethylhexahydrocarbazole (H-6) intermediate, which is the slower to convert to H-8. This is rationalized considering the nonaromatic structure of H-6, which would render desorption from the catalyst more difficult. Consistent with this notion, the H-6 → H-8 and H-10 → H12 transformations of "olefin-like" substrates have the highest apparent activation energies. Overall, the hydrogenation of NEC (in DEC) was completed to a greater than 95% selectivity to the desired NEDC (H-12) product with a less than 5% residue of the H-8 intermediate.

In an in-depth study of the hydrogenation of NEC, Eblagon and coworkers [198, 199] also provided valuable complementary insight not only on the catalysis but also on the stereochemistry of the NEDC reaction product. NEC was employed as a neat, molten substrate at the reaction temperatures of 120–170°C. A CSTR was used for the hydrogenation with provisions for taking small samples during the course of the reaction. These were analyzed by GC/MS and (at least for the final product) by NMR. Principally employed were Ru-based catalysts, which were found to be more active than Pd, Pt, and Ni composites.

For Ru/TiO$_2$ and Ru/Al$_2$O$_3$-catalyzed reactions at 130°C and 70 bar H$_2$, the intermediates found during the initial reaction stages were N-ethyltetrahydrocarbazole (H-4) and N-ethyloctahydrocarbazole (H-8) with low levels of the H-6 intermediate, as also reported by Sotoodeh et al. [188, 197]. A simpler consecutive pathway was suggested, consisting of a stepwise addition of two H atoms to NEC to give the H-4, H-6, and H-8 intermediates en route to the final formation of NEDC. First-order rate constants and apparent activation energies were derived for the Ru/Al$_2$O$_3$-catalyzed reaction at 120–170°C. The activation energy for consumption of NEC (58 kJ/mol) was lower than the cited 99.5 kJ/mol from Sotoodeh et al. [188] and was ascribed to "mass transfer limitations" related to having a molten reactant or simply due to different experimental conditions.

It was also found that the NEDC product was formed as a mixture of stereoisomers that were (in part) resolved as three fractions by GC/MS. In elegant subsequent analyses by NMR, it was discovered that one of the fractions consisted of two unresolved components. A total of four stereoisomers were identified by a combination of GC/MS and a variety of one-dimensional (1D) and two-dimensional (2D) NMR techniques. The provided structures of the six possible NMR-distinguishable stereoisomers of NEDC are reproduced in Figure 8.37.

Of these stereoisomers, four (labeled A to D) had an element of symmetry, while two (E and F) were asymmetric. An analysis (by both GC/MS and NMR) of the initial products of the hydrogenation reaction of NEC over a 5%Ru/Al$_2$O$_3$ catalyst displayed an initial preponderance of stereoisomer B and an E > B >> F distribution of the isomers at long reaction times when some conversion toward the thermodynamically more stable species had occurred. Gas-phase QM-based calculations predicted the ordering of NEC hydrogenation enthalpies to NEDC as E > F > B > A, where E would be the most stable species [203]. Also,

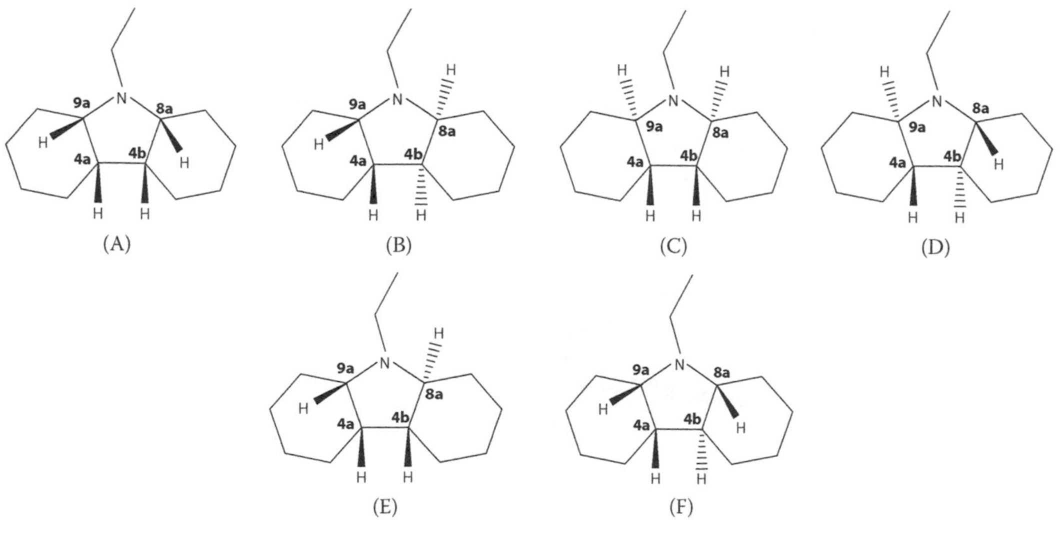

FIGURE 8.37
Chemical structures of all six possible N-ethyldodecahydrocarbazole (NEDC) stereoisomers that are potentially distinguishable by NMR. Structures A, B, E, and F were experimentally identified from hydrogenation product mixtures. E is the thermodynamically most stable and A the least stable. (Reproduced from Eblagon, K.M.; Rentsch, D.; Friedrichs, O.; Remhof, A.; Zuettel, A.; Ramirez-Cuesta, A.J.; Tsang, S.C.; *Int. J. Hydrogen Energy* 2010, *35*, 11609, with permission from the International Association of Hydrogen Energy.)

the structure of stereoisomer E corresponded to the optimized geometry for NEDC as calculated (using a different QM method) by Sotoodeh et al. [196].

The distribution of NEDC isomers was observed to vary greatly not only over the course of the reaction but also as a function of the employed hydrogenation catalyst. Figures 8.38a and 8.38b provide a direct comparison of the distribution of isomers A, E, and F (left vertical axis) and the ratio [A]/[total isomers] (on the right vertical axis) for reactions with a 5%Ru/TiO$_2$ and a Ru black catalyst, respectively.

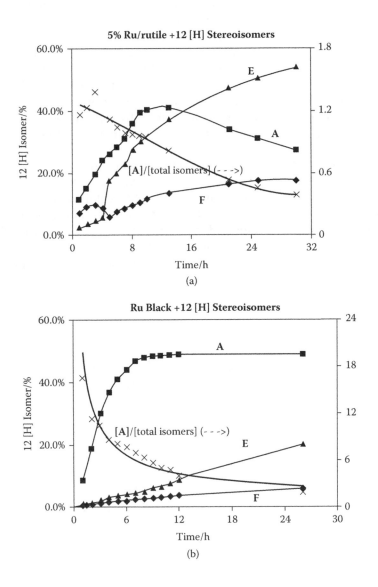

FIGURE 8.38

Concentration vs. time for the three N-dodecahydrocarbazole (H-12) stereoisomers A, E, and F produced during the hydrogenation of N-ethylcarbazole (NEC) over (a) 5 wt% Ru on rutile and (b) Ru black catalysts. Initially, the all-*cis* isomer predominates (particularly with Ru black), but eventually the reaction mixture trends toward the thermodynamically more stable E and F isomers. (Reproduced from Eblagon, K.M.; Tam, K.; Yu, K.M.K.; Zhao, S.-L.; Gong, X.-Q.; He, H.; Ye, L.; Wang, L.-C.; Ramirez-Cuesta, A.J.; Tsang, S.C.; *J. Phys. Chem. C* 2010, *114*, 9720, with permission from the American Chemical Society.)

From transmission electron microscopy (TEM) and CO chemisorption data, it was inferred that Ru/Al_2O_3 and Ru/TiO_2 catalysts consisted of small disordered metal particles with abundant low (metal atom-metal atom) coordination sites, such as edges and kinks. Conversely, Ru black has a preponderance of faceted sites and relatively fewer low-coordination sites.

Initially, for the Ru/TiO_2-catalyzed hydrogenation of NEC (Figure 8.38a), isomers with structures A and E were the main products, with the concentration of A decreasing with time after reaching a maximum. However, for the Ru black-catalyzed reaction (Figure 8.38b), isomer A was by far the predominant species, and the ratio of [A]/[total isomers] was very high. It is probable that the symmetric isomer A (having all the C-H bonds on one face) resulted from a concerted hydrogenation of an adsorbed NEC on the flat terraced sites of the catalyst. The more "structurally diverse" Ru/TiO_2 catalyst showed less selectivity for the symmetric species as the ratio [A]/[total isomers] was now much smaller and at longer reaction times E $\gg$ A > F. For both systems, however, there was eventually an isomerization to the thermodynamically more stable E and F species. In summary, it is recommended that the most productive catalyst should have a "proper combination" of flat terrace and low-coordination metal sites on small catalyst particles.

Ye et al. [200] investigated the kinetics of NEC hydrogenation over Raney Ni catalysts. Experiments were conducted using a CSTR at varying hydrogen pressures (20 to 60 bar) and at temperatures from 120°C to 240°C, with a maximum 91% conversion (5.3 wt% H_2 capacity) observed at 50 bar H_2 and 180°C. The apparent reaction rate was found to be of zero order with respect to NEC and first order in the concentration of hydrogen in the liquid phase (NEC melt). This is in contrast to the first- and zero-order dependency on reaction rate and H_2 pressure, respectively, as assumed by Sotoodeh et al. [188, 197] in modeling the hydrogenation of NEC (as a solution in DEC) using a 5% Ru/Al_2O_3 catalyst. It was concluded that when using Raney Ni in the hydrogenation, the process was totally determined by reactions at the solid-liquid interface. Accordingly, a kinetic model was provided for which the reaction rate showed only a dependency on temperature, the concentration of dissolved hydrogen, and the physical characteristics (i.e., mass, density, and size) of the catalyst particles.

Catalytic Dehydrogenation of N-Ethyldodecahydrocarbazole

In the early work of Adkins and coworkers [201], a complete dehydrogenation of NEDC to NEC was accomplished using a Ni/nickel chromite catalyst at 250°C, but only in the presence of benzene as an acceptor for the hydrogen. The dehydrogenation of NEDC was investigated at Air Products [184–187] by Pez and coworkers with the aim of devising a practical hydrogen storage system. The reaction was conducted both in a batch mode using a slurry catalyst in a CSTR and in a continuous flow system. The generated hydrogen was easily separated from the relatively involatile reaction components and in all cases was assayed as a flowing H_2 stream at realistic pressures of about 1 bar or higher. This is in contrast to the more common (and thermodynamically advantageous) practice of sweeping the hydrogen from the reactor with a flow of an inert gas.

The batchwise dehydrogenation of NEDC was performed in the presence of supported Pd, Pt, and bimetallic catalysts (e.g., Pt-Sn, Pt-Ir, Pt-Re) [184–186]. Typically, the reactor was ramped to a temperature of 200°C and held at that temperature to the desired level of conversion as gauged by the integrated H_2 flow and an analysis by GC/MS of the residual liquid. Analyses of the H_2 product from Pd-catalyzed dehydrogenation of NEDC showed part-per-million levels of CH_4 and C_2H_6 as the only identified impurities. Analyses of

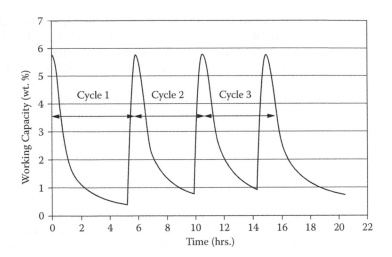

FIGURE 8.39
Sequential dehydrogenation and rehydrogenation of the N-ethyldodecahydrocarbazole/N-ethylcarbazole "molecule pair." (Reproduced from Cooper, A.; Abdourazak, A.; Cheng, H.; Fowler, D.; Pez, G.; Scott, A.; Design and development of new carbon-based sorbent systems for an effective containment of hydrogen. DOE Hydrogen Program FY 2005 annual progress report. http://www.hydrogen.energy.gov/pdfs/progress05/vi_b_3_cooper. pdf, accessed April 28, 2012.)

liquid samples taken during the course of the reaction suggested that the dehydrogenation proceeded in steps corresponding to the liberation of two moles of hydrogen from NEDC (H-12) to N-ethyloctahydrocarbazole (H-8), N-ethyltetrahydrocarbazole (H-4) to NEC. The latter (H-4) intermediate was clearly evident from NMR spectra of samples taken during the later stage of the reaction.

The reversibility of the NEDC/NEC + 6H$_2$ system was nicely illustrated by a sequential hydrogenation/dehydrogenation of the same sample of the molecule pair. A CSTR was charged with a 1:1 w/w mixture of 5% Pd/Al$_2$O$_3$ (dehydrogenation catalyst) and 5% Ru/Al$_2$O$_3$ (hydrogenation catalyst) and NEDC. As illustrated in Figure 8.39, the initial sample of NEDC was substantially dehydrogenated under about 1 bar H$_2$ at 200°C. With the application of hydrogen pressure to the dehydrogenated carrier (about 80 bar), a relatively fast rehydrogenation was realized. The sequence was repeated over three complete cycles with no apparent loss of hydrogen capacity as discerned by GC/MS analyses of the carrier medium [186, 213]. Additional cycling data were obtained using a continuous flow reactor (see the section on dehydrogenation reactor development).

The dehydrogenation of NEDC revealed an interesting dependence on the distribution of stereoisomers [186]. Hydrogenation of NEC over Ru/Al$_2$O$_3$ at 160°C (typical conditions) and 120°C (with 10% remaining N-ethyloctahydrocarbazole) resulted in products with a different distribution of stereoisomers. On additional heating of the latter lower-temperature product mixture in the presence of H$_2$ and catalyst at 160°C, a more stable (closer to equilibrium) distribution of isomers was obtained. Figure 8.40 displays a comparison of H$_2$ flow data for dehydrogenation of the 120°C reaction product ("L") and the 160°C product ("H"), containing the more stable distribution of stereoisomers. It is interesting that the L composition underwent significant H$_2$ desorption at temperatures as low as about 50°C.

Considering the data from Eblagon et al. [198, 199], it is tempting to speculate that the 120°C NEC hydrogenation product is rich in the symmetric, all-*cis* isomer, which they designated as A in Figure 8.37. By their calculations, A had a 22-kJ higher enthalpy of

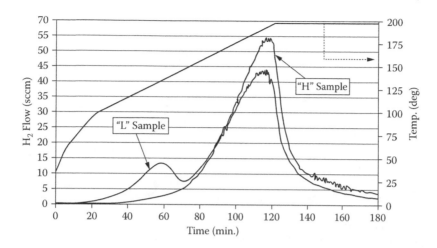

FIGURE 8.40

Hydrogen flow data for the dehydrogenation of two samples of N-dodecahydrocarbazole believed to consist of two different distributions of stereoisomers: the "L" and "H" samples, the former being the less stable and more reactive toward dehydrogenation. (Reproduced from Pez, G.P.; Scott, A.R.; Cooper, A.C.; Cheng, H.; U.S. Patent 7,429,372.)

formation than E, the most stable species. Thus, it would clearly be desirable to devise a high-yield stereoselective synthesis of isomer A, which should provide the lowest possible dehydrogenation temperature for the NEC/NEDC system.

Contributions by Sotoodeh, Smith, and coworkers [188, 197, 202, 203] and the surface science studies by the group of Sobota et al. [204], have provided valuable insight on the mechanism of NEDC dehydrogenation catalysis. In their first studies of this reaction, Sotoodeh et al. [188, 197] employed the hydrogenation product of a solution of NEC in DEC containing 5.8 wt% NEDC and less than 0.2% N-ethyloctahydrocarbazole (H-8). In the presence of a reduced 5% Pd/SiO$_2$ catalyst for 17 h at 170°C with hydrogen swept out using He flow, the NEDC dehydrogenation proceeded to full conversion to H-8 and N-ethyl-tetrahydrocarbazole (H-4) but also a minimal production of NEC, corresponding to only 4 wt% hydrogen recovery. From data of the distribution of NEDC, H-8, and H-4 in the reaction over time, a kinetics model was derived by which it was assumed that the dehydrogenation proceeded as two parallel reactions with activation energies of 67 kJ/mol and 144 kJ/mol for the production of H-8 and H-4, respectively.

In a significant advance ascribed to the development of an improved Pd/SiO$_2$ catalyst, it was possible to fully dehydrogenate NEDC within 1.6 h at 170°C under He sweep at 1 bar, with total selectivity to NEC and complete H$_2$ recovery [202, 203]. The more effective catalysts resulted from the use of an incipient wetness metal impregnation technique; the control of chloride, which was shown to have a deleterious effect; and catalyst structure-reactivity studies, which identified the optimum dispersion and dimension of the Pd nanoparticles (~6 nm). Product distributions over the course of the reaction were now interpreted in terms of a stepwise reaction from NEDC (H-12) → H-8 → H-4 → NEC.

From DFT modeling studies, it is envisaged that NEDC initially adsorbs parallel to a Pd(111) surface with the two H atoms of the C–H bonds adjacent to N in the central ring pointing directly at two Pd surface atoms (Figure 8.41).

This is consistent with the known weakening of the C–H bonds of carbon atoms in aliphatic amines that are adjacent to nitrogen (see the section on polyaromatic hydrocarbons with nitrogen and oxygen heterotoms). The arrangement is well poised for removal of the

Front View

Top View

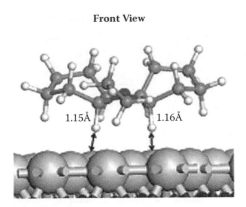
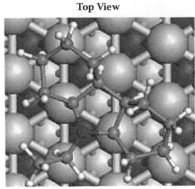

FIGURE 8.41 (See color insert.)
Calculated optimized adsorption geometries of a molecule of N-ethyldodecahydrocarbazole lying "flat" over a Pd(111) surface. Displayed is the interaction of the two H atoms of the C-H bonds in the central ring that are adjacent to nitrogen. See color insert Figure 8.41 for the location of the N atom. The other atoms are as follows: White, H; gray, C; light gray, first layer of Pd atoms; and black, second layer of Pd atoms. (Reproduced and adapted from Sotoodeh, F.; Smith, K.J.; *J. Catal.* 2011, *279*, 36, with permission from Elsevier.)

other two H atoms in this ring. On the less-well-packed structure of a Pd(110) surface, these two hydrogen atoms are too distant from Pd to directly interact, so the intermediate has to desorb and readsorb for further reaction. This is cited as an illustration of the structure sensitivity of the dehydrogenation catalysis.

Additional mechanistic insight was provided in complementary studies of an adsorption and reaction of NEDC on the surface of well-defined Pd nanoparticles on an alumina thin film in a ultra-high vacuum (UHV) system, using a combination of infrared reflection absorption spectroscopy (IRAS) and high-resolution photoelectron spectroscopy (HR-XPS) tools complemented by DFT calculations [204]. An ultrathin film of NEDC deposited on the Pd-Al_2O_3 surface warmed in stages from 110 K to 500 K, underwent adsorption and consecutive chemical reactions as characterized by the recorded spectra. Initially, NEDC adsorbs molecularly on both metal and support but then migrates selectively onto the Pd. Above 173 K, there is chemisorption with an abstraction of the two hydrogen atoms on the carbons that are adjacent to nitrogen, as suggested inter alia by the disappearance of their assigned ϑ(CH) vibrational modes. On further dehydrogenation, an intermediate displaying multiple binding of N and C atom centers to Pd is formed. On additional warming (>320 K), there is C-N bond scission, and surface-bound C_xH_y species are now evident (perhaps related to the experimental observation of trace CH_4, C_2H_4, and C_2H_6 by-products from dehydrogenation of NEDC over Pd catalysts at 200°C; see the section on dehydrogenation reactor development). Not surprisingly, at about 500 K, there is total fragmentation of the molecule, but the surface can be regenerated by oxidation.

The group of Ye et al. [205] who reported the Raney Ni-catalyzed hydrogenation of NEC to NEDC (see section on hydrogenation of N-ethylcarbazole) also investigated the possibility of a cyclic hydrogenation/dehydrogenation process using this catalyst for both conversions. In this study, they ran four cycles, each of 6-h duration, with H_2 absorption at 180–200°C/50-bar H_2 and desorption at 200°C and ambient pressure. From the reported H_2 absorption and desorption curves, it was clear that there was a large loss in capacity with cycling. This was ascribed to a limited dehydrogenation for each step, with remaining H-4 and H-8 intermediates, and a deactivation of the catalyst.

The aforementioned iridium pincer complex (see the section on DEC dehydrogenation catalysis) is known to function as a soluble catalyst for the aromatization of cycloalkanes to arenes, alkyl amines to imines and was evaluated by Wang et al. [206], for the dehydrogenation of NEDC. The reactions of a solution of $IrH_2[C_6H_3\text{-}2,6\text{-}(CH_2PBu^t_2)_2]$ and NEDC (initially in pentane) at 200°C generally provided a mixture of the N-ethyloctahydrocarbazole and N-ethyltetrahydrocarbazole. Only at very long reaction times was some of the completely dehydrogenated product observed.

Dehydrogenation Reactor Development

A practical liquid-carrier-based system would likely be required to deliver hydrogen as a continuous flow of gas. To this end, the dehydrogenation of NEDC as a liquid feed to a tubular continuous flow catalytic reactor was investigated. The reactor consisted of a vertically held 25-mm outside-diameter tube packed with 5% Pd/Al_2O_3 catalyst pellets [207]. Various engineering challenges were encountered in finding conditions for adequate operation of the three-phase, solid catalyst/liquid carrier/generated H_2 system.

One of the primary challenges was maintaining a "wet" catalyst surface while allowing an adequate flow of the generated gas [208]. Typically, the reactor was operated for long durations, providing a continuous flow of H_2 over several days. As an example (from Reference 186, Table 14), a 0.5-g/min inflow of NEDC liquid using 10 g catalyst at an operating temperature of 190°C resulted in a stable output H_2 flow of 160 standard cubic centimeters per minute (sccm) H_2. Similar to results for batch reactor operation, the output hydrogen contained part-per-million levels of CH_4 and C_2H_6. As a dehydrogenation/hydrogenation cycling demonstration, the packed bed reactor was employed for a continuous dehydrogenation of NEDC at 190°C to about 60% conversion, followed by batch rehydrogenation. Over six consecutive runs using the same batch of recycled feed, there was no significant loss in hydrogen capacity, as gauged by the H_2 flow rate during dehydrogenation [209]. In continuing reactor development work, Toseland et al. [210] employed catalyst-coated monolith systems for better precious metal utilization. More recently, research has been conducted on the use of microreactors [210, 211] with advantages of compactness, scalability (by sequential addition of basic components), and improved heat transfer for providing a controlled continuous flow of hydrogen from a catalytic dehydrogenation of liquid carriers.

Potential N- and O-Heterocyclic Molecule Carriers

Described in this section are selected perhydrogenated/dehydrogenated molecule pairs containing nitrogen or oxygen heteroatoms. While displaying H_2 reversibility, these carriers have evident limitations that may be mitigated by further research.

The Perhydro-4,7-Phenanthrolene/4,7-Phenanthrolene System

Six-membered ring N-heterocyclics have the potential to provide a 1:1 H/C or N atom ratio and therefore an about 7.2 wt% H_2 capacity. A fully hydrogenated carrier molecule was obtained in good yield and high purity by the hydrogenation of 4,7-phenanthrolene as a solution in THF (using preactivated 5% $Rh/LiAl_5O_8$ catalyst at 160°C, 55-bar H_2 pressure) [213a]. Samples were dehydrogenated in the presence of a 5% Pd/C catalyst. Beyond a

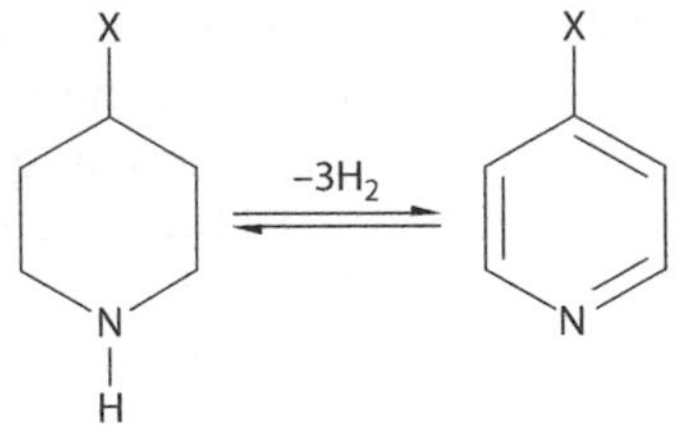

SCHEME 8.12
Dehydrogenation of piperidine to pyridine with release of hydrogen.

capacity of about 6 wt%, the dehydrogenation reaction became exceedingly slow, reaching a reversible 7 wt% H_2 loss only after 12 h at 250°C.

Piperidine and Octahydroindole Systems: Effects of Ring Substitution

As discussed in the section on polyaromatic hydrocarbons with nitrogen and oxygen heterotoms, ring substitution is expected to have a favorable influence on ΔH_D. To this end, Cui et al. [193] experimentally evaluated, with computational guidance, this hypothesis by investigating the effect of introducing substituents at designated positions of the title molecules. Considered first were the piperidine-pyridine molecule pair with substitution (by group X) at the 4 position, shown in Scheme 8.12.

From calculations and literature precedent, ΔH_D is expected to show a trend with more negative values of the Hammett σ paramater, as characteristic of more electron-donating groups favoring hydrogen loss. Dehydrogenation experiments were performed both in a closed system and using a nitrogen sweep. The nitrogen sweep procedure, for which a measure of kinetic rather than thermodynamic control may be expected, was the more discerning in terms of relative reactivity. In the following, the relative initial rate of dehydrogenation is related to the Hammett σ parameter:

X:	NMe$_2$	NH$_2$	C(O)NH$_2$	Alkyl	CN
σ:	−0.83	− 0.66	+0.36	−0.15	+0.66

There is an evident correlation between relative rate and σ, with the position of the amide group as an obvious anomaly. This observation of a higher relative dehydrogenation rate of piperidine-4-carboxylate is ascribed to the conjugated structure of the molecule. Utilizing such potentially conjugating ring substituent groups can thus be helpful, but as noted, the functional groups need to be inert under catalytic hydrogenation/dehydrogenation conditions. The 4-amino and the preceding amido-substituted piperidines with demonstrated 6.0 and 4.7 wt% hydrogen capacity were considered to be the more promising carrier systems for further study.

Indole-based two-ring systems with intended X = H, CH_3, OCH_3, and NH_2 were also investigated, as depicted in Scheme 8.13.

Based on calculations with X = H, an initial dehydrogenation of the five-membered ring is favored, but with an electron-donating group X (preferably NH_2), ΔH_D has almost the same value for the two rings. Partial dehydrogenation of 1-methyloctahydroindole at 170°C over Pd/SiO$_2$ was reported. As computationally predicted, the five-membered ring

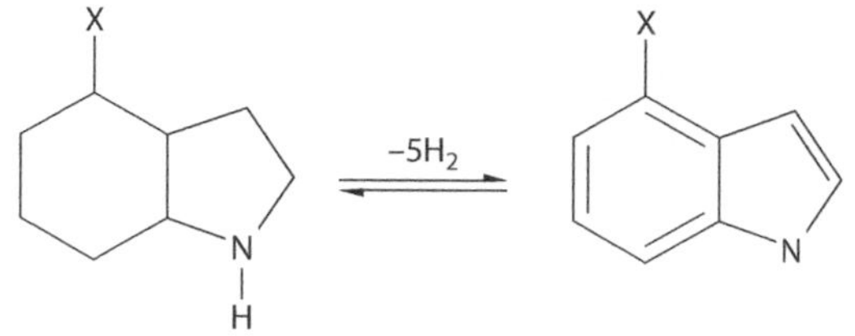

SCHEME 8.13
Indole-based hydrogen storage media.

with a relatively lower ΔH_D was dehydrogenated first. In these studies, problems were encountered with C-O and C-N bond cleavage, and the loss of substituents precluded finding viable H_2-reversible systems.

The Perhydrodibenzofuran-Dibenzofuran System

As a structural analogue of NEDC/NEC the perhydrodibenzofuran-Dibenzofuran system was investigated as a potentially higher-capacity carrier (6.5 vs. 5.8 wt% hydrogen). Dibenzofuran was hydrogenated using a 5% Ru/lithium aluminate catalyst under 60-bar H_2 at 100°C, giving a mixture of perhydrodibenzofuran (86%) and dodecahydrobiphenylene as shown in Scheme 8.14.

The perhydrodibenzofuran was separated by distillation and partially dehydrogenated over a Pd/C catalyst at long reaction times at 220°C with high selectivity (no indication of a rupture of the five-membered ring).

Wang et al. [214] reported on dehydrogenation studies of perhydrodibenzofuran and various perhydrogenated indoles and piperidines by means of the homogeneous $IrH_2[C_6H_3$-$2,6$-$(CH_2PBu^t)_2)_2]$, and related "pincer" complexes. The reaction of perhydrodibenzofuran with the $IrH_2[C_6H_3$-$2,6$-$(OPBu^t)_2]$ catalyst in a sealed flask at 200°C for 24 h resulted in only a 15% conversion to octahydrodibenzofuran. However, under the same conditions in the presence of *tert*-butylethylene as a hydrogen acceptor, a total conversion of perhydrodibenzofuran to a mixture of octahydrodibenzofuran (60%), tetrahydrodibenzofuran (32%), and dibenzofuran (8%) was achieved. With longer reaction times, the same level of dehydrogenation was achieved at 150°C, demonstrating that while the catalyst is active, the reaction is equilibrium limited. The authors cited thermodynamic bottlenecks in the dehydrogenation of perhydrodibenzofuran "as poignant examples of the dangers of selecting liquid organic carriers on the basis of the average ΔH_D/mole H_2 for converting the fully saturated to the fully unsaturated carrier." Instead, they recommend estimating ΔH_D/mol H_2 for each conceivable step where 1 mol H_2 is eliminated (see related comments at the end of the section on polyaromatic hydrocarbons with nitrogen and oxygen heterotoms).

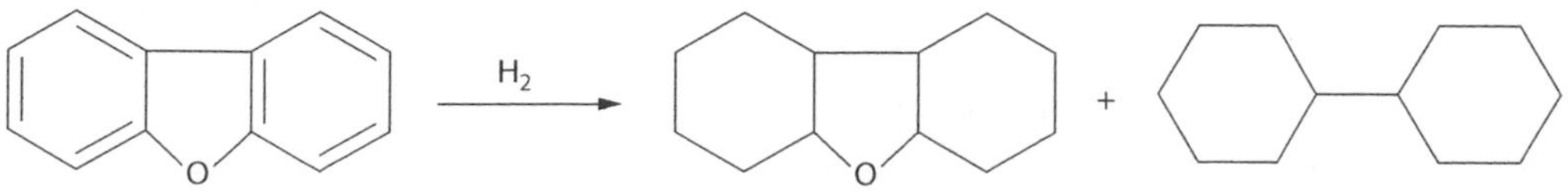

SCHEME 8.14
Hydrogenation of dibenzofuran, giving a mixture of perhydrodibenzofuran and dodecahydrobiphenylene.

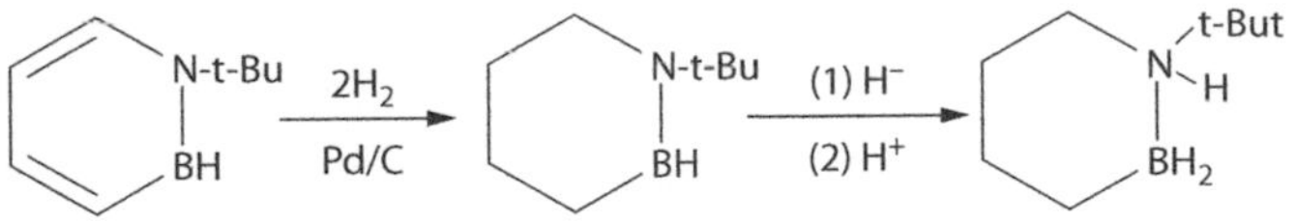

SCHEME 8.15
Hydrogenation of N-t-Bu-1,2-azaborine as a basis for H_2 storage.

Boron-Nitrogen Heterocyclics

The introduction of boron as a heteroatom in cycloalkanes has served to expand the scope of potential hydrogen-rich/hydrogen-poor molecule pairs, with molecules containing both B and N heteroatoms being of particular interest [215]. Campbell et al. [216] investigated the potential for a reversible hydrogenation of 1,2-dihydro-1,2-azaborine in which the overall hydrogenation is estimated to be close to thermoneutral. Experimentally, the more accessible N-t-Bu derivative with similar thermodynamics was used to successfully perform the first step of catalytically adding 2 equivalents of H_2, as shown in Scheme 8.15.

Catalytic hydrogenation across the B-N bond was not possible, but the desired perhydrogenated molecule was prepared by a two-step protonic reduction sequence for use as a model for dehydrogenation studies. In a later publication by the same group [217], an experimentally derived ΔH of –126(4) kJ/mol (which compares remarkably well with a G3(MP2) calculated value) was provided for the first step, consistent with the performed direct catalytic hydrogenation. The calculated *overall* ΔH was –105 kJ/mol (presumably for 3 H_2), implying that a direct addition of H_2 in the second step of the above sequence would be endothermic. Thus, dehydrogenation of the perhydrogenated molecule may be expected to proceed with loss of one H_2 from across the B-N bond but may encounter a "bottleneck" for a release of the two remaining H_2 equivalents for returning to the starting N-t-Bu-1,2-azaborine. Some means of an in situ energy coupling between these two steps may be needed for high conversion.

Autothermal Hydrogen Storage

The heat of dehydrogenation ΔH_D of a hydrogen carrier and the various means to supply this heat have been a recurring theme in this review. Ideally, the carrier system would supply its own heat at the temperature required for its dehydrogenation. There have been several approaches to such an autothermal storage of hydrogen, all consisting of a combination of thermally coupled chemical reactions selected for an overall thermoneutral release of hydrogen.

Coupled Dehydrogenation Reactions

Gelsey [218] describes a method and an apparatus for which the heat required for releasing hydrogen from a H_2-reversible metal hydride is provided by a concurrent exothermic catalytic hydrolysis of sodium borohydride to H_2 and sodium borate. In an attempt to provide these two functions in a single carrier, Wechsler et al. [219] investigated the possibility of catalyzed dehydrogenation and hydrolysis of the 4-aminopiperidine-borane adduct. All attempts at coupled reactions (using Pd/C and H_2O at 100°C) resulted in hydrogen production from the hydrolysis but no hydrogen release from dehydrogenation of the piperidine moiety. However, by employing the more easily dehydrogenated indoline

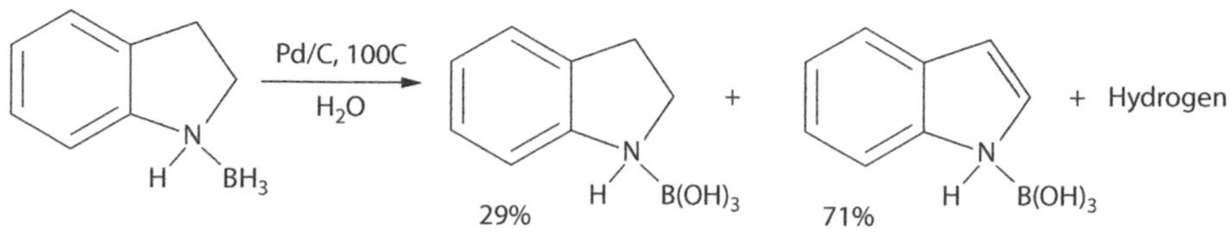

SCHEME 8.16
Partial dehydrogenation and complete hydrolysis of the indoline-borane adduct. (Adapted from Wechsler, D.; Cui, Y.; Dean, D.; Davis, B.; Jessop, P.G.; *J. Am. Chem. Soc.* 2008, *130*, 17195 with permission from the American Chemical Society.)

heterocycle, it was possible to achieve partial dehydrogenation and complete hydrolysis of the indoline-borane adduct, as depicted in Scheme 8.16.

Nonstoichiometric, physically mixed systems such as indoline/dimethylaminoborane and indoline/ammoniaborane at specific conditions provided hydrogen from both the hydrolysis and dehydrogenation reactions. There was some indication that in these reactions the borane component, once hydrolyzed, no longer inhibited the dehydrogenation of indoline. It was estimated that a 3:1 indoline/amine borane ($R_3N:BH_3$) combination should in theory provide a thermoneutral H_2 release if the two reactions had comparable rates [220]. For such a mixture at 100°C, hydrolysis is very fast, while the indoline dehydrogenation rate needs to be greatly enhanced. Various approaches to this included the use of catalysts other than Pd/C, adding 2,6-substituents on piperidine, and using higher reaction temperatures, all with apparently insufficient improvement in rate to render practical such an endothermic/exothermic "mixed fuel."

A second approach to thermoneutrality entails the use of a spontaneously H_2-releasing hydride rather than a hydrolysis reaction for providing the exotherm. In a patent application, Thorn et al. [221] cited the example of an ammoniaborane (NH_3BH_3)/DEC composite with a supported Pd or Pt catalyst, which when heated undergoes a spontaneous conversion to $(NH_2BH_2)_x$ or $(NHBH)_x$ materials with release of hydrogen. At greater than 250°C, the DEC is dehydrogenated to naphthalene with heat provided from the first reaction (which it also serves to moderate).

Wechsler et al. [222] attempted to convert the indoline-borane adduct under anhydrous conditions, hoping for a simultaneous two-site endothermic-exothermic dehydrogenation, as shown in Scheme 8.17. However, using various supported Pt group catalysts, there was no catalytic dehydrogenation of the adduct, and the BH_3 group was cleaved from the ring. With heat and no catalyst, hydrogen was produced but only from dehydrogenation of the aminoborane. From this investigation and following studies with now physically mixed indoline and dimethylaminoborane systems, it was concluded that the ring-dehydrogenation catalysis that would provide the endothermic component is somehow inhibited by the boron species, either by affecting the N-heterocycle substrate or deactivating the catalyst.

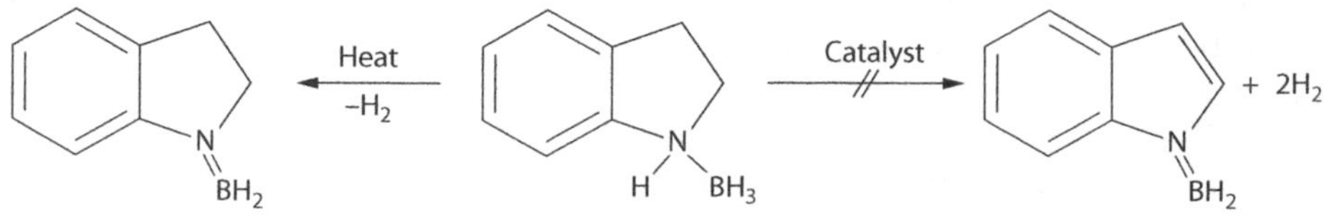

SCHEME 8.17
Attempts to dehydrogenate the indoline-borane adduct.

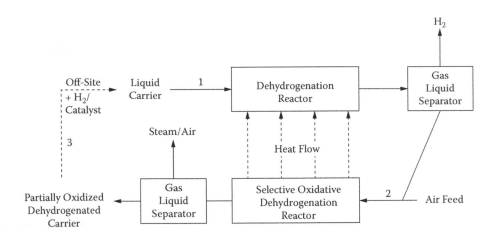

FIGURE 8.42
General schematic of an autothermal carrier cycle for which the fully loaded carrier is first dehydrogenated and
then undergoes a selective oxidation reaction that supplies the (stored) heat of dehydrogenation. Off-board the
vehicle, the spent carrier is entirely regenerated with hydrogen. (Reproduced from Pez, G.P.; Cooper, A.C.; Scott,
A.R.; U.S. Patent 8,003,073 (2011).)

Dehydrogenation Coupled with Selective Partial Oxidation

A serious drawback to using spontaneously H_2-releasing reactions (with $\Delta H_D \ll 0$) for heat
generation is that necessarily the "spent" materials are not directly reversible with hydrogen, although regeneration of the reagent is possible in multistep or electrolytic processes
(see the section on regeneration of AB from spent dehydrogenated products). This limitation led to the concept of devising an organic liquid carrier that could undergo the usual
(endothermic) catalytic ring dehydrogenation followed by heat-generating selective oxidation [223, 224]. These two chemical processes would be thermally coupled in an appropriate reactor system. The "spent" reagent would be regenerated off board the vehicle, solely
with molecular hydrogen, thus completing the cycle. A general schematic of one autothermal carrier cycle is shown in Figure 8.42.

The dodecahydrofluorene-fluorene-fluorenone system [223, 224] provides an example of an
autothermal carrier cycle, as described in Scheme 8.18. *Cis,cis*-dodecahydrofluorene (boiling
point 253°C) representing the fully loaded carrier in this cycle could be almost fully dehydrogenated to fluorene (6.65 wt% capacity) using a 5% Pt/Al_2O_3 catalyst in reaction at 235°C
for 6 h under 1 bar H_2. Significantly, dodecahydrofluorene samples consisting of a mixture of
stereoisomers were more difficult to fully dehydrogenate. This is reminiscent of the relatively
greater ease of dehydrogenation of *cis*-decalin vs. *trans*-decalin (see the section on DEC dehydrogenation catalysis) and the apparent isomer structure sensitivity during the dehydrogenation of NEDC (see the section on catalytic dehydrogenation of N-ethyldodecahydrocarbazole).

Sotoodeh et al. [225] reported on a kinetics study of dehydrogenation of dodecahydrofluorene as a 2.3 wt% solution in DEC over a Pd/C catalyst at 170°C. The reaction was
reported to be very slow; the two reaction intermediates, decahydrofluorene and hexa-
hydrofluorene, were produced with a high (95%) selectivity, but there was only a low
conversion (10.1% after 19 h) to the completely dehydrogenated product, fluorene. From
DFT calculations in their earlier reports, [188, 197] the binding energies for dodecahydrofluorene, NEDC, and dodecahydrocarbazole on a Pd (100) surface were –180.8, –95.0,
and –109.4 kJ/mol, respectively, with the dodecahydrofluorene molecule much closer
(parallel) to the surface. In view of this greater proximity and relatively tighter binding

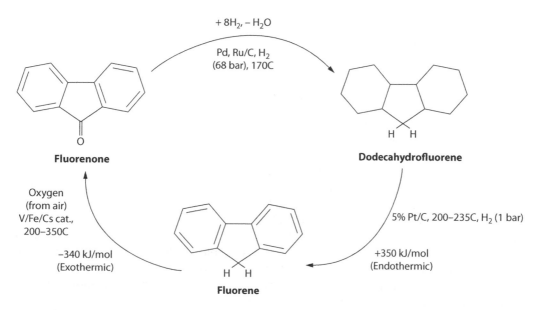

SCHEME 8.18
The dodecahydrofluorene-fluorene-fluorenone system as an example of an autothermal carrier cycle.

of dodecahydrofluorene as well as the observed high selectivity to fluorene (as the only ultimate product), it was suggested that dehydrogenation occurs via an abstraction of hydrogen from multiple carbon atoms. The higher adsorption energy of dodecahydro-fluorene vis-à-vis the isostructural N-heterocycles was suggested as the reason for the much slower rate of dehydrogenation.

In the autothermal carrier cycle sequence shown in Scheme 8.18, fluorene (melting point 116°C) is selectively oxidized with air in the vapor phase to fluorenone (melting point 84°C) and water, which was effected over a Cs-promoted V/Fe catalyst. Using data from computational modeling, the dehydrogenation and partial oxidation reaction enthalpies are approximately equal in magnitude but of opposite sign (in principle, resulting in thermoneutrality). During the spent carrier recycling step of the process, which would be conducted off board the vehicle, fluorenone is catalytically hydrogenated back to dodeca-hydrofluorene, with water as a by-product and molecular hydrogen providing (ideally) all the energy for the overall process. Of course, achieving this in practice will require the development of an on-board chemical reactor that is engineered for the diverse and thermally complementary catalytic chemistries to function as continuous processes. The carrier regeneration, however, could be practiced using relatively well-established industrial catalytic hydrogenation technologies.

Miscellaneous Potential Liquid Carriers

Sung and Kustov et al. [226] have investigated the terphenyl-tercyclohexane system for the storage of hydrogen. Hydrogenation of a commercial sample of terphenyl (presumably consisting of isomers of 1,2-diphenylbenzene) was selectively performed over a 10% Pt/C catalyst. Dehydrogenation of the tercyclohexane product over the same catalyst at 310–350°C resulted in nearly complete recovery of the hydrogen (7.25 wt% capacity). Trace levels of CH_4, C_2H_6, and C_3H_8 were observed in the H_2 effluent. In limited cycling studies, Pt-based catalysts were the most stable. The high dehydrogenation reaction temperatures

SCHEME 8.19
Proposed dehydrogenation of pentanethiol to 2-methyl-thiophene.

are expected for a ΔH_D that should be similar to that for cyclohexane-benzene. However, with terphenyl-tercyclohexane, there may be practical advantages for a system that is liquid at reaction conditions.

A measure of hydrogen storage was achieved by reversible hydrogenation of the phenyl ring in the 1-alkyl(aryl)-3-methylimidazolinium N-bis(trifluoromethanesulfonyl)imidate ionic liquid salt. Catalytic dehydrogenation using Pd/C to a moderate yield (60%) at 230–300°C under an Ar flow was reported [227].

As shown in Scheme 8.19, an alkyl-substituted thiophene liquid-carrier-based system for hydrogen storage was proposed by researchers at Asemblon [228] based on the discovery of a dehydrogenation of pentanethiol over a gold catalyst with cyclization to 2-methyl-thiophene.

A report by Zhao et al. [229] described work toward a catalytic hydrogenation and ring opening of thiophenes toward the development of a practical cycle.

There appears to be a renewed interest in storing hydrogen as formic acid, HCOOH (4.3 wt% H_2), which is derived from a catalytic hydrogenation of carbon dioxide. The reverse catalytic oxidation process provides H_2 and CO_2 [230]. The underlying chemistry is the reversible catalytic decomposition of formate salts to give bicarbonate and hydrogen [231].

$$HCO_2^- + H_2O \rightleftharpoons HCO_3^- + H_2 \tag{8.22}$$

The activation of hydrogen by nonmetal systems, for both reversible binding and catalysis, has captured intense interest in recent years [232]. A unique reactivity of H_2 is seen with Lewis acid–Lewis base combinations ("frustrated Lewis pairs" or FLPs), for which steric hindrance precludes the close encounter for a classical donor-acceptor complex. 2,6-Lutidine is sterically encumbered to the extent that it only displays a reversible interaction with tris(pentafluorophenyl)borane [233]. The system retains sufficient "FLP reactivity" to take up hydrogen with formation of the salt $[2,6\text{-}Me_2C_5H_3NH][HB(C_6F_5)_3]$. On heating, the salt loses H_2. While this FLP combination is clearly too "heavy" to be designated as a carrier, the new mode of reaction of hydrogen illustrated here may inspire new possibilities for hydrogen storage.

The loss of H_2 from the protic/hydridic salt ion pair is reminiscent of the discovery by Schwartz, Thorn, and coworkers [234, 235] of Pd-catalyzed hydrogen evolution from a benzimidazoline compound containing a carboxylic acid moiety. The hydridic proton of the C-H bond that is α to nitrogen and the acid reacts to produce H_2, but only in the presence of Pd (the role of the Pd catalyst is not clear). The reaction is exergonic ($\Delta G \ll 0$) and could not be reversed even with an applied pressure of hydrogen. Intended continuing efforts were aimed at devising more weakly acid, less hydridic, and overall lighter-weight combinations that might lead to a viable reversible reactivity for hydrogen storage.

Electrochemical "Virtual" Hydrogen Storage with Organic Liquid Carriers

Kariya, Ichikawa, et al. [179, 236, 237] from Hokkaido University, having worked with the reversible MCH-toluene and DEC-naphthalene systems (see the section on catalytic dehydrogenation under nonequilibrium conditions), also investigated using the cyclic hydrocarbons as direct fuels in a PEM fuel cell. The concept is that the "loaded" carrier could be spontaneously electrochemically oxidized in the fuel cell in conjunction with a parallel O_2 reduction reaction to generate electric power without the intermediacy of molecular hydrogen (and challenges with catalytic dehydrogenation). A schematic of the experimental setup and fuel cell design is provided in the cited references. Typically, the hydrocarbon vapor diluted with a humidified N_2 carrier gas was passed through the fuel cell anode compartment containing a Pt/C electrocatalyst. A humidified 1:4 O_2 to N_2 stream was led through the cathode. The operative reactions at the fuel cell anode and cathode were as follows [236]:

$$\text{Anode: } C_6H_{12} \leftrightarrow C_6H_6 + 6H^+ + 6e^- \qquad E_{300K} = -0.169 \text{ V} \qquad (8.23)$$

$$\text{Cathode: } O_2 + 4H^+ + 4e^- \leftrightarrow 2H_2O \qquad E_{300K} = 1.184 \text{ V} \qquad (8.24)$$

$$\text{Overall: } C_6H_{12} + 3/2O_2 \leftrightarrow C_6H_6 + 3H_2O \qquad \Delta E = 1.016 \text{ V} \qquad (8.25)$$

Fuel cell performance is conveyed by the polarization curve, a plot of the experimental output voltage vs. current density (Figure 8.43).

Key discerning features are the open circuit voltage (OCV), which characteristically diminishes with load due to mass transfer and other limitations of the cell [238] and the calculated power density. For cyclohexane, the OCV was 0.920 V (vs. an estimated 1.02 V), and a maximum power density of 15 mW cm^{-2} was reported. Putting these data in perspective, for a conventional hydrogen-air fuel cell, the power densities are much higher (e.g., 700 mW cm^{-2} at 0.68-V cell potential [238]), which means that the conversion of cyclohexane (Equation 8.25) in the experimental fuel cell must be quite small.

Ferrell et al. [239] reported an experimental survey of the use of cyclohexane, NEDC, dodecahydrofluorene, and hydroquinone as direct fuels in a PEM fuel cell. For cyclohexane electrooxidation was observed, but the OCV (400 mV) and current densities were lower than those as shown in Figure 8.43. It is suggested that the difference might arise from a cell design in the prior work that permitted better mixing of the fuel and water.

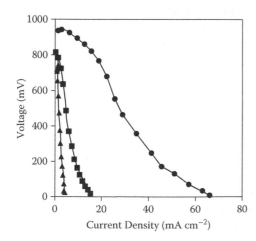 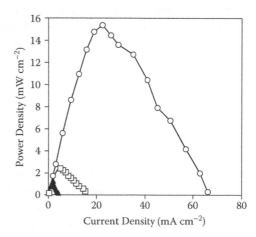

FIGURE 8.43

Current-voltage polarization curves (full circles) and current-power density plots (open circles) as performance indicators for direct cyclohexane fuel cells. Other data are for methylcyclohexane (squares) and cyclohexene fuels (triangles). Pt catalysts were used in the cathode and anode. This fuel cell was operated at 100°C with an oxygen flow at the cathode. (Reproduced from Kariya, N.; Fukuoka, A.; Ichikawa, M.; *Phys. Chem. Chem. Phys.* 2006, *8*, 1724, with permission of the PCCF Owner Societies, 2006 RSC Publishing.)

With NEDC at 120°C using a Pt/Ru catalyst, the oxidation currents were very small (<1 mA), but it was nevertheless possible to record a polarization curve. The OCV (>1.4 V) was unusually high as compared to a predicted 1.238 V for the equivalent cell with hydrogen as the fuel. This implies that the dehydrogenation, at least at very low conversions (i.e., NEDC to N-ethyldecahydrocarbazole), was spontaneous ($\Delta G_D < 0$). From the estimated thermodynamics for this reaction, a cell potential of 1.319 V was predicted. An electrooxidation of NEDC in acetonitrile was reported in a patent application by Soloveichik et al. [240]. Using a PEM FC with O_2 as the oxidant, an OCV of 340 mV was cited, increasing to 650 mV when using a carbon black/Ni/Pt electrode. Ferrell et al. [239] also found that under similar conditions (120°C, Pt/Ru catalyst), *cis,cis*-dodecahydrofluorene was reactive and likewise provided a very high OCV of greater than 1.2 V as compared to that for a practical H_2 fuel cell. The currents were very low, but always stable, and it was possible to record a polarization curve for the cell.

In their pioneering work in this area, Kariya et al. [179, 236, 237] also investigated MCH and 2-propanol as fuels. The latter fuel cell was also run in reverse with an applied potential generating acetone and oxygen from a simultaneous oxidation of water. Itoh [241] had previously demonstrated an electrolytic hydrogenation of benzene/water to cyclohexane and oxygen. The cyclohexane and 2-propanol PEM fuel cells were therefore referred to as "rechargeable (or regenerative) fuel cells," with the option that the regeneration of the fuel could be accomplished from either an electrolytic reduction or a conventional catalytic hydrogenation of the hydrogen-depleted fuel.

Crabtree [242], in a 2008 review article on "Hydrogen Storage in Liquid Organic Heterocycles," proposed this same concept of a direct use of a hydrogen storage material in a fuel cell, with regeneration in an electrocatalytic process, referring to it as "virtual H_2 storage." A recently reported anodic oxidation of a secondary amine in the presence of 2,3-dichloro-5,6-dicyanobenzoquinone as a electrocatalyst is presented as proof of principle for an organocatalytic CH-NH group dehydrogenation (as part of a cyclic amine carrier) and as a model system for virtual hydrogen storage [243].

At the time of writing (August 2012), there was a significant ongoing effort funded by the U.S. DOE Office of Science on this electrochemical hydrogen storage concept [244]. The research was being pursued at the multi-institutional Center for Electrocatalysis, Transport Phenomena, and Materials for Innovative Energy Storage (CETM) with General Electric Global Research as the lead institution [245].

Outlook for LOHC

In their recent "Perspective" article, Teichmann et al. [246] recast the conventional scenario of a hydrogen fuel economy in terms of a future energy supply that is based on employing LOHCs. They invoked as a generally accepted premise that energy will be increasingly derived from renewable resources, such as solar, wind, biomass, and geothermal sources. Since electricity production from many of these energy sources (at least wind and solar) is necessarily intermittent, they argued that there is a growing need for the means to store the generated power. As evidence for this (even present-day) need, they cited the occurrence of negative spot prices for electricity at the European Electricity Exchange that correlate with periods of relatively high production from wind plants. The negative prices are thus seen as evidence for a "lack of flexibility" in the electricity generation and distribution market that is likely to become even more problematic with an expected increasingly higher production from renewables (the potential role for hydrogen as a means of renewable energy storage was discussed in Chapter 1).

In this context, LOHC could be the medium for storing the intermittent electrical output of renewable energy generators, as well as for its transport to fixed and mobile sites of use. Hydrogen from electrolysis of water would be used to provide the "high-energy" form of the carrier via catalytic hydrogenation. The LOHC could easily be stored in large-scale tanks, similar to conventional hydrocarbon fuels, with (quantified) significant volumetric, gravimetric energy advantages vis-à-vis pumped hydroelectric power and high-pressure gaseous hydrogen.

Furthermore, the LOHC could be distributed in a similar fashion to hydrocarbons for use in mobile applications. On-board the vehicle, the fuel is catalytically dehydrogenated using the heat from the exhaust of an H_2 combustion engine to provide for the reaction endotherm. From "first considerations," it is felt that this quantity and level of waste heat should be sufficient, although not with currently available LOHC if a PEM fuel cell is the power source (see the following discussion). The "spent fuel" stored in the vehicle would be simultaneously replaced with fresh rehydrogenated carrier at the distribution station (a "dual-nozzle" design for this operation has been described [247]). The use of LOHC in the hydrogen infrastructure could be envisaged as a gradual transition from the existing fuels infrastructure to one based on hydrogen.

We have seen elements of the vision of an organic liquid-carrier-based hydrogen fuel infrastructure as conveyed in Figure 8.31 and the citations in the section on catalytic dehydrogenation under nonequilibrium conditions. Reference here is also made to reviews of the concept by the group of Biniwale et al., [248, 250] including a technoeconomic analysis [249]. Potential use for electrical energy storage (see the section on this topic) and vehicular applications including the development of a prototype truck using MCH as the hydrogen carrier (see the section on vehicular applications of the MTH cycle) have been discussed along with the vision of Ichikawa et al. [179] of a complex "hydrogen highway" interconnecting H_2 sources and customer sites with liquid carriers replacing power transmission lines (see the section on reactor development toward technology implementation). The present perspective of Teichmann and coauthors [246] is, however, particularly instructive because in addition it articulates the challenge of providing for the intermittence of now increasingly important renewable energy sources.

As was the case with AlH_3 and $LiAlH_4$ and AB previously in the chapter, an implementation of LOHC-based concepts for a future hydrogen-based energy supply clearly requires consideration of the entire system, including the reaction catalysts, chemical reactors, and associated technology for their use. Some of the significant issues/requirements expressed by Teichmann [246] and Crabtree [242] are as follows:

1. LOHC gravimetric and volumetric hydrogen capacity (U.S. DOE H_2 storage goals as a guide).
2. System well-to-tank (WTT) and on-board system energy efficiency.
3. Capability for heat integration, preferably with a PEM fuel cell.
4. Safety in use and environmental acceptance, particularly important requirements considering the large volumes of the fuel that would be involved.
5. Capital and recycling costs of the carrier.
6. The development of high power density reactors that can follow the varying power load of a vehicle.
7. Detailed engineering and economic assessments of how the existing fossil fuel infrastructure could be best employed for the distribution of recyclable liquid carriers.

Some of these issues/requirements were examined by Ahluwalia et al. [251] for hydrogen storage options for light-duty vehicles. These options include compressed gas, cryogenic H_2, cryogenic H_2 in an insulated pressure vessel, metal hydrides (e.g., AlH_3, $NaAlH_4$), "chemical storage" systems (e.g., $NaBH_4$, NH_3BH_3), and liquid organic carriers as exemplified by NEDC. For a model system [252] in which NEDC carrier is providing H_2 for a fuel cell, it is estimated that some 32% of the H_2 is combusted (68% on-board efficiency) to provide for the reaction heat. A higher efficiency would be possible with a lower ΔH_D carrier, and it could be 100% if ΔH_D were less than 40 kJ/mol H_2 (see the section on catalytic hydrogenation/dehydrogenation fundamentals) and the dehydrogenation catalysts were sufficiently active. Reaction temperatures could then be low enough to enable the utilization of the fuel cell's waste heat. The alternative approach is in adequately capturing the waste heat from an H_2 combustion engine (see Chapter 2). Since NEDC can be (exothermically) directly regenerated with hydrogen, it shows the highest WTT efficiency of the group, and it could be even higher if the heat were converted to electricity. An analysis has been conducted of an integrated production, storage, and delivery concept for a hydrogen fuel infrastructure using NEDC as a model carrier (but generally applicable to other liquids) [210]. A probability distribution of the cost for H_2 delivery was provided along with analysis of the contributing factors (e.g., cost of carrier, carrier losses, catalyst cost, and noble metal makeup). While not the "ideal carrier," NEDC is nevertheless seen as fulfilling many of the performance requirements and suitable as a test material for assessing the LOHC concept [246].

It is evident that the challenge of a carrier with a ΔH_D approaching 40 kJ/mol H_2 is still unmet, but what are the possibilities? Predictions of ΔH_D in this range for polycyclic molecules with two or more N heteroatoms are realistic and meaningful as long as the energetics along the envisaged reaction path are also considered (see the section on polyaromatic hydrocarbons with nitrogen and oxygen heterotoms). A target carrier molecule may be known or can be synthesized, but far less certain is finding catalysts that will readily promote practical rates of C-H and N-H bond breaking at PEM fuel cell temperatures of about 80°C. A case in point is the attempts to prepare carriers based on perhydro-β-carboline (ΔH_D = 44 kJ/mol), a heterocycle with two nitrogen atoms in the molecule; attempts were eventually successful in terms of preparing the compound. However, only minimal dehydrogenation was observed at temperatures up to 170°C in the presence of a Pt/C catalyst [224]. Autothermal H_2 delivery systems (see the section on autothermal hydrogen storage) do not necessarily have the approximately 80°C temperature limitation and deserve further investigation, particularly for energy storage applications, with the advantage of not requiring an energy input at the points of use. By employing the organic liquid carrier as the direct feed to a PEM-type fuel cell (see the section on electrochemical "virtual" hydrogen storage with organic liquid carriers), the thermodynamic limitation and equipment for an intermediate production of hydrogen are avoided. It is the most elegant approach, but as a challenge in electrocatalysis and requiring the development of new fuel cell technology, it is also the most daunting.

Acknowledgments

Much of the work on aluminum hydride included contributions from James Reilly and James Wegrzyn, and Jason Graetz is grateful to them for many years of collaboration and support. Jason also acknowledges funding from the U.S. DOE, Energy Efficiency and Renewable Energy (EERE), and the Office of Basic Energy Sciences. Lennie Klebanoff acknowledges helpful prior conversations about electrochemical generation of alane from Ragaiy Zidan at Savannah River National Laboratory and funding from the U.S. DOE EERE.

Guido Pez and Alan Cooper gratefully acknowledge funding provided by the U.S. DOE EERE through contracts DE-FC36-04GO14006 and DE-FG36-05GO15015 for the cited research on hydrogen storage using liquid carriers performed by Air Products and Chemicals, Inc. Guido and Alan are appreciative and deeply grateful to their colleagues for their dedicated efforts and creativity on this project, particularly the computational chemistry input of Hansong Cheng, which made possible a quantification of ideas; Bernard Toseland for engineering research; Sergei Ivanov, Michael Ford, Aaron Scott, Atteye Abdourazak, and Fred Wilhelm for their contributions in chemistry and catalysis; complemented by the invaluable technical assistance of Gian Muraro and Don Fowler.

References

1. Sandrock, G.; Reilly, J.; Graetz, J.; Zhou, W.-M.; Johnson, J.; Wegrzyn, J.; *Appl. Phys. A* 2005, *80*, 687.
2. Sandrock, G.; Reilly, J.; Graetz, J.; Zhou, W.-M.; Johnson, J.; Wegrzyn, J.; *J. Alloys Compd.* 2006, *421*, 185.
3. Graetz, J.; Reilly, J.J.; Yartys, V.A.; Maehlen, J.P.; Bulychev, B.M.; Antonov, V.E.; Tarasov, B.P.; Gabis, I.E.; *J. Alloys Compd.* 2011, *509S*, S517.
4. Brower, F.M.; Matzek, N.E.; Reigler, P.F.; Rinn, H.W.; Roberts, C.B.; Schmidt, D.L.; Snover, J.A.; Terada, K.; *J. Am. Chem. Soc.* 1976, *98*, 2450.
5. Paraskos, J.H.; Lund, G.; Insensitive Munitions and Energetic Materials Technology Symposium, Miami, FL, 2007.
6. Stecher, O.; Wiberg, E.; *Ber.* 1942, *75B*, 2003.
7. Finholt, A.E.; Bond, A.C.; Schlesinger, H.I.; *J. Am. Chem. Soc.* 1947, *69*, 1199.
8. Chizinsky, G.; *J. Am. Chem. Soc.* 1955, *77*, 3164.
9. Brinks, H.W.; Istad-Lem, A.; Hauback, B.C.; *J. Phys. Chem. B* 2206, *110*, 25833.
10. Ashby, E.C.; Yoon, N.M.; *J. Am. Chem. Soc.* 1966, *88*, 1464.
11. Heitsch, C.W.; Nordman, C.E.; Parry, R.W.; *Inorg. Chem.* 1963, 2, 508.
12. Dallacker, F.; Glombitza, K.W.; Lipp, M.; *Lieb. Ann. Chem.* 1961, *643*, 67.
13. Trevoy, L.W.; Brown, W.G.; *J. Am. Chem. Soc.* 1949, *71*, 1675.
14. Ashby, C.; Sanders, J.R.; Claudy, P.; Schwartz, R.; *J. Am. Chem. Soc.* 1973, *95*, 6485.
15. Finholt, A.E.; Barbaras, G.D.; Barbaras, G.K.; Urry, G.; Wartik, T.; Schlesinger, H.I.; *J. Inorg. Nucl. Chem.* 1955, *1*, 317.
16. Schwab, M.; Wintersberger, K.; *Z. Naturforsch.* 1953, *8*, 690.
17. Bulychev, B.M.; Golubeva, A.V.; Storozhenko, P.A.; Semenenko, K.N.; *Russ. J. Inorg. Chem.* 1998, *43*, 1141.
18. Bulychev, B.M.; Verbetskii, V.N.; Storozhenko, P.A.; *Russ. J. Inorg. Chem.* 2008, *53*, 1000.
19. Bulychev, B.M.; Storozhenko, P.A.; Fokin, V.N.; *Russ. Chem. Bull. Int. Ed.* 2009, *58*, 1817.
20. Turley, J.W.; Rinn, H.W.; *Inorg. Chem.* 1969, *8*, 18.
21. Brinks, H.W.; Istad-Lem, A.; Hauback, B.C.; *J. Phys. Chem. B* 2006, *110*, 25833.
22. Sinke, G.C.; Walker, L.C.; Oetting, F.L.; Stull, D.R.; *J. Chem. Phys.* 1967, *47*, 2759.
23. Graetz, J.; Reilly, J.; *J. Alloys Compd.* 2006, *424*, 262.
24. Orimo, S.; Nakamori, Y.; Kato, T.; Brown, C.; Jensen, C.M.; *Appl. Phys. A* 2006, *83*, 5.
25. Graetz, J.; Chaudhuri, S.; Lee, Y.; Vogt, T.; Reilly, J.J.; *Phys. Rev. B* 2006, *74*, 214114.
26. Ke, X.; Kuwabara, A.; Tanaka, I.; *Phys. Rev. B* 2005, *71*, 184107.

27. Sartori, S.; Opalka, S.M.; Løvvik, O.M.; Guzik, M.N.; Hauback, X.T.B.C.; *J. Mater. Chem.* 2008, *18*, 2361.
28. H.W. Brinks, W. Langley, C.M. Jensen, J. Graetz, J.J. Reilly, B.C. Hauback, *J. Alloys Compd.* 2007, *433*, 180.
29. Graetz, J.; Reilly, J.J.; Kulleck, J.G.; Bowman, R.C., Jr.; *J. Alloys Compd.* 2007, *271*, 446–447.
30. Yartys, V.A.; Denys, R.V.; Maehlen, J.P.; Frommen, Ch.; Fichtner, M.; Bulychev, B.M.; Emerich, H.; *Inorg. Chem.* 2007, *46*, 1051.
31. Brinks, H.W.; Brown, C.; Jensen, C.M.; Graetz, J.; Reilly, J.J.; Hauback, B.C.; *J. Alloys Compd.* 2007, *441*, 364.
32. Bulychev, B.M.; Storozhenko, P.A.; Fokin, V.N.; *Russ. Chem. Bull. Int. Ed.* 2009, *58*, 1817.
33. Niles, E.T.; Seaman, B.A.H.; Wilson, E.J.; U.S. Patent 3,869,544 (1975).
34. Cianciolo, A.D.; Sabatine, D.J.; Scruggs, J.A.; Trotz, S.I.; U.S. Patent 3,785,890 (1974).
35. Petrie, M.A.; Bottaro, J.C.; Schmitt, R.J.; Penwell, P.E.; Bomberger, D.C.; U.S. Patent 6,228,338 B1 (2001).
36. Kempa, P.B.; Thome, V.; Herrmann, M.; *Part. Part. Syst. Charact.* 2009, *26*, 132.
37. Wang, Y.; Palsson, G.K.; Raanaei, H.; Hjorvarsson, B.; *J. Alloys Compd.* 2008, *464*, L13.
38. Kato, S.; Bielmann, M.; Ikeda, K.; Orimo, S.-I.; Borgschulte, A.; Zuttel, A.; *Appl. Phys. Lett.* 2010, *96*, 051912.
39. Herley, P.J.; Irwin, R.H.; *J. Phys. Chem. Solids* 1978, *39*, 1013.
40. Herley, P.J.; Chrlstofferson, O.; Todd, J.A.; *J. Solid State Chem.* 1980, *35*, 391.
41. Herley, P.J.; Chrlstofferson, O.; Irwin, R.; *J. Phys. Chem.* 1981, *85*, 1874.
42. Herley, P.J.; Chrlstofferson, O.; *J. Phys. Chem.* 1981, *85*, 1882.
43. Herley, P.J.; Chrlstofferson, O.; *J. Phys. Chem.* 1981, *85*, 1887.
44. Graetz, J.; Reilly, J.J.; *J. Phys. Chem. B* 2005, *109*, 22181.
45. Maehlen, J.P.; Yartys, V.A.; Denys, R.V.; Fichtner, M.; Frommen, C.; Bulychev, B.M.; Pattison, P.; Emerich, H.; Filinchuk, Y.E.; Chernyshov, D.; *J. Alloys Compd.* 2007, *280*, 446–447.
46. Gabis; Dobrotvorski, M.; Evard, E.; Voit, A.; unpublished results, 2010.
47. Bulychev, B.M.; Verbetskii, V.N.; Sizov, A.I.; Zvukova, T.M.; Genchel, V.K.; Fokin, V.N.; *Russ. Chem. Bull. Int. Ed.* 2007, *56*, 1305.
48. Yartys, V.A.; Workshop CNES, Paris, November 2008.
49. Chopra, S.; Chaudhuri, S.; Veyan, J.-F.; Graetz, J.; Chabal, Y.J.; *J. Phys. Chem. C* 2011, *115*, 16701.
50. Bulychev, B.M.; Verbetsky, V.N.; unpublished results.
51. Graetz, J.; Catalyzed (de)hydrogenation reactions in Al hydrides, presented at the International Symposium on Metal-Hydrogen Systems, Reykjavik, Iceland, June 2008.
52. Tkacz, M.; Filipek, S.; Baranowski, B.; *Polish J. Chem.* 1983, *57*, 651.
53. Baranowski, B.; Tkacz, M.; *Z. Phys. Chem. N. F.* 1983, *135*, 27.
54. Konovalov, S.K.; Bulychev, B.M.; *Inorg. Chem.* 1995, *34*, 172.
55. Saitoh, H.; Machida, A.; Katayama, Y.; Aoki, K.; *Appl. Phys. Lett.* 2008, *93*, 151918.
56. Graetz, J.; Chaudhari, S.; Wegrzyn, J.; Celebi, Y.; Johnson, J.R.; Zhou, W.; Reilly, J.J.; *J. Phys. Chem. C* 2007, *111*, 19148.
57. Lacina, D.; Wegrzyn, J.; Reilly, J.; Johnson, J.; Celebi, Y.; Graetz, J.; *J. Phys. Chem. C* 2011, *115*, 3789.
58. Lacina, D.; Reilly, J.; Johnson, J.; Wegrzyn, J.; Graetz, J.; The reversible synthesis of bis(quinuclidine) alane, *J. Alloys Compd.* 2011, *509*, S654.
59. Lacina, D.; Wegrzyn, J.; Reilly, J.; Celebi, Y.; Graetz, J.; *Energy Environ. Sci.* 2010, *3*, 1099.
60. Hua, T.Q.; Ahluwalia, R.K.; *Int. J. Hydrogen Energy* 2011, *36*, 15259.
61. Clasen, H.; German Patent 1141 623 (1962).
62. Alpatova, N.M.; Dymova, T.N.; Kessler, Y.M.; Osipov, O.R.; *Russ. Chem. Rev.* 1968, *37*, 99.
63. Birnbaum, H.K.; Buckley, C.; Zeides, F.; Sirois, E.; Rozenak, P.; *J. Alloys Compd.* 1997, *260*, 253–254.
64. Zidan, R.; Garcia-Diaz, B.L.; Fenwox, C.S.; Stowe, A.C.; Gray, J.R.; Harter, A.G.; *Chem. Commun.* 2009, *25*, 3717.
65. Graetz, J.; Wegrzyn, J.; Reilly, J.J.; *J. Am. Chem. Soc.* 2008, *130*, 17790.

66. Liu, X.; McGrady, G.S.; Langmi, H.W.; Jensen, C.M.; *J. Am. Chem. Soc.* 2009, *131*, 5032.
67. (a) Stephens, F.H.; Pons, V.; Baker, R.T.; *Dalton Trans.* 2007, 2613–2626; (b) Hamilton, C.W.; Baker, R.T.; Staubitz, A.; Manners, I.; *Chem. Soc. Rev.* 2009, *38*, 279–293; (c) Staubitz, A.; Robertson, A.P.M.; Manners, I.; *Chem. Rev.* 2010, *110*, 4079–4124; (d) Bowden, M.; Autrey, T.; *Curr. Opin. Solid State Mater. Sci.* 2011, *15*, 73–79.
68. (a) Orimo, S.-I.; Nakamori, Y.; Eliseo, J.R.; Züttel, A.; Jensen, C.M.; *Chem. Rev.* 2007, *107*, 4111–4132; (b) Jain, I.P.; Jain, P.; Jain, A.; *J. Alloys Compd.* 2010, *503*, 303–339.
69. (a) Hughes, E.W.; *J. Am. Chem. Soc.* 1956, *78*, 502–503; (b) Lippert, E.L.; Lipscomb, W.N.; *J. Am. Chem. Soc.* 1956, *78*, 503–504.
70. (a) Bowden, M.E.; Gainsford, G.J.; Robinson, W.T.; *Aust. J. Chem.* 2007, *60*, 149–153; (b) Yang, J.B.; Lamsal, J.; Cai, Q.; James, W.J.; Yelon, W.B.; *Appl. Phys. Lett.* 2008, *92*, 091916–091916-3.
71. Hess, N.J.; Schenter, G.K.; Hartman, M.R.; Daemen, L.L.; Proffen, T.; Kathmann, S.M.; Mundy, C.J.; Hart, M.; Heldebrant, D.J.; Stowe, A.C.; Autrey, T.; *J. Phys. Chem.* 2009, *113*, 5723–5735.
72. Klooster, W.T.; Koetzler, T.F.; Siegbahn, P.E.M.; Richardson, T.B.; Crabtree, R.H.; *J. Am. Chem. Soc.* 1999, *121*, 6337–6343.
73. (a) Bondi, A.; *J. Phys. Chem.* 1964, *68*, 441; (b) Nyburg, S.C.; Faerman, C.H.; *Acta Crystallogr. B* 1985, *41*, 274–279.
74. Stowe, A.C.; Shaw, W.J.; Linehan, J.C.; Schmid, Autrey, T.; *Phys, Chem. Chem. Phys.* 2007, *9*, 1831–1836.
75. Bowden, M.; Heldebrant, D.J.; Karkamkar, A.; Proffen, T.; Schenter, G.K.; Autrey, T.; *Chem. Commun.* 2010, *46*, 8564–8566.
76. Filinchuk, Y.; Nevidomskyy, A.H.; Chernyshov, D.; Dmitriev, V.; *Phys. Rev. B: Condens. Matter Mater. Phys.* 2009, *79*, 214111/1.
77. Heldebrant, D.J.; Karkamkar, A.; Hess, N.J.; Bowden, M.; Rassat, S.; Zheng, F.; Rappe, K.; Autrey, T.; *Chem. Mater.* 2008, *20*, 5332–5336.
78. Jacquemin, D.; Perpete, E.A.; Wathelet, V.; Andre, J.-M.; *J. Phys. Chem. A* 2004, *108*, 9616–9624.
79. Li, J.; Kathmann, S.M.; Schenter, G.K.; Gutowski, M.; *J. Phys. Chem. C* 2007, *111*, 3294–3299.
80. (a) Trefonas, L.M.; Mathews, F.S.; Lipscomb, W.N.; *Acta Crystallogr.* 1961, *14*, 273–278; (b) Jaska, C.A.; Temple, K.; Lough, A.J.; Manners, I.; *J. Am. Chem. Soc.* 2003, *125*, 9424–9434.
81. Hu, M.G.; Geanangel, R.A.; Wendlandt, W.W.; *Thermochim. Acta* 1978, *23*, 249–255.
82. Sit, V.; Geanangel, R.A.; Wendlandt, W.W.; *Thermochim. Acta* 1987, *113*, 379–382.
83. Geanangel, R.A.; Rabalais, J.W.; *Inorg. Chim. Acta* 1985, *97*, 59–64.
84. Wolf, G.; Baumann, J.; Baitalow, F.; Hoffmann, F.P.; *Thermochim. Acta* 2000, *343*, 19–25.
85. Baitalow, F.; Wolf, G.; Grolier, J.P.E.; Dan, F.; Randzio, S.L.; *Thermochim. Acta* 2006, *445*, 121–125.
86. Nylén, J.; Sato, T.; Soignard, E.; Yarger, J.L.; Stoyanov, E.; Häussermann, U.; *J. Chem. Phys.* 2009, *131*, 104506/1.
87. Palumbo, O.; Paolone, A.; Rispoli P.; Cantelli, R.; Autrey, T.; *J. Power Sources* 2010, *195*, 1615–1618.
88. Smith, R.S.; Kay, B.D.; Schmid, B.; Li, L.; Hess, N.J.; Gutowski, M.; Autrey, T.; *Prep. Pap. Am. Chem. Soc., Div. Fuel Chem.* 2005, *50*, 112–113.
89. Shaw, W.J.; Bowden, M.; Karlamkar, A.; Howard, C.J.; Heldebrant, D.J.; Hess, N.J.; Linehan, J.C.; Autrey, T.; *Energy Environ. Sci.* 2010, *3*, 796–804.
90. Wolstenholme, D.J.; Traboulsee, K.T.; Hua, Y.; Calhoun, L.A.; McGrady, G.S.; *Chem. Commun.* 2012, *48*, 2597–2599.
91. Wang, J.S.; Geanangel, R.A.; *Inorg. Chim. Acta* 1988, *148*, 185–190.
92. Shaw, W.J.; Linehan, J.C.; Szymczak, N.K.; Heldebrant, D.J.; Yonker, C.; Camaioni, D.M.; Baker, R.T.; Autrey, T.; *Angew. Chem. Int. Ed.* 2008, *47*, 7493–7496.
93. Ryschkewitsch, G.E.; Wiggins, J.W.; *Inorg. Chem.* 1970, *9*, 314–317.
94. Chen, X.; Bao, X.; Zhao, J.-C.; Shore, S.G.; *J. Am. Chem. Soc.* 2011, *133*, 14172–14175.
95. (a) Li, Q.S.; Zhang, J.; Zhang, S.; *Chem. Phys. Lett.* 2005, *404*, 100–106; (b) Zhang, J.; Zhang, S.; Li, Q.S.; *Theochem* 2005, *717*, 33–39.
96. Nguyen, M.T.; Nguyen, V.S.; Matus, M.H.; Gopakumar, G.; Dixon, D.A.; *J. Phys. Chem. A* 2007, *111*, 679–690.
97. Weismiller, M.R.; van Duin, A.C.T.; Lee, J.; Yetter, R.A.; *J. Phys. Chem. A* 2010, *114*, 5485–5492.

98. Nguyen, V.S.; Matus, M.H.; Grant, D.J.; Nguyen, M.T.; Dixon, D.A.; *J. Phys. Chem. A* 2007, *111*, 8844–8856.
99. Miranda, C.R.; Ceder, G.; *J. Chem. Phys.* 2007, *126*, 184703/1.
100. Jaska, C.A.; Temple, K.; Lough, A.J.; Manners, I.; *Chem. Commun.* 2001, 962–963.
101. (a) Fazen, P.J.; Remsen, E.E.; Carroll, P.J.; Beck, J.S.; McGhie, A.R.; Sneddon, L.G.; *Chem. Mater.* 1995, *7*, 1942–1956; (b) Wideman, T.; Fazen, P.J.; Lynch, A.T.; Su, K.; Remsen, E.E.; Sneddon, L.G.; *Inorg. Synth.* 1998, *32*, 232.
102. Denney, M.C.; Pons, V.; Hebden, T.J.; Heinekey, D.M.; Goldberg, K.I.; *J. Am. Chem. Soc.* 2006, *128*, 12048–12049.
103. Böddeker, K.W.; Shore, S.G.; Bunting, R.K.; *J. Am. Chem. Soc.* 1966, *88*, 4396–4401.
104. Paul, A.; Musgrave, C.B.; *Angew. Chem. Int. Ed.* 2007, *46*, 8153–8156.
105. Keaton, R.J.; Blacquiere, J.M.; Baker, R.T.; *J. Am. Chem. Soc.* 2007, *129*, 1844–1845.
106. (a) Yang, X.; Hall, M.B.; *J. Am. Chem. Soc.* 2008, *130*, 1798–1799; (b) Zimmerman, P.M.; Paul, A.; Zhang, Z.; Musgrave, C.B.; *Angew. Chem. Int. Ed.* 2009, *48*, 2201–2205; (c) Zimmerman, P.M.; Paul, A.; Musgrave, C.B.; *Inorg. Chem.* 2009, *48*, 5418–5433.
107. Tang, C.T.; Thompson, A.L.; Aldridge, S.; *J. Am Chem. Soc.* 2010, *132*, 10579.
108. Tang, C.T.; Thompson, A.L.; Aldridge, S.; *Angew. Chem. Int. Ed..* 2010, *49*, 921.
109. Alcaraz, G.; Vendier, L.; Clot, E.; Sabo-Etienne, S.; *Angew. Chem. Int. Ed.* 2010, *49*, 918–920.
110. Alcaraz, G.; Sabo-Etienne, S.; *Angew. Chem. Int. Ed.* 2010, *49*, 7170–7179.
111. Kelly, H.C.; Marriott, V.B.; *Inorg. Chem.* 1979, *18*, 2875–2878.
112. Stephens, F.H.; Baker, R.T.; Matus, M.H.; Grant, D.J.; Dixon, D.A.; *Angew. Chem. Int. Ed.* 2007, *46*, 746–749.
113. Bluhm, M.E.; Bradley, M.G.; Butterick, R., III; Kusari, U.; Sneddon, L.G.; *J. Am. Chem. Soc.* 2006, *128*, 7748–7749.
114. Freemantle, M.; *An Introduction to Ionic Liquids*, RSC, Cambridge, UK, 2010.
115. Gutowska, A.; Li, L.; Shin, Y.; Wang, C.M.; Li, X.S.; Linehan, J.C.; Smith, R.S.; Kay, B.D.; Schmid, B.; Shaw, W.; Gutowski, M.; Autrey, T.; *Angew. Chem. Int. Ed.* 2005, *44*, 3578–3582.
116. Wang, L.-Q.; Karkamkar, A.; Autrey, T.; Exarhos, G.J.; *J. Phys. Chem. C* 2009, *113*, 6485–6490.
117. Paolone, A.; Palumbo, O.; Rispoli, P.; Autrey, T.; Karkamkar, A.; *J. Phys. Chem. C* 2009, *113*, 10319–10321.
118. Feaver, A.; Sepehri, S.; Shamberger, P.; Stowe, A.; Autrey, T.; Cao, G.; *J. Phys. Chem. B* 2007, *111*, 7469–7472.
119. Sepehri, S.; Feaver, A.A.; Shaw, W.J.; Howard, C.J.; Zhang, Q.; Autrey, T.; Cao, G.; *J. Phys. Chem. B* 2007, *111*, 14285–14289.
120. Servoss, R.R.; Clark, H.M.; *J. Chem. Phys.* 1957, *26*, 1179–1184.
121. Xiong, Z.; Chua, Y.; Wu, G.; Wang, L.; Wong, M.W.; Kam, Z.M.; Autrey, T.; Kemmitt, T.; Chen, P.; *Dalton Trans.* 2010, *39*, 720–722.
122. Chua, Y.S.; Chen, P.; Wu, G.; Xiong, Z.; *Chem. Commun.* 2011, *47*; 5116–5129.
123. Xiong, Z.; Yong, C.K.; Wu, G.; Chen, P.; Shaw, W.; Karkamkar, A.; Autrey, T.; Jones, M.O.; Johnson, S.R.; Edwards, P.P.; David, W.I.F.; *Nat. Mater.* 2008, *7*, 138–141.
124. Xiong, Z.; Wu, G.; Chua, Y.S.; Hu, J.; He, T.; Xu, W.; Chen, P.; *Energy Environ. Sci.* 2008, *1*, 360–363.
125. Fijalkowski, K.J.; Grochala, W.; *J. Mater. Chem.* 2009, *19*, 2043–2050.
126. Diyabalanage, H.V.K.; Nakagawa, T.; Shrestha, R.P.; Semelsberger, T.A.; David, B.L.; Scott, B.L.; Burrell, A.K.; David, W.I.F.; Ryan, K.R.; Jones, M.O.; Edwards, P.P.; *J. Am. Chem. Soc.* 2010, *132*, 11836–11837.
127. Lee, T.B.; McKee, M.L.; *Inorg. Chem.* 2009, *48*, 7564–7575.
128. Kim, D.Y.; Singh, N.J.; Lee, H.M.; Kim, K.S.; *Chem.-Eur. J.* 2009, *15*, 5598–5604.
129. Luedtke, A.T.; Autrey, T.; *Inorg. Chem.* 2010, *49*, 3905–3910.
130. (a) Dietrich, B.L.; Goldberg, K.I.; Heinekey, D.M.; Autrey, T.; Linehan, J.C.; *Inorg. Chem.* 2008, *47*, 8583–8585; (b) Miller, A.J.M.; Bercaw, J.E.; *Chem. Commun.* 2010, *46*, 1709–1711.
131. Wolstenholme, D.J.; Titah, J.T.; Che, F.N.; Traboulsee, K.T.; Flogeras, J.; McGrady, G.S.; *J. Am. Chem. Soc.* 2011, *133*, 16598–16604.
132. Ramachandran, P.V.; Gagare, P.D.; *Inorg. Chem.* 2007, *46*, 7810–7817.

133. Hausdorf, S.; Baitalow, F.; Wolf, G.; Mertens, F.O.R.L.; *Int. J. Hydrogen Energy* 2008, *33*, 608–614.
134. Sneddon, L.G.; Amineborane based chemical hydrogen storage, DoE hydrogen annual merit review, 2007. http://www.hydrogen.energy.gov/pdfs/review07/st_27_sneddon.pdf.
135. Taylor, F.M.; Dewing, J.; U.S. Patent 3,103,417, 1963.
136. Davis, B.L.; Dixon, D.A.; Garner, E.B.; Gordon, J.C.; Matus, M.H.; Scott, B.; Stephens, F.H.; *Angew. Chem. Int. Ed.* 2009, *48*, 6812–6816.
137. Sutton, A.D.; Davis, B.L.; Bhattacharyya, K.X.; Ellis, B.D.; Gordon, J.C.; Power, P.P.; *Chem. Commun.* 2010, *46*, 148–149.
138. Sutton, A.D.; Burrell, A.K.; Dixon, D.A.; Garner, E.B., III; Gordon, J.C.; Nakagawa, T.; Ott, K.C.; Robinson, J.P.; Vasiliu, M.; *Science* 2011, *331*, 1426–1429.
139. Hughes, A.M.; Corruccini, R.J.; Gilbert, E.C.; *J. Am. Chem. Soc.* 1939, *61*, 2639–2642.
140. Liu, X.F.; McGrady, G.S.; Langmi, H.W.; Jensen, C.M.; *J. Am. Chem. Soc.* 2009, *131*, 5032–5033.
141. Langmi, H.W.; McGrady, G.S.; Liu, X.F.; Jensen, C.M.; *J. Phys. Chem. C* 2010, *114*, 10666–10669.
142. Liu, X.F.; Langmi, H.W.; Beattie, S.D.; Azenwi, F.F.; McGrady, G.S.; Jensen, C.M.; *J. Am. Chem. Soc.* 2011, *133*, 15593–15597.
143. Wu, C.; Wu, G.; Xiong, Z.; Han, X.; Chu, H.; He, T.; Chen, P.; *Chem. Mater.* 2010, *22*, 3–5.
144. *Basic Research Needs for the Hydrogen Economy*; U.S. DoE Report, May 2003.
145. Bélanger, G.; Cyclohexane, a liquid hydride; hydrogen energy progress-IV, *Proceedings of the 4th World Energy Conference*, Vol. 3; Pergamon Press, Oxford, UK, 1982, pp. 1335–1343.
146. Touzani, A.; Klvana, D.; Bélanger, G.; *Int. J. Hydrogen Energy* 1984, *9*, 929.
147. Cacciola, G.; Giordano, N.; Restuccia, G.; *Int. J. Hydrogen Energy* 1984, *9*, 411.
148. Itoh, N.; Watanabe, S.; Kawasoe, K.; Sato, T.; Tsuji, T.; *Desalination* 2008, *234*, 261.
149. Sultan, O.; Shaw, H.; Study of automotive storage of hydrogen using recyclable liquid chemical carriers. Technical Report TEC75/003, 1975. http://www.osti.gov/energycitations/product.biblio.jsp?osti_id = 5000657.
150. Alhumaidan, F.; Cresswell, D.; Garforth, A. *Energy Fuels* 2011, *25*, 4217.
151. Shukla, A.A.; Gosavi, P.V.; Pande, J.V.; Kumar, V.P.; Chary, K.V.R.; Biniwale, R.B.; *Int. J. Hydrogen Energy* 2010, *35*, 4020.
152. Ali, J.K.; Newson, E.J.; Rippin, D.W.T.; *Chem. Eng. Sci.* 1994, *49*, 2129.
153. Ali, J.K.; Baiker, A.; *Appl. Catal. A* 1997, *155*, 41.
154. Oda, K.; Akamatsu, K.; Sugawara, T.; Kikuchi, R.; Segawa, A.; Nakao, S.; *Ind. Eng. Chem. Res.* 2010, *49*, 11287.
155. Kerleau, P.; Swesi, Y.; Meille, V.; Pitault, I.; Heurtaux, F.; *Catal. Today* 2010, *157*, 321.
156. Taube, M.; Taube, P.; *Adv. Hydrogen Energy* 1981, *2*, 1077.
157. Taube, M.; Rippin, D.W.T.; Cresswell, D.L.; Knecht, W.; *Int. J. Hydrogen Energy* 1983, *8*, 213.
158. Taube, M.; Rippin, D.W.T.; Knecht, W.; Hakimfard, D.; Milisavljevic, B.; Gruenenfelder, N.; *Int. J. Hydrogen Energy* 1985, *10*, 595.
159. Gruenenfelder, N.; Schucan, T.H.; *Int. J. Hydrogen Energy* 1989, *14*, 579.
160. Scherer, G.W.H.; Newson, E.; Wokaun, A.; *Int. J. Hydrogen Energy* 1999, *24*, 1157.
161. Weitkamp, A.W.; *Adv. Catal.* 1968, *18*, 1.
162. Rautanen, P.A.; Lylykangas, M.S.; Aittamaa, J.R.; Krause, A.O.I.; *Ind. Eng. Chem. Res.* 2002, *41*, 5966.
163. Newson, E.; Optimization of seasonal energy storage in stationary systems with liquid hydrogen carriers, decalin and methylcyclohexane; Hydrogen Energy Progress XII; Proceedings of the 12th World Hydrogen Energy Conference 1998, 2, 935.
164. Wang, B.; Goodman, D.W.; Froment, G.F.; *J. Catal.* 2008, *253*, 229.
165. Wang, Y.; Shah, N.; Huggins, F.E.; Huffman, G.P.; *Energy & Fuels* 2006, *20*, 2612.
166. Gupta, M.; Hagan, C.; Kaska, W.C.; Cramer, R.E.; Jensen, C.M.; *J. Am. Chem. Soc.* 1997, *119*, 840.
167. Loutfy, R.O.; Veksler, E.M.; Investigation of Hydrogen Storage in Liquid Organic Hydrides; HYFORUM 2000: The International Hydrogen Energy Forum 2000; EFO Energie Forum, Bonn, Germany, pp. 335–340.
168. Rahimpour, M.R.; Mirvakili, A.; Paymooni, K.; *Int. J. Hydrogen Energy* 2011, *36*, 6970.
169. Hodoshima, S.; Arai, H.; Saito, Y.; *Int. J. Hydrogen Energy* 2003, *28*, 197.
170. Hodoshima, S.; Arai, H.; Takaiwa, S.; Saito, Y.; *Int. J. Hydrogen Energy* 2003, *28*, 1255.

171. Shinohara, C.; Kawakami, S.; Moriga, T.; Hayashi, H.; Hodoshima, S.; Saito, Y.; Sugiyama, S.; *Appl. Catal. A* 2004, *266*, 251.
172. Hodoshima, S.; Saito, Y.J.; *Chem. Eng. Jpn.* 2004, *37*, 391.
173. Shono, A.; Hashimoto, T.; Hodoshima, S.; Satoh, K.; Saito, Y.; *J. Chem. Eng. Jpn.* 2006, *39*, 211.
174. Hodoshima, S.; Takaiwa, S.; Shono, A.; Satoh, K.; Saito, Y. *Appl. Catal. A* 2005, *283*, 235.
175. Saito, Y.; Aramaki, K.; Hodoshima, S.; Saito, M.; Shono, A.; Kuwano, J.; Otake, K.; *Chem. Eng. Sci.* 2008, *63*, 4935.
176. Hodoshima, S.; Shono, A.; Saito, Y.; *Energy & Fuels* 2008, 22, 2559.
177. Kariya, N.; Fukuoka, A.; Ichikawa, M.; *Appl. Catal. A* 2002, *233*, 91.
178. Kariya, N.; Fukuoka, A.; Utagawa, T.; Sakuramoto, M.; Goto, Y.; Ichikawa, M.; *Appl. Catal. A* 2003, *247*, 247.
179. Ichikawa, M.; Organic liquid carriers for hydrogen storage, in *Solid State Hydrogen Storage: Materials and Chemistry*, Woodhead, Cambridge, UK, 2008, p. 500.
180. Biniwale, R.B.; Ichikawa, M.; *Chem. Eng. Sci.* 2007, *62*, 7370.
181. (a) Sebastián, D.; Bordejé, E.G.; Calvillo, L.; Lázaro, M.J.; Moliner, R.; *Int. J. Hydrogen Energy* 2008, *33*, 1329. (b) Lázaro, M.P.; Garcia-Bordejé, E.; Sebastián, D.; Lázaro, M.J.; Moliner, R.; *Catal. Today* 2008, *138*, 203.
182. Meng, N.; Shinoda, S.; Saito, Y.; *Int. J. Hydrogen Energy* 1997, 22, 361.
183. Kobayashi, I.; Yamamoto, K.; Kameyama, H.; *Chem. Eng. Sci.* 1999, *54*, 1319.
184. Pez, G.P.; Scott, A.R.; Cooper, A.C.; Cheng, H.; U.S. Pat. Appl. Publ. 2004/0223907 A1, 2004.
185. Pez, G.P.; Scott, A.R.; Cooper, A.C.; Cheng, H.; U.S. Patent 7,101,530, 2006.
186. Pez, G.P.; Scott, A.R.; Cooper, A.C.; Cheng, H.; Wilhelm, F.C.; Abdourazak, A.H.; U.S. Patent 7,351,395, 2008.
187. Pez, G.P.; Scott, A.R.; Cooper, A.C.; Cheng, H.; U.S. Patent 7,429,372, 2008.
188. Sotoodeh, F.; Zhao, L.; Smith, K.; *J. Appl. Catal. A* 2009, *362*, 155.
189. Cheng, H.; Chen, L.; Cooper, A.C.; Sha, X.; Pez, G.P.; *Energy Environ. Sci.* 2008, *1*, 338.
190. Moores, A.; Poyatos, M.; Luo, Y.; Crabtree, R.H.; *New J. Chem.* 2006, *30*, 1675.
191. Clot, E.; Eisenstein, O.; Crabtree, R.H.; *Chem. Commun.* 2007, 2231.
192. Lu, R.-F.; Boëthius, G.; Wen, S.-H.; Su, Y.; Deng, W.-Q.; *Chem. Commun.* 2009, 1751.
193. Cui, Y.; Kwok, S.; Bucholtz, A.; Davis, B.; Whitney, R.A.; Jessop, P.G.; *New J. Chem.* 2008, *32*, 1027.
194. Perkin, W.H., Jr.; Plant, S.G.P.; *J. Chem. Soc. Trans.* 1924, *125*, 1503.
195. Adkins, H.; Coonradt, H.L.; *J. Am. Chem. Soc.* 1941, *63*, 1564.
196. Nishimura, S.; *Handbook of Heterogeneous Catalytic Hydrogenation and Organic Synthesis*, Wiley, New York, 2001, pp. 501–502.
197. Sotoodeh, F.; Smith, K.J.; *Ind. Eng. Chem. Res.* 2010, *49*, 1018.
198. Eblagon, K.M.; Tam, K.; Yu, K.M.K.; Zhao, S.-L.; Gong, X.-Q.; He, H.; Ye, L.; Wang, L.-C.; Ramirez-Cuesta, A.J.; Tsang, S.C.; *J. Phys. Chem. C* 2010, *114*, 9720.
199. Eblagon, K.M.; Rentsch, D.; Friedrichs, O.; Remhof, A.; Zuettel, A.; Ramirez-Cuesta, A.J.; Tsang, S.C.; *Int. J. Hydrogen Energy* 2010, *35*, 11609.
200. Ye, X.; An, Y.; Xu, G.; *J. Alloys Compd.* 2011, *509*, 152.
201. Adkins, H.; Lundsted, L.G.; *J. Am. Chem. Soc.* 1949, *71*, 2964.
202. Sotoodeh, F.; Smith, K.J.; *J. Catal.* 2011, *279*, 36.
203. Sotoodeh, F.; Huber, B.J.M.; Smith, K.J.; *Int. J. Hydrogen Energy* 2012, *37*, 2715.
204. Sobota, M.; Nikiforidis, I.; Amende, M.; Zanón, B.S.; Staudt, T.; Höfert, O.; Lykhach, Y.; Papp, C.; Hieringer, W.; Laurin, M.; Assenbaum, D.; Wasserscheid, P.; Steinrück, H.P.; Görling, A.; Libuda, J.; *Chem. Eur. J.* 2011, *17*, 11542.
205. Ye, X.-F.; An, Y.; Xu, Y.-Y.; Kong, W.-J. Xu, G.-H.; *Huaxue Gongcheng/Chemical Engineering (China)* 2011, *39*, 29.
206. Wang, Z.; Tonks, I.; Belli, J.; Jensen, C.M.; *J. Organomet. Chem.* 2009, *694*, 2854.
207. Pez, G.P.; Toseland, B.; Reversible liquid carriers for an integrated production, storage, and delivery of hydrogen. DOE Hydrogen Program annual review, May 2005. http://www.hydrogen.energy.gov/pdfs/review05/pd34_pez.pdf (accessed February 25, 2012).
208. Danckwerts, P.V.; *Gas-liquid Reactions*, McGraw-Hill, New York, 1970.

209. Pez, G.P.; Toseland, B., Reversible liquid carriers for an integrated production, storage, and delivery of hydrogen. DOE Hydrogen program FY 2005 progress report. http://www.hydrogen.energy.gov/pdfs/progress05/v_b_1_pez.pdf (accessed February 25, 2012).
210. Toseland, B.; Pez, G.P.; Cooper, A.; Scott, A.; Fowler, D. (combined authors); Reversible liquid carriers for an integrated production, storage, and delivery of hydrogen. DOE Hydrogen Program FY 2007 to FY 2010 annual progress reports. http://www.hydrogen.energy.gov.
211. Toseland, B.A.; Pez, G.P.; Puri, P.S.; U.S. Patent 7,485,161, 2009.
212. Cooper, A.; Pez, G.; Abdourazak, A.; Scott, A.; Fowler, D.; Wilhelm, F.; Monk, V.; Cheng, H.; Design and development of new carbon-based sorbent systems for an effective containment of hydrogen. DOE Hydrogen Program FY 2006 annual progress report. http://www.hydrogen.energy.gov/pdfs/progress06/iv_b_3_cooper.pdf (accessed February 25, 2012).
213. (a) Cooper, A.C.; Campbell, K.M.; Pez, G.P.; An Integrated Hydrogen Storage and Delivery Approach Using Organic Liquid-Phase Carriers; Proceedings of the 2006 World Hydrogen Energy Conference, June 13–16, 2006, Lyon, France. (b) Cooper, A.; Abdourazak, A.; Cheng, H.; Fowler, D.; Pez, G.; Scott, A.; Design and development of new carbon-based sorbent systems for an effective containment of hydrogen. DOE Hydrogen Program FY 2005 annual progress report. http://www.hydrogen.energy.gov/pdfs/progress05/vi_b_3_cooper.pdf (accessed April 28, 2012).
214. Wang, Z.; Belli, J.; Jensen, C.M.; *Faraday Discuss.* 2011, *151*, 297.
215. Matus, M.H.; Liu, S.-Y.; Dixon, D.A.; *J.Phys. Chem.* A 2010, *114*, 2644.
216. Campbell, P.G.; Zakharov, L.N.; Grant, D.J.; Dixon, D.A.; Liu, S.-Y.; *J. Am. Chem. Soc.* 2010, *132*, 3289.
217. Campbell, P.G.; Abbey, E.R.; Neiner, D.; Grant, D.J.; Dixon, D.A.; Liu, S.-Y.; *J. Am. Chem. Soc.* 2010, *132*, 18048.
218. Gelsey, J.; U.S. Patent 7,108,933 (2006).
219. Wechsler, D.; Cui, Y.; Dean, D.; Davis, B.; Jessop, P.G.; *J. Am. Chem. Soc.* 2008, *130*, 17195.
220. Dean, D.; Davis, B.; Jessop, P.G.; *New J. Chem.* 2011, *35*, 417.
221. Thorn, D.L.; Tumas, W.; Ott, K.C.; Burrell, A.K.; U.S. Pat. Appl. Publ. 2007/0183967 A1, 2007.
222. Wechsler, D.; Davis, B.; Jessop, P.G.; *Can. J. Chem.* 2010, *88*, 548.
223. Pez, G.P.; Cooper, A.C.; Scott, A.R.; U.S. Patent 8,003,073 (2011).
224. Cooper, A.; Scott, A.; Fowler, D.; Cunningham, J.; Ford, M.; Wilhelm, F.; Monk, V.; Cheng, H.; Pez, G.; Hydrogen storage by reversible hydrogenation of liquid-phase hydrogen carriers. DOE Hydrogen Program FY 2008 annual progress report. http://www.hydrogen.energy.gov/pdfs/progress08/iv_b_2_cooper.pdf (accessed February 25, 2012).
225. Sotoodeh, F.; Huber, B.J.M.; Smith, K.J.; *Appl. Cat A: Gen.* 2012, *419*, 67.
226. Sung, J.S.; Choo, K.Y.; Kim, T.H.; Tarasov, A.L.; Tkachenko, O.P.; Kustov, L.M.; *Int. J. Hydrogen Energy* 2008, *33*, 2721.
227. Stracke, M.P.; Ebeling, G.; Cataluña, R.; Dupont, J.; *Energy & Fuels* 2007, *21*, 1695.
228. Ratner, B.D.; Naeemi, E.; U.S. Pat. Appl. Publ. 2007/0003476 A1, 2007.
229. Zhao, H.Y.; Oyama, S.T.; Naeemi, E.D.; *Catal. Today* 2010, *149*, 172.
230. Enthaler, S.; von Langermann, J.; Schmidt, T.; *Energy Environ. Sci.* 2010, *3*, 1207.
231. Zaidman, B.; Wiener, H.; Sasson, Y.; *Int. J. Hydrogen Energy* 1986, *11*, 341.
232. Stephan, D.W.; *Chem.Commun.* 2010, *46*, 8526.
233. Geier, S.J.; Stephan, D.W.; *J. Am. Chem. Soc.* 2009, *131*, 3476.
234. Schwarz, D.E.; Cameron, T.M.; Hay, P.J.; Scott, B.L.; Tumas, W.; Thorn, D.L.; *Chem. Commun.* 2005, 5919.
235. Thorn, D.L.; Tumas, W.; Hay, J.P.; Schwarz, D.E.; Cameron, T.M.; PCT Int. Appl. 2006009630 A2 (2006).
236. Kariya, N.; Fukuoka, A.; Ichikawa, M.; *Chem. Commun.* 2003, 690.
237. Kariya, N.; Fukuoka, A.; Ichikawa, M.; *Phys. Chem. Chem. Phys.* 2006, *8*, 1724.
238. Gasteiger, H.A.; Kocha, S.S.; Sompalli, B.; Wagner, F.T.; *Appl. Catal.* B 2005, *56*, 9.
239. Ferrell, J.R., III; Sachdeva, S.; Strobel, T.A.; Gopalakrishnan, G.; Koh, C.A.; Pez, G.; Cooper, A.C.; Herring, A.M.; *J. Electrochem. Soc.* 2012, *159*, B371.
240. Soloveichik, G.L.; Zhao, J.-C.; U.S. Pat. Appl. US2008/0248345, 2008.
241. Itoh, N.; Xu, W.C.; Hara, S.; Sakaki, K.; *Catal. Today* 2000, *56*, 307.

242. Crabtree, R.H.; *Energy Environ. Sci.* 2008, *1*, 134.
243. Luca, O.R.; Wang, T.; Konezny, S.J.; Batista, V.S.; Crabtree, R.H.; *New J. Chem.* 2011, *35*, 998.
244. U.S. Department of Energy, Energy Frontier Research Centers. http://www.science.energy. gov/bes/efrc (accessed February 25, 2012).
245. General Electric Global Research. http://ge.geglobalresearch.com/ (accessed February 25, 2012).
246. Teichmann, D.; Arlt, W.; Wasserscheid, P.; Freymann, R.; *Energy Environ. Sci.* 2011, *4*, 2767.
247. Bagzis, L.R.; Appleby, J.B.; Pez, G.P.; Cooper, A.C.; U.S. Pat. Appl. 2005/0013767 A1, 2005.
248. Biniwale, B.B.; Rayalu, S.; Devotta, S.; Ichikawa, M.; *Int. J. Hydrogen Energy* 2008, *33*, 360.
249. Pradhan, A.U.; Shukla, A.; Pande, J.V.; Karmarkar, S.; Biniwale, R.B.; *Int. J. Hydrogen Energy* 2011, *36*, 680.
250. Shukla, A.; Karmakar, S.; Biniwale, R.B.; *Int. J. Hydrogen Energy* 2012, *37*, 3719.
251. Ahluwalia, R.K.; Hua, T.Q.; Peng, J.K.; *Int. J. Hydrogen Energy* 2012, *37*, 2891.
252. Ahluwalia, R.K.; Hua, T.Q.; Peng, J.; Kumar, R.; System level analysis of hydrogen storage options. DOE Hydrogen Program FY 2007 annual progress report. http://www.hydrogen. energy.gov/pdfs/progress07/iv_f_1_ahluwalia.pdf (accessed March 2, 2012).

Section III

Engineered Hydrogen Storage Systems

Materials, Methods, and Codes and Standards

9

Engineering Properties of Hydrogen Storage Materials

Daniel Dedrick

CONTENTS

Introduction 331
Packing Density..332
Thermal Properties 333
Enhancement Strategies 338
    Gas Pressure..338
    Increased Particle Diameter..339
    Densification 339
    Gas Flow 339
    High-Conductivity Coatings 340
    High-Thermal-Conductivity Composite Alloying..340
    High-Thermal-Conductivity Structures..341
    High-Thermal-Conductivity Additives 341
Flow Properties...343
Future Outlook 344
References..345

Introduction

Previous chapters have reviewed the properties of materials that directly affect their abilities to release or absorb hydrogen reversibly. This chapter examines the properties of materials that most impact their practical implementation in real hydrogen storage systems. These "engineering properties" are defined as the characteristics of heat transfer, mass transfer, and mechanical stress within a physical arrangement of storage material particles consisting of amorphous structures, single crystals, or aggregates. This arrangement of particles is called a *bed*. Engineering properties considered in this chapter include packing density, thermal conductivity, expansion, and permeability.

In general, hydrogen storage materials are high-surface-area materials consisting typically of transition metals, alkali metals, and other light elements such as B, N, or C. Characteristic particle sizes tend to be small. Beds may sinter or decrepitate with thermal or hydrogen cycling. Decrepitation is the systematic disintegration of bulk particles into smaller particles or "fines." The morphology of the bed may vary significantly as a function of hydrogen content. Also, for reversible materials, significant changes can occur during the life cycle of the bed, leading to order-of-magnitude changes in engineering properties.

The engineering properties, coupled with the chemical kinetics, define the dimension and engineering characteristics of the storage system. In general, packing density controls the overall volume of the system and is arguably the most influential parameter. Thermal conductivity controls the rate at which heat can be added or removed from the bed and helps define the characteristic thermal path in the bed. Permeability controls the rate at which hydrogen can be added or removed from the bed. Mechanical expansion forces during sorption of hydrogen control packing density and vessel design (e.g., strength).

In this chapter, packing density is discussed first as it has an impact on all subsequent engineering properties. This is followed with a discussion of thermal transport and enhancement as these are the most impactful engineering properties on the bed design. Finally, the flow properties and permeability of the characteristic bed are described. The focus is on metal hydride materials, although the concepts apply to the engineering properties of many hydrogen storage materials.

Packing Density

The packing density of particles and powders has been studied and characterized by industries such as food processing, chemical packaging, oil and gas, and geotechnology [1, 2]. The packing density of a bed depends on a large number of particle characteristics, including size distribution, shape, plasticity, and interparticle friction. A convention for the engineering packing density is referred to as the *tap density*, which is simply defined as the bulk density resulting from manually compacting the bed with mechanical shock. The process of developing tap density relies on developing the optimal arrangement of close-packed particles in the absence of physical pressure being directly applied to the particles or bed. Although tap densities can vary tremendously, 50% of single-crystal density is common.

Exceeding tap density typically requires the process of continuous mechanical vibration, melting, or hydraulic/mechanical bed packing. Mechanical packing is the most straightforward of packing processes and typically is performed mechanically or hydraulically. Other material-specific methods have been considered, including melting the materials to form a bed (heating the materials to a point that a phase transition occurs from solid to liquid). Care must be taken when performing this process as phase segregation or insufficient permeability can result.

Packing density has a significant impact on each of the other engineering properties. Increased packing density has a positive impact on thermal properties (as described in the following section) by improving solid fraction and decreasing interfacial resistance. The resulting thermal properties can then result in less system complexity. Conversely, high packing density decreases permeability by both increasing the tortuousity and restricting the flow area through the bed. This can result in challenges to adding or removing hydrogen. For some materials, such as the AB_2 or AB_5 metal hydrides discussed in Chapter 5, the increased packing density can result in excess mechanical pressures due to swelling and contraction of the matrix during hydrogenation and dehydrogenation, respectively. The presence of this excess pressure can require overdesign to accommodate the added stress to avoid safety impacts such as vessel structural failure.

Thermal Properties

The thermal properties of hydrogen storage beds directly influence the efficiency, performance, and cost of hydrogen storage systems. Thermal design is perhaps the most challenging aspect of hydride-based automotive hydrogen storage system design due to the thermodynamic characteristics of hydrogen storage materials that can operate in the pressure and temperature regime of conventional proton exchange membrane (PEM) fuel cells. Enthalpies of hydrogenation in the best cases are about -30 to -40 kJ/mol H_2, requiring approximately 200 MJ of heat transfer for each hydrogen fill-and-delivery cycle when considering an automotive-scale system. Hydrogen uptake (refueling) is the most thermally challenging operational state as the reaction is exothermic, and on-board automotive refueling scenarios require rapid refueling rates. During rapid hydrogen uptake of an automotive-scale system, nearly 0.5 MW of cooling can be required to maintain a constant system temperature.

These intensive heat flux conditions demand a detailed understanding and optimization of the thermal transfer properties present within a hydrogen storage system. High thermal conductivity of the metal hydride and low thermal resistance at the wall mitigate the effects of thermally limited reactions, enabling the use of larger diameters and a reduction in the number of parts required for efficient operation. Since thermal resistance at the vessel wall can generally be accommodated during absorption with the temperature and flow characteristics of the coolant fluid, the effective thermal conductivity of the storage material is of specific interest as this can limit the rate of the absorption reaction most significantly.

As discussed, hydrogen storage beds generally take the form of packed particles immersed in hydrogen gas. There are three different regimes of effective thermal conductivity within a packed particle bed. The low-pressure regime (typically subatmospheric) is characterized by molecular transport where gas pressure does not influence the thermal conductivity of the bed. The intermediate-pressure regime is characterized by transition between molecular and continuum transport within the void spaces where thermal conductivity is significantly affected by the gas pressure. The high-pressure regime is characterized predominantly by continuum transport within the void spaces, and thus pressure no longer affects the thermal conductivity of the bed. The inflection point between transition and continuum transport is called the *critical pressure*. The boundaries of each of these regimes depend on the characteristic void space pore size within the bed. The intermediate- and high-pressure regimes are of specific interest for typical hydride-based hydrogen storage systems.

There are a total of six modes of heat transfer within the bulk metal hydride packed particle structure as described by Oi [3] in Figure 9.1. The six modes of heat transfer are:

1. Heat conduction at the particle contacts
2. Heat conduction through a thin "film" of hydrogen
3. Heat radiation between the particles
4. Heat conduction through the particles
5. Heat conduction through hydrogen in larger void spaces
6. Heat radiation between vacant spaces

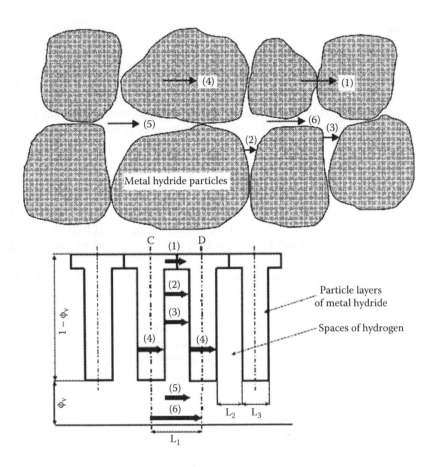

FIGURE 9.1
Schematic and resistance model of six heat transfer modes within a hydrogen storage bed. (Reproduced from T. Oi, K. Maki, and Y. Sasaki, *Journal of Power Sources* **125** (2004) 52–61 with permission of Elsevier.)

Models have been constructed describing the complete heat transfer mechanisms within the bed. Yagi and Kunii [4] developed generalized resistance models for packed beds, which others adapted for application to metal hydride beds [3, 5–7]. For lower- and moderate-temperature applications of these models, radiation heat transfer can be neglected [3, 5, 6, 8, 9]. In general, the resistance model of effective thermal conductivity of a packed metal hydride bed can be described as

$$K_{th} = K_{H2} \left[\varphi_v + \frac{1 - \varphi_v}{\gamma + \frac{2}{3}\left(\frac{K_{H2}}{K_S}\right)} \right] \tag{9.1}$$

where K_{H2} is the thermal conductivity of the hydrogen gas, K_S is the thermal conductivity of the metal hydride solid, φ_v is the porosity of the bed. The constant γ is defined as the ratio of the effective length of solid relating to conduction and the average hydride particle diameter. This ratio is difficult to quantify and is a function of particle contact angle, shape, and roughness. For metal hydrides, typical values fall between 0.01 and 0.1 (see Figure 5 of Reference 10). Using this resistance model, Oi [3] predicted the thermal conductivity of a

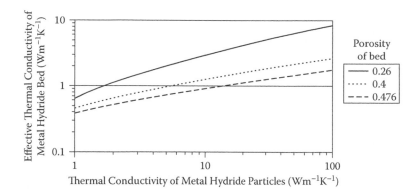

FIGURE 9.2
Resistance model results for effective thermal conductivity of a metal hydride bed as a function of particle thermal conductivity and void fraction. (Reproduced from T. Oi, K. Maki, and Y. Sasaki, *Journal of Power Sources* **125** (2004) 52–61 with permission of Elsevier.)

typical metal hydride bed as a function of metal hydride particle thermal conductivity for three different porosities, as shown in Figure 9.2.

Resistance models are useful for making quick estimations of packed bed effective thermal conductivities, yet are subjective and lack accuracy. Other types of models have been developed, including numeric methods [11]. The most useful and apparently accurate model was developed by Zehner, Bauer, and Schlünder and reported in Tsotsas [12] and adapted by Sanchez [13]. It consists of a unit cell of two particle halves of equivalent shape encased in a cylinder of fluid (hydrogen). The model presented here neglects radiation and assumes spherical particles. The effective thermal conductivity of the bed K_{th} is calculated by

$$K_{th} = K_g \left\{ (1 - \sqrt{1 - \Psi}) \cdot \Psi \cdot \frac{1}{\Psi - 1 + \dfrac{1}{K_g}} + \sqrt{1 - \Psi} \cdot \left[\phi \cdot K_p + K_{SO} \cdot (1 - \phi) \right] \right\} \tag{9.2}$$

In this expression, K_g is the thermal conductivity of hydrogen, Ψ is the porosity, ϕ is a flattening coefficient that defines contact quality, K_p is the particle thermal conductivity, and K_{SO} is an expression for the thermal conductivity at the particle-gas-particle interface. The effective thermal conductivity is highly influenced by K_{SO} as it describes the contribution of the fluid to the particle thermal contact quality. For a complete description of the model, refer to the work of Sanchez [13].

Using the Sanchez model, effective thermal conductivities of sodium alanate (a prototypical complex metal hydride) were calculated as a function of hydrogen pressure, particle thermal conductivity, particle diameter, and void fraction. The results are shown in Figure 9.3.

Ranges selected for these variables were chosen based on typical metal hydride packed bed characteristics. Other model inputs include the deformation factor and contact-flattening coefficient of the particle—indicating the area of contact between the particles. These factors are difficult to calculate or predict and are usually estimated from experimental results. Overall, the effective conductivity of the bed is less sensitive to the

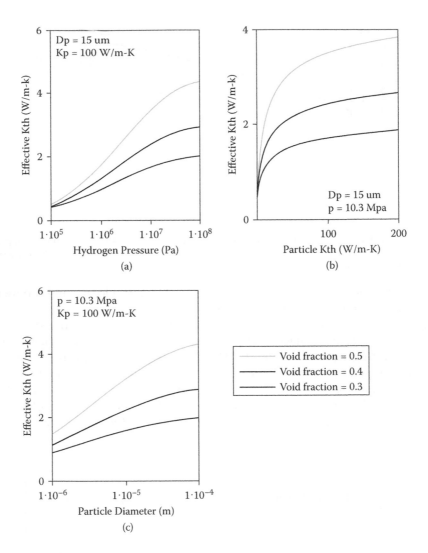

FIGURE 9.3 (See color insert.)
Calculated effective thermal conductivity of a packed particle bed as a function of void fraction, hydrogen pressure, and particle characteristics.

deformation factor and contact-flattening coefficient compared to the other variables. For these calculations, typical values of these coefficients were used as in the work of Sanchez [13].

Effective thermal conductivity as a function of hydrogen pressure and void fraction is described in Figure 9.3(a). In general, higher pressures and lower void fractions lead to higher conductivities up to the critical pressure. The critical pressure for these calculated geometries is near 1×10^8 Pa (1000 bar), as observed by a reduction in rate of effective thermal conductivity increase with pressure. The effective thermal conductivity as a function of particle thermal conductivity is shown in Figure 9.3(b). In general, particle thermal conductivities above 50 W/m-K do not significantly influence bed effective thermal conductivity due to dominance in this regime by porosity and the particle-to-particle thermal

FIGURE 9.4
(Left) The sintered solid resulting from hydrogen cycling of sodium alanates. (Right) A scanning electron micrographic (SEM) image of the sintered solid.

contact. The effective thermal conductivity as a function of particle diameter is presented in Figure 9.3(c). Larger particle diameters lead to higher effective conductivities due to the reduction of particle-to-particle contacts and an increase in continuum regime gas transport relative to transition regime gas transport.

The calculations presented here are consistent with many models and measurements described in the literature [5–8, 11, 14, 15]. Models and measurements indicate that the effective thermal conductivity of particles loaded in a packed bed is generally limited to values below about 5 W/m-K, even with significant increases in the particle thermal conductivity, as shown in Figure 9.3(b). More clever methods must be employed to enhance thermal conductivity to levels above 5 W/m-K. In addition, the models discussed have been developed for distinct particles typical of classic/interstitial hydride materials. These beds are generally characterized as unsintered powders, while complex hydrides such as sodium alanate ($NaAlH_4$) can become porous sintered solids as seen in Figure 9.4. Sintered solids are formed when a powder is heated (but not melted) and made to coalesce into an agglomerated porous mass. Packed particle models have not been directly applied to sintered solid materials.

Measurements at Sandia National Laboratories, California, of sodium aluminum hydride sintered solids have demonstrated that the effective conductivities are similar to other packed particle beds [16]. The thermal conductivities of stoichiometric sodium alanates compacted at 40% of the single-crystal density were found to vary between 0.5 and 1.0 W/m-K depending on cycle, hydrogen content, and gas pressure. Effective thermal conductivities as a function of gas pressure for fully cycled stoichiometric sodium alanates are shown in Figure 9.5.

As with packed particle beds, low thermal conductivities of sodium alanates present an engineering challenge when integrated within a system. Although the physical appearance of a sintered sodium alanate is dissimilar to a bed of close-packed spheres, the thermal transport behavior of both cases are similar. Both cases contain a characteristic thermal path length that influences the thermal transport within the bed when compared to the mean free path of the gas. In addition, the sintered sodium alanate bed can be modeled as packed spheres (or islands) of material with improved thermal contact between each sphere or "island." Given the similarities in fundamental mechanisms of heat transport of

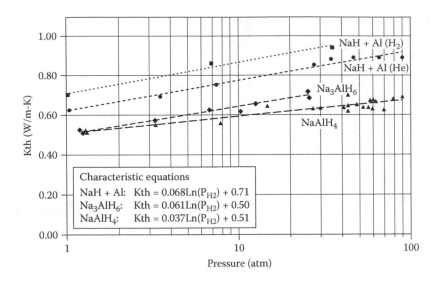

FIGURE 9.5
Effective thermal conductivity as a function of gas pressure for fully cycled stoichiometric sodium alanates. (Reproduced from D.E. Dedrick, M.P. Kanouff, W.C. Replogle, and K.J. Gross, *Journal of Alloys and Compounds* **389** (2005) 299–305 with permission of Elsevier B.V.)

each case, we can assume that many of the same thermal conductivity enhancement strategies will work for both a packed particle and a sintered solid bed.

As shown, additional means may need to be undertaken to attain higher effective thermal conductivities. A variety of approaches are explored next.

Enhancement Strategies

Several authors have described the importance of thermal conductivity enhancement to provide improved performance of metal-hydride-based hydrogen storage systems [17–19]. Lower thermal resistances enable the use of larger thermal length scales and increased rate of hydrogen uptake. Many methods have been proposed and some tested for enhancing the thermal conductivity of hydrogen storage materials. Each of the avenues are explored separately and their application to sodium-alanate-based systems is established in the next sections.

Gas Pressure

Increasing gas pressure of the surrounding hydrogen up to the critical pressure improves the overall effective conductivity of the bed as shown in the previous calculations (Figure 9.3(a)). The critical pressure is defined as the point at which the mean free path of the gas is significantly less than a critical thermal path length within the bed. Further application of gas pressure to thermal property enhancement to sodium alanates is limited as the gas pressures required for absorption are near to or exceed the *critical pressure* of the bed (note the small increase in effective thermal conductivity of sodium alanates between 50 and 100 atm as shown in Figure 9.5).

Increased Particle Diameter

Larger particles or grains will significantly enhance thermal conductivity by reducing the number of particle-to-particle contacts and an increase in continuum-regime gas transport relative to transition regime gas transport [13, 20]. Application of this enhancement mode to complex hydrides is not straightforward as particle size and morphology vary significantly as a function of hydrogen content, and larger grains lead to a reduction of reaction surface area. If the grain size could be controlled with some undefined method, the performance improvement due to thermal conductivity enhancement would need to be balanced with the required reaction surface area.

Densification

Densification and reduced porosity improve particle contact and volumetric solid fraction, thus enhancing heat transport within the bed. Upper limits of densification are governed by requirements for gas diffusion within the bed. Experience has proven that further densification may lead to gas-diffusion-limited reactions of metal hydride systems.

Gas Flow

Flow through the porous bed enhances the radial effective or apparent thermal conductivity of packed beds [4, 20]. Winterberg and Tsotsas [21] developed models and heat transfer coefficients for packed spherical particle reactors that are invariant with the bed-to-particle-diameter ratio. The radial effective thermal conductivity is defined as the summation of the thermal transport of the packed bed K_{bed} and the thermal dispersion caused by fluid flow K_{flow}, or

$$K_{bed+flow} = K_{bed} + K_{flow} = K_{bed} + X_1 \cdot Pe_0 \cdot \frac{u_c}{u_{ave}} \cdot f(r) \cdot K_{gas} \qquad (9.3)$$

The coefficient X_1 is a correlation function that describes the rate of increase of the effective thermal conductivity with flow velocity, Pe_0 is the Péclet number that describes the contribution of forced convection relative to hydrogen heat conduction, u_c is the velocity at the centerline of the bed, u_{ave} is the average velocity, $f(r)$ describes the radial variation in dispersion, and K_{gas} is the thermal conductivity of the fluid (hydrogen). For our purposes, u_c and u_{ave} are assumed to be near equivalent due to the large vessel-to-particle-diameter ratio. The radial dispersion variation $f(r)$ is assumed to be unity for similar reasons. For a complete description of the model and correlations, refer to the work of Winterberg and Tsotsas [21]. Although the model is promoted to be invariable to the ratio of the vessel diameter to the particle diameter, the correlations were built using experimental data with ratios from 5.5 to 65. In the case of the sodium alanate bed, this ratio may be as high as 1000 or more.

Using the model described, added effective thermal conductivity was calculated for a 2-cm-diameter sodium alanate bed. Properties for hydrogen at 100°C and 120 atm were utilized, and the velocity along the centerline was assumed to be equivalent to that of the average. Significant enhancements in thermal conductivity may be experienced at modest flow velocities of 20–30 m/s (this corresponds to Reynolds numbers Re_0 of ~1000 and a volumetric flow rate of ~2 L/s). Figure 9.6 describes the relationship of increased effective conductivity to hydrogen flow.

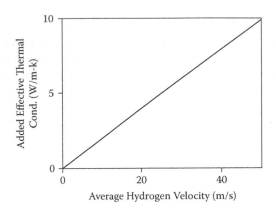

FIGURE 9.6
The increase in the effective thermal conductivity of a sodium alanate bed as a function of average flow.

Additional effort will be required to (a) determine the application of this model to vessel-to-particle-diameters ratios above 65 and (b) determine the flow resistance properties of a specific metal hydride bed. Estimations of the pressure drop per unit length associated with gas flow through a packed bed of spheres can be calculated by

$$\frac{P}{L} = 150 \cdot \frac{\mu \cdot v_0}{D_P^2} \cdot \frac{(1-\varepsilon)^2}{\varepsilon^3}$$

where μ is the viscosity of the gas, v_0 is the average velocity of the gas, D_P is the particle diameter, and ε is the porosity [22]. The pressure drop within a sodium alanate bed with Reynolds numbers of about 1000 could be very high due to the dependence on the inverse of diameter squared (>1000 psi). If present, added pressure requirements of this magnitude could be detrimental to the energy densities of hydrogen storage systems.

High-Conductivity Coatings

Researchers have experimented with enhancing heat transport by coating metal hydride pellets with high-conductivity, high-ductility metals. Copper has been frequently used for this purpose [23–26]. Kim et al. [24] reported conductivities as high as 9 W/m-K for higher packing densities of 25% by weight for copper-coated LaNi$_5$ powders (thermal conductivity of uncoated powders typically less than 1 W/m-K). This corresponds well with results published by Kurosaki and coworkers [23], seen in Figure 9.7, which describes effective thermal conductivity of LaNi$_5$ as a function of copper mass percentage.

This approach may not be the most favorable method of enhancing the thermal properties of sodium alanate or other high-energy-density metal hydride systems, as significant copper mass is required to attain moderate increases in thermal conductivity. In addition, gas and solid mass transport may be adversely affected when this technology is applied.

High-Thermal-Conductivity Composite Alloying

Various high-conductivity materials have been alloyed with metal hydrides to form enhanced heat transport composite materials. Eaton et al. [27] experimented with various

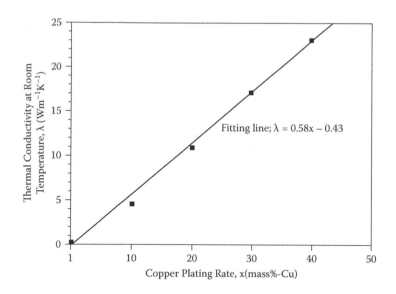

FIGURE 9.7

Thermal conductivity of LaNi$_5$ powders at approximately 60% packing density as a function of copper mass percentage. (Reproduced from K. Kurosaki, T. Maruyama, K. Takahashi, H. Muta, D. Uno, and S. Yamanaka, *Sensors and Actuators* **A 113** (2004) 118–123 with permission of Elsevier S.A.)

alloyed metal additives, including copper, aluminum, lead, and lead-tin. The samples documented in this reference were alloyed at elevated temperature (200–600°C) and cycled. In many samples, cycling resulted in the separation and fracture of the alloy and thus a reduction in composite thermal conductivity. Sintered aluminum structures of 20% solid fraction have been integrated with LaNi$_5$ hydride materials with success, resulting in effective thermal conductivities of 10 to 33 W/m-K [28–30]. Temperatures required for this process and added mass and volume may exclude application to some materials such as complex metal hydrides.

High-Thermal-Conductivity Structures

High-thermal-conductivity structures have been used to enhance thermal conductivity, including copper wire matrices [31], periodic plates [19], nickel foams [26], and aluminum foams [32]. Nagel et al. [31] designed and integrated a 90% porous corrugated copper wire matrix with a MmMi$_{4.46}$Al$_{0.54}$ hydride bed and improved the overall conductivity by a modest 15%. Aluminum and nickel foams have been used with some success [26]. Practically, metal foams tend to be costly and are a significant challenge to load with metal hydrides in ways that result in low void fractions. Thermal analysis of a sodium alanate bed with periodic structures indicated that thermal conductivities near 20 W/m-K may be possible as shown in Figure 9.8.

High-Thermal-Conductivity Additives

Work at Sandia has demonstrated the use of excess aluminum powders to increase the average particle thermal conductivity and enhance the effective conductivity of the bed. Thermal conductivity was enhanced threefold (from 0.5 to 1.5 W/m-K) by adding 12% by mass of 20-μm aluminum to stoichiometric sodium alanates and compacting the bed to

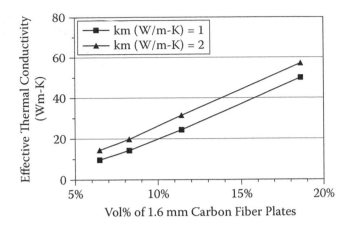

FIGURE 9.8
The relationship between the effective conductivity and the volume fraction of 1.6-mm carbon fiber plates installed in a 10-cm-diameter metal hydride bed for two metal hydride effective conductivities.

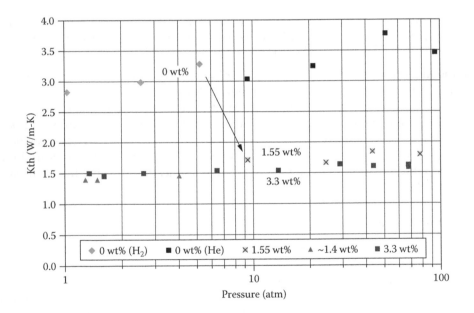

FIGURE 9.9 (See color insert.)
Effective thermal conductivity of a fully cycled NaAlH$_4$ bed, with 12% by mass added Al, as a function of gas pressure and H$_2$ capacity.

approximately 60% of the single-crystal density. Measurements were made using the thermal probe method as described by Dedrick [16]. Effective thermal conductivity as a function of gas pressure is shown in Figure 9.9. The improvement in bed thermal conductivity can be seen by comparing Figure 9.9 (with Al additive) to Figure 9.5 (without Al additive).

Carbon fibers have also been implemented by Sandia to enhance the thermal conductivity of sodium alanate beds. A total of 9% by mass of woven carbon fibers was found to enhance effective thermal conductivity by at least an order of magnitude. Results for

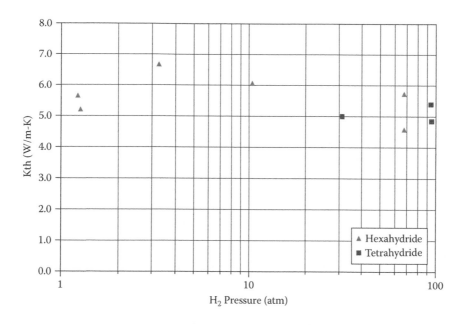

FIGURE 9.10
Thermal conductivity of NaAlH$_4$ bed enhanced by an order of magnitude via the use of 9 wt% carbon fiber.

effective conductivity measurements are shown in Figure 9.10. The conductivities experienced exceed the designed measurement range of the developed thermal probe method, so it should be noted that these measurements are approximate and should be considered minimums of the conductivity enhancement. High-conductivity additives are considered one of the most promising methods for enhancing thermal conductivities.

Expanded natural graphite fibers have been described as appropriate additives due to their characteristic high thermal conductivity, porosity, dispersibility, and low cost [13, 33, 34]. Expanded natural graphite (ENG) fibers are produced from natural graphite that is soaked in sulfuric acid and heated to high temperatures, thereby expanding to very fine flakes. Thermal conductivities of approximately 20 W/m-K are attainable with volumetric fractions as low as 10%. Experimentally, thermal conductivities as high as 10 W/m-K were attained with mass fractions as low as 5%, although significant scatter is present within the measured data. Predictions of thermal conductivity enhancement resulting from the combination of sodium alanates with ENG fibers using volumetric weighting as in the work of Sanchez [13] is shown in Figure 9.11.

Observed thermal conductivities of sodium aluminum compacts combined with ENG fibers are lower than predicted. This could be due to either poor alignment of the ENG fibers during the compaction or poor thermal contact between the metal hydride and ENG flakes. The compaction process could potentially be optimized to better align the ENG fibers.

Flow Properties

The permeability of a packed bed is well understood through many experimental and analytical studies in the 20th century. Generally, the metal hydride bed consists of

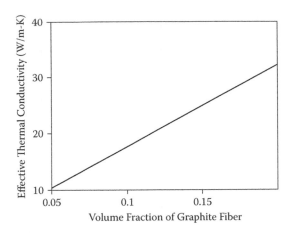

FIGURE 9.11
Prediction of effective thermal conductivity of sodium alanates and ENG fiber compacts with 5% voids as a function of volume fraction of ENG fiber.

high-surface-area close-packed particles. Models used to described flow through a packed bed generally need to accommodate Knudsen effects due to the small characteristic particle size. Experimental studies by Voskuilen et al. [35] of permeability characteristics of sodium aluminum hydride beds indicated that a variety of conventional permeability models can be utilized for predicting flow through a metal hydride bed. In addition, the researchers found that flow along walls can play a significant part in the effective permeability of a metal hydride bed. Estimations for the permeability enhancement due to wall effects indicated that the effective permeability at the wall could be as much as 15 times greater compared to the bulk properties.

The flow properties of a metal hydride bed can be enhanced through the integration of flow channels in the bed. Voskuilen et al. [35] analyzed a variety of configurations to determine the optimal approach to flow enhancement.

Future Outlook

In general, the engineering properties of metal hydride beds follow the behavior of generic packed beds. This will likely be true for new materials that are developed that have similar physical characteristics, such as high surface area and small particle sizes. As new reversible materials are developed with lower enthalpies of hydrogen desorption and rehydrogenation, thermal properties may become less important due to reduced thermal duty. Conversely, as materials with higher hydrogen capacities are developed, flow properties become more important to enable high fluxes in and out of the metal matrix.

For many applications, life-cycle properties such as capacity durability will become important to understand and manage. Life-cycle durability may be impacted by contamination or morphology changes as experienced during the cyclic loading of the matrix. These properties and characteristics vary entirely depending on the specifics of the material chemistry and physical characteristics of the developed materials.

References

1. Y. Wanibe and T. Itoh, *New Quantitative Approach to Powder Technology*, Wiley, New York, March 1999.
2. D. McGlinchey, *Bulk Solids Handling*, Wiley, New York, April 2008.
3. T. Oi, K. Maki, and Y. Sasaki, *Journal of Power Sources* **125** (2004) 52–61.
4. S. Yagi and S. Kunii, *AIChE Journal* **3**, No. 3 (1957) 373–381.
5. D. Sun, *International Journal of Hydrogen Energy* **15**, No. 5 (1990) 331–336.
6. Y. Ishido, M. Kawamura, and S. Ono, *International Journal of Hydrogen Energy* **7**, No. 2 (1982) 173–182.
7. E. Suissa, I. Jacob, and Z. Hadari, *Journal of the Less-Common Metals* **104** (1984) 287–295.
8. A. Isselhorst, *Journal of Alloys and Compounds* **231** (1995) 871–879.
9. J. Kapischke and J. Hapke, *Experimental Thermal and Fluid Science* **17** (1998) 347–355.
10. D. Kunii and S. Smith, *AIChE Journal* **6**, No. 1 (1960) 71–79.
11. Y. Asakuma, S. Miyauchi, T. Yamamoto, H. Aoki, and T. Miura, *International Journal of Hydrogen Energy* **29** (2004) 209–216.
12. E. Tsotsas and H. Martin, *Chemical Engineering and Processing* **22** (1987) 19–37.
13. A. Rodriguez Sanchez, H.-P. Klein, and M. Groll, *International Journal of Hydrogen Energy* **28** (2003) 515–527.
14. S. Suda, N. Kobayashi, and K. Yoshida, *International Journal of Hydrogen Energy* **6**, No. 5 (1981) 521–528.
15. A. Kempf and W.R.B. Martin, *International Journal of Hydrogen Energy* **11**, No. 2 (1986) 107–116.
16. D.E. Dedrick, M.P. Kanouff, W.C. Replogle, and K.J. Gross, *Journal of Alloys and Compounds* **389** (2005) 299–305.
17. M. Gopal and S. Murthy, *Chemical Engineering and Processing* **32** (1993) 217–223.
18. M. Nagel, Y. Komazaki, and S. Suda, *Journal of the Less-Common Metals* **120** (1986) 35–43.
19. D. Sun, *International Journal of Hydrogen Energy* **17**, No. 12 (1992) 945–949.
20. M. Pons and P. Dantzer, *International Journal of Hydrogen Energy* **19**, No. 7 (1994) 611–616.
21. M. Winterberg, E. Tsotsas, A. Krischke, and D. Vortmeyer, *Chemical Engineering Science* **55** (2000) 967–979.
22. R. Bird, W.E. Stewart, and E.W. Lightfoot, *Transport Phenomena*, 2nd edition, Wiley, New York.
23. K. Kurosaki, T. Maruyama, K. Takahashi, H. Muta, D. Uno, and S. Yamanaka, *Sensors and Actuators* A **113** (2004) 118–123.
24. K.J. Kim, G. Lloyd, A. Razani, and K.T. Feldman, Jr., *Powder Technology* **99** (1998) 40–45.
25. H. Ishikawa, K. Oguro, A. Kato, H. Suzuki, and E. Ishii, *Journal of the Less-Common Metals* **120** (1986) 123–133.
26. Y. Chen, C.A.C. Sequeira, C. Chen, X. Wang, and Q. Wang, *International Journal of Hydrogen Energy* **28** (2003) 329–333.
27. E. Eaton, C.E. Olsen, H. Sheinberg, and W.A. Steyert, *International Journal of Hydrogen Energy* **6**, No. 6 (1981) 609–623.
28. M. Ron, D. Gruen, M. Mendelsohn, and I. Sheft, *Journal of the Less-Common Metals* **74** (1980) 445–448.
29. E. Bershadsky, Y. Josephy, and M. Ron, *Journal of the Less-Common Metals* **153** (1989) 65–78.
30. Y. Josephy, Y. Eisenberg, S. Perez, A. Ben-David, and M. Ron, *Journal of the Less-Common Metals* **104** (1984) 297–305.
31. M. Nagel, Y. Komazaki, and S. Suda, *Journal of the Less-Common Metals* **120** (1986) 35–43.
32. W. Supper, M. Groll, and U. Mayer, *Journal of the Less-Common Metals* **104** (1984) 219–286.
33. K.J. Kim, B. Montoya, A. Razani, and K.-H. Lee, *International Journal of Hydrogen Energy* **26** (2001) 609–613.
34. H.-P. Klein and M. Groll, *International Journal of Hydrogen Energy* **29** (2004) 1503–1511.
35. T. Voskuilen, D. Dedrick, and M. Kanouff, System level permeability modeling of porous hydrogen storage materials, Sandia Report, SAND2010-0254, January 2010.

References

10

Solid-State H_2 Storage System Engineering: Direct H_2 Refueling

Terry Johnson and Pierre Bénard

CONTENTS

Introduction 348
Elements of System Design...348
    Operating Environment 348
    Sorption Materials..349
    Classic Metal Hydrides 351
    Complex Metal Hydrides..352
General Design Concepts..352
General Considerations of Thermal Management...355
    Heat Conduction 355
    Heat Exchange 357
    Internal Heat Exchangers 358
    External Heat Exchangers 359
    Insulation..359
Computational Simulation: Heat and Mass Transport...363
    Heat Exchanger Design 363
    Storage Vessel Design 363
    Mass Conservation..364
    Momentum Conservation...364
    Energy Conservation 365
    Species Conservation 366
    Physical Property Models 366
Structural Design..367
    Pressure Vessel Design 367
    Heat Exchanger Design 369
    Structural Materials Selection..369
Balance of Plant 371
System Example: Storing Hydrogen in an Engineered $NaAlH_4$ Bed 372
Safety..377
Performance Relative to Targets 378
Future Outlook 379
References..381

Introduction

An efficient mode of hydrogen storage has to satisfy a set of criteria based on net energy density, weight, safety, cost, charging-discharging kinetics, cycle life, and other considerations for a particular application. Although having the largest heat of combustion per unit of mass, hydrogen energy systems suffer from performance issues when volumetric considerations are factored in due to the low density of hydrogen gas at ambient temperature and pressure conditions. For the demanding automotive application, significant progress has been achieved recently. However there is as of yet no single storage material or method that satisfies all desired automotive design criteria. To date, it has been most convenient to store hydrogen in high-pressure tanks or in large cryogenic storage systems.

Going beyond compression or liquid storage will require relying on the interactions of molecular and atomic hydrogen with other atoms. Previous chapters have described in great detail the nature of hydrogen bonding in interstitial metal hydrides (Chapter 5), complex metal hydrides (Chapters 6 and 8), physisorption materials (Chapter 7), ammonia borane (Chapter 8), and liquid organic hydrogen carriers (Chapter 8). Here, we take a deep dive into the engineering aspects of how the various material properties can be managed to produce a functioning hydrogen storage system and give a description of a complete system that was developed at Sandia National Laboratories in Livermore, California. In Chapter 11, Bob Bowman, Ned Stetson, and Don Anton give a broad systems engineering overview of how these different materials can form the basis of engineered automotive systems with different storage properties.

Elements of System Design

Operating Environment

The operating environment of hydrogen storage systems includes primarily the temperature and pressure required for the system to store and deliver hydrogen. These operating conditions, along with the storage material properties, are the primary drivers for the design. For example, system geometry and structural material selection are primarily determined by the operating temperature, hydrogen compatibility, and hydrogen pressure required for refueling, which determines the required material strength.

Chao and Klebanoff in Chapter 5 reviewed how the van't Hoff equation relates the changes in reaction enthalpy ΔH and entropy ΔS to the equilibrium pressure existing above a hydrogen storage material at a given temperature T. These immutable thermodynamic quantities determine the ball field in which engineers play to exploit the hydrogen storage properties of materials in real systems. However, the van't Hoff equilibrium expression may not describe the actual pressure measured if the approach to equilibrium is slow (poor kinetics) and true equilibrium is not achieved. Both thermodynamics and kinetics are very important for the real engineered hydrogen storage systems.

The hydrogen refueling of a "spent material" typically occurs at elevated pressure to provide fast kinetics. The refueling pressure is thus the pressure limit for design since hydrogen desorption takes place at pressures below the material equilibrium pressure at similar temperatures to refueling. The operating environment for refueling is typically set to maximize the capacity while refueling in a reasonable time, 3 to 10 min perhaps. For some systems, an optimization for pressure and temperature can be performed based on the material kinetics and thermodynamics.

In contrast, the operating conditions for hydrogen delivery are determined by the fuel cell and application requirements. For example, for light-duty vehicles, the U.S. Department of Energy (DOE) has defined targets of 3-bar delivery pressure and 0.02-g/s/kW hydrogen flow rate [1]. Many classic interstitial metal hydrides can satisfy these requirements at or near room temperature, while complex metal hydrides typically require elevated (often excessive) temperature due to thermodynamics, kinetics, or both. In this case, the desorption temperature required to release hydrogen at a sufficient rate defines the maximum operational temperature. Some sorption materials will not release their entire capacity at 3 bar unless heated. This will affect system design and overall system delivery efficiency.

As described in the prior chapters, on-board regenerable systems include hydrogen storage materials that span a broad range of temperatures from 77 to 600 K. In addition, while generally an elevated pressure is required for a high capacity, pressures range from tens of bar up to several hundred bar for solid-state hydrogen storage systems. However, each class of storage material has more tightly grouped properties.

Several system design considerations are based on the operating environment. For high-pressure compressed hydrogen tanks, expensive carbon fiber composite vessels are used for the extremely high specific strength of this material. However, at some lower pressure level the specific strength of carbon fiber is unnecessary because the wall thickness decreases to a level that is impractical to fabricate. For metal hydrides, system design can be optimized by trading refueling pressure against vessel wall thickness. To some limit, increasing the refueling pressure for metal hydrides will increase the absorption rate, assuming that temperature is controlled. However, the greater the refueling pressure, the thicker the vessel walls must be to maintain a sufficient factor of safety against failure. Since thicker walls result in a heavier vessel, increasing the refueling pressure creates competing effects that result in an optimum pressure for system gravimetric efficiency. The pressure-wall-thickness trade-off was also seen in the discussion of cryocompressed hydrogen storage systems given in Chapter 4.

Temperature can be a driver for material selection as well. For example, cryogenic temperatures may require different vessel materials than operation at 500 K. Low-temperature embrittlement can be a problem for some steels, for instance, and these materials should not be used for cryogenic service. In addition, sealing materials can be problematic at both high temperatures (>200°C) and cryogenic temperatures. Many elastomers are not suitable for either of these temperature regimes.

Sorption Materials

In Chapter 7, Purewal and Ahn described the nature of molecular hydrogen adsorption in physisorption systems such as metal organic frameworks (MOFs), activated carbons, and zeolites. Here, we revisit these adsorption materials and their environment from an engineering perspective.

Since sorption materials rely on physisorption, they are typically only effective near cryogenic temperatures. Most storage capacities are listed at 77 K, for instance, although recent work has pushed toward operation at temperatures closer to ambient (see Chapter 7). Figure 10.1 reports the hydrogen adsorption excess isotherms on the activated carbon AX-21 (which is equivalent to the activated carbon MaxSorb) as a function of temperature from 30 to 298 K [2].

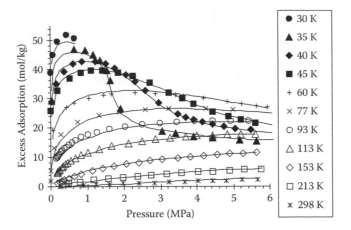

FIGURE 10.1
Adsorption isotherms of hydrogen on the activated carbon AX-21. (Reproduced from M.A. Richard, P. Bénard P,
R. Chahine, Gas adsorption process in activated carbon over a wide temperature range above the critical point.
Part 1: Modified Dubinin-Astakhov model, *Adsorption* **15**, 43–51 (2009). With permission.)

A simplified version of this figure appears in Chapter 7, Figure 7.7. The adsorption isotherms n_a shown in Figure 10.1 represent the amount of hydrogen adsorbed per unit mass
of the adsorbent as a function of pressure and temperature in excess of the amount that
would be present in the dead volume of the adsorbent. The isotherms shown in Figure 10.1
indicate that significant excess densities (about half the liquid density of hydrogen) can
be achieved at pressures on the order of 10–30 bars for temperatures lower than 77 K,
whereas negligible quantities of hydrogen can be found in the porous structure at ambient conditions in excess of the bulk gas density. Although researchers are working to find
a sorption material that can function closer to room temperature, such a material has to
date remained elusive. The operating environment of the physisorption storage unit thus
requires cryogenic conditions, and the storage efficiency is expected to drop rapidly as
temperature is increased.

Most sorption studies have considered moderate hydrogen pressures. However, for
many materials, increasing pressure has limited positive impact on capacity. In fact, it is
release pressure that limits these materials. Many cannot release their entire capacity at
the pressure required for delivery to a fuel cell.

The recoverable hydrogen is the net storage density of a sorption-based hydrogen storage
unit. It corresponds to the amount of hydrogen stored that can be practically recovered from
the storage system by reducing the pressure to a minimum set pressure, which could be
the operating pressure of a hydrogen fuel cell. The recoverable hydrogen density depends
on the application (which sets the minimum set pressure required) and the thermodynamic conditions during discharge. Discharge can be isothermal or adiabatic or performed
by allowing a temperature swing in which the temperature of the reservoir is allowed to
rise during discharge. The residual storage density, which is the density adsorbed in the
tank at the minimum pressure, becomes larger as the temperature is lowered. The residual
density can be partially recovered by heating the system using heat from the environment
or a heater. In the latter case, the storage system is then considered "active."

Ahluwalia has examined the recoverable hydrogen issue [3] with results presented in
Figure 10.2. From Figure 10.2, we see that the recoverable hydrogen is a strong function of

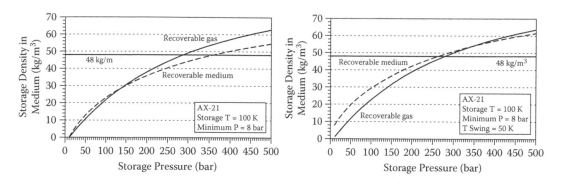

FIGURE 10.2
Recoverable hydrogen as a function of storage pressure using isothermal discharge (left) or by allowing a temperature swing of 50 K (right). The minimum pressure is 8 bars. The storage density in the adsorbent material required to achieve a recoverable system storage density of 36 kg/m³ is indicated as a constant line (48 kg/m³). (Reproduced from R.K. Ahluwalia, J.K. Peng, Automotive hydrogen storage system using cryo-adsorption on activated carbon, *Int. J. Hydrogen Energy* **34**, 5476 (2009) with permission of Elsevier.)

storage pressure using either isothermal discharge (Figure 10.2, left) or by allowing a temperature swing of 50 K (Figure 10.2, right). The minimum pressure is 8 bars. The storage density in the adsorbent material required to achieve a recoverable system storage density of 36 kg/m³ is indicated as a constant line (48 kg/m³).

Classic Metal Hydrides

As described in Chapter 5, classic interstitial metal hydride hydrogen storage materials generally operate at near-ambient temperatures and pressures below 70 bar. Alloying can tailor these materials to release hydrogen at 3 bar at low temperatures (see Figure 5.10 of Chapter 5). Refueling can be performed with hydrogen pressure as low as 10–20 bar as long as the material temperature is sufficiently controlled. The main problem with the classic interstitial materials is their poor gravimetric capacity, leading to engineered H₂ storage systems that are too heavy and carry too little hydrogen for some applications, particularly for light-duty vehicles. This deficiency led to the development of complex metal hydrides as high-capacity storage materials, as reviewed in Chapter 6. In contrast, interstitial metal hydrides have relatively high volumetric storage capacity due to the high density of the metal alloys.

To take advantage of the high volumetric capacity of metal hydrides and the high gravimetric capacity of high-pressure composite tanks, Toyota has proposed the design of a hybrid hydrogen storage system that combines the two. Mori and his team have combined a gaseous hybrid vessel (rated at 350 bars) with an AB₂ Laves phase Ti-Cr-Mn alloy for a fuel cell vehicle [4]. The alloy has a desorption hydrogen capacity of 1.9 wt% at 298 K (25°C). Four high-pressure Type III tanks (total volume of 180 L) like that shown in Figure 10.3 were able to store 7.3 kg of hydrogen, which is about 2.5 times more than that of a 350-bar compressed vessel system with the same volume. Through the use of an internal heat exchanger, these tanks could be filled rapidly, reaching 80% capacity in 5 min. A similar high-pressure system was developed and tested at Purdue University through partial funding by General Motors [5, 6]. An internal finned-tube heat exchanger was developed and optimized to allow for a 5-min fill of the Ti₁.₁CrMn hydride material.

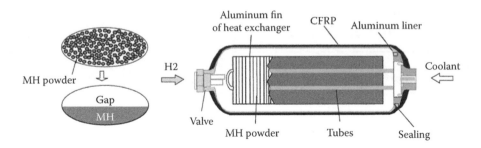

FIGURE 10.3
High-pressure metal hydride hybrid tank concept. (Reproduced from D. Mori, K. Hirose, Recent challenges of hydrogen storage technologies for fuel cell vehicles, *Int. J. Hydrogen Energy*, **34**, 4569 (2009) with permission of Elsevier.)

Complex Metal Hydrides

Complex metal hydrides generally require high pressure (>100 bar) for refueling along with elevated temperature for both refueling and delivery, due primarily to kinetic limitations. Sodium alanate ($NaAlH_4$), which has been the most studied complex metal hydride, requires pressures of 100–140 bar and temperatures of 120–150°C for rapid refueling to full capacity. Most other complex hydrides require even more extreme pressures and temperatures [7, 8]. Chapter 6 reviewed the significant recent effort to reduce the pressure required for hydrogen uptake in complex metal hydrides. Material researchers are at the same time trying to reduce the hydrogen release temperatures of these materials. Yet, as with pressure, significant temperatures up to 300°C are required for some complex metal hydrides to absorb and desorb hydrogen rapidly. An example of the engineering challenges that complex metal hydrides pose is discussed further in this chapter regarding a description of an engineered $NaAlH_4$ hydrogen storage system.

General Design Concepts

All hydrogen storage systems must contain an elevated pressure of hydrogen for delivery to a fuel cell or hydrogen internal combustion engine (ICE), whether the pressure is 10 or 700 bar. As a practical matter, since bottles pressurized to 150 bar are the most convenient and widely available form of merchant hydrogen, ambient temperature systems must be able to withstand at least 150 bar if they are to be refueled from compressed gas cylinders. In general, to withstand elevated hydrogen pressure, the hydrogen storage vessel is constrained to either a spherical or a cylindrical shape to provide uniform stress distribution, thus minimizing the wall thickness and the weight of the vessel. Any flat-sided shape would require much thicker walls, resulting in a much more massive vessel. In addition, a cylindrical shape is much more convenient than a sphere with regard to tank packaging, fabrication, and material loading. Thus, most hydrogen storage systems consist of one or more cylindrical vessels.

Some attempts have been made to design conformable or semiconformable hydrogen storage vessels. Conformability is highly desirable so that the storage tank can fit into spaces between other vehicle components like gasoline tanks do today. Hydrogen storage tanks can be made in odd or noncylindrical shapes, but it is difficult to do so without greatly increasing the weight for a given pressure rating. The U.S. DOE has funded research in this area at Lawrence Livermore National Lab [9]. Semiconformable fiber wound pressure vessels were designed and tested as well as internally structured vessels with a macrolattice of metal rods. A more easily manufactured approach using internal structure with repeated structural members was also conceptualized [9]. One advantage of an internally structured vessel in addition to conformability would be the potential to use the structure to enhance heat transfer within the vessel. However, the complexity of such a vessel has thus far eluded a practical design.

Turning back to cylindrical shapes, the number and diameter of the cylinders that make up the system are largely determined by the thermal management design. For hydrogen storage solutions that do not require thermal management, such as gaseous storage, a single large cylinder is typically optimal. However, most, if not all, advanced storage materials will require heat removal during hydrogen refueling and heat addition during hydrogen delivery. This heat transfer is governed by the material reaction enthalpy. For example, regenerating $NaAlH_4$ from spent $NaH + Al$ with hydrogen releases 40 kJ/mol H_2 [10]. In other words, the enthalpy ΔH of hydrogenation is –40 kJ/mol H_2. So, by absorbing 6 kg of hydrogen (a reasonable amount for a fuel cell vehicle) in 5 min, 400 kW of heat is produced by the system. Figure 7.2 of Chapter 7 shows an enthalpy of hydrogenation of –40 kJ/mol H_2 which corresponds to enough heat to bring to boil 80 gal water, a considerable release of heat that must be managed.

This heat must be removed or the hydrogen storage material will heat to a temperature above the optimum fill temperature, which will slow or even stop the absorption of hydrogen. This is a nice "self-regulating" feature of solid-state hydrogen storage, but care must be taken that the maximum temperature reached does not damage the surrounding tank material. To affect this heat removal or, in the case of hydrogen delivery, heat addition, heat exchange must be designed into the storage system. Broadly, the heat exchangers for hydrogen storage vessels can be lumped into two categories: external or internal. External heat exchange involves circulating a heat transfer fluid over the outside of the storage cylinders. With multiple cylinders, this design takes on the form of a shell-and-tube heat exchanger. Internal heat exchange involves circulating a heat transfer fluid through the inside of the cylinders via cooling tubes. With internal heat exchange, the hydrogen storage cylinders are typically larger in diameter than with external heat exchange. In either case, the thermal conductivity of the hydrogen storage material may require enhancement by some means to achieve a practical or optimal design. This was explored in Chapter 9 by Daniel Dedrick.

For vessel length, longer is better for mass and volume efficiency since end caps will have a fixed size independent of length. To illustrate this, consider two extremes: (1) a system with a single tube long enough to hold all of the required metal hydride and (2) a system with hundreds of short tubes to hold the same amount of hydride. In the first case, only two end caps are required, while the second case requires hundreds of end caps. Longer tubes are also better for cost since part count is reduced. Actual tube length will be limited based on packaging constraints for the application. For the automotive market, cylinder lengths are probably limited to a meter or less for transverse mounting or 3 m for longitudinal mounting.

Modularity is a concept developed to maximize the energy efficiency and performance of an advanced hydrogen storage system. The differences between a bulk bed and a modular system are shown in Figure 10.4. The same mass of storage media M is used, but it is distributed to n smaller vessels; the exact number of modules depends on the properties of the

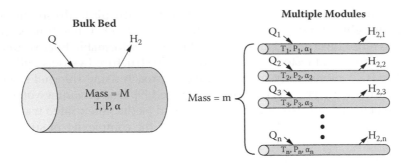

FIGURE 10.4
Comparison of bulk bed to modular system configurations. Here, Q is heat added to the system, and α is the hydrogen concentration in the solid species.

storage media, the details of the control strategy, and the vehicle size. Each of these smaller vessels is individually controlled, with its own temperature T_n, pressure P_n, and composition.

The main benefit of modular systems is an overall increase in storage system energy density due to three enabling aspects: efficiency in thermal management, reducing parasitic heating losses; increased hydrogen flow rate capability at all fill states; and rapid startup. During desorption, if only a portion of the system is required for operation, only that segment is heated, reducing parasitic losses. When starting up a cold system, a single module can be rapidly heated to the operating temperature required to deliver hydrogen. Finally, only the modules that have hydrogen left in them are heated. This is especially important in operation over many days because continually heating spent storage media wastes energy. In addition, if certain portions of the system are reacting at different rates (due to composition, temperature, etc.), the capabilities of each module can be coordinated to provide the appropriate flow for the immediate system demands. This can reduce heat use as well as ensure maximum hydrogen flow rate when the storage system is nearly empty.

Beyond the efficiency improvement, there are additional system advantages of a modular approach. For one, contamination control and quarantine are made possible. With individual modules, the amount of material exposed to air or water would be minimized if only one module of a 10-module system were damaged. Another advantage is conformability. Many small modules could conform to an odd shape or could be distributed throughout a vehicle. Third, counting modules as they are emptied could be used as a coarse fuel gauge.

To summarize, the basic concept for an advanced hydrogen storage system consists of one or more high-aspect-ratio cylindrical pressure vessels with external or internal heat exchange. The details of the number and size of the vessels are driven primarily by the operating conditions and properties of the storage material. Trade-offs between cost and mass and volume efficiency may also shape the design space. An example of these choices is given further in this chapter when an $NaAlH_4$ storage system is described in detail.

For material-based systems that required hydrogen desorption temperatures greater than the waste heat available from a fuel cell (~85°C for a proton exchange membrane [PEM] fuel cell), another design concept that is required is an on-board heater. The heating method that is most practical for a fuel cell vehicle is hydrogen combustion since hydrogen is stored on-board and can be converted to heat much more efficiently than electrical energy. Heat can be produced and distributed to the hydrogen storage system using a hydrogen burner coupled to a heat exchanger or with a catalytic heater. A catalytic heater produces heat using low-temperature catalytic reaction of hydrogen and oxygen. Concepts for burners and catalytic heaters have been developed for sodium alanate systems, for example [11–13].

General Considerations of Thermal Management

Thermal management is the most challenging aspect of the design of advanced hydrogen storage systems. For these material-based systems, the movement of hydrogen to and from the storage system is often accompanied by the movement of significant amounts of heat. The challenge is transferring heat efficiently to and from solid storage materials, which are generally characterized by poor thermal transport properties due to either the inherent material properties or the morphology of a packed powder or highly porous solid. The dependence of thermal transport properties on material microstructure was examined in Chapter 9.

The reaction enthalpy of the storage material coupled with the material thermal conductivity determines the degree of the challenge. For materials with high reaction enthalpy and low thermal conductivity, such as most complex metal hydrides, the challenge is greatest. During hydrogen absorption, the high reaction enthalpy results in a large amount of heat generation. Low thermal conductivity prevents this heat from moving easily through the material to be coupled to the tank cooling loop. This combination results in high peak temperatures, large temperature gradients, and poor thermal control, all of which are undesirable. Peak temperatures that are too high will prevent hydrogen absorption and actually promote hydrogen desorption from the material. The ideal material from a thermal management perspective would have a low reaction enthalpy and high thermal conductivity. This material would produce little heat during absorption and require little heat to desorb hydrogen. High thermal conductivity would result in uniform temperatures that could easily be maintained at the optimum value.

For most advanced hydrogen storage systems, the thermal management design will be driven by requirements for hydrogen refueling rather than delivery. This is because the rate of hydrogen sorption is typically much higher for refueling where the entire storage capacity must be replaced in a few minutes. High hydrogen absorption rates equate to high heating rates.

Taking absorption as the limiting case, the hydrogen storage vessels must be cooled to control the storage material temperature for the maximum refueling rate. With external heat exchange, a heat transfer fluid is circulated over the outer surface of the storage vessel or vessels. With internal heat exchange, a heat transfer fluid is circulated through one or more cooling tubes inside the vessel(s). In either case, to design the heat exchanger requires analyzing the conduction heat transfer through the storage material and the coupled convective heat transfer to the heat transfer fluid. The goal of the design is to minimize the added mass and volume of the heat exchanger while achieving the desired temperature distribution in the storage vessel.

Heat Conduction

Heat conduction within the hydrogen storage material is governed by Fourier's law, which states that heat flux q (watts per square meter) is proportional to the negative of the local temperature gradient: $q = -k*dT/dx$. The proportionality constant k is the thermal conductivity in watts per meter per degree Kelvin (W/mK) of the material. The thermal conductivity of hydrogen storage materials was discussed in detail in Chapter 9, including the effect of porosity and gas pressure. Here, the various contributions to heat transfer are lumped together into an effective thermal conductivity of the material, and we consider the effect of this material property on the design of the storage vessel. Fourier's law defines

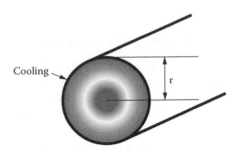

FIGURE 10.5
Externally cooled cylindrical storage vessel.

a relationship between the effective thermal conductivity, heat flux, and the temperature difference over a unit length of material. Assuming that a maximum temperature difference can be defined, this relationship can be used to determine a characteristic length for the design. This length is the maximum allowable distance to a temperature-controlled surface whether it is the radius of a vessel with external cooling or the distance between cooling tubes in an internally cooled system.

As an example, consider an externally cooled cylindrical storage vessel as shown in Figure 10.5. The material in the center of the cylinder is hottest, while the material next to the cooled cylinder wall is coolest.

For this simple geometry and with some simplifying assumptions, the maximum vessel radius can be defined. The primary assumption is that there is constant and uniform volumetric heat generation due to the hydrogen absorption reaction. The volumetric heat generation Q (watts per cubic meter) is equal to the hydrogen absorption rate per unit volume multiplied by the absorption enthalpy ΔH. Typically, $\Delta H < 0$ for absorption. The hydrogen absorption rate per unit volume can be defined as the gravimetric capacity w divided by the time to reach that capacity t multiplied by the hydride packing density ρ_p divided by the molecular weight of hydrogen M. Thus, Q can be defined as shown in Equation 10.1.

$$Q = -\frac{w \cdot \rho_p \cdot H}{M \cdot t} \tag{10.1}$$

If it is then assumed that the thermal conductivity is not a strong function of temperature, an analytical solution to the one-dimensional, steady-state heat transfer equation exists as shown in Equation 10.2.

$$T_i - T_o = \frac{Q}{4k} \cdot r_o^2 \tag{10.2}$$

Here, T_i is the temperature at the centerline, and T_o is the temperature at the cooled outer radius. Rearranging Equation 10.2 shows that the vessel radius can be calculated as

$$r_o = \sqrt{(T_i - T_o) \cdot \frac{4k}{Q}} \tag{10.3}$$

The temperature difference is chosen based on the sensitivity of the hydriding kinetics to temperature. A gradient in temperature can affect the kinetics of the absorption reaction and thus the effective capacity. So, the temperature gradient must be correlated with the effective capacity from Equation 10.1. Ideally, the temperature difference would be zero, and the entire volume of the material would be at the optimum refueling temperature. However, a zero-temperature difference is impractical for a real material, and a temperature difference must be found that results in an acceptable reduction in the absorption rate.

The characteristic length determines much about the heat exchanger geometry and is essentially fixed by the packing density, enthalpy, and thermal conductivity of the storage material and the desired refueling rate. However, rather than be limited by this intrinsic characteristic length, in some cases it is desirable to increase the characteristic length through thermal conductivity enhancement. Thermal conductivity enhancement is the addition of another material or structure to the hydrogen storage material that results in an increased effective thermal conductivity of the mixture. Thermal conductivity enhancement of hydrogen storage materials was examined in detail in Chapter 9. Unfortunately, since the enhancement material typically does not also absorb hydrogen, this technique lowers the gravimetric and volumetric efficiencies of the storage material. Adding thermal conductivity enhancement is often done to improve the practicality or simplicity of the design, to reduce cost, or to enable another design aspect that improves mass and volume efficiency.

Heat Exchange

Heat exchange is required for most, if not all, materials-based hydrogen storage systems. The enthalpy of reaction and rate of absorption or desorption determine the required heat removal or heat addition rates, respectively. To transfer heat to and from the storage vessel or vessels, these systems need an internal or external heat exchanger and a circulating heat transfer fluid. Certain sorption systems may be the exception.

For sorption materials, low reaction enthalpy may allow for thermal management schemes that only include the flow of precooled or liquid hydrogen during refueling. For example, Richard et al. [14] considered adiabatic filling of an activated carbon reservoir using hydrogen precooled to 80 K. The initial pressure and temperature of the adsorbent bed were set to 80, 200, and 298 K with a residual pressure of 2.5 bar. However, only with the initial bed temperature of 80 K could the required 5 kg of hydrogen be stored. Delivery under isothermal conditions can be hampered by the residual density of the sorption storage unit, resulting in lower accessible storage densities, depending on the temperature. This can be initiated by heating.

These authors also considered adiabatic filling of the reservoir using a mixture of gaseous and liquid hydrogen at 20 K for the initial conditions mentioned. They concluded that 5 kg of hydrogen could be stored without additional cooling if the system was adiabatically filled with liquid hydrogen, assuming, in the case of an initial temperature of 298 K, that the adsorbent was properly insulated from the container.

Adiabatic filling with liquid hydrogen was also considered by Ahluwalia [3] to eliminate the need for an on-board coolant. The hydrogen was pumped at 350 bar. However, an in-tank heat exchanger was still required to assist in performing temperature swings for desorption.

Paggiaro et al. studied a strategy in which the heat of adsorption was removed from activated carbon using a recirculation strategy [15]. In this approach, the system uses the hydrogen flowing into the system to cool the adsorbent bed, as shown in Figure 10.6. Hydrogen is charged at a constant flow rate (5 kg per 3 min in their example) and allowed to flow out of the system through a valve once a maximum pressure of the storage unit is reached. They showed that 90% of the maximum amount of hydrogen could be stored

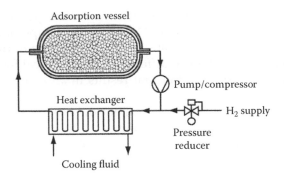

FIGURE 10.6
Recirculation cooling strategy for filling. (Reproduced from R. Paggiaro, F. Michl, P. Benard, W. Polifke, Cryo-adsorptive hydrogen storage on activated carbon. II: Investigation of the thermal effects during filling at cryogenic temperatures, *Int. J. Hydrogen Energy* **35**, 648 (2010) with permission of Elsevier.)

in the reservoir in less than 10 min, compared to over an hour in the case of a vessel immersed in a liquid nitrogen bath.

Beyond these solid-state physisorption exceptions, system design concepts for materials-based storage largely include heat exchange for thermal management during refueling and hydrogen delivery. Heat exchangers internal to the storage vessel consist of a heat transfer fluid circulated through tubing that may be finned to increase the heat transfer area. External heat exchange takes place in a shell around a single storage vessel or bank of vessels, as in a shell-and-tube heat exchanger. Fluid flow in this case may be parallel to the vessel axis or axes or, with the use of baffles, in cross flow. Examples of internal and external heat exchangers for hydrogen storage systems are shown in Figure 10.7 [16].

Internal Heat Exchangers

Numerous examples of internal heat exchangers can be found in the literature [17–31]. Internal heat exchangers consist of one or more coolant tubes that provide fluid flow straight through the vessel or more commonly are looped so that the inlet and outlet are at the same end of the vessel. Recently, some researchers have suggested the use of spiral or helical cooling tubes as the best way to provide an even distribution of heat transfer [21–26]. Often, the tubes are finned to provide more heat transfer surface area. The fins can be either radial or longitudinal. In all cases, there is a trade-off between minimizing the mass and volume of the heat exchanger and maximizing the temperature uniformity of the bed.

For internal heat exchangers, the convective heat transfer coefficient can be calculated using a correlation for coolant flow in a round tube. These can be found in any heat transfer textbook [32, 33]. Since the heat transfer coefficient for laminar flow is quite low (50–200 W/m² K) and independent of flow velocity, turbulent flow may be required to achieve the required cooling rate. With a Reynolds number in the 5000 to 10,000 range (turbulent flow), a factor of 10 increase in the heat transfer coefficient can be achieved over the laminar case. However, the increased heat transfer rate comes at a cost. The higher coolant flow rate will require more pump work. Pressure drop through the heat exchanger can be calculated with correlations as well and must be taken into account in the overall design.

Ahluwalia et al. [3] examined the pressure and temperature variations present in an adsorption system during a refueling scenario in which hydrogen was precooled to 100 K

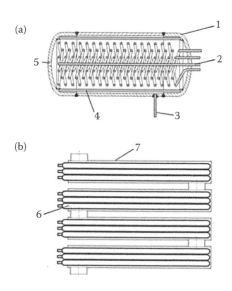

FIGURE 10.7

(a) Large hydride vessel with internal heat exchanger (1) shell, (2) coiled tube for heat transfer fluid, (3) hydrogen nozzle, (4) peripheral hydrogen supply, (5) opening for filling with active hydrogen storage material. (b) External heat exchanger concept (6) tubular reactor element, (7) heat transfer fluid shell. (Reproduced from C. Na Ranong, M. Höhne, J. Franzen, J. Hapke, G. Fieg, M. Dornheim, N. Eigen, J.M. Bellosta von Colbe, O. Metz, et al., Concept, design and manufacture of a prototype hydrogen storage tank based on sodium alanate, *Chem. Eng. Technol.* **32**, 1154 (2009) with permission of John Wiley and Sons.)

and supplied at 25% overpressure. Off-board liquid nitrogen was circulated through the internal heat exchanger in the tank to remove the heat of adsorption. The refueling system they proposed is shown in Figure 10.8.

External Heat Exchangers

The other option for thermal management for hydrogen storage systems is to perform convective heat transfer on the outside of the storage vessel using an external heat exchanger. For a single vessel, the heat exchanger could be an annular shell in which heat transfer fluid flows around the vessel. More commonly, an external heat exchanger configuration takes the form of a shell-and-tube heat exchanger where multiple hydrogen storage vessels make up the tube bank and fluid flows within a surrounding shell [16, 34–45]. The fluid flow can be along the vessel axes, or baffles can be used to create multiple cross-flow passes over the bank of tubes. The latter case produces higher heat transfer rates for a given fluid flow rate but also results in higher pressure drop.

An example of an external heat exchanger for a metal hydride system is shown in Figure 10.9. This system was designed by Ergenics and built for a fuel-cell-powered mine loader [45]. The metal hydride powder was contained within the small diameter (15.9-mm) tubes, which were manifolded together and placed inside the rectangular shell of the heat exchanger.

Insulation

For hydrogen storage systems that operate at temperatures significantly higher or lower than the ambient environment, insulation can be an important design aspect. As previously

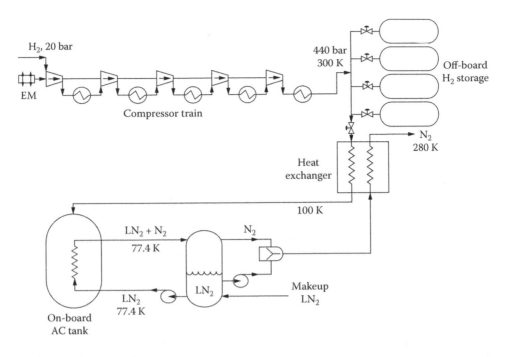

FIGURE 10.8
Refueling system with off-board liquid nitrogen cooling. (Reproduced from R.K. Ahluwalia, J.K. Peng, Automotive hydrogen storage system using cryo-adsorption on activated carbon, *Int. J. Hydrogen Energy* **34**, 5476 (2009) with permission of Elsevier.)

FIGURE 10.9
External heat exchanger example. Tube manifolds containing hydride alloy powder are stacked within a stainless steel shell. Water ethylene glycol circulates through the shell and around the manifolds to remove the heat of formation of hydrogen-metal bonds during recharge. (Reproduced from A.R. Miller, D.H. DaCosta, M. Golben, Reversible metal-hydride storage for a fuelcell mine loader, presented at Intertech-Pira's Hydrogen Production and Storage Forum, Vancouver, Canada, September 11–13, 2006, with permission.)

discussed, complex metal hydrides typically are operated at temperatures much higher than ambient. These systems must be insulated to maintain their operating temperature without significant heat input. Adsorption systems typically are operated at temperatures much lower than ambient and must be insulated to minimize heating from the environment. Both of these cases are considered now in more detail.

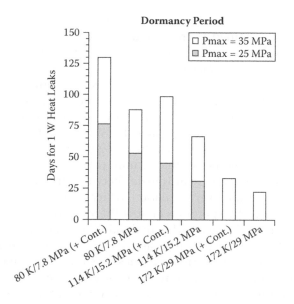

FIGURE 10.10

Predicted dormancy for a 150-L adsorption storage unit filled with AX-21 as a function of storage conditions and maximum pressure settings. A 1-W heat leak is assumed. (Reproduced from M.-A. Richard, D. Cossement, P.-A. Chandonia, R. Chahine, D. Mori, K. Hirose, Preliminary evaluation of the performance of an adsorption-based hydrogen storage system, *AIChE J.* **55**(11), 2985–2996 (2009) with permission of John Wiley and Sons.)

Cryogenic storage systems are invariably subject to heat leaks from the environment, the degree of which is set by the quality of the insulation of the storage system. The heat leaks result in heating the storage unit, which causes a release of hydrogen from the dense phase (either liquid or adsorbed) of the stored hydrogen, resulting in the gradual increase of the pressure of the gaseous hydrogen in thermodynamic equilibrium with the stored phase. In the case of a cryocompressed storage unit, heat leaks will result in an increase of temperature inside the reservoir and an increase in pressure. When the level of pressure reaches the maximum storage pressure, the gas must be vented to prevent the pressure to build up to unsafe levels. The time required to reach this point is defined as the dormancy of the storage system. The dormancy depends linearly on the intensity of the heat leak. Dormancy was discussed in Chapter 4 for cryocompressed systems and was shown to be a strong function of the pressure of the cryocompressed tank and the profile with which hydrogen is withdrawn from the system during use.

For a sorption-material hydrogen storage tank operating near 77 K, the dormancy depends highly on the tank design. Figure 10.10 shows the dormancy for a 1-W heat leak for the AX-21 sorption storage unit considered by Richard et al. in Reference 14. Figure 10.10 shows the results for a 150-L storage unit filled with 5 kg of hydrogen with maximum tank pressures set to 25 MPa (250 bar) and 35 MPa (350 bar). Due to linear dependence of the dormancy with the heat leaks, the dormancy can be estimated by dividing the dormancy with the thermal power of the heat leak. If the heat leaks can be limited to 1 W, the dormancy can reach several weeks.

Ahluwalia et al. [3] also examined the boil-off of solid-state cryogenic storage materials operating via physisorption and estimated a value of 18 watt-days for initial storage conditions of 350 bar and 100 K, which is similar to the results obtained by Richard and coworkers at the higher-pressure range they considered. Ahluwalia noted that the value they obtained was comparable to the dormancy of cryocompressed systems operating under similar conditions. They obtained a venting rate of 0.8 g/h/W, which decreased over time. They also noted that cryoadsorption systems have longer dormancy than conventional low-pressure liquid hydrogen storage systems.

Heat leaks from the environment must be minimized through proper insulation to maximize dormancy and maintain the temperature as close as possible to the optimal operating temperature of the storage unit, which for hydrogen is on the order of 100 K. This is typically done through superinsulation (also known as multilayer vacuum insulation), which consists of a vacuum space between the cryogenic vessel containing the adsorbent and the environment. The presence of the vacuum reduces heat transfer by gaseous conduction and convection. Radiative heat transfer between the inner vessel and the external shell is minimized by multilayer insulation, which consists of layers of low-emittance radiation shields supported by spacers with low thermal conductivity. Experimental test beds generally use a liquid nitrogen bed to maintain the temperature close to 77 K.

High-temperature hydrogen storage systems (i.e., complex metal hydride systems) experience a challenge similar to dormancy. These systems must be maintained at an elevated temperature to overcome thermodynamic or kinetic limitations to rapidly absorb and release hydrogen. The energy required to heat and maintain these systems at elevated temperature is a parasitic loss to the hydrogen storage capacity. For example, if a hydrogen burner or catalytic heater is used to provide heat to the system, a fraction of the stored hydrogen must be used for this purpose rather than for fuel. To minimize this parasitic loss, these systems should be well insulated.

Multilayer vacuum insulation can be used for these systems, just as for adsorption systems. However, it can be costly and bulky, and in some cases lower-performance alternatives may be acceptable. One such alternative is the use of vacuum insulation panels (VIPs). As shown in Figure 10.11, most VIPs consist of a micro- or nanoporous material enclosed in a film barrier that is evacuated to about 1 mbar. Fumed silica and aerogels are the most widely used core materials. At vacuum, thermal conductivities are reduced by a factor of 5 or more so that room temperature conductivities of less than 0.005 W/mK are achieved. This is equivalent to an R value (thermal resistance R = temperature difference/heat flux) of 29 per inch of insulation. Common household insulation has an R value of around 4 per inch, so VIPs are more than seven times better insulators.

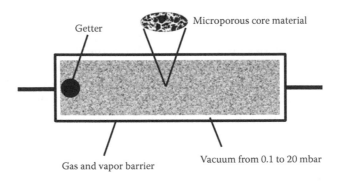

FIGURE 10.11
Typical construction of a vacuum insulation panel.

Computational Simulation: Heat and Mass Transport

Detailed computational models are required to optimize the design of advanced hydrogen storage systems. The hydrogen refueling and delivery performance of these systems are influenced by the highly coupled processes of heat transfer, mass transfer, and, in some cases, chemical reaction. Minimizing, or even eliminating, process bottlenecks during refueling and hydrogen release requires the solution to a complex set of coupled equations that must be solved numerically. There are two primary aspects of hydrogen storage system design that may require detailed computational simulation of heat and mass transfer: heat exchanger design and storage vessel design.

Heat Exchanger Design

We have previously discussed that heat exchange for hydrogen storage systems may take place in internal or external heat exchangers. For internal heat exchangers, coolant flows within tubes inside the storage vessels, and for external heat exchangers, coolant flows on the outside of the vessels. In either case, heat transfer calculations for flow in tubes or for shell-and-tube heat exchangers can be made using empirical correlations, which can be found in standard heat transfer textbooks [32, 33]. The correlations are used to compute an average heat transfer coefficient based on geometry and fluid properties. Pressure drop can also be estimated. However, for more accurate predictions or cases of complex geometry, a higher-fidelity model may be required. In these cases, fluid dynamics models must be developed and coupled to conjugate heat transfer models to more accurately predict local heat transfer rates, temperature variations, and fluid pressure drop.

For complex heat exchanger manifolds, flow uniformity and pressure drop are calculated using computational fluid dynamics (CFD) models that solve the Navier-Stokes equations for momentum and mass transport of the heat transfer fluid, including turbulent kinetic energy and dissipation rate (*k-e* model), when appropriate. For complex heat exchanger designs, the energy transport equation is also solved for the conjugate heat transfer problem. These heat transfer solutions can then be directly or indirectly coupled to heat transfer calculations within the storage vessels to provide the overall storage system thermal management solution. Heat transfer calculations within the storage vessels are covered in the following discussion.

Storage Vessel Design

Advanced hydrogen storage materials are often comprised of sintered and fine powders that undergo significant morphological changes during hydrogen sorption. Efficient transport of hydrogen and heat into and out of porous beds is required to attain rapid refueling and dehydriding performance demanded by the fuel cell application. Due to energy density goals, system designers must specify high-solid fractions [46], which, in the case of mass transfer, can cause transport bottlenecks of hydrogen gas through the bed. In addition, packed particle beds typically possess low thermal conductivities due to a complex set of competing heat transfer modes, including interface effects [47] (see Figure 9.1 from Chapter 9). Because the fundamental kinetics of hydriding and dehydriding are only applicable to a localized surface of a particle within a well-characterized environment (temperature and pressure), the processes within a packed-bed reactor become highly coupled and complex.

If heat and mass transport processes are not understood, the local temperature and pressure environment of each particle is unknown and subject to averaging uncertainties. The result of this uncertainty is poor prediction of hydriding and dehydriding rates. In the case of rapid hydriding in packed-bed reactors, as reactants are consumed, the intrinsic porous nature of the bed limits the rate of hydrogen replenishment, and the effective rate of reaction is limited not by local kinetics, but rather by the depletion of available reactants. In the case of dehydriding processes, an inability to remove low-pressure hydrogen rapidly from the solid matrix results in rate limitations.

Predicting the state (temperature, pressure, and species concentrations) of these hydrogen storage vessels during hydrogen absorption and desorption requires the simultaneous solution to the conservation equations for mass, momentum, energy, and species [16, 31, 48]. Each of these conservation equations in turn requires material property models for closure.

Mass Conservation

For all material-based hydrogen storage systems, hydrogen will be either physically or chemically bound to the solid. Thus, all such systems require a mass balance for hydrogen to account for the conversion from gas phase to solid. A general form of mass conservation is shown in Equation 10.4:

$$\frac{\partial \rho}{\partial t} + \nabla \cdot (\rho v) = R \tag{10.4}$$

where ρ and v are the hydrogen density and velocity, respectively. The right-hand side of the mass conservation equation is nonzero to account for absorption or desorption of hydrogen to and from the solid phase. This source term R is calculated from an equation or set of equations that describes the sorption kinetics for the material.

Momentum Conservation

Solid storage materials are typically packed powders or porous solids. Thus, momentum conservation is expressed for gas flow through a porous medium. A number of different expressions for gas flow through porous media have been developed and reported in the literature. Some of the most commonly used models include the Ergun model [49] and the Darcy and the Darcy-Brinkman models [50, 51]. The Ergun model has several variations, including the Carman-Kozeny and Blake-Kozeny equations.

The Ergun model was originally introduced by Ergun in 1952 [49]. The Ergun model, shown in Equation 10.5, treats the porous medium as a series of capillary flow channels that induce viscous and inertial pressure losses similar to those resulting from porous media.

$$\frac{P}{L} = A\mu v + B\rho v^2 \tag{10.5}$$

Here, the gradient of pressure P is related to the velocity v, where μ is dynamic viscosity and ρ is the fluid density. Ergun also went on to derive the values of the A and B coefficients for beds of packed spheres as follows:

$$A = 150\frac{(1-\phi)^2}{\phi^3 d_p{}^2}, \quad B = 1.75\frac{(1-\phi)}{\phi^3 d_p}$$

Here, ϕ is the porosity and d_p is the diameter of the spheres.

Darcy's law describes a linear relationship between flow through porous media and the pressure gradient. The constant of proportionality is the material permeability K divided by the fluid dynamic viscosity μ. The Darcy-Brinkman equation combines Darcy's law with the Stokes equation to produce the following relationship [50, 51]:

$$\mu_e \,{}^2 v - \, p - \frac{\mu}{K}v = 0 \tag{10.6}$$

This relationship allows for modeling of gas flow through multidimensional porous media using finite element analysis. It does, however, require knowledge of the permeability of the porous media and the effective viscosity μ_e. It has been shown experimentally that the effective viscosity term is required to reconcile experimental data with the model [52–54]. However, often it is assumed that the effective viscosity is equal to the "real" viscosity μ.

For a transient analysis, Brinkman's equation results in the following momentum equation:

$$\frac{\rho \partial v}{\phi \partial t} + \frac{\rho}{\phi}v \cdot \, u = - \, p + \, \cdot \left[\frac{\mu}{\phi}\left(\, v + \, v^T\right)\right] - \frac{\mu}{K}v \tag{10.7}$$

In Equation 10.7, the hydrogen superficial velocity v is equal to the seepage velocity u multiplied by the porosity. Similar equations for momentum conservation have been used in the literature for flow through hydrogen storage materials [16, 31]. For high-flow rates, an additional viscous drag term is sometimes included on the right-hand side. This Forchheimer term is proportional to the square of the fluid velocity $-\beta|u|u$.

Energy Conservation

One form of the energy equation for hydrogen storage materials is given by Equation 10.8:

$$\left(\rho c_p\right)_m \frac{\partial T}{\partial t} + \left(\rho c_p\right)_g v \cdot \, T = \, \cdot \left(k_m \, T\right) + R \, H \tag{10.8}$$

In addition to the variables defined previously, c_p is specific heat capacity of the storage material m and the hydrogen gas g. In most cases, the hydride material temperature T is assumed to be in thermal equilibrium with the gas temperature. The term k_m is the thermal conductivity of the solid. The first term in Equation 10.8 is the rate of change of temperature of the solid, while the second term is the hydrogen advection term. Note that the advection term uses the density and specific heat of the advecting phase (i.e., hydrogen), while the thermal inertia term uses the density and specific heat of the solid phase, which is the dominant source of inertia. The first term on the right-hand side accounts for heat conduction through the solid, where an effective conductivity of the hydride bed is used (see Chapter 9). The second term on the right accounts for the chemical reaction rate R between the hydride and gas phase, which can be either exothermic (hydrogenation) or endothermic (dehydrogenation), where ΔH is the enthalpy of reaction. Note that terms accounting for pressure variation work and viscous dissipation are neglected here because they are much smaller (by three or more orders of magnitude) than the other terms.

Species Conservation

For all materials-based hydrogen storage systems, a detailed analysis must keep track of the free hydrogen, typically gaseous, and the hydrogen bound by the solid storage material. In chemically reacting systems such as complex metal hydrides, several solid species must be accounted for as well. Thus, species conservation equations are a necessary part of the analysis. Here, we consider the most complicated case of a multispecies complex metal hydride.

In this general case, species conservation for each solid, immobile species is given by

$$\frac{\partial c_j}{\partial t} = \dot{R}_j \tag{10.9}$$

where j stands for the reacting species. For sodium alanates, for instance, this would include NaH, Na_3AlH_6, $NaAlH_4$, and Al (see Chapter 6). The solid species production rates $\dot{R}_j$ are given by combinations of the reaction rates according to the stoichiometry of the reactions.

Physical Property Models

The conservation equations rely on the solution to the chemical rate equations to determine mass source terms for hydrogen. To this end, chemical kinetics models must be used, many of which have been developed for a number of hydrogen storage materials. For instance, a number of kinetic models of sodium alanate have been developed [16, 55–59]. These models primarily take the following form:

$$\dot{R}_j = Ae^{Q/RT} \cdot f(P) \cdot f(c_j) \tag{10.10}$$

where there is an Arrhenius temperature dependence multiplied by a function of the hydrogen pressure and a function of the hydrogen concentration in the solid phase. The functions of pressure vary but always contain a reference to the equilibrium pressure of the storage material. This guarantees that the sign of the rate is in agreement with the thermodynamics; that is, hydrogen is absorbed when the pressure is greater than the equilibrium pressure and desorbed when the reverse is true. Often, the equilibrium pressures are modeled as an Arrhenius expression themselves based on the van't Hoff equation (see Chapter 5). Functions of solid-phase concentration also vary but enforce that the rate approaches zero when the solid is saturated.

In addition to the chemical kinetics model, several other properties are required to solve the system of equations. To solve the momentum equation, hydrogen transport through the porous media must be described. This solution requires a permeability model for the material. One such model developed by Young and Todd [54] captures the dependence on porosity ϕ, the particle diameter d_p, the tortuosity τ, and the Knudsen number Kn as shown in Equation 10.11. This model spans both continuum (viscous) and free molecular flow regimes (Knudsen flow). The porosity K can be determined from measurements of the storage material mass loading, volume of the vessel, and knowledge of the single-crystal density of the storage material. The pore size and tortuosity can be determined from microscopy or, as is more likely, fit using experiments carried out on samples of the material. These experiments involve flowing gas through the samples while measuring the mass flow rate and pressure drop across the sample.

$$K = \frac{\phi d_p^{\ 2}}{\tau^2}\left(\frac{1}{32} + \frac{5}{12}Kn\right) \tag{10.11}$$

To solve the energy equation, the thermal properties of the storage material are needed. In particular, the thermal conductivity of the porous solid is required. Ideally, this property will be measured for the storage material in the form and density that it will take within the hydrogen storage system since bulk properties do not apply for small particles at packing densities below the solid density. Chapter 9 contains an in-depth discussion of the thermal conductivity of hydrogen storage materials. If physical measurements cannot be made, models exist that can predict the thermal conductivity of porous materials. Since thermal conductivities of packed particle beds are inherently low, one might try to enhance the thermal conductivity of these materials with additives (Chapter 9). In addition to thermal conductivity, the specific heat capacity of the storage material must be known. A simple combination of the weighted bulk specific heats is often used.

Structural Design

Pressure Vessel Design

All advanced hydrogen storage systems must be designed as hydrogen pressure vessels due to the fact that at some point in the refueling or delivery of hydrogen an elevated pressure of hydrogen must be contained. As previously discussed, this constrains most storage vessels to a cylindrical geometry. To design the vessels for safe operation, a structural analysis is required to determine the thickness of the vessel walls. This includes the cylinder wall thickness t_w and the thickness of the cylinder ends t_e, which may be different. For simple geometries, these thicknesses can be estimated using equations for thin-walled pressure vessels, Equations 10.12 and 10.13. The quantities t_w and t_e both depend on the applied hydrogen pressure P, the vessel material allowable stress σ, and Poisson's ratio v. For more complex geometries, as with a detailed final design, finite element models may be required. Also, these equations assume isotropic material properties. Composite vessels, even if simple cylinders, may require more complex design calculations due to anisotropic material properties inherent in these structures.

$$t_w = \frac{P \cdot r_o}{(\sigma - 0.6P)} \tag{10.12}$$

$$t_e = \sqrt{\frac{3 \cdot P \cdot r_0^{\ 2} \cdot \sqrt{v^2 - v + 1}}{4\sigma}} \tag{10.13}$$

The hydrogen pressure used in these calculations should be the highest pressure to which the vessel will be subjected, which is likely to be the refueling pressure. In most cases, higher pressure improves hydriding kinetics but also requires thicker-walled pressure vessels. Since the system gravimetric efficiency includes the weight of the vessel as well as the storage material, an optimum pressure can be found that maximizes the gravimetric efficiency of the system for a given refueling time. At a hydrogen refueling pressure lower than the optimum, slower refueling is dominant over the reduced weight of the vessel. At a hydrogen refueling pressure higher than the optimum, the increased vessel weight dominates the improvement in refueling rate. The optimum refueling pressure is thus a function of the pressure dependence of the storage material chemical kinetics.

In addition to the maximum operating pressure, the structural design of some hydrogen storage systems must take into account density changes of the storage material during hydrogen uptake and release. Some materials, notably interstitial and complex metal hydrides, undergo significant density changes during hydrogen absorption. In Chapter 5, volume expansion of interstitial metal hydrides of up to 40% was noted. This expansion must be taken into account in the design of the storage vessel. If sufficient space is not provided to accommodate the material expansion with hydrogenation, then a significant pressure can be developed in addition to the hydrogen pressure. This additional pressure must be understood and accounted for in the structural analysis of the vessels.

In addition to material expansion, decrepitation, which is the breakdown of metal hydrides into smaller particles due to hydrogenation cycles, can pose structural design challenges. Decrepitation can lead to density gradients in a storage vessel due to the movement of smaller particles due to gravity. With a large density gradient, the local expansion of the material in the high-density region due to hydrogen absorption can cause such extreme pressures that vessel failure can result.

The allowable stress σ used in design calculations may be based on material property data or code requirements, depending on the application. For instance, Section VIII of the ASME Boiler and Pressure Vessel Code provides guidance for the design of pressure vessels. The design rules in Division 1 use an allowable stress, which is provided for a number of common pressure vessel materials in Section II of the code. It is important to consider the allowable stress at the operating temperature since most materials have reduced strength at elevated temperature. In addition, Section VIII, Division 3, contains alternative design rules for high-pressure vessels, including Article KD-10, which describes special requirements for hydrogen pressure vessels.

If finite element analysis is required to determine the pressure vessel geometry, the allowable stress for the material is normally compared to the calculated maximum von Mises stress under the maximum pressure load. The geometry is then adjusted until the maximum von Mises stress is everywhere lower than the allowable stress.

Hydrogen storage vessels must be designed to meet the requirements of safety codes and standards for each application, such as Society of Automotive Engineers (SAE) J2579 for hydrogen vehicles. This document defines design, construction, operational, and maintenance requirements for hydrogen storage systems for on-road vehicles. It includes performance-based requirements and test protocols that hydrogen storage systems must meet that are not prescriptive in terms of design specifics such as materials or vessel wall thickness. It is up to the storage system designer to ensure that the vessel design can meet these requirements. The hydrogen storage and technology safety codes and standards are discussed in Chapter 12 by Chris Sloane. Other codes and standards that apply to hydrogen storage systems include those from the International Organization for Standardization (ISO), NASA, and American Institute of Aeronautics and Astronautics (AIAA) [60–63].

Heat Exchanger Design

In addition to the pressure vessels, the heat exchanger may require a structural analysis for design. In the case of external heat exchange, the shell must contain the heat transfer fluid and withstand the operating pressure as the fluid is circulated through the heat exchanger. So, a structural analysis is required to define the shell thickness and design any reinforcement that may be required. In this case, the shell material does not have the requirement of hydrogen compatibility, so the material space is significantly expanded. A low-cost, high-strength material should be used. A finite element structural model may be required to evaluate the minimum thickness to withstand the force, including a factor of safety from the yield strength.

With an internal heat exchanger, the heat transfer fluid is circulated through the hydrogen storage vessel through tubing. This tubing must withstand the internal fluid pressure along with the external hydrogen pressure. In addition, the tubing must be made of a hydrogen-compatible material, limiting material choice. The thin-walled pressure vessel equation shown previously can be used for sizing this tubing.

Structural Materials Selection

A key part in the design of hydrogen storage systems is structural material selection. Current compressed H_2 vessels for vehicle applications, as discussed in Chapters 3 and 4, are Type III or Type IV pressure vessels that rely on high-strength carbon fiber composite outer layers for strength. Type III vessels have aluminum inner liners, while Type IV have plastic liners, typically high-density polyethylene (HDPE). The liners are fairly thin, a few millimeters thick, and serve only to prevent hydrogen permeation. The material choices for these vessels are based on hydrogen compatibility and low weight. Aluminum and HDPE are essentially inert in hydrogen, and carbon fiber composites can obtain extremely high specific strength. However, this specific strength comes at a high cost, which is one of the primary limitations of current compressed H_2 systems.

The materials used in compressed H_2 systems may be appropriate for some advanced hydrogen storage systems, but operating environments must be taken into account. The 700-bar compressed H_2 tanks require the high specific strength of composites to minimize weight, but many materials-based systems operate at much lower pressures. For low-pressure systems, a lower-strength material may provide a low-cost alternative. In addition, compressed H_2 tanks operate over a range of temperatures from −40°C to 85°C. Adsorption systems that operate much colder or metal hydride systems that operate much hotter may require different materials. For instance, many composites may not withstand temperatures greater than 150°C, where some complex metal hydrides operate. Thus, each material operating environment must be considered when choosing the best structural materials.

That being said, one of the primary drivers in the choice of structural materials for hydrogen storage systems is hydrogen compatibility. Many structural metals are embrittled by hydrogen [64, 65]. So, while titanium alloys may have good specific strength, they are mostly a poor choice for a hydrogen storage vessel due to the high permeability of hydrogen in titanium and subsequent formation of titanium hydride. Hydrogen compatibility

TABLE 10.1

Comparison of Structural Materials Preferred for Hydrogen Storage Vessels

Material	Advantages	Disadvantages
A286	High specific strength Good thermal stability Engineering experience in H_2 Published data in H_2	Expensive Difficult to weld
316 stainless steel	Lower cost Can be welded Engineering experience in H_2 Published data in H_2	Low specific strength
2219 aluminum	Good specific strength Can be welded Lower cost	Microstructure stability at 200°C Poor corrosion resistance
Pipeline steels (carbon steels)	Engineering experience in H_2 Fracture resistance in H_2 Low cost	Limited property database in H_2 Low strength (YS ~ 35–52 ksi) Temperature effects not known
Pressure vessel steels (low-alloy steels)	Engineering experience in H_2 Fracture resistance in H_2 Low cost	Limited property database in H_2 Temperature effects not known No experience with welds in H_2
Carbon fiber composite	Highest specific strength Opportunity for future cost reduction	Limited temperature (250°C max) Expensive manufacturing process Limited to larger-diameter vessels

coupled with high specific strength and ductility leads to a fairly narrow choice of materials that should be used for hydrogen storage vessels. A listing of some of these preferred materials is shown in Table 10.1. A summary of research in this area can be found in the book by Gangloff and Somerday [65].

Of the materials listed in Table 10.1, the pipeline steels and pressure vessel steels offer the possibility of greatly reduced cost with little loss in specific strength since higher-strength pipeline steels are being developed [66–68]. Pipeline steels have been successfully utilized for hydrogen service in the past, so there is considerable engineering experience with them. However, this experience is primarily with low-strength materials (yield strengths of 35–52 ksi) used in applications with high safety factors. These applications are also at ambient temperature, so there is not much experience at elevated temperature. At higher temperature, hydrogen precipitation at hydrogen trapping sites during cooling might occur. There is limited knowledge of properties in hydrogen gas, and potentially susceptible microstructures are not known. Also, resistance to fatigue in pipeline steels is not well established. For joining pipeline steels, welds can be more sensitive to environmental effects than base materials. Also, postweld heat treating is required. All of these facts suggest that while a pipeline steel could be a good hydrogen storage vessel candidate, more information is needed to be certain.

A similar statement can be made for pressure vessel steels. These are quenched and tempered low-alloy steels that, in addition to the issues with pipeline steels, can be susceptible to hydrogen-assisted fracture, especially at higher strengths and higher operating stress. Figure 10.12 shows the effect of hydrogen on a number of materials, including some carbon and low-alloy steels. Note that compared to these materials, 316L and A286 are nearly impervious to hydrogen. Coupled with cost and weldability, this is why 316L is the most common structural material used for high-pressure hydrogen components such as tubing, fittings, valves, regulators, and so on.

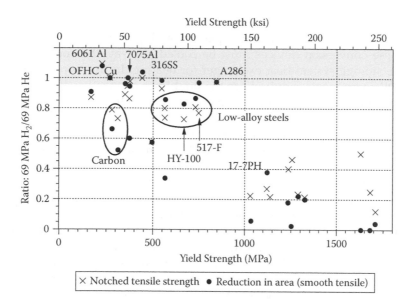

FIGURE 10.12
Effect of hydrogen on strength and ductility of structural metals, relative to a control measurement using helium.

Balance of Plant

Hydrogen storage systems require a number of ancillary components in addition to the primary storage vessel itself. These balance-of-plant (BOP) components will almost universally include the following: tubing, valves, pressure regulators, pressure relief devices, and pressure transducers. In addition, many advanced storage systems will require temperature sensors for operation and control.

Components that are "wetted" by pressurized hydrogen will have the same compatibility requirements as the structural materials of the storage vessels. Most BOP components used in H$_2$ storage systems today are made of 316L stainless steel for reasons previously discussed. In fact, the Japanese government currently only allows H$_2$ storage system components to be constructed from 316L stainless steel. However, as with vessel structural materials, research is ongoing to find or develop alternative materials for these components that will meet or exceed the performance of stainless steel to lower system costs. For example, the U.S. DOE has recently funded research to reduce the cost of BOP components for H$_2$ storage applications.

In addition to the gas system components, advanced H$_2$ storage systems are likely to include some form of heat exchanger. Typically, the heat exchanger is an integral part of the storage vessel and is not considered separately here. However, additional BOP components will be required for heat exchange. For refueling, a common assumption is that the cooling fluid that is used for heat removal from the storage system is supplied and circulated by the refueling station. Thus, the only additional BOP components required for refueling are the coolant lines that connect the heat exchanger to the refueling interface. However, for those systems that require heat for H$_2$ delivery to the conversion device, the heat exchange must take place while the system is in use. For these systems, heat of sufficient quality (temperature) must be generated and distributed to the H$_2$ storage vessel on-board the vehicle.

For sorption systems and low-temperature metal hydrides, the heat from a PEM fuel cell or an H_2 ICE is available in sufficient quantity and quality for the purpose of desorption. Since these conversion devices must be cooled anyway, the coolant can be routed directly to the storage system heat exchanger for heating. In this configuration, the only additional BOP components required are the fluid lines connecting the two. An example of this system configuration was implemented by Ovonic Hydrogen Solutions in a hybrid Prius vehicle as discussed in Chapter 3.

With this symbiotic relationship, the radiator that would be used to reject the conversion device heat to ambient air can be reduced in size or removed completely depending on the energy balance between the heat production from consuming hydrogen and the heat consumption required for hydrogen desorption from the storage system. While the radiator may be reduced in size, the coolant pump may have to grow in size to handle the additional pressure drop through the storage system heat exchanger.

Waste heat from a fuel cell or ICE is available in large enough quantities, but it is at a lower temperature than most complex hydrides require. Thus, a heater is required as part of the BOP of a complex metal hydride system. The most efficient way to produce the required heat is by burning some of the stored hydrogen in a burner or reaction of H_2 and O_2 in a catalytic heater. Heat is produced by the oxidation of hydrogen as described by the chemical reaction, Equation 10.14.

$$H_2 + \tfrac{1}{2}O_2 \rightarrow H_2O, \qquad\qquad (10.14)$$

which is an exothermic reaction producing 242 kJ of heat per mole of H_2 (LHV). Uncatalzyed, this reaction occurs in the gas phase only at high temperatures (above 1000°C). At lower temperatures, a catalyst is required to promote the oxidation process; that is, the reaction takes place on the surface of a precious metal (e.g., palladium). Consequently, to oxidize hydrogen without producing excessively high temperatures, catalytic heaters use the flow of hydrogen and air through catalyst-coated channels to promote the reaction.

Whether a hydrogen burner or a catalytic heater is used, the system will require a flow rate of air and hydrogen commensurate with the desired heat output. Depending on the enthalpy of the storage material and the operating temperature, this may be tens of kilowatts of heating power. The hydrogen can be supplied by the storage system, but the air may require a separate blower. Also, tubing, fittings, and valves will be required for the gases as well as the heat transfer fluid. These components must all be considered part of the overall hydrogen storage system.

System Example: Storing Hydrogen in an Engineered NaAlH₄ Bed

In the previous sections, the engineering of advanced hydrogen storage systems has been treated generally with a few brief examples to illustrate various aspects of the process. In this section, an example system is described in some detail to show how all of the elements are brought together. In 2003, General Motors partnered with Sandia National Laboratories to develop an advanced hydrogen storage system based on sodium alanate (NaAlH₄). Sodium alanate is a prototypical complex metal hydride that was described in some detail in Chapter 6. While NaAlH₄ does not possess sufficient capacity to supply fuel for light-duty vehicles with the desired 300 mile range, it serves as a well-studied surrogate with many of the properties anticipated for a suitable complex hydride, including low thermal conductivity, kinetics that require catalysts, significant heat release during hydrogen filling, and high reactivity. The goals of the project were to design, build, and test a hydrogen storage system for use with complex hydrides and to assess system performance. Specifically, performance goals included the ability to supply a fuel cell in simulated driving scenarios, to reduce tank costs as much as possible while maintaining rapid filling and reliable performance, and to achieve good energy content on a mass or volume basis. Several papers have been published previously describing the details of the GM/Sandia hydrogen storage system [11, 12, 44, 56]. A summary is given here.

The GM/Sandia hydrogen storage system is to date the first and only full-scale complex metal hydride system built and optimized to meet the performance requirements of the dynamic automotive platform. The system was designed to be refueled in approximately 10 min and to deliver hydrogen at up to 2.0 g/s. For refueling, a circulating heat transfer fluid (a synthetic mineral oil) was used to remove the heat of rehydrogenation from the sodium alanate, as much as 60 MJ. The design was based on the assumption that this heat is removed from the vehicle and rejected at the refueling station. However, for hydrogen delivery to the fuel cell, heat must be provided to the hydrogen storage system in situ. The GM/Sandia system did not incorporate a fuel cell (or its waste heat), so a hydrogen catalytic heater was used to supply this energy [12]. The same heat transfer fluid used for cooling was circulated between the heater and the hydrogen storage system. Hydrogen from the storage system was mixed with air and pumped through the catalytic heater, producing heat that was transferred to the circulating fluid.

Modularity was a key concept in the design of the GM/Sandia hydrogen storage system. Multiple hydrogen storage modules, rather than a single large tank, enabled the use of hydrogen delivery control logic that significantly increased the efficiency of the system. The hydrogen storage system consisted of four identical modules. Each storage module was a shell-and-tube heat exchanger with sodium alanate stored in tubular vessels enclosed within the heat exchanger shell. Twelve vessels arranged in a staggered 4 by 3 array made up each module. The module vessels were 316L stainless steel and 2.25 inches (0.057 m) in outer diameter and about 36 inches (0.914 m) long. The vessels were designed for a maximum working pressure of 138-bar hydrogen. Each vessel was packed to a density of 1 g/cc (1000 kg/m³) with 1.79 kg of a sodium alanate-graphite mixture so that each module contained about 21.5 kg of material.

Figure 10.13 shows close-up views of a vessel, a bank of vessels prior to shell assembly, a completed module with the heat exchanger shell attached, and the four-module system installed in the test cell. The heat exchanger was divided into sections by the baffles shown around the tube bank. This was to provide for fluid flow across the tube bank for cooling and heating. The completed module included fluid flow distribution manifolds and a hydrogen distribution manifold. The modules were insulated with VIPs and assembled in a test cell. Each hydrogen storage module was designed to hold approximately 0.75 kg of hydrogen for a total storage system capacity of 3.0 kg.

The completed system was the result of an engineering effort that relied heavily on experimentally validated computational simulations. As described, these modules consisted of sodium alanate vessels arranged inside a heat exchanger shell. Two primary modeling efforts were carried out: one for the hydride vessels and one for the heat exchanger. For the hydride vessels, a one-dimensional chemical kinetics and heat transfer model was used to simulate refueling and delivery scenarios and to optimize the effective thermal

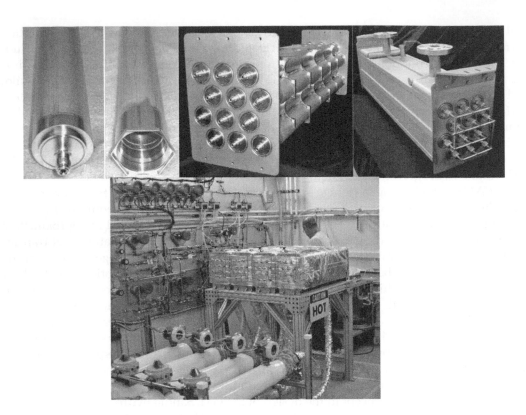

FIGURE 10.13
GM/Sandia hydrogen storage system. Close-up views of one vessel, a bank of vessels prior to shell assembly, a completed module with the heat exchanger shell attached, and the four module system installed in the test cell.

conductivity, operating pressure, and operating temperature for the hydrogen storage modules. A second, more detailed, two-dimensional finite element model was used for understanding the performance of the GM/Sandia hydrogen storage system.

The one-dimensional model was developed to predict radial variations of temperature and chemical composition within a hydrogen storage vessel. The model included coupled chemical kinetics and heat transfer; hydrogen pressure was assumed to be uniform. Given an initial sodium alanate composition profile, the rate equations were integrated using a specified hydrogen pressure and a computed temperature profile. The temperature profile was calculated from an initial condition based on conduction heat transfer within the alanates. A thermal conductivity model was used for these calculations along with heat flux or convective boundary conditions on the outer vessel surface and a volumetric heat source term due to the chemical reaction rate.

Despite the relative simplicity of the one-dimensional model, it was used extensively in the design of the GM/Sandia system. For refueling, the model was used to optimize vessel diameter given constraints on refueling pressure and cooling temperature. It was also used to develop optimum refueling pressure profiles for specified vessel diameter, cooling temperature, and alanate thermal conductivity. Cooling loads for refueling the hydrogen storage system were also calculated with this model. For hydrogen delivery, the model was used to predict delivery rates for specified temperature and pressure conditions. It was also used to determine the heat required to meet specified hydrogen delivery requirements. This information was used, in turn, to size the heating system for the GM/Sandia system.

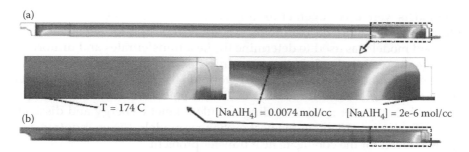

FIGURE 10.14 (See color insert.)

Two-dimensional axisymmetric model of GM/Sandia vessel containing sodium alanate: (a) phase distribution and (b) temperature distribution. The dotted regions in each distribution of (a) and (b) are blown up in the middle figure. The figure shows a snapshot in time during the second step of a hydrogen absorption simulation. Figure 10.14(a) shows the instantaneous concentration of NaAlH₄ and Figure 10.14(b) shows the temperature distribution. Note that the hydrogen gas cools the bed at the entrance on the right-hand side while the absorption reaction heats the bed producing peak temperatures along the axis. The temperature gradients produce concentration gradients as hydrogen is absorbed at different rates depending on local temperature.

For a more detailed representation of the key physics involved in the demonstration system operation, a two-dimensional axisymmetric model of a hydrogen storage vessel was developed. This model was built using Comsol Multiphysics [69]. The model included the complete geometry of a single module vessel represented in two dimensions as shown in Figure 10.14. It included the thermocouple well used in the experiments to access alanate temperatures, as well as the hydrogen inlet geometry.

All of the relevant sodium alanate properties were represented in the model. The chemical rate equations were integrated with heat transfer and mass transfer calculations. For heat transfer, both radial and axial temperature gradients were calculated using conduction heat transfer within the alanates with the phase-dependent thermal conductivity. A convection boundary condition [i.e., $kdT/dr = h(T_o - T)$] for the energy equation was used on the outer surface of the vessel to simulate the effect of the oil flow. Here, T_o is the temperature at the outer radius of the vessel, and T is the bulk oil temperature. The heat transfer coefficient was determined first from a heat exchanger model, then from experimental data. This heat transfer coefficient was assumed to be uniform over the outer surface of each vessel.

Mass transfer of hydrogen was modeled such that radial and axial pressure gradients were calculated based on boundary conditions and the alanate permeability model. The permeability model was included in the momentum equation to calculate the two-dimensional pressure profile. A pressure boundary condition for the momentum equation was applied to the vessel inlet to calculate the flow of hydrogen into the vessel. A constant or time-dependent pressure boundary condition could be used.

The design of the hydrogen storage modules was primarily an effort to maximize the heat exchange between the storage vessels and the heat transfer fluid used to maintain the system at the optimum temperature for both hydrogen refueling and delivery. Based on sizing calculations and given geometric constraints, the modules were configured as rectangular shell-and-tube heat exchangers with a 4 × 3 arrangement of storage vessels as the tubes. The goal for the module-level modeling effort was to develop a fluid flow distribution design that balanced heat transfer against pressure drop. Uniform fluid flow across all storage vessels with high heat transfer would promote optimum hydrogenation and dehydrogenation conditions. The trade-off was high-pressure drop, which would reduce the overall system efficiency.

Several models were used together to fully define the module design. A CFD model of the module inlet and outlet oil manifolds was used to design for adequate flow uniformity. A second model was used to determine the heat transfer rates and uniformity in the heat exchanger as well as the fluid pressure drop through the heat exchanger. A similar, but higher-fidelity, model was then used to determine the effects of the flow pattern from the inlet manifold. In each of these CFD models, momentum and mass transport were solved for the heat transfer fluid, including turbulent kinetic energy and dissipation rate (k-e model) where appropriate. In the heat exchanger models, the energy transport equation was also solved for the conjugate heat transfer problem.

To provide heat to the storage system for hydrogen delivery, a catalytic heater was developed and demonstrated [11, 12]. The catalytic heater was designed to transfer up to 30 kW of heat from the catalytic reaction to the storage system through an intermediate heat transfer fluid. The fluid, a synthetic mineral oil, was circulated between the catalytic heater and the hydrogen storage system to transfer the heat. The design of the catalytic heater consisted of three main parts: (1) the reactor, (2) the gas heat recuperator, and (3) oil and gas flow distribution manifolds. The reactor was a brazed aluminum assembly of 10 alternating layers of finned channels for gas and oil flow. The gas heat recuperator was designed to maximize heat exchange with minimum pressure drop and size. A three-layer finned plate design was used for this purpose. The recuperator was tightly integrated with the reactor to minimize the overall footprint, and it served as a cover for one end of the reactor gas channels. The catalyst coating used with the catalytic heater consisted of a carbon-supported palladium powder mixed with a high-temperature paint and lacquer thinner. At full power, the heater was able to catalytically combust a 10% hydrogen/air mixture flowing at over 80 ft³/min and transfer 30 kW of heat to a 30-gal/min flow of oil over a temperature range from 100°C to 220°C. The total efficiency of the catalytic heater, defined as the heat transferred to the oil divided by the inlet hydrogen chemical energy, was determined to exceed the design goal of 80% for oil temperatures from 60°C to 165°C. The system is shown in Figure 10.15.

During the 15 months of testing the GM/Sandia system, 40 refueling and delivery cycles were performed, including reference cycles to track system capacity, refueling experiments

FIGURE 10.15
Fully integrated 30-kW catalytic heater.

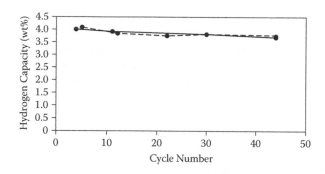

FIGURE 10.16

NaAlH$_4$ storage bed capacity variation as a function of cycle number.

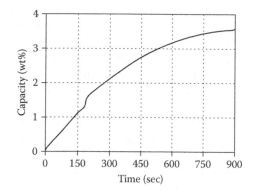

FIGURE 10.17

Example hydrogen absorption data from the GM/Sandia NaAlH$_4$ storage system.

to optimize rate and simulate real-world scenarios, and hydrogen delivery experiments to characterize delivery rate, compare efficiency of control methods, demonstrate startup and peak rates, and simulate realistic fuel cell drive cycles. Figure 10.16 shows the measured hydrogen capacity of the system taking into account only the mass of the stored NaAlH$_4$. An impressive 4.0 wt% capacity was achieved for the full-scale system, which is essentially the same as that observed in gram-level laboratory quantities of NaAlH$_4$. In addition, rapid refueling was demonstrated as shown in Figure 10.17; a capacity of 3.2 wt% was reached in 10 min.

Safety

With regard to safety, there are a number of hazards associated with a materials-based hydrogen storage system. Along with the hazards of a pressurized combustible gas, these systems may include pyrophoric or water-reactive material and either high-temperature or cryogenic thermal hazards. From a practical standpoint, these hazards must be able to be managed safely for such a technology to be commercialized. From the system engineering perspective, it would be required not only to be able to handle the materials safely but also to do so economically in a production manufacturing environment. For some of the more hazardous materials, that could pose a significant challenge.

From the perspective of qualification, an advanced hydrogen storage system would need to meet the codes and standards requirements that govern the application in which it would be used. For vehicular applications, codes and standards are currently being developed for hydrogen storage systems and their refueling, as discussed in the structural design section of this chapter and discussed in more detail in Chapter 12.

Performance Relative to Targets

Through the FreedomCAR and Fuel Partnership, the U.S. DOE, USCAR, energy companies, and utility partners have developed performance, cost, durability, and safety targets for hydrogen storage systems for light-duty vehicles [70]. These targets have evolved since they were first introduced in 2003 and have become the standard by which hydrogen storage systems are measured. The targets for automotive systems have proven very difficult to meet due to the desire to provide the consumer with the same attributes as a fuel tank for a gasoline ICE. The DOE is currently developing similar targets for other hydrogen storage markets that will potentially be relaxed in some areas of performance relative to the automotive requirements.

The three rows of Table 10.2 show the current gravimetric and volumetric density targets along with older cost targets for hydrogen storage systems for light-duty vehicles developed by the DOE. The current cost targets are to be determined (TBD). The current system density targets are given for the years 2010, 2017, and for "ultimate full fleet," which represents full vehicle penetration into the market. The 2010 target for gravimetric energy density of a vehicle hydrogen storage system is 4.5 wt%. Thus, a system that weighs 100 kg must store 4.5 kg of hydrogen to meet this target. The 2010 target for volumetric density is 28 gH$_2$/L of storage system volume. Applying this target to the previous example, the 100-kg system must be equal to or less than 160.7 L in volume. To meet the 2010 target for cost, that system would have to cost less than \$599 (4.5 kg × \$133/kg). The 2017 targets are more difficult to meet at 5.5 wt%, 40 gH$_2$/L, and \$67/kg H$_2$. Finally, the ultimate full fleet targets are 7.5 wt% and 70 gH$_2$/L, with a cost target that is yet to be determined.

To compare these targets to the current state of the art, Figures 10.18 to 10.20 (reproduced from [71]) show the weight, volume, and projected costs of a number of hydrogen storage systems. The values given for the various systems were taken from the DOE-sponsored analysis of these systems performed by Argonne National Labs and TIAX, LLC, and

TABLE 10.2

Energy Density and Cost Targets for Hydrogen Storage Systems for Light-Duty Vehicles

Target	Gravimetric Density (wt%)	Volumetric Density (gH$_2$/L)	Capital Cost (\$/kWh)[a]
DOE light-duty vehicle targets, 2010	4.5	28	4[b]
DOE light-duty vehicle targets, 2017	5.5	40	2[c]
DOE light-duty vehicle targets, ultimate	7.5	70	TBD

[a] From M.A. Richard, P. Bénard, R. Chahine, Gas adsorption process in activated carbon over a wide temperature range above the critical point. Part 1: Modified Dubinin-Astakhov model. *Adsorption* **15**, 43–51 (2009).
[b] Old 2010 cost target. Current target to be determined.
[c] Old 2015 cost target. Current 2017 target to be determined.

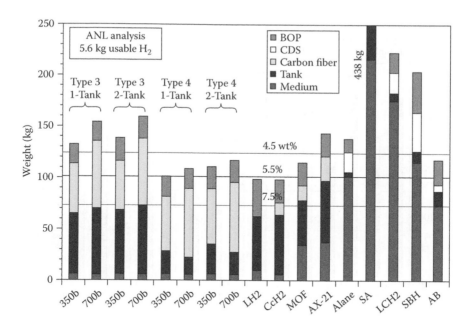

FIGURE 10.18 (See color insert.)
System weight for 5.6-kg H$_2$ storage for various storage options. Definitions: SA—sodium alanate, LCH2—liquid organic carrier, SBH—sodium borohydride, AB—ammonia borane. (Reproduced from R.K. Ahluwalia, T.Q. Hua, J.K. Peng, On-board and off-board performance of hydrogen storage options for light-duty vehicles, *Int. J. Hydrogen Energy* **37**, 2891 (2012) with permission of Elsevier.)

presented at the DOE annual merit review meetings from 2005 to 2010 [72–75]. The storage system options listed include high-pressure gaseous hydrogen storage, liquid H$_2$, cryo-compression, metal hydrides, sorption systems, and several off-board regenerable systems. Of the systems that can be refueled on-board with H$_2$, only cryocompressed (CcH2 in Figures 10.18 to 10.20) and the MOF adsorption systems were found capable of meeting both the 2010 targets for gravimetric and volumetric density, although the 700-bar com-pressed and the liquid H$_2$ systems were close. Furthermore, as it currently stands, only the cryocompressed system appears capable of meeting the 2017 performance targets, and no current system meets the ultimate performance. Perhaps more important, Figure 10.20 shows that none of the currently pursued systems comes close to even the old 2010 cost target shown on this graph at $4/kWh. The LH$_2$ system, at twice the cost of the 2010 target, is the closest. The hybrid tank concept [4–6] was not analyzed, but would have a very high system weight (offscale in Figure 10.18) due to the use of heavy interstitial metal hydrides.

Future Outlook

The implication of Figures 10.18 to 10.20 is that significant advancements enabling improved performance and cost reductions are required if hydrogen storage systems are going to meet the DOE targets for light-duty vehicles. To that end, the DOE funded the Hydrogen Storage Engineering Center of Excellence (HSECoE) beginning in 2009. Part of the activ-ity of the HSECoE is to further develop and assess system-level designs for advanced

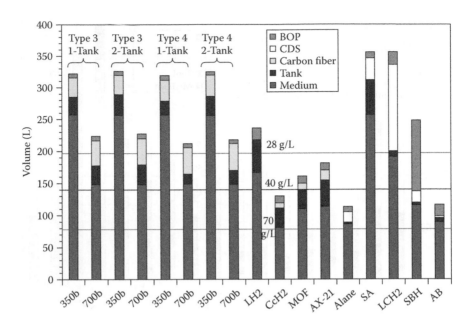

FIGURE 10.19 (See color insert.)
System volume for 5.6-kg H₂ storage for various storage options. (Reproduced from R.K. Ahluwalia, T.Q. Hua, J.K. Peng, On-board and off-board performance of hydrogen storage options for light-duty vehicles, *Int. J. Hydrogen Energy* **37**, 2891 (2012) with permission of Elsevier.)

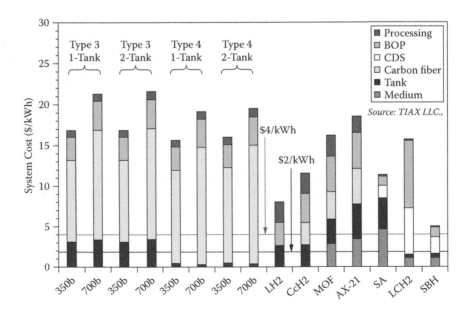

FIGURE 10.20 (See color insert.)
Projected cost of hydrogen storage systems at high-volume manufacturing. (Reproduced from R.K. Ahluwalia, T.Q. Hua, J.K. Peng, On-board and off-board performance of hydrogen storage options for light-duty vehicles, *Int. J. Hydrogen Energy* **37**, 2891 (2012) with permission of Elsevier.)

hydrogen storage systems. Some of that engineering assessment work is reported in the next chapter. Both new material development/discovery and improved systems engineering must take place, it is hoped in concert, for significant advances to be made in this field.

It is important to emphasize that the comparison to targets made in Figures 10.18 to 10.20 applies only to the problem of storing hydrogen for light-duty vehicles. For other (nonvehicular) applications of hydrogen storage, different storage targets would likely apply. Sandia recently conducted a study [76] to assess the energy storage and density requirements for nonvehicular applications for which fuel cell technology (and hydrogen storage) could be applied. The purpose of the study was to enable the DOE to develop hydrogen storage targets for these different applications, such as construction equipment, aviation ground support equipment, and portable electronics [76]. It remains to be seen how the current hydrogen storage technology will compare to those nonmotive targets. All applications of hydrogen storage would benefit from improvements in the thermodynamic and kinetic properties of these materials.

References

1. U.S. Department of Energy, The Hydrogen, Fuel Cells and Infrastructure Technologies Program 2006 Multi-year Research Demonstration and Development Plan (MYRDDP). http://www1.eere.energy.gov/hydrogenandfuelcells/mypp/.
2. M.A. Richard, P. Bénard, R. Chahine, Gas adsorption process in activated carbon over a wide temperature range above the critical point. Part 1: Modified Dubinin-Astakhov model, *Adsorption* **15**, 43–51 (2009).
3. R.K. Ahluwalia, J.K. Peng, Automotive hydrogen storage system using cryo-adsorption on activated carbon, *Int. J. Hydrogen Energy* **34**, 5476 (2009).
4. D. Mori, K. Hirose, Recent challenges of hydrogen storage technologies for fuel cell vehicles, *Int. J. Hydrogen Energy* **34**, 4569 (2009).
5. M. Visaria, I. Mudawar, T. Pourpoint, Enhanced heat exchanger design for hydrogen storage using high-pressure metal hydride. Part 1: Design methodology and computational results, *Int. J. Heat Mass Transfer* **54**, 413 (2011).
6. M. Visaria, I. Mudawar, T. Pourpoint, Enhanced heat exchanger design for hydrogen storage using high-pressure metal hydride. Part 2. Experimental results, *Int. J. Heat Mass Transfer* **54**, 424 (2011).
7. I.P. Jain, P. Jain, A. Jain, Novel hydrogen storage materials: a review of lightweight complex hydrides, *J. Alloys Compd.* **503**, 303 (2010).
8. B. Sakintuna, F. Lamari-Darkrim, and M. Hirscher, Metal hydride materials for solid hydrogen storage: a review, *Int. J. Hydrogen Energy* **32**, 1121 (2007).
9. S.M. Aceves, G.D. Berry, A.H. Weisberg, F. Espinosa-Loza, S.A. Perfect, Advanced concepts for vehicular containment of compressed and cryogenic hydrogen, Proceedings of WHEC 16, 2006. Lyon, France.
10. B. Bogdanovic, M. Schwickardi, Ti-doped alkali metal aluminium hydrides as potential novel reversible hydrogen storage materials, *J. Alloys Compd.* **1**, 253 (1997).
11. T.A. Johnson, M.P. Kanouff, Performance characterization of a hydrogen catalytic heater, Sandia Technical Report, SAND2010-2474, 2010, Sandia National Laboratories, Albuquerque, NM.
12. T.A. Johnson, M.P. Kanouff, Development of a hydrogen catalytic heater for heating metal hydride hydrogen storage systems, *Int. J. Hydrogen Energy* **37**, 2304 (2011).
13. S. Lasher, Analyses of hydrogen storage materials and on-board systems, Presented at the 2005 DOE annual merit review. May 2005. http://www.hydrogen.energy.gov/pdfs/review05/st19_lasher.pdf.

14. M.A. Richard, D. Cossement, P.A. Chandonia, R. Chahine, D. Mori, K. Hirose, Preliminary evaluation of the performance of an adsorption-based hydrogen storage system, *AIChE J.* **55**(11), 2985–2996 (2009).

15. R. Paggiaro, F. Michl, P. Benard, W. Polifke, Cryo-adsorptive hydrogen storage on activated carbon. II: Investigation of the thermal effects during filling at cryogenic temperatures, *Int. J. Hydrogen Energy* **35**, 648 (2010).

16. C. Na Ranong, M. Höhne, J. Franzen, J. Hapke, G. Fieg, M. Dornheim, N. Eigen, J.M. Bellosta von Colbe, O. Metz, et al., Concept, design and manufacture of a prototype hydrogen storage tank based on sodium alanate, *Chem. Eng. Technol.* **32**, 1154 (2009).

17. F. Askri, M. Ben Salah, A. Jemni, S. Ben Nasrallah, Heat and mass transfer studies on metal-hydrogen reactor filled with MmNi$_{4.6}$ Fe$_{0.4}$, *Int. J. Hydrogen Energy* **34**, 6705 (2009).

18. C.A. Chung, C.S. Lin, Prediction of hydrogen desorption performance of Mg$_2$Ni hydride reactors, *Int. J. Hydrogen Energy* **34**, 9409 (2009).

19. A. Chaise, P. de Rango, P. Marty, D. Fruchart, S. Miraglia, R. Olives, et al. Enhancement of hydrogen sorption in magnesium hydride using expanded natural graphite, *Int. J. Hydrogen Energy* **34**, 8589 (2009).

20. T.L. Pourpoint, V. Velagapudi, I. Mudawar, Y. Zheng, T.S. Fisher, Active cooling of a metal hydride system for hydrogen storage, *Int. J. Heat Mass Transfer* **53**, 1326 (2010).

21. S. Mellouli, F. Askri, H. Dhaou, A. Jemni, S. Ben Nasrallah, A novel design of a heat exchanger for a metal-hydrogen reactor, *Int. J. Hydrogen Energy* **32**, 3501 (2007).

22. S. Mellouli, F. Askri, H. Dhaou, A. Jemni, S. Ben Nasrallah, Numerical study of heat exchanger effects on charge/discharge times of metal-hydrogen storage vessel, *Int. J. Hydrogen Energy* **34**, 3005 (2009).

23. H. Dhaou, A. Souahlia, S. Mellouli, F. Askri, A. Jemni, S. Ben Nasrallah, Experimental study of a metal hydride vessel based on a finned spiral heat exchanger, *Int. J. Hydrogen Energy* **35**, 1674 (2010).

24. S. Mellouli, F. Askri, H. Dhaou, A. Jemni, S. Ben Nasrallah, Numerical simulation of heat and mass transfer in metal hydride hydrogen storage tanks for fuel cell vehicles, *Int. J. Hydrogen Energy* **35**, 1693 (2010).

25. M. Raju, S. Kumar, Optimization of heat exchanger designs in metal hydride based hydrogen storage systems, *Int. J. Hydrogen Energy* (2011), doi:10.1016/j.ijhydene.2011.06.120.

26. G. Mohan, M.P. Maiya, S.S. Murthy, Performance simulation of metal hydride hydrogen storage device with embedded filters and heat exchanger tubes, *Int. J. Hydrogen Energy* **32**, 4978 (2007).

27. A. Freni, F. Cipitı, G. Cacciola, Finite element-based simulation of a metal hydride-based hydrogen storage tank, *Int. J. Hydrogen Energy* **34**, 8574 (2009).

28. D. Mosher, S. Arsenault, X. Tang, D. Anton, Design, fabrication and testing of NaAlH$_4$ based hydrogen storage systems, *J. Alloys Compd.* **446–447**, 707 (2007).

29. D. Anton, D. Mosher, T. Xia, R. Brown, S. Arsenault, A. Saitta, et al., *High Density Hydrogen Storage System Demonstration Using NaAlH$_4$ Based Complex Compound Hydrides, Final Report*, U.S. Department of Energy, Washington, DC, 2007.

30. B.J. Hardy, D.L. Anton, Hierarchical methodology for modeling hydrogen storage systems. Part I: Scoping models, *Int. J. Hydrogen Energy* **34**, 2269 (2009).

31. B.J. Hardy, D.L. Anton, Hierarchical methodology for modeling hydrogen storage systems. Part II: Detailed models, *Int. J. Hydrogen Energy* **34**, 2992 (2009).

32. A.F. Mills, *Heat Transfer*, Irwin, Homewood, IL, 1992.

33. F.P. Incropera, D.P. Dewitt, *Fundamentals of Heat and Mass Transfer*, 5th edition, Wiley, New York, 2001.

34. A. Isselhorst, Heat and mass transfer in coupled hydride reaction beds, *J. Alloys Compd.* **231**, 871 (1995).

35. M.R. Gopal, S.S. Murthy, Studies on heat and mass transfer in metal hydride beds, *Int. J. Hydrogen Energy* **20**, 911 (1995).

36. M.Y. Ha, I.K. Kim, H.D. Song, S. Sung, D.H. Lee, A numerical study of thermo-fluid phenomena in metal hydride beds in the hydriding process, *Int. J. Heat Mass Transfer* **47**, 2901 (2004).

37. C.A. Krokos, D. Nikolic, E.S. Kikkinides, M.C. Georgiadis, A.K. Stubos, Modeling and optimization of multi-tubular metal hydride beds for efficient hydrogen storage, *Int. J. Hydrogen Energy* **34**, 9128 (2009).

38. J. Ye, L. Jiang, Z. Li, X. Liu, S. Wang, X. Li, Numerical analysis of heat and mass transfer during absorption of hydrogen in metal hydride based hydrogen storage tanks, *Int. J. Hydrogen Energy* **35**, 8216 (2010).

39. Y. Wang, F. Yang, X. Meng, Q. Guo, Z. Zhang, I.S. Park, et al., Simulation study on the reaction process based single stage metal hydride thermal compressor, *Int. J. Hydrogen Energy* **35**, 321 (2010).

40. B. MacDonald, A. Rowe, A thermally coupled metal hydride hydrogen storage and fuel cell system, *J. Power Sources* **161**, 346 (2006).

41. B. MacDonald, A. Rowe, Impacts of external heat transfer enhancements on metal hydride storage tanks, *Int. J. Hydrogen Energy* **31**, 1721 (2006).

42. F. Askri, M. Ben Salah, A. Jemni, S. Ben Nasrallah, Optimization of hydrogen storage in metal-hydride tanks, *Int. J. Hydrogen Energy* **34**, 897 (2009).

43. Y. Kaplan, Effect of design parameters on enhancement of hydrogen charging in metal hydride reactors, *Int. J. Hydrogen Energy* **34**, 2288 (2009).

44. T. Johnson, S. Jorgensen, D. Dedrick, Performance of a full-scale hydrogen-storage tank based on complex hydrides, *Faraday Disc.* **151**, 385–397 (2011) doi:10.1039/C0FD00017E.

45. A.R. Miller, D.H. DaCosta, M. Golben, Reversible metal-hydride storage for a fuelcell mine loader, presented at Intertech-Pira's Hydrogen Production and Storage Forum, Vancouver, Canada, September 11–13, 2006.

46. T.A. Johnson, D.E. Dedrick, Effects of metal hydride properties on the performance of hydrogen storage systems, Proceedings from the 2007 MS&T Conference, Detroit, MI.

47. D.E. Dedrick, Solid-state hydrogen storage system design, in *Solid-State Hydrogen Storage: Materials and Chemistry*, edited by G. Walker, Woodhead, University of Nottingham, UK, 2008, Chap. 4.

48. T.A. Johnson, M.P. Kanouff, D.E. Dedrick, G.H. Evans, S.W. Jorgensen, Model-based design of an automotive-scale, metal hydride hydrogen storage system, *Int. J. Hydrogen Energy* **37**, 2835 (2012).

49. S. Ergun, Fluid flow through packed columns, *Chem. Eng. Prog.* **48**, 89 (1952).

50. L.J. Durlofsky, J.F. Brady, Analysis of the Brinkman equation as a model for flow in porous media, *Phys. Fluids* **30**, 3329 (1987).

51. M. Parvazinia, V. Nassehi, R.J. Wakeman, M.H.R. Ghoreshy, Finite element modeling of flow in a porous medium between two parallel plates using the Brinkman equation, *Transport Porous Media* **63**, 71 (2006).

52. F.J. Valdes-Parada, J.A. Ochoa-Tapia, J. Alvarez-Ramirez, On the effective viscosity for the Darcy-Brinkman equation, *Physics A* **385**, 69 (2007).

53. H. Lui, P.R. Patil, U. Narusawa, On Darcy-Brinkman equation: viscous flow between two parallel plates packed with regular square arrays of cylinders, *Entropy* **9**, 118 (2007).

54. J.B. Young, B. Todd, Modeling of multi-component gas flows in capillaries and porous solids, *Int. J. Heat Mass Transfer* **48**, 5338 (2005).

55. B.J. Hardy, Integrated hydrogen storage system model, Report, WSRC-TR-20007-00440, rev. 0., Savannah River National Laboratory. November 16, 2007. http://www1.eere.energy.gov/hydrogenandfuelcells/pdfs/bruce_hardy_srnl-2007-0043_part2.pdf#h2_storage.

56. D.E. Dedrick, M.P. Kanouff, R.S. Larson, T.A. Johnson, S.W. Jorgensen, Heat and mass transport in metal hydride based hydrogen storage systems, HT2009-88231, Proceedings of ASME Summer Heat Transfer Conference, 2009, San Francisco, CA.

57. W. Luo, K. Gross, A kinetics model of hydrogen absorption and desorption in Ti-doped NaAlH₄, *J. Alloys Compd.* **385**, 224 (2004).

58. G.A. Lozano, C. Na Ranong, J.M. Bellosta von Colbe, R. Bormann, G. Fieg, J. Hapke, M. Dornheim, Empirical kinetic model of sodium alanate reacting system (I). Hydrogen absorption, *Int. J. Hydrogen Energy* **35**, 6763 (2010).

59. G.A. Lozano, C. Na Ranong, J.M. Bellosta von Colbe, R. Bormann, G. Fieg, J. Hapke, M. Dornheim, Empirical kinetic model of sodium alanate reacting system (II). Hydrogen desorption, *Int. J. Hydrogen Energy* **35**, 7539 (2010).

60. NASA–NSS1740.16: 1997 safety standard for hydrogen and hydrogen systems—guidelines for hydrogen system design, materials selection, operations, storage and transportation. 1997.

61. AIAA G-095: 2004 guide to safety of hydrogen and hydrogen systems. 2004. National Aeronautics and Space Administration, American Institute of Aeronautics and Astronautics, Reston, VA.

62. ISO/TR 15916: 2004 basic considerations for the safety of hydrogen systems. 2004. International Organization for Standardization, Geneva, Switzerland.

63. ISO 16111: 2008 transportable gas storage devices—hydrogen absorbed in reversible metal hydride. 2008.

64. B.P. Somerday, C. San Marchi, Technical reference on hydrogen compatibility of materials, Sandia National Laboratories Technical Report No. SAND2008-1163, Albuquerque, NM.

65. R.P. Gangloff and B.P. Somerday, *Gaseous Hydrogen Embrittlement of Materials in Energy Technologies: Volume 1: The Problem, Its Characterisation, and Effects on Particular Alloy Classes*, Woodhead, Cambridge, UK, 2012.

66. K.A. Nibur, C. San Marchi, B.P. Somerday, Fracture and fatigue tolerant steel pressure vessels for gaseous hydrogen (PVP2010–25827), ASME Pressure Vessels and Piping Division Conference, July 18–22, 2010, Bellevue WA.

67. C. San Marchi, B.P. Somerday, K.A. Nibur, D.G. Stalheim, T. Boggess, S. Jansto, Fracture and fatigue of commercial grade pipeline steels in gaseous hydrogen (PVP2010–25825), ASME Pressure Vessels and Piping Division Conference, July 18–22, 2010, Bellevue, WA.

68. K.A. Nibur, B.P. Somerday, C. San Marchi, J.W. Foulk, III, M. Dadfarnia, P. Sofronis, G.A. Hayden, Measurement and interpretation of threshold stress intensity factors for steels in high-pressure hydrogen gas, Sandia Technical Report SAND2010-4633, Sandia National Laboratories, Livermore, CA, July 2010.

69. Comsol, Inc., Comsol Multiphysics, Comsol, Palo Alto, CA.

70. U.S. Department of Energy, Targets for onboard hydrogen storage systems for light-duty vehicles, U.S. Department of Energy, Office of Energy Efficiency and Renewable Energy and the FreedomCAR and Fuel Partnership, September 2009. Revision 4.0, p. 9.

71. R.K. Ahluwalia, T.Q. Hua, J.K. Peng, On-board and off-board performance of hydrogen storage options for light-duty vehicles, *Int. J. Hydrogen Energy* **37**, 2891 (2012).

72. S. Lasher, K. McKenney, Y. Yang, M. Hooks, Analysis of hydrogen storage materials and on-board systems—cryocompressed and liquid hydrogen system cost assessments; 2008. Presentation at DOE Hydrogen Program annual review, Arlington, VA. 2008. http://www.hydrogen.energy.gov/annual_review08_storage.html#test.

73. S. Lasher, K. McKenney, J. Sinha, P. Chin, Analysis of hydrogen storage materials and on-board systems—compressed and liquid hydrogen carrier system cost assessments; 2009. Presentation at DOE Hydrogen Program annual review, Arlington, VA. 2009. http://www.hydrogen.energy.gov/annual_review09_storage.html.

74. S. Lasher, Analysis of hydrogen storage materials and onboard systems—updated cryogenic and compressed hydrogen storage system cost assessments; 2010. Presentation at DOE Hydrogen Program annual review, Washington, DC. http://www.hydrogen.energy.gov/annual_review10_storage.html.

75. R.K. Ahluwalia, T.Q. Hua, J.K. Peng, R. Kumar, System level analysis of hydrogen storage options; 2007–2010. Presentation at DOE Hydrogen Program annual review, Arlington, VA.

76. L.E. Klebanoff, J.W. Pratt, T.A. Johnson, M. Arienti, L. Shaw, M. Moreno, Final report for analysis of H_2 storage needs for early market non-motive fuel cell applications. Sandia Technical Report SAND2012-1739, 2012, Sandia National Laboratories, Albuquerque, NM.

11

Engineering Assessments of Condensed-Phase Hydrogen Storage Systems

Bob Bowman, Don Anton, and Ned Stetson

CONTENTS

Introduction 385
Assessments of Automotive Hydrogen Storage Systems 386
    Performance Objectives for Storage Systems..387
    Integrated Vehicle Systems Modeling..387
    Specifications for Drive Cycles...389
    Complex Hydride Systems..389
        Materials Properties and Behavior..389
        System Descriptions..392
        Performance Predictions 392
    Chemical Hydride Storage Systems 394
        Material Properties and Behavior 394
        System Description..394
        Performance Attributes...396
    Adsorbent-Based Hydrogen Storage Systems 396
        Material Descriptions and Behavior 396
        Storage System Description 397
        Performance Attributes...397
    Summary of Assessments 397
Ranges of Applicability for Hydrogen Storage Systems 399
Status and Outlook 400
Acknowledgments...401
References...401

Introduction

The development of efficient and safe hydrogen storage systems includes the design and fabrication of containment and auxiliary components (i.e., the so-called engineering issues). Whether hydrogen storage is via physical or chemical means, this hardware must address initial and long-term performance requirements for the desired application [1–5]. Furthermore, the engineering process will impact overall cost and system-level performances [6, 7]. Since H_2 gas is highly flammable, all storage systems need to be configured and constructed to minimize its leakage into any confined spaces, such as engine or passenger compartments within vehicles or the structures where the hydrogen-powered devices are kept or operated [8]. The chemical reactivity of candidate chemical storage media with the atmosphere and common environmental materials (e.g., water, alcohols, lubrication oils, etc.) should be investigated and addressed via appropriate engineered safeguards for normal operation and plausible accident scenarios [9–11]. To optimize the weight and volume storage densities of condensed-phase chemical storage media, these materials are often highly compacted powders [12–14]. Vessel designs for these compacts must accommodate the expansion/contraction cycles during the hydrogen absorption/ desorption process without generating excessive stresses on their structural walls [15]. Substantial engineering efforts are necessary to meet these often-conflicting requirements.

Probably the most commonly used (and frequently misrepresented) parameter in hydrogen storage technology is *gravimetric capacity*, which is the ratio of hydrogen (H_2) mass stored to either the mass of the hydrogenated sorbent material itself or to the mass of the complete storage system [16]. Researchers and proponents for improved or "advanced" condensed-phase storage media often assign gravimetric capacity to the theoretical (or idealized) total hydrogen content of a chemical system rather than the actual amount available during its discharge from a storage container under realistic operational conditions. Hence, their performance claims can be greatly exaggerated or outright misleading. In a recent paper, Demirci and Miele [16] thoroughly discussed the various aspects for defining gravimetric capacity with respect to either material-level or system-level viewpoints. While using the materials-based criteria can be useful for screening candidates, only system-level capacities that include the storage media as well as initially stored hydrogen along with all vessel and auxiliary components are meaningful for representing the performance level for the complete storage system in any specific application. Furthermore, the contributions from the storage vessel and components will exceed the weight of the condensed-phase storage media (i.e., combination of hydrogen and host materials themselves) by at least 50–60% even for the most aggressive (i.e., minimal weight/volumes and highly integrated components) designs and under the most favorable operating scenarios. The actual weights and volumes of past and current hydrogen storage systems built and demonstrated under laboratory or field conditions have weighed more than the storage media by factors greater than two and sometimes as much as an order of magnitude [1, 2, 4]. These increases reflect use of commercially available hardware components, efforts to minimize manufacturing and fabrication costs, and high safety factors to account for high pressure and temperature during testing. Significant refinements in engineering designs and analyses are needed to permit meeting all performance and safety targets with minimal weight and volume for the hydrogen storage systems.

In Chapter 10, Terry Johnson and Pierre Bénard provided a general overview of engineering approaches and issues with hydrogen storage systems. They addressed structural and thermal designs; material properties and selection, including hydrogen compatibility and safety concerns, analytical modeling of heat, and mass transfers for absorption and adsorption storage media; and testing of prototype storage devices. In particular, they illustrated many of these features with the design, modeling, fabrication, and laboratory testing of a full-scale (e.g., nominal 3-kg H_2 storage capacity) sodium alanate prototype storage unit.

Assessments of Automotive Hydrogen Storage Systems

This section describes the engineering status of condensed-phase hydrogen storage delivery systems for automotive and transport applications [17], which have been heavily studied in recent years. These include adsorption systems, chemical hydride systems, and metal/complex hydride systems. Broad descriptions are given of storage systems and their anticipated performance against a set of engineering technical targets developed by the U.S. Department of Energy (U.S. DOE) [18]. One prototypical material was selected for each storage material type. These prototypical material selections were made due to their relatively high degree of data availability and their availability commercially in large quantities. These materials are not intended to be a selection of recommended candidates but rather used to identify the current state of the art for storage system technologies. Similarly, the system descriptions and their associated balance of plant (BOP) were selected as starting points to determine if the technical targets can be met or to what extent the performance is deficient. System descriptions and results are drawn from the combined work of the U.S. DOE Hydrogen Storage Engineering Center of Excellence (HSECoE) [19]. Descriptions of the systems and their components are brief, and large amounts of secondary BOP components such as valves, pressure relief devices, and the like are not described. These system concepts are rudimentary, and details need to be further developed for specific applications. Descriptions are given only to serve as a guide in determining the technical barriers, which need to be addressed to make these systems viable.

Performance Objectives for Storage Systems

All engineering systems have performance metrics, which need to be met. The U.S. DOE has established such metrics for automotive hydrogen storage and regularly updates these metrics as new vehicle data and specifications are brought to light. The current set of these technical metrics [18] is given in Table 11.1. The table gives the 2017 and Ultimate "Nonquantified," "Mandatory," and "Desirable" targets, with current values listed for three solid-state storage systems. The metal hydride material listed is $NaAlH_4$; the chemical hydride listed is for 10 wt.% ammonia borane (AB) in the ionic liquid 1-butyl-3-methylimidazolium chloride (bmimCl) and the adsorbent material is the activated carbon AX-21. These material properties will be discussed more in the pages to follow. The "nonquantified" targets cannot be represented by a number, and relate to very stringent safety requirements. The "mandatory" metrics are associated with storage system function, such as hydrogen delivery temperature and pressure. The "desirable" targets in Table 11.1 define the level of system performance and will differentiate system and vehicle operation. These attributes include gravimetric and volumetric density, fill time, and fuel purity. These characteristics will need to be balanced and traded against each other to meet the requirements of the application and customer.

Integrated Vehicle Systems Modeling

To determine the current state-of-the-art system performance, preliminary system architectures were designed and used in a unified modeling system run under a MatLab-SimuLink environment developed by the HSECoE [20]. This SimuLink model included the hydrogen storage system, fuel cell system, and vehicle system models.

A number of the system targets described in Table 11.1 are not easily determined by component performance alone but need to be evaluated in the context of the entire vehicle system. These targets include the amount of available hydrogen, transient response times, start time to full flow rate, and dormancy. To quantify system performance against these targets, an integrated system model is required. This model needs to be used in parallel with component development to set the operating space for the hydrogen storage system. Such a model

TABLE 11.1

Summary of Automotive Hydrogen Storage Technical Targets [18] and System Performance Parameters Used in HSECoE Analyses

Technical Target	Units	2017	Ultimate	Metal Hydride	Chemical Hydride	Adsorbent
Nonquantified						
Permeation and leakage	scc/h	[a]	[a]	[b]	[b]	[b]
Toxicity		[a]	[a]	[b]	[b]	[b]
Safety		[a]	[a]	[b]	[b]	[b]
Mandatory						
Minimum delivery temperature	°C	−40	−40	−40	−40	−40
Maximum delivery temperature	°C	85	85	85	85	85
Minimum delivery pressure (PEM)	bar	5	3	5	5	5
Maximum delivery pressure	bar	12	12	12	12	12
Minimum operating temperature	°C	−40	−40	−30	—	−30
Maximum operating temperature	°C	60	60	50	50	50
Desirable						
Gravimetric density	kgH_2/kg system	0.055	0.075	0.012	0.038	0.039
Minimum full flow rate	$[gH_2/s]$/kW	0.02	0.02	0.02	0.02	0.02
System cost[c]	$/kWh net	2	TBD	49.0	25.6	18.5
On-board efficiency	%	90	90	78	97	95
Volumetric density	kgH_2/l	0.040	0.070	0.012	0.034	0.024
Cycle life	N	1500	1500	1000	1000	1000
Fuel cost[c]	$/gge	2–6	2–3	7.3[e]	—	4.8[e]
Loss of usable hydrogen	$[gH_2/h]/kgH_2$	0.05	0.05	0.1	0.1	0.44
WPP efficiency[d]	%	60	60	44.1	37.0	40.1
Fuel purity	%	99.97	99.97	99.97	99.97	99.99
Transient response	s	0.75	0.75	0.75	0.49	0.75
Start time to full flow (−20°C)	s	15	15	15	1	15
Fill time	min	3.3	2.5	10.5	5.4	4.2
Start time to full flow (20°C)	s	5	5	5	1	5

Note: TBD, to be determined as DOE and industry are currently reassessing appropriate value in light of technical progress.

[a] Undefined.

[b] Satisfactory.

[c] Previous values.

[d] WPP: Wheel to power plant.

[e] Value from Reference [20].

required integration of vehicle performance, fuel cell performance, and storage system performance. This has been accomplished through the development of an *integrated system model* run under a MatLab-SimuLink environment developed by the HSECoE partners [20–22].

This integrated system model uses lumped-parameter models for the thermal integration of the storage system and fuel cell BOP feeding into the vehicle requirements. The vehicle demand is dictated by a series of well-established drive cycles described in the following section. An outline of this integrated model is given in Figure 11.1. Each system component (vehicle, fuel cell, or storage system) has a series of inputs or demands placed on it and outputs or the results of these demands. These results are calculated from the lumped-parameter models embedded in the code, which simulates the systems operation. Many of the tools developed by the HSECoE are now available on the Internet [23] for general use and evaluation.

Specifications for Drive Cycles

A number of drive cycles were implemented to identify storage system performance. This is particularly relevant in the case of transient system performance, where the targets given in Table 11.1 have not identified the duration for which the targets need to be met. In many instances, engineering solutions have been implemented to meet the targets not directly related to the storage system media used. A summary of the drive cycles and the targets determined by them are given in Table 11.2. These drive cycles were input parameters into the integrated system model described previously and used to predict the performance against the technical targets [20].

Complex Hydride Systems

Materials Properties and Behavior

Conventional interstitial metal hydrides (presented in Chapter 5) and complex metal hydrides (discussed in Chapter 6) are chemically similar from the storage system viewpoint and are considered together here. These materials are rechargeable inside their container with only the addition of hydrogen gas and the removal of heat. These materials typically have enthalpies of hydrogen absorption of −30 to −60 kJ/mol H_2. Thus, during charging, the system will need to dissipate this amount of heat during the charging time. In addition, as hydrogen is delivered, the same amount of heat will need to be supplied to system. Additional information on relevant engineering aspects of these storage media is included in Chapter 10 by Johnson and Bénard.

For automotive applications, the ultimate DOE target is to load 5 kg of hydrogen (i.e., 2500 mol) in 2.5 min [18]. For a compound with rehydrogenation enthalpy of −40 kJ/mol H_2, this requires a heat dissipation of 100 MJ in 2.5 min or 40 MJ/min. This typical thermal transfer requirement necessitates the use of substantial heat exchange equipment within the storage vessel. A number of recent studies [24–27] have been published to determine the required heat exchanger characteristics as a function of media-specific heat, thermal conductivity, enthalpy, and absorption kinetics. A very important study [28] has developed a simple relationship between the technical target, the media characteristics, and the heat exchanger design. This relationship is given as

$$\left(\frac{1}{L^2}\right)\left(\frac{k\ M_{Hyd\_eff}}{-\ H_{overall}}\frac{T}{\rho_{Hydride}}\right) = \frac{1}{mM_{H2}}\frac{m_{H2}}{t} \tag{11.1}$$

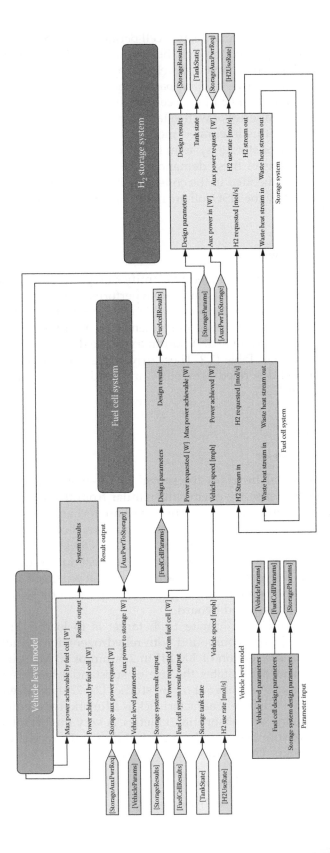

FIGURE 11.1 (See color insert.)
Integrated system model: hierarchy of modules that was developed and utilized by the Hydrogen Storage Engineering Center of Excellence [HSECoE] [Reference 20].

TABLE 11.2

Drive Cycles Utilized in System Target Status Assessments [20]

Drive Cycle	Test Schedule	Cycle	Description	Target	Temperature (°C)
1	**Ambient drive cycle** — Repeat the EPA FE cycles from full to empty and adjust for 5-cycle post-2008	UDDS	Low speeds in stop-and-go urban traffic	System size	24
		HWFET	Free-flow traffic at highway speeds		24
2	**Aggressive drive cycle** —Repeat from full to empty	US06	Higher speeds; harder acceleration and braking	Minimum flow rate and transient response	24
3	**Cold drive cycle** —Repeat from full to empty	FTP-75 (cold)	FTP-75 at colder ambient temperature	Start time to full flow rate (−20°C)	−20
4	**Hot drive cycle** —Repeat from full to empty	SC03	Air conditioning use under hot ambient conditions	Start time to full flow rate (20°C)	35
5	**Dormancy test**	n/a	Static test of the storage system: 31 days	Dormancy	35

Note: Definitions to table: EPA: Environmental Protection Agency. FE: Fuel Economy. UDDS: Urban Dynamometer Drive Schedule. HWFET: Highway Fuel Economy Cycle. US06: Supplemental Federal Test Procedure (Condition 6). FTP-75: Federal Test Procedure Version 75. SCO3: Supplemental Federal Test Procedure (Condition 3).

where L is the characteristic length between heat transfer components, k is the media thermal conductivity, M_{hyd_eff} is the mass of hydride required to meet the system capacity target, ΔT is the temperature difference allowable in the media to meet the absorption target, $\Delta H_{overall}$ is the total enthalpy of hydrogen absorption, $\rho_{Hydride}$ is the density of hydride media, m is a geometric constant, M_{H2} is the molar mass of hydrogen, and $\Delta m_{H2}/\Delta t$ is the charging rate. This equation can be broken down into a system design component $1/L^2$, media characteristics, and a critical system design target as the filling time $\Delta m_{H2}/\Delta t$.

The implications of this relationship are discussed in greater detail in [28], but it is immediately apparent that the critical system design element (in this case the cooling element spacing) and a set of media-specific characteristics can be expressed as a given technical target [28]. Equation 11.1 shows the fundamental relationship between technical target of hydrogen charging, the media characteristics, and fundamental heat exchanger spacing. This relationship illustrates the importance of modifying specific media characteristics to increase heat exchanger spacing. This is illustrated by the necessity of increasing thermal conductivity. To minimize the heat exchanger mass and cost, thermal conductivity enhancements may be added to the storage media, as described previously by Dedrick in Chapter 9. This addition will allow for increased heat exchanger surface spacing. In addition to the tank and internal heat exchanger, one needs to supply heat to aid in the hydrogen discharge. This heat supply is more than an order of magnitude slower, however, since driving a vehicle takes significantly longer than refueling.

Many conventional interstitial metal hydrides considered for storage applications have very rapid hydrogen adsorption and desorption characteristics. Thus, they are not kinetically limited. They are, however, thermally limited in that during charging the evolved heat needs to be removed at a fast enough rate to allow for further charging. Complex hydrides, however, are usually kinetically limited. The slow kinetics is a major technical hurdle for complex hydrides that needs to be overcome in their potential implementation.

The baseline material used in this simulation was sodium aluminum hydride, $NaAlH_4$, ball milled with 3 mol% $TiCl_3$ as a catalyst. The performance data for this material were taken from unpublished work developed within the HSECoE.

System Descriptions

Complex and metal hydride storage systems are comprised of a storage tank, which is typically a pressure vessel. The utilization of a pressure vessel is important in meeting both the DOE volumetric target as well as the charging target. The free space within the storage tank not filled with media can be used to store compressed hydrogen gas. This compressed gas also serves as a buffer, supplying hydrogen under high-demand circumstances but prior to the delivery of heat to the system. Since heat needs to be added to the system continuously to release hydrogen as noted, a heat source is needed. This can be accomplished by either burning some of the discharged hydrogen or supplying heat from a waste heat source or electrical resistive heating. Hydrogen combustion is the most efficient when used in conjunction with a waste heat source. The amount of hydrogen combusted depends on the quality of the waste heat stream and the ability to deliver this heat into the storage system. These issues were examined in Chapter 10.

For many conventional metal hydrides, water can be used as the heat transfer medium since these materials are benign to potential water exposure from an inadvertent leak. In using complex hydrides, one must first understand the risk of inadvertent water contact and assess the risk of using this material as a heat transfer fluid [9–11]. For a number of the complex hydrides under consideration, such as $NaAlH_4$ and $LiNH_2$, water contact could potentially liberate significant hydrogen in an exothermic reaction. Thus, prudent system design would use a nonreacting heat transfer fluid such as any number of heat transfer oils as was implemented in the GM/Sandia $NaAlH_4$ storage sysem described in Chapter 10. The use of any heat transfer fluid will necessitate the inclusion of a pump and external heat exchanger.

With all of these considerations taken into account, the minimal system design for a potential automotive application is given in Figure 11.2a [14, 22]. In this specific case, the system is composed of a Type III (i.e., carbon-fiber-wrapped metal liner) $NaAlH_4$ tank operational at 150 bar with an accompanying hydrogen catalytic burner [29, 30]. The internal heat exchange system is a conventional tube/fin design.

Performance Predictions

This system meets 10 of the DoE/FreedomCAR technical targets for 2017 that are shown in Table 11.1, is above 50% of these targets for 4 targets, and below 50% for 6 targets. The summary of these attributes is given in the spider diagram of Figure 11.2b, where full realization of the target is denoted by shading to the outer diameter. Those targets posing the greatest technical challenges for deployment of metal or complex hydride systems are gravimetric density, cost, volumetric density, well-to-wheels efficiency, and fill time. The gravimetric target is missed substantially, primarily due to the lack of an on-board reversible hydride medium having a gravimetric capacity greater than 12 wt%. With the inclusion of a high-pressure Type IV graphite-reinforced composite tank to benefit the gravimetric target, the projected cost is significantly greater than the DOE target. Even though the medium itself holds hydrogen in a crystallographically dense manner, when these compounds are processed into particles of micrometers and smaller in size, the

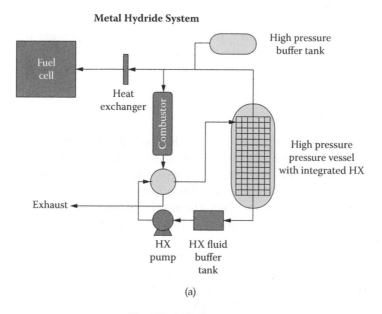

(a)

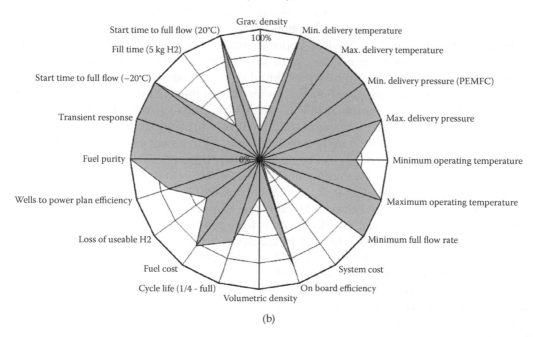

(b)

FIGURE 11.2
(a) Idealized reversible metal hydride system configuration with combustor. HX, heat exchange. (b) Spider chart depicting metal hydride system performance characteristics against the 2017 technical targets.

packing density of these materials drops to 40%, the primary cause of missed volumetric density. Since significant heat is required for hydrogen discharge for the 42 kJ/mol NaAlH$_4$, the well-to-power plant efficiency is also severely impacted. The ultimate fill time target of 2.5 min cannot be reached with current materials due to slow absorption kinetics. Although greater heat exchange can typically be added at the penalty of mass, inherent kinetics is a severe limitation to complex hydrides with no solution in sight.

Chemical Hydride Storage Systems

Material Properties and Behavior

Chemical hydrides differ from metal and complex hydrides in that after hydrogen release the storage media must be removed from the system and regenerated at a separate off-board chemical processing facility. The most significant technical hurdle involved in the development of chemical hydride material is the efficiency by which the spent material can be regenerated. The "off-board-reversible" chemical hydrides AlH$_3$, LiAlH$_4$, and NH$_3$BH$_3$ (ammonia borane, AB) and liquid organic hydrogen carriers (LOHCs) were examined in depth in Chapter 8.

The chemical hydrides can be either exothermic or endothermic discharge materials. A prototypical exothermic material is ammonia borane (AB) [31–33], which has a 20 wt% hydrogen density if fully discharged but with an exothermic –20 kJ/mol H$_2$ discharge enthalpy. This large enthalpy works to both decrease the overall well-to-wheels efficiency [6, 34] since this is nominally wasted energy and to increase the system mass since heat exchange BOP must be added to the system to remove the heat released during hydrogen desorption. Aluminum hydride (AlH$_3$), on the other hand, is a typical endothermic chemical hydride having an enthalpy of 10 kJ/mol H$_2$ [35]. Endothermic media will require continuous feed of heat to maintain the discharge reaction, reducing its wells-to-wheels efficiency.

The chemicals themselves can be either liquid or solid materials at ambient conditions both before and after dehydrogenation. It is preferred that they maintain their original form throughout the reaction to minimize BOP complexity. Liquids are most attractive since they are easily pumped, and heat transfer is facilitated. A hybrid approach is to use the solids mixed with a nonreactive liquid into pumpable slurry. Slurries are often problematic, tending to separate over time, and they can be quite abrasive. In addition, the carrier liquid needs to maintain chemical inactivity and maintain a low vapor pressure at the dehydrogenation temperatures in the reactor. Either liquids or slurries will need to maintain a viscosity, at all operational temperatures, below 1500 centipoise to maintain pumpability with minimal impact on efficiency. Significant work yet needs to be performed to ensure slurry stability.

Solid materials are generally difficult to transport both on and off the vehicle during filling and through the reactor during discharge [21, 36]. For this reason, for the current analysis, solid transport was not considered.

The baseline material used in this engineering assessment was liquid ammonia borane composed of pure NH$_3$BH$_3$ dissolved in bmimCl to a mass fraction of 10% AB. The performance data for this material were taken from unpublished work developed within the HSECoE [37].

System Description

Chemical hydride reactors can be conceived in many different forms [6, 21, 36, 38, 39], but a simple flow-through reactor is simple, with the fewest components. A typical chemical

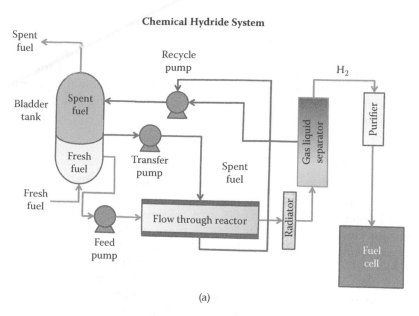

Chemical Hydride System

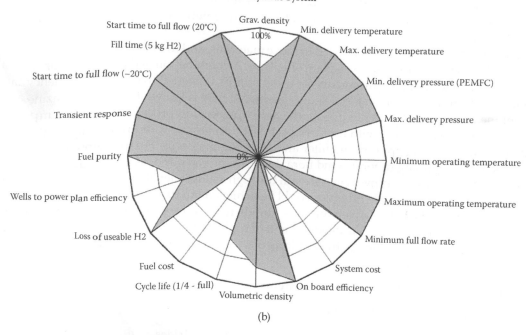

Chemical Hydride System

(b)

FIGURE 11.3

(a) Idealized liquid/slurry media chemical hydrogen storage system. (b) Spider chart depicting chemical hydride system performance characteristics against the 2017 technical targets.

hydride system incorporating a flow-through reactor is depicted in Figure 11.3a [37, 39]. It is composed of a bladder tank and pump feeding a flow-through reactor followed by a gas liquid separator (GLS), a pump with heat exchange loop, and purifier. The bladder tank holds both the fresh and the spent fuel in a single volume, reducing volumetric capacity to its minimum. A pump is included to maintain both fresh and spent fuel flow, with a portion of the spent fuel recirculated to absorb excess heat generated at the reactor. The flow-through reactor transfers heat away from the medium as it passes inside, while the GLS separates the evolved hydrogen from the liquid stream and returns the spent fuel to the bladder tank. The reactor and GLS work at high pressure, and the high-pressure hydrogen stream is stored in a GLS for system transit operation. Finally, prior to hydrogen discharge to the fuel cell, the gas is purified, if necessary, to remove potential contaminants such as diborane [32, 33].

Performance Attributes

This system meets 13 of the FreedomCAR/Fuel technical targets for 2017 as shown in Table 11.1; where 4 of the targets are above 50%, 1 is below 50%, and 2 are undefined since off-board regeneration is required, and minimum operating temperatures have not yet been determined [21, 40, 41]. The summary of these attributes is given in the spider diagram of Figure 11.3b, in which full realization of the target is denoted by shading to the outer diameter. Those targets not met include gravimetric density, volumetric density, wells-to-power-plant efficiency, hydrogen purity, and fill time. Missing the gravimetric density target is the result of the large number of pumps in the system, which has not been optimized for automobile applications. The low solubility of ammonia borane in bmimCl resulted in this system design missing the volumetric target. Identification of liquids or slurries with higher gravimetric capacities would overcome this obstacle.

The large number of pumps needed to move materials, and their inherent parasitic losses, resulted in the low tank-to-power-plant efficiency. A combination of these pumps or system designs alleviating the requirements for these pumps is needed to meet the efficiency target. The ammonia borane material being considered here is known to release borazine and diborane along with hydrogen during the dehydrogenation process [37]. These impurity releases may be mitigated with improved catalyst development, and remaining impurities can be trapped using appropriate filter and trapping materials. Further work needs to be conducted on trapping diborane and borazine contaminates and reduction in evolution of these contaminants to the hydrogen stream through catalyst development. In the projected filling scenarios, reasonable fill times could be achieved but did not meet the DOE targets. Significant quantities of relatively viscous liquids need to be transported simultaneously both on and off the vehicle. The development of higher-capacity fluids would significantly alleviate this drawback along with simultaneous filling and emptying scenarios.

Adsorbent-Based Hydrogen Storage Systems

Material Descriptions and Behavior

Adsorbent materials, similar to metal and complex hydrides, are in situ rechargeable materials. Adsorption-based hydrogen storage was examined in Chapter 7 by Justin Purewal and Channing Ahn. The significant difference is that the enthalpy of adsorption of these materials, which is also known as the isosteric heat of adsorption, is typically -4 to -10 kJ/mol H_2. This requires maintenance of the adsorbent at cryogenic temperatures, typically 77 K or below. Both superactivated carbon and framework materials [42–44] have been considered for automotive storage applications [45]. These materials typically hold 6–8 wt% hydrogen at 77 K. They are high-surface-area materials, holding hydrogen on the surface through induced dipole interactions. These high surface areas, measured at about

3000 m^2/g, result in rather voluminous powders, somewhat hindering volumetric capacity. In Chapter 10, Johnson and Bénard discussed a number of properties and issues of hydrogen storage systems utilizing adsorbents.

The baseline material selected for the present HSECoE analysis was a superactivated carbon commercially available and designated as AX-21. The performance data for this material were taken from studies by Richard et al. [46, 47].

Storage System Description

The adsorbent system used for this effort is given in Figure 11.4a. It is composed of a 200-bar high-pressure Type III vessel enclosed in a multilayer vacuum-insulated jacket with a 5-W heat leak at 80 K [48]. Charging is achieved via *flow-through cooling* [49, 50] with an initial bed temperature of 80 K. The flow-through cooling design continuously passes chilled hydrogen gas at 77 K through the tank. As hydrogen is adsorbed, the storage medium is heated. As more chilled hydrogen passes over the medium, the gas is heated and the media cooled. The exiting heated hydrogen gas, which now has an average temperature of 160 K, is returned to the fueling station for either rechilling or other uses. Conversely, heat needs to be introduced into the tank for discharging. This is achieved utilizing an in-tank electrical resistance heater. Finally, an external heat exchanger is utilized to heat the released hydrogen to ambient temperatures for use by the fuel cell. The heat exchanger requires significant technical development since the outflowing gas is at cryogenic temperatures, and the ambient atmosphere with its moisture content will ice over conventional heat exchangers.

Performance Attributes

This adsorbent system meets 14 of the FreedomCAR/Fuel technical targets for 2017; 4 of the targets are above 50% and 2 below 50%. The summary of these attributes is given in the spider diagram of Figure 11.4b, in which full realization of the target is denoted by shading to the outer diameter. The primary deficiencies of this system design are gravimetric density, system cost, volumetric density, loss of usable hydrogen, and well-to-power-plant efficiency. The low gravimetric density is due in part to the use of a high-pressure vessel and enveloping multilayer vacuum insulation (MLVI). The former increases volumetric density, and the latter increases dormancy time. Incorporation of MLVI and a Type III pressure vessel increases the cost of these systems beyond the DOE target. The low volumetric density of the superactivated carbon drives the volumes up, resulting in the missed volumetric density target. The MLVI, while projected to perform well, still did not meet the very difficult dormancy target. The well-to-power-plant efficiency was not met due to the relative inefficient use of hydrogen energy through conversion to electricity (~50%) to run the resistance heating elements.

Summary of Assessments

A comparison of the simulated system performance vs. the FreedomCAR/Fuel targets is given in Table 11.1. A number of overall conclusions can be drawn. The gravimetric density, volumetric density, cost, fill time, and efficiency are common weaknesses of all three condensed-phase hydrogen storage systems for the analyzed configurations and storage material considered. Gravimetric density shortcoming is a direct result of current media not having sufficient hydrogen storage capacities. The volumetric density shortfall is related to an inability to densely pack fine powders into relatively fragile heat exchange devices in the cases of adsorbents and

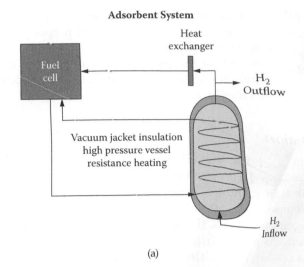

(a)

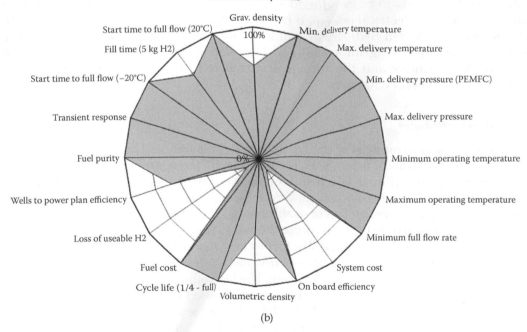

(b)

FIGURE 11.4
(a) Idealized adsorption hydrogen storage system. (b) Spider chart depicting adsorbent system performance characteristics against the 2017 technical targets.

complex hydrides. Well-to-wheel efficiencies are limited due to an overall limitation in current ability to manage waste heat and use it efficiently. This leads to the conclusion that while a great deal of progress has been made in developing solid-state hydrogen storage materials and engineered systems, even more research needs to be done to make these systems commercially attractive for light-duty vehicles. This materials science and engineering work should be done in parallel so that the strengths of the various media can be utilized, while their deficiencies are focused on and overcome by the various engineering tools available.

It is important to emphasize that the comparison to targets made in Figures 11.2b, 11.3b, and 11.4b and the conclusions discussed apply to the implementation of these materials for the particular application of light-duty vehicles. For other (nonvehicular) applications of hydrogen storage, different storage targets would likely apply, with different conclusions drawn. Indeed, Klebanoff et al. have conducted a recent study to assess the energy storage requirements for a sizable range of applications for which fuel cell technology (and hydrogen storage) could be applied for the purpose of the DOE eventually developing hydrogen storage targets for these different applications, such as construction equipment, aviation ground support equipment, and portable electronics [51]. It is fair to say, however, that all applications would benefit from improvements in the thermodynamic and kinetic properties of these materials.

Ranges of Applicability for Hydrogen Storage Systems

Figure 11.5 shows the approximate ranges in hydrogen capacity that storage systems may find initial acceptance. The total capacity range spans from 1 g to 100,000 kg of hydrogen and is divided into roughly three segments: portable, transportable, and stationary. For use in this discussion, *portable* is defined as being carried easily by one human, such as would be the case for using small electronic equipment of less than 1 kW as categorized by Friedrich and Buchi [52]. *Transportable* is meant to cover mobile applications such as small through heavy-duty vehicles for which energy consumption is 1–200 kW. *Stationary storage* is defined as greater than 100 kg of hydrogen is envisioned for distributed and centralized energy storage with generating capacity greater than 500 kW or launching spacecraft [53]. These definitions are arbitrary and may be revised or adapted as new technologies become available, but they should serve as a helpful guide.

Gaseous compressed hydrogen is bound at the lower limit of about 1 kg by the mass of the conventional Type I (i.e., all-metal) tank required to meet government-mandated safety conditions. More exotic tanks incorporating graphite reinforcement have been estimated [54] to be expensive for most commercial light-duty vehicles. Replacement of graphite fibers with glass or aramid fibers could be useful in driving down this cost and thus lower

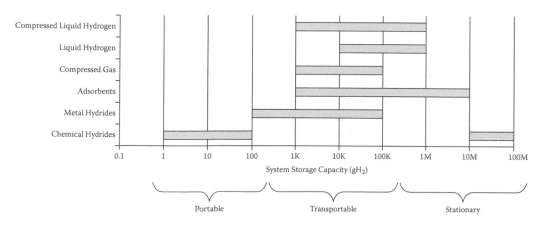

FIGURE 11.5
Approximate range of storage system hydrogen content for the various system types.

limits for compressed gas acceptability in this and other applications. The upper bound on compressed hydrogen is about 100 kg and is primarily governed by the size of the tanks. Gaseous hydrogen is not volumetrically efficient, and large storage capacities will require large tanks and high expense. Some stationary applications may be able to take advantage of relatively cheap storage, such as underground reservoirs either naturally occurring or made specifically for gas storage.

Cryogenic liquid and compressed liquid hydrogen storage vessels are anticipated to be viable above 1 but less than 1000 kg of storage hydrogen for most power applications. Conventional liquid hydrogen storage systems have limited dormancy time and are thus unsuitable for most stationary applications except for when usage will be in relatively short times such as rocket fuel [53]. Compressed liquid storage is foreseen for use at larger transportable applications, including aviation [55, 56] and small stationary applications where larger size and resultant lower surface area to volume will reduce the heat loss of the cryogenic systems.

Adsorbents, similarly limited to cryogenic temperature operation, will probably retain many of the same disadvantages as compressed liquid hydrogen. This behavior may limit their storage capacity range to 1 to 1000 kg.

Metal and complex hydrides do not have the temperature limitations of adsorbents and compressed liquid hydrogen but still require significant tankage and heat exchange. These requirements will likely limit their size to roughly 0.1–100 kg of stored hydrogen.

The chemical hydrides are different from most of the other media. Regeneration at a central facility will affect the storage capacity of these systems. At small capacities, ranging from 1 g to 1 kg, these materials provide viable options for high-value functions, including military systems. The chemical storage media can be saved for recycling or discarded as economics dictates. Large-scale uses of chemical hydrides for light-duty transportation systems will require development of an efficient and cost-effective infrastructure to handle interchangeable tanks with hydrogen capacities of 1–10 kg along with central reprocessing/refilling. Requiring a large stationary facility for regeneration would allow chemical hydrides to also find application in large storage systems, greater than 10,000 kg, where the reprocessing facility can be custom built specifically for that storage system's needs. For the LOHCs, this concept was described in Chapter 8 for the cyclohexane-toluene process [57] to store seasonal hydroelectric-generated hydrogen for use throughout the year.

Status and Outlook

In summary, numerous physical and chemical methods have been explored and developed over the past century, with the evolutionary process continuing. When a storage material or system can match the technical requirements for a given application, it has usually been exploited, often with commercial products becoming available. Examples include compressed gas cylinders for academic and industrial laboratories, large-scale liquid hydrogen Dewars at space launch facilities [53], and devices using various hydrides for portable and backup power sources [58]. For space and many military applications, performance and reliability usually trump the cost issues. The reverse will be true for mass-market usage, such as passenger vehicles, for which cost is perceived to be the dominating factor, although performance and safety targets must be met as well to satisfy both customers and governmental regulatory agencies. The previous chapters in this book have provided in-depth treatment of various engineering issues and materials now being investigated and developed for hydrogen storage.

Acknowledgments

We would like to acknowledge the discussion and use of information developed by contributors to the HSECoE (in alphabetical order): K. Brooks, R. Chahine, C. Corgnali, K. Day, M. Devarakonda, K. Drost, S. Garrison, B. Hardy, J. Holladay, D. Kumar, D. Mosher, T. Motyka, N. Newhouse, J.-M. Pasini, S. Rassot, A. Raymond, J. Reiter, E. Ronnebro, T. Semelsberger, D. Siegel, K. Simons, L. Simpson, A. Sudik, D. Tamburello, M. Thornton, B. van Hassel, and M. Veenstra.

References

1. Züttel, A., M. Hirscher, K. Yvon, et al. 2008. Hydrogen storage. In *Hydrogen as a Future Energy Carrier*, ed. A. Züttel, A. Borgschulte, L. Schlapbach, Chapter 6:165–264. Weinheim, Germany: Wiley-VCH.
2. Dantzer, P. 1997. Metal-hydride technology: a critical review. In *Hydrogen in Metals III*, ed. H. Wipf, 279–340. Berlin: Springer.
3. Felderhoff, M., C. Weidenthaler, R. von Helmolt, U. Eberle. 2007. Hydrogen storage: the remaining scientific and technological challenges. *Phys. Chem. Chem. Phys.* **9**:2643–2653.
4. Mori, D., K. Hirose. 2009. Recent challenges of hydrogen storage technologies for fuel cell vehicles. *Int. J. Hydrogen Energy* **34**:4569–4574.
5. Hirscher, M. 2010. *Handbook of Hydrogen Storage*. Weinheim: Wiley-VCH.
6. Paster, M. D., R. K. Ahluwalia, G. Berry, A. Elgowainy, S. Lasher, K. McKenney, M. Gardiner. 2011. Hydrogen storage technology options for fuel cell vehicles: well-to-wheel costs, energy efficiencies, and greenhouse gas emissions. *Int. J. Hydrogen Energy* **36**:14534–14551.
7. Jepsen, J., J. M. Bellosta von Colbe, T. Klassen, M. Dornheim. 2012. Economic potential of complex hydrides compared to conventional hydrogen storage systems. *Int. J. Hydrogen Energy* **37**:4204–4214.
8. Ekoto, I. W., E. G. Merilo, D. E. Dedrick, M. A. Groethe. 2011. Performance-based testing for hydrogen leakage into passenger vehicle compartments. *Int. J. Hydrogen Energy* **36**:10169–10178.
9. Tanaka, H., K. Tokoyoda, M. Matsumoto, Y. Suzuki, T. Kiyobayashi, N. Kuriyama. 2009. Hazard assessment of complex hydrides as hydrogen storage materials. *Int. J. Hydrogen Energy* **34**:3210–3218.
10. Lohstroh, W., M. Fichtner, W. Breitung. 2009. Complex hydrides as solid storage materials: first safety tests. *Int. J. Hydrogen Energy* **34**:5981–5985.
11. James, C. W., J. A. Cortes-Concepcion, D. A. Tamburello, D. L. Anton. 2012. Environmental reactivity of solid-state hydrogen storage systems; fundamental testing and evaluation. *Int. J. Hydrogen Energy* **37**:2885–2890.
12. Lozano, G. A., J. M. Bellosta von Colbe, R. Bormann, T. Klassen, M. Dornheim. 2011. Enhanced volumetric hydrogen density in sodium alanate by compaction. *J. Power Sources* **196**:9254–9259.
13. Lozano, G. A., C. N. Ranong, J. M. Bellosta von Colbe, R. Bormann, J. Hapke, G. Fieg, T. Klassen, M. Dornheim. 2012. Optimization of hydrogen storage tubular tanks based on light weight hydrides. *Int. J. Hydrogen Energy* **37**:2825–2834.

14. van Hassel, B. A., D. Mosher, J. M. Pasini, M. Gorbounov, J. Holowczak, X. Tang, R. Brown, B. Laube, L. Pryor. 2012. Engineering improvement of NaAlH₄ system. *Int. J. Hydrogen Energy* **37**:2756–2766.
15. Okumura, M., K. Terui, A. Ikado, Y. Saito, M. Shoji, Y. Matsushita, H. Aoki, T. Miura, Y. Kawakami. 2012. Investigation of wall stress development and packing ratio distribution in the metal hydride reactor. *Int. J. Hydrogen Energy* **37**:6686–6693.
16. Demirci, U. B., P. Miele. 2011. Chemical hydrogen storage: "material" gravimetric capacity versus "system" gravimetric capacity. *Energy Environ. Sci.* **4**:3334–3341.
17. Satyapal, S., J. Petrovic, C. Read, G. Thomas, G. Ordaz. 2007. The U.S. Department of Energy's National Hydrogen Storage Project: progress towards meeting hydrogen-powered vehicle requirements. *Catal. Today* **120**:246–256.
18. DOE targets for onboard hydrogen storage systems for light-duty vehicles (Revision 4.0). 2009. http://www1.eere.energy.gov/hydrogenandfuelcells/storage/pdfs/targets_onboard_hydro_storage.pdf.
19. Hydrogen Storage Engineering Center of Excellence Web site. http://hsecoe.srs.gov/index.html.
20. Thornton, M. 2011. System design, analysis, modeling, and media engineering properties for hydrogen energy storage. Proceedings of the 2011 DOE annual merit review. http://www.hydrogen.energy.gov/pdfs/review11/st008_thornton_2011_o.pdf.
21. Devarakonda, M., K. Brooks, E. Rönnebro, S. Rassat. 2012. Systems modeling, simulation and material operating requirements for chemical hydride based hydrogen storage. *Int. J. Hydrogen Energy* **37**:2779–2793.
22. Pasini, J. M., B. A. van Hassel, D. A. Mosher, M. J. Veenstra. 2012. System modeling methodology and analyses for materials-based hydrogen storage. *Int. J. Hydrogen Energy* **37**:2874–2884.
23. HSECoE models and analysis download Web page. http://hsecoe.srs.gov/models.html.
24. Zhang, J., T. S. Fisher, P. V. Ramachandran, J. P. Gore, I. Mudawar. 2005. A review of heat transfer issues in hydrogen storage technologies. *J. Heat Transfer* **127**:1391–1399.
25. Visaria, M., I. Mudaward, T. Pourpoint, S. Kumar. 2010. Study of heat transfer and kinetics parameters influencing the design of heat exchangers for hydrogen storage in high-pressure metal hydrides. *Int. J. Heat Mass Transfer* **53**:2229–2239.
26. Hardy, B., D. L. Anton. 2009. Hierarchical methodology system for hydrogen storage systems. Part I: scoping models. *Int. J. Hydrogen Energy* **34**:2269–2277.
27. Hardy, B., D. L. Anton. 2009. Hierarchical methodology system for hydrogen storage systems. Part II: detailed models. *Int. J. Hydrogen Energy* **34**:2992–3004.
28. Corgnale, C., B. J. Hardy, D. A. Tamburello, S. L. Garrison, D. L. Anton. 2012. Acceptability envelope for metal hydride-based hydrogen storage systems. *Int. J. Hydrogen Energy* **37**:2812–2824.
29. Raju, M., J. P. Ortmann, S. Kumar. 2010. System simulation models for high-pressure metal hydride storage systems. *Int. J. Hydrogen Energy* **35**:8742–8754.
30. Raju, M., S. Kumar. 2011. System simulation modeling and heat transfer in sodium alanate based hydrogen storage systems. *Int. J. Hydrogen Energy* **36**:1578–1591.
31. Stephens, F. H., V. Pons, R. T. Baker. 2007. Ammonia–borane: the hydrogen source par excellence? *Dalton Trans.* **25**:2613–2626.
32. Smythe, N. C., J. C. Gordon. 2010. Ammonia borane as a hydrogen carrier: dehydrogenation and regeneration. *Eur. J. Inorg. Chem.* 509–521.
33. Staubitz, A., A. P. M. Robertson, I. Manners. 2010. Ammonia-borane and related compounds as dihydrogen sources. *Chem. Rev.* **110**:4079–4124.
34. Ahluwalia, R. K., T. Q. Hua, J. K. Peng. 2012. On-board and off-board performance of hydrogen storage options for light-duty vehicles. *Int. J. Hydrogen Energy* **37**:2891–2910.
35. Graetz, J., J. J. Reilly. 2006. Thermodynamics of the α, β and γ polymorphs of AlH₃. *J. Alloys Compounds* **424**:262–265.
36. Holladay, J., D. Herling, K. Brooks, K. Simmons, E. Rönnebro, M. Weimar, S. Rassat. 2011. Systems engineering of chemical hydride, pressure vessel, and balance of plant for on-board hydrogen storage. Proceedings of the 2011 DOE annual merit review. http://www.hydrogen.energy.gov/pdfs/review11/st005_holladay_2011_o.pdf.

37. Semelsberger, T. A., T. Burrell, T. Rockward, E. Brosha, J. Tafoya, G. Purdy, T. Nakagawa, B. Davis. 2011. Chemical hydride rate modeling, validation, and system demonstration. Proceedings of the 2011 DOE annual merit review. http://www.hydrogen.energy.gov/pdfs/review11/st007_semelsberger_2011_o.pdf.

38. Aardahl, C. L., S. D. Rassat. 2009. Overview of systems considerations for on-board chemical hydrogen storage. *Int. J. Hydrogen Energy* **34**:6676–6683.

39. Ahluwalia, R. K., J. K. Peng, T. Q. Hua. 2011. Hydrogen release from ammonia borane dissolved in an ionic liquid. *Int. J. Hydrogen Energy* **36**:15689–15697.

40. Deverakonda, M., J. Holladay, K. P. Brooks, S. Rassat, D. Herling. 2010. Dynamic modeling and simulation based analysis of an ammonia borane (AB) reactor system for hydrogen storage. *Electrochem. Soc. Trans.* **33**:1959–1972.

41. Brooks, K. P., M. Devarakonda, S. Rassat, D. A. King, D. Herling. 2010. Systems modeling of ammonia borane bead reactor for on-board regenerable hydrogen storage in PEM fuel cell applications. *Proc. ASME 2010 8th Fuel Cell Sci. Eng. Technol. Conf.* **1**:729–734.

42. Thomas, K. M. 2007. Hydrogen adsorption and storage on porous materials. *Catalysis Today* **120**:389–398.

43. Panella, B., M. Hirscher. 2010. Physisorption in porous materials. In *Handbook of Hydrogen Storage*, ed. M. Hirscher, 39–62. Weinheim, Germany: Wiley-VCH.

44. Meisner, G. P., Q. Hu. 2009. High surface area microporous carbon materials for cryogenic hydrogen storage synthesized using new template-based and activation-based approaches. *Nanotechology* **20**:204023.

45. Ahluwalia, R. K., J. K. Peng. 2009. Automotive hydrogen storage system using cryo-adsorption on activated carbon. *Int. J. Hydrogen Energy* **34**:5476–5487.

46. Richard, M.-A., P. Bénard, R. Chahine. 2009. Gas adsorption process in activated carbon over a wide temperature range above the critical point. Part 1: modified Dubinin-Astakhov model. *Adsorption* **15**:43–51.

47. Richard, M.-A., P. Bénard, R. Chahine. 2009. Gas adsorption process in activated carbon over a wide temperature range above the critical point. Part 2: conservation of mass and energy. Adsorption **15**:53–63.

48. Tamburello, D. 2011. Personal communication to D. L. Anton.

49. Kumar, S., M. Raju, V. S. Kumar. 2012. System simulation models for on-board hydrogen storage. *Int. J. Hydrogen Energy* **37**:2862–2873.

50. Hardy, B., C. Corgnale, R. Chahine, M.-A. Richard, S. Garrison, D. Tamburello, D. Cossement, D. Anton. 2012. Modeling of adsorbent based hydrogen storage systems. *Int. J. Hydrogen Energy* **37**:5691–5705.

51. Klebanoff, L. E., J. W. Pratt, T. A. Johnson, M. Arienti, L. Shaw, M. Moreno. 2012. Analysis of H_2 storage needs for early market non-motive fuel cell applications, Sandia Technical Report SAND2012-1739. Sandia National Laboratories, Livermore, CA.

52. Friedrich, K. A., F. N. Buchi. 2008. Fuel cells using hydrogen. In *Hydrogen as a Future Energy Carrier*, ed. A. Zuettel, A. Borgschulte, L. Schlapbach, Chap. 8.1, 356. Weinheim, Germany, Wiley-VCH.

53. Bowman, R. C., Jr. 2006. Roles of hydrogen in space explorations. In *Hydrogen in Matter*, ed. G. R. Myneni, B. Hjorvarsson, 175–199. New York: American Institute of Physics.

54. Hua, T. Q., R. K. Ahluwalia, J.-K. Peng, M. Kromer, S. Lasher, K. McKenney, K. Law, J. Sinha. 2011. Technical assessment of compressed hydrogen storage tank systems for automotive applications. *Int. J. Hydrogen Energy* **36**:3037–3049.

55. Haglind, F., A. Hasselrot, R. Singh. 2006 Potential of reducing the environmental impact of aviation by using hydrogen. Part I: background, prospects and challenges. *Aeronaut. J.* **110**:533–540.

56. Verstraeta, D., P. Hendrick, P. Pilidis, K. Ramsden. 2010. Hydrogen fuel tanks for subsonic transport aircraft. *Int. J. Hydrogen Energy* **35**:11085–11098.

57. Cacciola, G., N. Giordano, G. Resticcia. 1984. Cyclohexane as a liquid phase carrier in hydrogen storage and transport. *Int. J. Hydrogen Energy* **9**:411–419.

58. McWhorter, S., C. Read, G. Ordaz, N. Stetson. 2011. Materials-based hydrogen storage: attributes for near-term, early market PEM fuel cells. *Curr. Opin. Solid State Mater. Sci.* **15**:29–38.

12

Codes and Standards for Hydrogen Storage in Vehicles

Christine Sloane

CONTENTS

Introduction 405
Historical Perspective..408
Management of Risk: Minimum Performance Criteria.......................................409
Performance Requirements: Qualification Testing..411
    Qualification Test Protocols 411
        Baseline Performance..411
        Expected Operation..412
        Extreme Service Durability 414
        Fire 418
    Acceptance Criteria...418
        Criteria for Permeation and Leakage 419
        Criteria for End-of-Life Strength..420
Temperature Specifications in Qualification Tests 421
Validation of Qualification Requirements for Hydrogen Storage....................421
What Makes the New Performance-Based Requirements More Stringent Than
Earlier Approaches?...422
New Requirements for Test Facilities...423
Are We Done Yet? What Additional Changes in Requirements Are under Discussion?.....423
Conclusion 424
Acknowledgments...425
References..425

Introduction

Previous chapters have described the need, methods, and engineering approaches for developing hydrogen storage systems for a variety of emerging hydrogen-powered applications, such as fuel cell light-duty vehicles, portable power, and construction equipment. The field testing and deployment of these systems have required the establishment of hydrogen technology "codes and standards." Codes and standards developed by industry enable the coordinated commercial deployment of a technology. They specify interfaces and expectations for operation so that components and systems made by diverse suppliers will work together. They also specify requirements for safe design, operation, and test standards for certifying performance. Regulations developed by governments specify legally enforced safety requirements for use of the technology. Regulations often are developed from or directly cite the requirements found in industry codes and standards.

Hydrogen has been used commercially for decades (e.g., agriculture, refining, scientific laboratories, and hospitals). It is transported over roadways in trucks and moves through pipelines to refineries. Industry codes and standards and government regulations for the transport and use of gaseous and liquid hydrogen are well established [1–3].

This chapter focuses on the more recent development of codes and standards for hydrogen use as a propulsion fuel in vehicles, specifically on-road vehicles (e.g., cars, trucks, and buses) [4–22]. International vehicle standards have been under development primarily at the Society of Automotive Engineers International (SAE) and the International Organization for Standardization (ISO). A global technical regulation (GTR) is under development at the United Nations as a voluntary basis for global harmonization of national regulations for the safety of hydrogen-fueled passenger vehicles.

Industry standards for fuel cell vehicles have been developed before the vehicles have been commercialized by retail sales in the marketplace, but are in the advanced precommercialization phase, during which fleets of prototype vehicles are in evaluation in on-road service. The development of industry safety standards and government regulations for fuel cell vehicles well before commercialization has been driven by interest in achieving global harmonization of regulatory requirements to enable vehicles to be deployed internationally. With global harmonization higher initial volumes can be realized, and, thereby, the environmental benefits of these vehicles can be achieved more rapidly on a global scale.

Industry codes and standards are being established for the following:

1. Installation/operation of vehicle fueling stations (National Fire Protection Association [NFPA] 2) [1]
2. Station fueling nozzle and vehicle fuel receptacle (SAE J2600) [4]
3. Fueling station protocol for fuel transfer (e.g., flow rate and pressure cutoff) (SAE J2601) [5]
4. Vehicle fuel system leak tightness (on road and in crash) (SAE J2578) [6]
5. Hydrogen fuel storage on-board vehicles (SAE J2579) [7]

This chapter focuses on the current development of standards for the hydrogen storage system within on-road vehicles. Although the focus here is on vehicles, the codes and standards have a broader relevance since the emerging fuel-cell-based construction equipment items (mobile lights, portable generators, etc.) are designed to be compliant with the vehicular hydrogen codes and standards.

The safety standards for the hydrogen storage systems in vehicles contain best-practice guidelines for their design and installation. They also specify the performance-based test procedures that are used to formally qualify storage systems for on-road service. At present, compressed gaseous hydrogen is the type of storage system receiving the primary focus because of its wide use by vehicle manufacturers of fuel cell prototype vehicles and because the rupture risk from the high pressure of compressed gaseous storage merits specific attention. From a safety perspective, high-pressure on-board hydrogen storage is the feature that most distinguishes fuel cell vehicles from battery-powered electric vehicles. Battery and fuel-cell powered vehicles use similar high-voltage electric power trains; their safety is ensured by requirements like those that apply to hybrid electric vehicles.

Hydrogen-fueled vehicles are required to pass crash safety tests with leakage limits equivalent (on an energy basis) to gasoline-fueled vehicles.

In recent years, the development of industry standards has focused on performance-based requirements instead of prescriptive requirements that focus on known failures of previous technologies. There are two key reasons for the focus on performance-based standards. First, on-road safety is a high priority. Therefore, requirements must be developed comprehensively and in anticipation of extremes of on-road service learned from decades of experience to avert adverse on-road experiences. Second, they provide stringent requirements to verify safe performance of new technologies with environmental benefits under broad conditions of on-road use. Vehicle crash tests are an example of performance-based requirements—the crash conditions relate to statistics for on-road crash conditions. They are applied uniformly to all vehicles without differentiation for the materials used in a vehicle's structure, for its structural design, or for the techniques used in its manufacture, that is, without preknowledge of specific failure modes. Vehicles must pass crash tests to qualify for on-road service. Likewise, hydrogen storage systems must pass additional performance-based tests based on extreme conditions of on-road usage to qualify for on-road service.

A typical hydrogen storage system is illustrated in Figure 12.1. It consists of a high-pressure containment vessel equipped with three components: a check valve, a temperature-activated pressure relief device (TPRD), and an automatic shutoff valve. The check valve is installed on the containment vessel at the connection to the fuel line from the fueling receptacle. It prevents backflow of hydrogen gas. The TPRD opens when activated by fire. It releases hydrogen gas in a controlled manner (hydrogen gas then dissipates rapidly due to its high buoyancy). The automatic shutoff valve is installed in the fuel line to the power system (e.g., the fuel cell system). It is equipped to close automatically in the event of an accident, when the vehicle is not operating, or other specified conditions. It is important to note that the high-pressure part of the hydrogen storage system shown in Figure 12.1 does not incorporate a pressure-activated pressure relief device (PRD). As described further in this chapter, the responsibility to prevent overpressurization during refueling lies with the hydrogen fueling station. Hydrogen storage systems are generally pressurized to either 350 or 700 bar when fully filled. Leak-tight performance (leak less than 3.6 normal cubic centimeters per minute (Ncc/min)) is required throughout service life. This leak rate specification is described in this chapter.

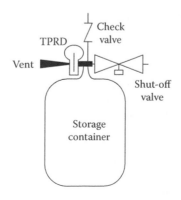

FIGURE 12.1
Typical gaseous hydrogen storage system.

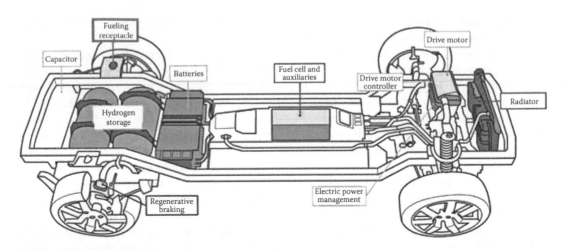

FIGURE 12.2
Typical mounting of hydrogen storage in a passenger car.

Several hydrogen storage systems are generally mounted in a single vehicle to provide sufficient fuel for expected driving distances. In passenger cars, they are generally mounted under the rear end of the vehicle, as illustrated in Figure 12.2. In buses, they are often mounted on the roof.

Historical Perspective

Compressed natural gas (CNG) vehicles provide a useful historical reference for compressed gaseous fuel storage up to 300 bar. Perhaps surprisingly, storage standards for natural gas vehicles are not sufficiently general to handle all safety issues specific to hydrogen storage. Four aspects lead to unique safety considerations: (1) chemical and physical properties of hydrogen (e.g., Joule Thompson temperature effect, rates of permeability and dispersion, and hydrogen embrittlement); (2) higher pressures (greater than 300 bar); (3) broader operating conditions; and (4) fuel quality constraints.

Many of the previous historical requirements for the storage of natural gas in vehicles can be seen as a list of tests that screen for on-road failures that have occurred in CNG tanks operating at less than 300 bar made with materials and structures developed prior to 1990. Recorded failures (rupture) of CNG tanks in vehicles have been traced to fire, to abuse/damage, to chemical exposure, or to the durability of metal and glass-fiber-composite tanks after extended static and cycling pressure. Each circumstance was addressed in historical standards with a test replicating the extremity of conditions causing the failure. In several cases, the resulting requirements are prescriptive, rather than performance based, in that tests are applied differently for systems with different materials or construction design. Such tests target the specific failure mechanisms that have occurred with CNG tanks with historical materials/constructions operating at pressures under 300 bar. Hence, the prescriptive requirements are reactive (reflecting past failures) rather than proactive (screening for future failures that may occur under different combinations of on-road conditions, new materials, and new constructions).

Performance-based standards do not assume failure mechanisms are known or that materials and constructions can be fully anticipated but instead seek to identify comprehensively worst-case, stressful on-road conditions to be managed by a roadworthy system. The breadth of conditions under which CNG metal and glass-fiber tanks failed when operated at pressures below 300 bar is not necessarily sufficient to encompass risk conditions for hydrogen systems made with new materials and constructions or operating at different working pressures and fueling conditions.

Another difference from the CNG history has been the involvement of a broader community of experts early in the process of developing safety requirements for hydrogen storage. The early CNG community drew primarily on the expertise of pressure vessel and component manufacturers and aftermarket converters of gasoline vehicles into CNG vehicles. The broader community engaged in development of hydrogen requirements for vehicles includes vehicle designers and manufacturers, who bring broader experience with on-road vehicle safety and have experience with management of risk when a product is in use by a diverse population and in highly variable on-road conditions. In addition, experts from the emerging fueling and repair/service infrastructures were engaged to coordinate requirements for fueling and defueling. Component manufacturers were engaged to specifically coordinate requirements for component reliability, durability, and performance so that those features would be consistent with the expected performance of safety for the storage systems into which they would be incorporated. And finally, independent test laboratories (including national laboratories) have evaluated material properties and developed test procedures.

Management of Risk: Minimum Performance Criteria

The development of qualification tests for on-road service and other specifications for hydrogen storage systems have been based on systematic risk management. Risk = (Probability of an adverse event occurring) × (Severity of an adverse event). Risk is managed by reducing the probability or severity of adverse events. The unintended releases of fuel by leakage or rupture are the primary adverse events of concern with high-pressure hydrogen storage systems in fuel cell vehicles.

Figure 12.3 illustrates the approach used in SAE J2579. The probability of an adverse event is given by the overlap of two probability distribution functions: the probability distribution for stressful conditions that occur in varying severities during service life and the probability distribution for storage systems to leak/rupture when subjected to various stressful on-road conditions. If the overlap is insignificant, then the risk is minimal.

The severity of stresses applied in qualification testing is represented by the "performance qualification criteria" in Figure 12.3, which correspond to severe stress conditions at or beyond the extreme for on-road lifetime service. Setting these criteria correctly—at worst-case extremes of on-road conditions—is the key to establishing robust design qualification tests.

Design qualification testing is designed to manage both leak and rupture risk. The risk associated with leakage is primarily a fire risk; that risk is mitigated by the rapid dispersion of hydrogen away from a vehicle and by vehicle controls that shut down the vehicle well before hydrogen leakage can reach its lower flammability limit (LFL). The probability

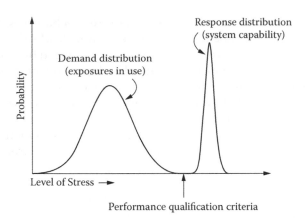

FIGURE 12.3
Management of risk.

of vessel leakage or rupture is specifically managed with requirements for durability, residual structural strength, and stability at the end of service.

It is not enough for qualified (tested) prototype storage system units to demonstrate their ability to sustain comprehensive on-road stresses. In addition, the production units must demonstrate tight correspondence to the capabilities of the validated tested units to ensure that production units have the required performance capability.

Three features make qualification tests established in this manner more demanding than historical CNG vessel tests:

1. Compounded stresses,
2. Full-system performance, and
3. Comprehensive on-road stress conditions.

Compound stresses refer to the application of on-road stresses simultaneously or in sequence to a single storage system unit. For example, in real-world on-road use, a vessel that is subjected to chemical exposures may also have been dropped during installation and could be in service for decades and thereby experience a lifetime of fueling pressure cycles. So, to verify on-road performance, these stresses must be applied to a single vessel, each in its worst-case extreme. This compounded test criterion is more severe than historical requirements, for which each stress is applied to a separate vessel. And, it corresponds more closely to the extreme of real-world potential risk. In another example, internal temperature stresses exacerbated by the presence of hydrogen gas and extreme ambient temperatures during fueling/defueling are produced by the simultaneous application of these factors during testing.

Full-system performance refers to the requirement that the full storage system (defined as all surfaces in direct contact with high-pressure hydrogen in the storage vessel) must demonstrate reliable containment of hydrogen without rupture or unacceptable leak. The key feature is that interfaces between components (e.g., shutoff valves, check valves, TPRDs, and connecting O-rings) and the storage tanks are validated as leak tight and not susceptible to blowout during extremes of vehicle service. In addition, components responsible for enclosure of high-pressure hydrogen (shutoff valves, check valves, and TPRDs) are expected to be individually qualified for reliable and durable operation with high-pressure hydrogen.

And finally, comprehensive on-road stress conditions refer to a comprehensive evaluation of all anticipated sources of stress during service life, not just those that have been linked to failures of earlier tank designs or materials. Therefore, if new designs/materials/manufacturing processes are subject to failure mechanisms that differ from those seen in the past, the qualification testing is designed to assess performance under extreme conditions where those failure mechanisms might occur in on-road service.

Performance Requirements: Qualification Testing

Performance standards designed to verify safety in on-road vehicle service need two elements:

- Qualification test protocols that produce worst-case stress conditions of service life
- Acceptance criteria

Qualification Test Protocols

In setting test conditions, awareness of historical failure mechanisms is helpful but not sufficient. Instead, a comprehensive approach to defining extreme conditions of on-road vehicle service is required. Storage systems must function under stresses of

1. baseline performance,
2. expected (normal) vehicle operation,
3. extreme service conditions (externally imposed stresses and conditions of extreme usage), and
4. fire.

Verification testing for normal, or expected, vehicle operation and performance in fire (items 2 and 4) are performed pneumatically with hydrogen gas to replicate real-world factors potentially leading to leakage/permeation as well as rupture. Verification testing for baseline and extreme conditions (items 1 and 3) requires the application of physical stresses potentially leading to structural failure and rupture; these pressure forces can be applied hydraulically.

Baseline Performance

Baseline performance metrics provide reference points for subsequent design qualification testing and for comparison of production units with the prototypes that undergo formal design qualification testing. They also establish that design prototypes are sufficiently similar to one another to be representative of design properties and are sufficiently capable to merit further testing.

For example, SAE J2579 requires three types of baseline system performance tests. The first are material tests that establish reference points for comparison of production units with the formally qualified (tested) design prototypes and establish that metals in contact with hydrogen are not subject to embrittlement. The second are initial burst pressure (BP_0) tests that establish that three randomly selected design prototype units are similar to one another (within ±10%) and that the BP_0 of each exceeds 200% nominal working pressure (NWP). NWP is the maximum (full-fill) pressure at 15°C, the rated pressure for fueling station fills, and is equivalent to the service pressure of gas cylinders.

The third verifies minimal pressure cycle life (number of full-fill pressure cycles until failure) of three vessels and assesses the variability in their pressure cycle life. If variability in pressure cycle life is greater than ±25%, then more than one vessel must be subjected to the item 3 extreme service testing.

Expected Operation

SAE J2579 provides an example of robust performance requirements that comprehensively address expected (normal) vehicle operations. The Expected Service (Pneumatic) Performance Test is illustrated in Figure 12.4. Storage systems are subjected to a high incidence of expected worst-case conditions and are required to perform without rupture or unacceptable leak/permeation. In addition, they must demonstrate residual strength after lifetime-equivalent durations of exposure to extreme stresses of vehicle service.

Expected vehicle operations for storage systems include fueling, parking, and driving (defueling). In each case, stress is applied internally by high-pressure hydrogen. Performance must be demonstrated with hydrogen gas because the unique properties of hydrogen can create additional internally imposed stresses. For example, because hydrogen is a small molecule, pressure changes may drive greater infusion-into/withdrawal-from interstices within materials and sealing interfaces than would occur with larger

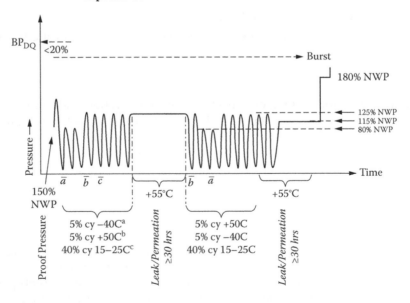

Expected-Service Performance Verification Test

a Fuel/defuel cycles @ −40°C with initial system equilibration @ −40°C, 5 cycles with +20°C fuel; 5 cycles with <−35°C fuel.
b Fuel/defuel cycles @ +50°C with initial system equilibration @ +50°C, 5 cycles with <−35°C fuel.
c Fuel/defuel cycles @ 15–25°C with service (maintenance) defuel rate, 50 cycles.

FIGURE 12.4
Expected service performance test.

molecules, like natural gas. The combination of stress factors that occur in normal operation could promote previously unrecognized failure mechanisms (e.g., temperature-amplified, hydrogen-induced embrittlement of metals or contraction and rigidity in sealing materials) that should be elicited in qualification testing rather than on-road service. Therefore, qualification testing must include compounded and realistic extreme conditions of fueling, parking, and defueling (driving) with hydrogen.

Fueling/Defueling (Driving) Performance Requirement

The following features are key to achieving comprehensiveness with regard to the stressful fueling/defueling conditions that are tested:

- Rate of internal pressure change. Maximum temperature and pressure shock occur with rapid fueling and defueling, so rapid 3-min fuelings to 125% NWP (using a constant-pressure ramp rate according to SAE J2601) are required, as are defuelings at the maximum rate (maximum vehicle fuel consumption rate) to 20 bar. If the defueling rate used for vehicle maintenance/service is higher, then systems must demonstrate at least 50 of the more rapid defuelings in addition to normal defuelings associated with driving.

- Peak fueling pressure. Maximum stress is applied with fueling from 20 bar to 125% NWP at ambient (20°C) and high (+50°C) external temperatures, and fueling to 80% NWP at cold (−40°C) external temperatures. Fueling protocols (e.g., SAE J2601) do not permit fueling above 80% NWP at −40°C to ensure that systems are not overpressured if subsequently exposed to warmer temperatures (e.g., moving inside or in diurnal warming). A precision of ±10 bar on pressure cycle extremes ensures correspondence with vehicle usage and that appropriate pressure controls are used in the testing facility.

- Tank internal temperature and pressure at onset of fueling. Maximum thermal shock is produced using both cold-soak and hot-soak test conditions. A −40°C-cold-soaked, fully filled system is subjected to rapid defueling (to create the coldest possible interior temperature conditions) followed by rapid heating from fueling with 25°C fuel (indoor fueling without temperature control) and also with fuel at −35°C or less (expected at public fueling stations). In addition, a +50°C hot-soaked, empty (20-bar) system is subjected to the impact of −40°C fuel under fueling.

- External temperature. Fueling/defueling cycles are conducted under both −40°C and +50°C extremes of environmental (external) temperature. The percentage of extreme temperature pressure cycles corresponds to twice the percentage of occurrence in populated extreme latitudes. The remaining pressure cycles are conducted at nominal 20°C ambient temperature.

- Number of expected fueling/defueling pressure cycles. The maximum number of high-stress full fueling/defueling cycles has been determined from (Lifetime vehicle range)/(Driving range per full tank) = L/R. Under no circumstances is this expected service qualification test cycle number allowed to be less than 500 empty-to-full fuelings with hydrogen gas (500 = 150,000 lifetime miles of driving/300 miles per fueling). The extremity of this number derives from the low probability that all vehicle fuelings through a vehicle's life would be under the maximum stress condition of an empty-to-full fill. In addition, potential buildup of impurities or charge from multiple fuelings is managed by fuel quality require-ments (SAE J2719 and ISO 14687-2).

- Interaction between fueling/defueling pressure/temperature cycles and the static stress associated with parking. Sensitivity to fatigue mechanisms induced by the interplay and cumulative impact of cyclic and static pressure [10] is elicited by sequential exposure to cyclic and static high pressure [11].

Parking

Parking is associated with prolonged exposures to static high-pressure hydrogen gas dur-ing which physical force is applied to the vessel structure by the internal pressure; in addition, small hydrogen molecules are forced into, and possibly through, the vessel walls (permeation). The worst-case extreme condition for applied physical force is addressed in verification testing for extreme service durability (discussed in the next section). The worst-case condition for permeability (including leakage) occurs when the vessel interior, which generally provides a sealing or gas-tight interface with the hydrogen gas, has expe-rienced fatigue from extremes of pressure cycling and when the extent of hydrogen migra-tion has risen to a steady-state level under prolonged high temperature and pressure.

The maximum allowable discharge due to permeation (including leakage) from a hydro-gen storage system is established to prevent a buildup to 25% LFL (the LFL of hydrogen in air is 4%) in a confined space, such as a home garage. The leakage limit is established to prevent the smallest possible pinhole flame on the surface of the storage vessel.

Extreme Service Durability

In SAE J2579, qualification for lifetime service requires a robust test of the durability of ves-sels to survive exposure to externally imposed stresses and thereafter to perform reliably through a service lifetime of extreme usage.

Externally Imposed Stresses

Storage systems used in on-road vehicles can be expected to encounter four types of externally imposed stresses: impact (drop during installation, road wear, and crash dam-age), fire, fueling station malfunction (overpressurization), and environmental exposures. Crash performance is evaluated in government-required vehicle crash testing, for which fuel integrity requirements are established to manage fire risk due to leakage (e.g., SAE J2578). For exposure to external fire sources, the probability of rupture is reduced by the requirement to demonstrate release without rupture (discussed in the following section on fire). The remaining external factors—drop, road wear, and environmental exposure (chemicals combined with temperature/humidity exposure), and fueling station malfunc-tion (overpressurization)—can be aggressive assaults on the vessel structure and hence present rupture risks. SAE J2579 provides an example of robust performance require-ments that comprehensively address these externally imposed stresses. The Durability (Hydraulic) Performance Test: Extreme Conditions and Extended Usage requirement is illustrated in Figure 12.5.

Since a vessel dropped during installation would thereafter be exposed to road wear (e.g., stone chips and abrasion from vessel bindings) and to environmental factors, the application of these stresses in series is required. The worst-case factors to be applied include drop (e.g., from a forklift), expected in-service wear, and concentrated application of chemicals to punctures through external coatings with refreshed exposure throughout

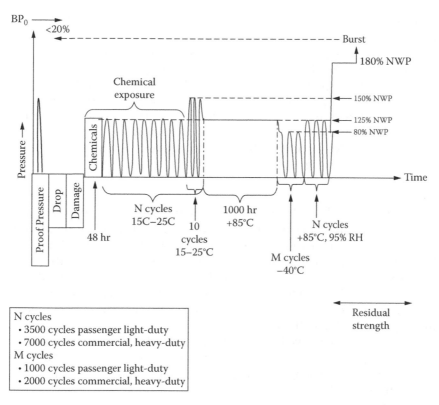

FIGURE 12.5
Durability performance test.

48 h of high humidity with the system at full pressure, followed by a demonstration of the capability to sustain extreme usage in vehicle operations that include fueling, parking, and driving (defueling). Stresses exerted during fueling and driving (defueling) are replicated by hydraulic pressure cycles. Stresses exerted during parking are replicated by sustained hydraulic pressure.

Fueling and Defueling (Driving)

Surveys of gasoline passenger vehicles [8] (not buses) showed the extreme lifetime in miles has been about 360,000. Since vehicles currently exceed 200 miles of travel with a full fuel tank, 360,000/200 = 1800 exceeds the number of full fuelings in current vehicles. As the driving range per full fueling rises, the maximum number of full fuelings decreases. As an added assurance of residual strength for rupture prevention, consideration of additional numbers of fuelings in the likelihood of partial fillings is accommodated in the qualification testing by considering a lifetime of fuelings at only one-third capacity, thereby requiring 3 × 1830 = 5500 pressure cycles to 125% NWP; this is beyond worst case since partial fuelings have been shown to cause considerably less fatigue than full fuelings. As a corroboration, it has been shown [6] that a six-sigma projection (~10^{-9} probability) from the distribution of recorded lifetime vehicle miles traveled [8] would also support about 5500 pressure cycles as the statistically extreme worst case. These pressure cycles, which serve to accelerate the fatigue of vessel wrap material, can be applied either pneumatically or hydraulically. For commercial heavy-duty vehicles, such as buses, 15,000 pressure cycles are required to accommodate higher lifetime usage.

Requirements for fueling stations have been established (CSA and ISO have standards under development; the NFPA has established requirements) that limit the fueling pressure to 125% NWP and require monitoring/intervention to limit a potential overpressurization to less than 150% NWP. Thus, it is the responsibility of the hydrogen fueling station to prevent dangerous overpressure events, hence the requirement for the fueling stations to install PRDs to manage pressure. As a worst case, exposure to overpressurization due to fueling station failure is assumed to occur 10 times during service life prior to extreme parking (static pressure) and extreme temperature refuelings. In the absence of reports of overpressurization at CNG fueling stations, this frequency is expected to be extreme. In addition, a final end-of-service overpressurization is applied.

Parking

Parking is associated with prolonged exposures to high pressure, which could cause fatigue and stress rupture. Parking cannot be directly replicated in qualification testing because the relevant cumulative time period is years to decades, which is too long for a practical qualification test. Therefore, a performance verification test that amplifies the physical stress to accelerate resulting stress fatigue is needed. The amplified stress should not, however, induce failure modes that could not occur in on-road service.

The worst-case exposure to static high pressure (full-fill parking) has been selected as 25 years under full-fill conditions (100% NWP). Vessel strength derives from properties of the structures, which have been metal in past portable applications and have been composite (resin-impregnated fiber) in modern vehicle applications. A robust approach for accelerating the effects of prolonged high-pressure exposure is to set requirements appropriate to verifiable properties.

Experimental findings on tensile stress failure of representative strands used in composite wraps of pressure vessels are shown in Figures 12.6a and 12.6b [12–15]. (Strands are fibers [10,000–15,000 filaments] coated in resin and cured.)

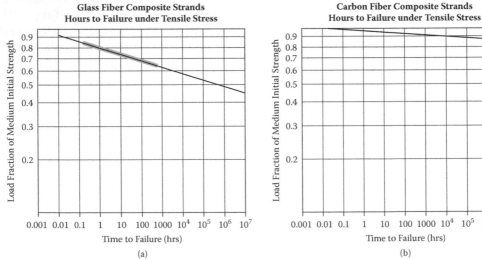
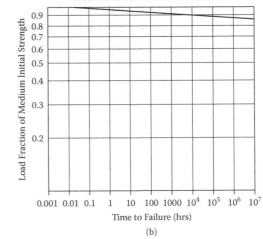

FIGURE 12.6
Composite strand experimental data: hours to failure under sustained tensile stress. (a) glass fiber composite strands; (b) carbon fiber composite strands.

In each case, failures were shown to be well represented by Weibull distributions [12], which have the characteristic that failures that occur with the same probability fall on a straight line in log-log plots of the load fraction (load as a fraction of the median initial strength) plotted against time to failure when held under that load. The plots show the median (0.5 probability), and the shaded area indicates the region of data used for the Weibull fit. By looking at the slopes, it is clear that the time to failure increased when the load was reduced. Carbon strands showed greater resistance to stress rupture than glass strands in that a small reduction in the applied load resulted in a greater increase in time to failure for the carbon strands than for the glass strands. Carbon strands appeared highly resistant to stress rupture. In contrast, glass strands fatigued more rapidly (failure at a much lower load fraction for a specific time to failure, or stated another way, much shorter time to failure for a given load fraction). For carbon strands, a reduction in load of just 3% resulted in a lengthening time to failure by a factor of 100; for glass strands, a reduction in load of 18% was required for the same improvement in time to failure.

It is important to note that the failure probability shown in Figure 12.6 is not the probability for vessel failure; it is the estimated probability for a single-point strand failure. Vessel fiber-wrap patterns and resin adhesion accommodate individual strand segment failures through load sharing. This is consistent with the observation that multiple adjacent strand failures transferring large load-sharing burdens are needed for a broader failure [16]. Vessel measurements have shown longer times to failure for a given load fraction than strand data.

The *slopes* of the lines in Figure 12.6 have relevance to vessels, however, since the fatigue response of strands to load stress is the main underlying source of vessel fatigue. Using the slope of lines in Figure 12.6(a) one can conclude that for vessels with glass-fiber-composite strand wraps, the probability of failure for 25 years under tensile stress imparted by 100% NWP was equivalent to 1000 h under tensile stress imparted by 122% NWP. Consequently, a glass-composite vessel would have the same probability of failing (or passing) a performance test where it is subjected to 100% NWP for 25 years as for a performance test where it is subjected to 122% NWP for 1000 h. Given the difference in the practicality of the different test times, the 1000-h test has been selected. From a test facility perspective, a shorter test conducted at a higher pressure would be even more convenient, but since even short-term parking pressures cannot exceed the fueling maximum pressure, 125% NWP, that could potentially drive failure modes that would not occur in real-world vehicle service.

Figure 12.5 shows that SAE J2579 includes 1000 h of exposure to 125% NWP as the performance test for survival of parking stresses. An elevated temperature of 85°C, which is an extreme recorded under-hood temperature for a parked vehicle, is applied to account for systems subject to heat-accelerated deterioration. Exposure to ultraviolet light is not included as SAE J2579 explicitly requires ultraviolet shielding of vehicle storage systems.

Confirmation of Parking (Stress Rupture) Criteria

The parking performance test verifies sufficient resistance to stress rupture for full-fill parking lasting 25 years for a worst-case (such as glass-fiber-reinforced composite) where (for a given failure probability) the time to failure shortens times 10^{-2} when the sustained pressure increases by 18%. To ensure that this is the worst-case condition for future fiber technologies, SAE considered adding the requirement that vessels demonstrate capability beyond this expected worst case by requiring that vessels be held at 75% BP_0 for 1000 h without rupture. In addition to verifying that the qualification testing applies for 25 years of parking at 100% NWP, it also ensures capability to sustain 115% NWP for 10 years, which is a lifetime of overnight parking after full fills. Experiments on vehicles fueled and parked overnight at temperatures over 20°C showed internal pressures within carbon fiber tanks with insulating plastic liners dropped by 10% within 5 hr, so a criterion to demonstrate capability equivalent to holding 115% NWP for 10 years (12 hr per day over 20 yr of service) would be conservative. And it would ensure that fibers with higher sensitivity to stress rupture than glass fibers would not be qualified for on-road service.

More recently, SAE has undertaken consideration of an alternative criterion that vessels have capability to sustain parking at 150% NWP for 25 years. This would be intended to correspond to the even more conservative extreme of a tank subjected to maximum over-pressurization due to service station malfunction followed by a lifetime of exposure to an ultra-extreme elevated temperature of 85°C. If this criterion is adopted, then tanks would be required to demonstrate the capability to sustain pressures above 175% NWP for 1000 hrs without failure, and fiber-reinforced composite tanks would be limited in construction to carbon or glass fiber. In that case, it would be expected that generic performance-based requirements for vessel qualification would be developed in the future that would allow the qualification of future vessels using novel fibers.

In either case, because extensive experimental data and records of established on-road service are available for glass-fiber and carbon-fiber reinforced composite vessels, they would be allowed to be qualified more simply by demonstrating 330% NWP and 200% NWP initial burst pressures, BP_0, respectively.

Fire

Reliable rupture-free performance is the paramount requirement for on-road service. In the event of a vehicle fire, rupture could be caused by fire damage of a storage vessel or by increased pressure of the contained gas. Therefore, storage systems are required to vent hydrogen in a controlled manner through a TPRD when exposed to fire. To verify this capability, storage systems are exposed to fire conditions that replicate worst-case conditions (temperature and timing) determined from studies of vehicle fires. An assortment of localized vehicle fires have been examined; the duration and temperature are considered to be relevant only when the localized fire makes contact with the storage system above 300°C (the temperature at which resin reaction is noted). The test protocol in SAE J2579, which is illustrated in Figure 12.7, remains under discussion and hence subject to revision. It begins with a localized fire exposure that progresses to an engulfing fire condition. The localized fire is applied at 600°C, an extremely damaging condition not expected to remain localized as long in an on-road vehicle fire as in the test before transitioning to an engulfing fire; therefore, it is considered to be an extreme-case condition. A storage system may be tested with vehicle componentry designed to reduce localized fire intrusions, in which case the storage system alone is also tested with exposure to an immediately engulfing bonfire to ensure basic capability for the isolated storage system.

Acceptance Criteria

The performance metrics for storage are the absence of rupture or unacceptable leak throughout service life. So, both must be evaluated during testing and at the end of test sequences to verify the performance capability is retained at end of life.

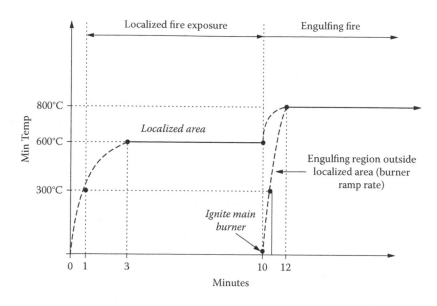

FIGURE 12.7
Fire performance test.

Criteria for Permeation and Leakage

In the pneumatic test (Figure 12.4), permeation (including leakage) is evaluated. The permeation limit is set at 150 Ncc/min, so that the flammability within a tight space cannot exceed 25% LFL (LFL of hydrogen in air is 4%). Consistent with defining worst-case requirements, *tight* means 30.4 m³ (close to nominal car dimensions) and having the minimum air exchange per hour (ACH) recorded for a garage (0.03 ACH). For vehicles of unconventional size, the permeation limit is apportioned to the minimum size of garage where it will fit. Hence, even at the end of a worst-case stressful service life, the system could be parked safely in confined spaces.

The maximum allowable discharge due to leakage and permeation from a hydrogen storage system was originally established at 150 Ncc/min for standard passenger vehicles to prevent a buildup to 25% LFL in a "very tight" 30.4-m³ garage with the lowest recorded 0.03 air changes per hour. The value is obtained from

$$C_\% = 100\, Q_{H2}/(Q_{air} + Q_{H2})$$

where $C_\%$ is the steady-state hydrogen concentration (percentage in air), Q_{air} is the airflow rate, and Q_{H2} is the hydrogen permeation rate.

Consideration of tight garage volumes for vehicles of unconventional size led to the requirement that the maximum allowable permeation/leakage discharge for systems be $A \cdot 150$ Ncc/min, where $A = (V_{width} + 1) \cdot (V_{height} + 0.5) \cdot (V_{length} + 1)/30.4$ m³ and V_{width}, V_{height}, and V_{length} are the vehicle width, height, length (m), respectively. Equivalently, if the total water capacity of the storage system is less than 330 L, then the system may qualify to 46 mL/L/h at 55°C and 115% NWP. The temperature for the leak/permeation test was selected to be at least 55°C as an extreme condition for long-term parking in a tight garage. The full-fill pressure at 55°C is 115% NWP. These requirements are consistent with conclusions of the European Union HySafe project [20].

In addition to the permeation requirement, localized leakage is evaluated by bubble testing and is limited to 3.6 Ncc/min. This is the smallest leakage that could support a pinhole flame [19]. Thus, a localized hydrogen leak cannot sustain a flame that could subsequently degrade surface material and cause a loss of containment. Per SAE 2008-01-0726, "Flame Quenching Limits of Hydrogen Leaks," the lowest flow of H_2 that can support a flame is 20 Ncc/min from a typical compression fitting, and the lowest leak possible from a miniature burner configuration is 3.6 Ncc/min. Since the miniature burner configuration is considered a conservative worst case, the maximum leakage criterion was selected as 3.6 Ncc/min. It is expected that visual detection of unacceptable leakage should be feasible. When using standard leak test fluid, the bubble size is expected to be approximately 1.5 mm in diameter. For a localized rate of 3.6 Ncc/min, the resultant allowable rate of bubble generation is about 2030 bubbles per minute. Even if much larger bubbles are formed, the leak should be readily detectable. For example, the allowable bubble rate for 6-mm bubbles would be approximately 32 bubbles per minute.

Criteria for End-of-Life Strength

End-of-Life strength, as used here, refers to residual strength to resist rupture after having survived compounded worst-cast extremes of service life. The robustness of the test sequences (expected operation and extreme service durability) has been quantified. For example, extremes of environmental temperature are taken from climate data. Also, the number of stressful fueling-induced pressure cycles was linked to 10^{-9} probability of occurrence from broad surveys of vehicle lifetime mileage. Further, worst-case parking stress was linked through stress rupture data for composite strands to 25 years of parking under full-fill conditions.

However, given the severe consequences of rupture, two additional quantifiable metrics for residual, or end-of-life, strength have been developed:

1. Residual strength: end-of-life capability
2. Residual strength: stability

Residual Strength: End-of-Life Capability

As illustrated in Figures 12.4 and 12.5, vessels are required to demonstrate the capability to sustain a 150% NWP overpressurization from fueling station failure at the very end of simulated extreme service life and have the capability to sustain the overpressurization for 10 h. This requirement is linked through strand data (Figure 12.6) to the test criterion of 180% NWP for more than 4 min without rupture.

Residual Strength: Stability

The vessel BP measured at the end of life must be within 20% of the median BP of fresh vessels. Therefore, the mean virgin burst pressure BP_0 must be verified as part of design qualification. Once again, *end of life* is defined as the end of the test sequences shown in Figures 12.4 and 12.5. This requirement is consistent with the expectation that the vessel has not reached a condition for which the primary failure mechanism has advanced sufficiently to produce pronounced loss of strength. This expectation is consistent with traditional vehicle safety assurance, which precludes use of materials whose properties change substantially over vehicle life, even if excess capacity were built into newly manufactured units.

Temperature Specifications in Qualification Tests

Temperature specifications in performance tests designed to confirm function during expected service (Figure 12.5) are linked to worst-case on-road requirements as follows:

- Temperatures of −40°C and +50°C during pressure cycling correspond to nominal extremes in global ambient temperature records that could be encountered repeatedly during vehicle service.

- Temperatures of 55°C or greater during permeation testing correspond closely to the global maximum recorded temperature. Since temperature stabilization of a storage system requires many hours, a peak temperature below the global short-term high would be an extreme condition for garage parking.

- A temperature of 85°C during 1000 h at 125% NWP has three elements of justification:

   a. The maximum temperatures known to have occurred within vehicles parked in direct sunlight on dark asphalt are in the range 80–85°C.

   b. Full-fill fuelings (SAE J2601 when completed) may commonly result in interior tank temperatures of 85°C. For 150,000-mile lifetime mileage and 200 miles/fueling, a vehicle would experience 750 lifetime fuelings. Allowing time for cool down, 500–1000 h of exposure of the interior liner to 85°C is a possibility. Time spent in the test is intended to show that the materials can withstand the maximum expected temperatures without degrading or creeping under load.

   c. Elevated temperatures have been used to accelerate stress rupture testing, although a test of a single cylinder or limited number of cylinders may not have the statistical significance that has been associated with planned studies involving a larger number of cylinders. The acceleration of testing is associated with the Arrhenius rate equation, in which rate of reaction varies with temperature. It has been observed [17, 18] that for many composites, the rate of reaction doubles with a 10°C increase in temperature. To this extent, 85°C represents a 2^7-fold time acceleration of these factors (1000 h $\times$ 2^7 > 14 years) and can thereby emulate stress rupture in an accelerated manner.

Validation of Qualification Requirements for Hydrogen Storage

There are two ways to validate qualification requirements for on-road service. One is retroactive: If mass-produced storage systems qualified using the requirements provide safe performance over 15 years or more of vehicle service, then the requirements have been sufficient (and possibly, but not necessarily, overly conservative). The other is proactive: (1) The requirements must be capable of execution by qualified test facilities; (2) they must screen out vessels known to fail in on-road service; (3) they must address known extremes of on-road service; and (4) they must only fail vessels having good on-road service if a previously unrecognized vulnerability to failure is revealed. The SAE J2579-2008 requirements were submitted to validation according to these four criteria at Powertech Laboratories [21] in Vancouver, Canada, with positive results.

What Makes the New Performance-Based Requirements More Stringent Than Earlier Approaches?

Performance requirements as described here, with SAE J2579 used as an example, differ from historical requirements in several respects that impose additional stringency on storage system performance. Highlights are the following:

- Most notably, historical requirements apply to individual vessels rather than full storage systems. It is known that early 700-bar systems were withdrawn from on-road service because leaks developed at interfaces. Those systems would not have qualified under comprehensive system-level performance tests specified in SAE J2579.

- In addition, the compounding of stresses that could be simultaneously experienced in service adds to the severity of stresses imposed on the storage systems that are evaluated for on-road suitability. For example, some warm prototype liners have failed when exposed to high-pressure hydrogen for a sufficient time for liner permeation to equilibrate before stressful pressure cycling—a condition that can only be evaluated when both temperature extremes and full-stress pressure cycling with hydrogen gas occur simultaneously.

- Pneumatic testing with hydrogen is required to simulate the full range of internal temperatures that can occur during defueling. These temperatures are linked to boss embrittlement, liner durability, and leak tightness of valves. Leakage from prototype systems tested with hydrogen under realistic fueling conditions has been discovered with this test protocol (from a September 2008 Powertech Laboratories report to the SAE Safety Work Group [21]).

- The requirement for manufacturing quality control to ensure that the BP of new (unused) vessels is controlled to 90% or greater of the median BP of vessels tested in design qualification (and equal to or greater than the minimum BP requirement) ensures that manufactured vessels correspond to those qualified for service. Also, the requirement that the more than 10 vessels presented to undergo design qualification testing must have BP_0's within +10% of their median BP_0 provides assurance that tested vessels are representative of manufactured units. The employment and documentation of formal quality control procedures are additional requirements.

- The test to confirm the parking (stress rupture) criteria provides confirmation of minimal vessel structural stability over time. The criteria primarily ensure performance consistent with historical expectations for fiber/resin composite and metal vessels.

- Finally, the new requirements of end-of-life residual strength will likely prove challenging to some systems that previously qualified only the strength of new vessels. The new end-of-life requirements require demonstration of residual strength (>4 min at 180% NWP) and BP stability (BP > 80% of nominal virgin BP). Also, BP measurements require a slower rate of pressure increase (3-min constant pressure ramp rate to 150% NWP followed by slower increase), which effectively stiffens the requirement over previous methods of BP evaluation. Further elements of improvement over use of historical BP ratios are documented [22].

Some requirements may appear to be less stringent than those used previously, but in reality they are not because real risk is not increased with these requirements. For example, some earlier testing protocols called for up to 45,000 hydraulic pressure cycles. Considering that a single full-fill cycle represents 200–300 miles of driving, 5500 cycles corresponds to more than 1–2 million miles of driving, which is already far beyond the lifetime range of on-road passenger vehicles. In another example, CNG vessels are often tested for 1000 pneumatic cycles to allow for interior electrical charge accumulation; however, for hydrogen systems, that risk is addressed at its origin: Hydrogen fuel quality requirements limit water and particulate content of the fuel (hence, charge-carrying aerosol).

New Requirements for Test Facilities

The primary new facility requirements for conducting these performance tests pertain to the pneumatic test sequence (Figure 12.5). This test sequence requires environmental temperature control (i.e., the surroundings) for the storage system being tested during pressure cycling. It also requires that the pressure cycling be conducted with hydrogen gas applied under realistic fueling conditions (e.g., SAE J2601), which means that the capability to reliably control the temperature of the hydrogen gas at the system inlet to −35°C or less is required. Several fueling stations and several storage qualification test centers have installed sufficient capability for temperature control of hydrogen fuel at the point of delivery to the storage system. The Powertech report to the SAE Safety Work Group [21] contains details of experience in refitting a storage test facility to be able to conduct the test sequence illustrated in Figure 12.5. That report recommended conducting the test sequence shown in Figure 12.4 before conducting the test sequence shown in Figure 12.5 as a means of verifying the rupture resistance of the system before conducting the pneumatic pressure tests and thereby avoid putting the facility at risk of vessel rupture during the more time-intensive pneumatic testing.

The length of time required to conduct the testing required by the sequences shown in Figures 12.4 and 12.5 is expected (from the testing done by Powertech Laboratories for SAE [21]) to be approximately 3–4 weeks longer than the time required to conduct more traditional CNG vessel qualification. The primary uncertainty is the time required for the defueling phases during pressure cycling of 700-bar systems.

Are We Done Yet? What Additional Changes in Requirements Are under Discussion?

The development of a performance standard for gaseous hydrogen on vehicles is not complete at SAE or ISO. In addition to the topics discussed, issues that will likely merit further consideration include the following:

- Inspection/Recertification

  At present, periodic in-service inspection requirements are not specified within SAE J2579. There are two primary reasons: First, individual vehicle owners cannot be compelled effectively to comply with inspection requirements in all cultures where vehicles are deployed; hence, reliance on inspection for public safety has not yet been accepted as a global strategy. Second, nondestructive evaluation (NDE) methods of inspection are useful but not yet sufficiently capable of reliable indication of damage or stress rupture potential. Third, invasive interior inspection techniques that require removal of vessels from vehicles may be impractical to apply, especially when very large numbers of vehicles are deployed, and may result in added risk from vessel damage during removal and reinstallation. More robust NDE methods are under development and may provide future opportunities for inspection/recertification.

Recertification requirements for reuse of storage systems removed from a vehicle (e.g., after a crash) may need to be updated for the higher-pressure systems used for hydrogen storage. Therefore, in the current SAE J2579, it is specified that a storage system cannot be transferred from one vehicle to another for continued use.

- Hydrogen Embrittlement

  Test procedures to qualify metal alloys for use in contact with hydrogen in hydrogen storage systems have recently been developed within SAE J2579 to include extreme conditions of $-50°C$ or less and 900 bar, which occur in hydrogen storage systems in on-road service. Steel alloys found to be qualified include UNS S31600 and UNS S31603 having 12.5% or greater nickel composition and 0.1% or less magnetic phases by volume and used in conditions in which operational stress is less than 67% of yield strength. Aluminum alloys found to be qualified include A6061-T6, A6061-T62, A6061-T651, A6082-T6, A6082-T62, A6082-T651, and A6082-T6511. Additional metal alloys may be qualified either by testing vessels with specified pressure cycling using hydrogen gas under extreme temperature conditions or by qualifying the alloys with specified extreme-temperature slow-strain rate tests, fatigue life tests, and fatigue crack growth rate (compact tension) tests. SAE test procedures were developed with assistance from experimental research and expertise at Sandia National Laboratories and Fukuoka University.

- Storage Systems Other Than Compressed Gaseous Hydrogen

  If storage systems other than gaseous hydrogen are developed for vehicles, such as the solid-state storage systems described throughout this book, those qualification test procedures will be adapted to encompass appropriate extreme conditions for fueling and defueling and static containment using those technologies. Complementary standards for fueling stations would also be required. Corresponding test requirements for liquid hydrogen storage are contained in SAE J2579 and within the U.N. global technical regulation.

Conclusion

The recent efforts to develop requirements for vehicle hydrogen storage have drawn on the rich history with CNG vehicle storage and from the broad community of experts. That community has included pressure vessel and component manufacturers, experts from the emerging fueling and repair/service infrastructures, and the vehicle designers/manufacturers, who bring broad experience with on-road vehicle safety and management of risk when a product is used by a diverse population in uncontrolled conditions.

The more fundamental shifts in perspective that have occurred during these recent efforts have pertained to the data-driven approach to developing performance-based requirements that clearly address simultaneous exposures to different stresses and stress rupture susceptibility that could develop over vehicle life. The end-of-life requirements for residual strength (resistance to rupture and stability of ultimate burst strength) focus on verifying actual characteristics of storage systems after severe exposures to on-road stresses.

The resultant shift to performance-based requirements for compressed storage of hydrogen on vehicles provides a proactive, rather than reactive, approach to preventing on-road failures. It focuses on the breadth and severity of on-road service conditions that all storage systems must survive with full performance. Comprehensive on-road worst-case conditions encompass circumstances of past failures, while they apply more generally to future technologies. The requirements are linked to the severity of on-road stresses that can be experimentally measured to substantiate the specification of performance requirements with data. By these means, the requirements are designed to provide a higher level of safety assurance.

Acknowledgments

I am pleased to acknowledge discussions with developers of the SAE J2579 requirements for hydrogen storage, the SAE Fuel Cell Vehicle Safety Work Group, specifically Glenn Scheffler, Marcus Jung, Ian Sutherland, Yoshio Fujimoto, Mike Veenstra, Akihiro Sakakida, Gini Sage, Wolfgang Oelerich, Livio Gambone, Mark McDougal, Thorsten Michler and Brian Somerday.

References

1. National Fire Protection Association (NFPA) 2, Hydrogen Technologies Code (2011) as incorporated in individual state and municipal fire codes.
2. *Code of Federal Regulations*, Title 49 (49 CFR), parts 171–180 and parts 190–199 (pipelines).
3. *Code of Federal Regulations*, Title 49 (49 CFR), part 571, Federal Motor Vehicle Safety Standards.
4. Society of Automotive Engineers (SAE), J2600, Technical Information Report for Compressed Hydrogen Surface Vehicle Fueling Connection Devices (2012 revision in progress), SAE International, Warrendale, PA.
5. Society of Automotive Engineers (SAE), J2601, Technical Information Report for Fueling Protocols for Light Duty Gaseous Hydrogen Surface Vehicles (2012 revision in progress), SAE International, Warrendale, PA.
6. Society of Automotive Engineers (SAE), J2578, Recommended Practice for General Fuel Cell Vehicle Safety (2012 revision in progress), SAE International, Warrendale, PA.
7. Society of Automotive Engineers (SAE), J2579, Technical Information Report for fuel systems in fuel cell and other hydrogen vehicles (2012 revision in progress), SAE International, Warrendale, PA.
8. Sierra Research Report Number SR2004-09-04, Review of the August 2004 Proposed CARB Regulations to Control Greenhouse Gas Emissions from Motor Vehicles: Cost Effectiveness for the Vehicle Owner or Operator—Appendix C to the Comments of The Alliance of Automobile

Manufacturers. Austin, Thomas C., Thomas R. Carlson, James M. Lyons, Francis Di Genova, Garrett D. Torgerson as prepared for the Alliance of Automobile Manufacturers, September 2004; Sierra Research Inc., Sacramento CA.

9. Informal presentation to SAE Fuel Cell Safety Work Group (2008) Troy, MI by Michael Veenstra; Ford Motor Co, Dearborn, MI.

10. J.F. Mandell, MIT Report RR-R81-2, Fatigue behavior of fiber-resin composites (1981) MIT, Cambridge, MA; also published as chapter 4 in Developments in Reinforced Plastics, 2nd ed, G. Pritchard; Applied Sci Publ.

11. S.L. Phoenix, P. Schwartz, H.H. Robinson IV, *Composites Sci. Technol.* (1988) 32, 81.

12. E.Y. Robinson, NASA Aerospace Report No. ATR 92(2743)-1, Design prediction for long-term stress rupture service of composite pressure vessels (1991); The Aerospace Corp, El Segundo, CA.

13. R.E. Glaser, R.L. Moore, T.T. Chiao, *Composites Tech. Rev.* (1983) 5, 21.

14. T.T. Chiao, C.C. Chiao, R.J. Sherry, UCRL-78467, Lifetimes of fiber composites under sustained tensile loading (1976); Lawrence Livermore National Laboratory, Livermore, CA.

15. J.T. Shaffer, Stress rupture of carbon fiber composite materials, 18th Int. SAMPE Technical Conference (1986); Seattle, WA.

16. S. Blassiau, A.R. Bunsell, A. Thionnet, *Proc. R. Soc. A* (2007) 463, 1135.

17. C.C. Chiao, R.J. Sherry, N.W. Hetherington, Lawrence Livermore Laboratory Report UCRL-78496, Experimental Verification of an Accelerated Test for Predicting the Lifetime of Organic Fiber Composites (1976); Lawrence Livermore Laboratory. Livermore, CA.

18. G.M. Ecord, NASA EM2-96-006, Stress rupture failure of a Kevlar overwrapped pressure vessel during stress rupture testing (1996); L B Johnson Space Center, Houston, TX.

19. M.S. Butler, R.L. Axelbaum, C.W. Moran, P.B. Sunderland, SAE Technical Paper 2008-01-0726, Flame quenching limits of hydrogen leaks; SAE World Congress, Detroit MI; SAE International, Warrendale, PA.

20. P. Adams, A. Bengaouer, B. Cariteau, V. Molkov, A. Venetsanos, *Int. J. Hydrogen Energy* (2011); 36(3) 2742–2749.

21. M. McDougal, SAE J2579 Validation testing program powertech: final report, National Renewable Energy Laboratory Report No. SR-5600-49867. (Note: downloads available at NREL publications database http://www.nrel.gov/publications/.)

22. C. Sloane, SAE Technical Paper 2009-01-0012 (2009), *SAE Int. J. Passenger Cars Mech. Syst* (2009); 2(1), 193–205.

Editor's Epilogue and Acknowledgments

Now that you have seen what we have written, I'd like to offer an "Editor's Epilogue," offering my own personal views of "where things stand" and some thoughts about the road ahead.

We made the case in Chapter 1 that if we are going to solve our fuel resource insecurity, political insecurity and environmental sustainability problems surrounding our current fossil-fuel-based energy infrastructure, we are going to need to turn to hydrogen. In particular, environmental sustainability will only be achieved through a zero-carbon energy solution that hydrogen technology can ultimately provide. Unless we have a carbon-free solution, none of the benefits accrued to a new infrastructure (such as reduction in GHG emissions) can survive growth in either population, or growth in the intensity with which technology uses energy. So, the need is clear for hydrogen-based power technology, and it needs to be a zero-carbon technology. The time-scales for technology change discussed in Chapter 1, along with the ~50 year energy horizon that we have, indicates we have to start the conversion to a zero carbon technology-like hydrogen now, and we need to be going much faster than we are.

In Chapter 2, we gave a description of the hydrogen energy conversion devices (H_2 ICEs and fuel cells) that are required to convert hydrogen into electrical and shaft power. Hydrogen fuel cells are commercially available, whereas the H_2 ICEs are "commercial ready." PEM fuel cells in particular are already finding use in the first fuel cell vehicles, and also commercial use in portable power, backup power, material handling equipment, and construction equipment (e.g., the Fuel Cell Mobile Light). H_2 ICEs have already been demonstrated in H_2 ICE light-duty vehicles, with even small production runs made for the BMW Hydrogen 7 vehicle. Fuel cells still need reductions in cost for widespread market infusion, but there is every reason to believe these price reductions would occur with the economies of a larger manufacturing scale. Hydrogen ICE's, being modifications of the existing combustion engines, can be produced for nearly the same cost as the conventionally fueled version of the technology. Of course improvements in the durability of hydrogen conversion technologies are desirable. That's true for any technology. However, the technical status of hydrogen fuel cells and ICEs is sufficient *now* to begin the conversion to a hydrogen-based energy infrastructure *today*.

Hydrogen is a chemical commodity that is widely available. As described in Chapter 3, hydrogen has found use in manufacturing, science, and technology over the years because its use offered a distinct advantage in certain applications. For chemical manufacturing and laboratory uses of hydrogen, hydrogen is conveniently stored as compressed hydrogen gas in appropriately sized cylinders (steel, aluminum, or composite) at pressures up to ~200–300 bar. All of these high-pressure tanks are safe and reliable means of storing the quantity of hydrogen for which they were designed. The composite tank manufacturers in particular deserve a lot of credit for producing a high-quality product. For spacecraft applications (i.e., as a rocket propellant) requiring very large quantities of hydrogen, hydrogen is stored in cryogenic LH_2 containment vessels in large stationary storage facilities and also as part of the launch vehicle itself. The first hydrogen-powered vehicles have used both compressed gas and cryogenic methods of hydrogen storage.

The search for new ways of storing hydrogen comes from a desire to take advantage of hydrogen's properties in new and different applications that challenge the existing storage methods (high-pressure cylinders, LH_2 dewars). Use of small amounts of LH_2 by everyday consumers would place demands on the poor dormancy performance associated with LH_2 storage in conventional dewars. As discussed in Chapter 4, this problem can be technically addressed through the use of cryo-compressed LH_2 storage techniques, or via adsorption-based systems as discussed in Chapters 7 and 10. If a large-scale infrastructure appears based on LH_2, storing small amounts (~5 kg) of hydrogen for long periods of time (weeks) can now be contemplated due to the dormancy increases afforded by both of these storage technologies. The cryo-compressed storage is more advanced from an engineering perspective than the adsorption hydrogen storage technology. As adsorption materials progress, there is potential for further improvements in dormancy, and a reduction in the average pressure within the hydrogen storage system. The state of adsorption system engineering will get a strong boost from the efforts of the U.S. DOE Hydrogen Storage Engineering Center of Excellence (HSECoE), as the HSECoE is in the process of designing, building, and testing a full-scale adsorption-based hydrogen storage system.

It may very well be that a large-scale LH_2 infrastructure does arise. Although it typically requires about 30% of the hydrogen fuel energy to liquefy it, it might be worth it (in cost and energy) because once liquefied, LH_2 can be transported and dispensed very easily without the need for (currently) expensive pressure vessels and dispensing hardware rated for high pressure. It's possible that the overall cost reductions for transporting LH_2 might make it the physical state of choice for a hydrogen energy infrastructure. I personally believe cost will drive the infrastructure rather than the eventual physical state of hydrogen that ends up being stored on a particular application. If a substantial LH_2 infrastructure appears (perhaps alongside a compressed gas infrastructure), then the advanced LH_2 storage methods discussed in Chapters 4, 7, and 10 will be ready. I see no "show-stopping" problems in the engineering implementation of either cryo-compressed or solid-state adsorption hydrogen storage systems, although clearly important engineering work remains for both approaches.

The desire to turn to other solid-state hydrogen storage methods comes about because in some applications (such as automobiles), improvements in volumetric and gravimetric density are desired. In addition, refueling using ambient-temperature low-pressure hydrogen, or at least pressures available with merchant hydrogen (~150 bar), would be convenient, as LH_2 is currently not as available as compressed hydrogen and it does remain to be seen in what physical state (gas, liquid) the hydrogen energy infrastructure will develop. Summarizing this motivation for solid-state storage, there exists automotive desire for more gravimetrically and volumetrically efficient hydrogen storage systems that can be refueled readily from compressed hydrogen at ambient temperature.

Chapter 5 reviewed the development of the first generation of such solid-state materials, namely interstitial metal hydrides. These are really remarkable materials. They store and release hydrogen with excellent kinetics at ambient temperatures, have good volumetric storage density, have low operating pressures (~15 bar), and can be refueled with gaseous hydrogen. For a number of applications such as hydrogen storage for stationary applications where one needs to store energy for grid applications, or applications where the weight of the storage system is an advantage, such as in forklift trucks, the performance of interstitial metal hydrides is excellent. Solid-state hydrogen storage via interstitial metal hydrides has been demonstrated to be a technically sound approach even for light-duty vehicles, as shown in Figure 3.6, although there clearly is room for performance improvement and the cost of hydrogen storage tanks based on interstitial metal hydrides needs to come down.

The important point here is that whether it be high-pressure compressed storage of hydrogen, cryo-compressed storage of hydrogen, or storage in a low-gravimetric capacity interstitial metal hydrides, there exist diverse ways now of storing hydrogen in support of the early deployment of hydrogen powered technology (vehicles, equipment, portable power). Technically, there is no hydrogen storage impediment to the initial deployment of a hydrogen energy infrastructure. Although currently expensive, the methods already demonstrated would lend themselves to economies of manufacturing scale, and costs would eventually come down. As with any technology, there is room for material and method improvement, and this is where the most recent work on solid-state methods of hydrogen storage arises and this book springs forth.

The recent drive to develop even more advanced hydrogen storage materials came from a desire to optimize hydrogen storage materials for use in light-duty vehicles with long driving range. As described in Chapters 6, 7, and 8, there are three broad classes of solid-state storage: on-board reversible complex metal hydrides, on-board reversible hydrogen adsorption materials, and off-board reversible hydrogen storage materials. The terms "on-board" and "off-board" are vehicle-centric, but all applications of hydrogen storage must consider the nature of the storage material regeneration. Assessments of the "state-of-the-art" and research needs in these advanced materials can be found throughout Chapters 6–8. A few thoughts are offered here.

As described above, one can think of these materials relative to the type of hydrogen infrastructure that might arise and the mode of refueling. If there is an expansion of the existing compressed gas infrastructure (compressed to the ~70–200 bar range typical of merchant and hydrogen pipeline gas), and if one assumes the current refueling paradigm of a hydrogen vehicle pulling up to a H_2 station, and quickly being refueled with ambient-temperature gaseous hydrogen, then one is in the realm of the "on-board reversible complex metal hydride" hydrogen storage materials. Enormous progress has been made in finding new materials and understanding the chemical properties of complex metal hydride materials. However, that one material that combines all the attributes of high gravimetric and volumetric capacities, full reversibility, and good thermodynamics and kinetics remains to be discovered. It is my view that this truly optimal material is within reach, and may reside amongst the borohydrides.

As discussed in Chapter 6, many borohydride materials have been examined in recent years as reversible hydrogen storage materials. An interesting feature of the borohydrides is that they are typically high gravimetric and volumetric density materials whose hydrogen storage properties can vary widely with regard to thermodynamics, kinetics, the purity of the hydrogen released, and reversibility. In many ways, the optimal solid-state hydrogen storage material lies *within* the chemistry boundaries observed for the borohydrides, not outside those boundaries. In a sense, finding the right borohydride is a problem of interpolation, not extrapolation, and interpolation is fundamentally a much less risky proposition. With further fundamental understanding of the chemical variations seen in the borohydrides, a truly remarkable borohydride-based hydrogen storage material may be discovered.

The use of "off-board reversible" hydrogen storage materials would require a paradigm shift in refueling concepts on the part of the automobile manufacturers. The shift involves the concept of removing "spent" storage material from the vehicle and replacing it with fresh hydrogenated material, with the spent material sent to a central facility for rehydrogenation. Although this paradigm is currently a foreign concept for automotive uses, the idea of an irreversible power source is not foreign to many non-vehicular applications.

Disposable batteries are used in man-carry portable electronics and indeed materials such as AlH_3 are currently being contemplated for such applications. If such a paradigm shift in the automotive industry were to be made, then recent R&D has produced several materials that could serve as high performance hydrogen storage materials in this "off-board reversible" world. The materials themselves (AlH_3, $LiAlH_4$, NH_3BH_3, and liquid organic carries) need to have much more work demonstrating robust dehydrogenation/rehydrogenation cycling with good energy efficiency, and this is particularly true for AlH_3, $LiAlH_4$, and NH_3BH_3. The liquid organic carriers need more research improving the thermodynamics of the hydrogenation/dehydrogenation. However, all of these materials have demonstrated limited reversibility in the laboratory, with more R&D needed to demonstrate robust cycling.

As described in Chapter 9, many of the materials-based problems associated with operating practical hydrogen storage systems based on metal hydrides have been identified and analyzed. Chapter 10 described the real engineered and operating solid-state hydrogen storage system that was built and tested by GM/Sandia using $NaAlH_4$. Additional engineering work would be desirable for the N-based complex hydride materials, such as the ($2LiNH_2 + MgH_2$) material described in Chapter 6, since the material properties of this material differ significantly from the sodium alanate material for which engineering learning has already taken place. Engineering studies of first the N-based materials then the B-based materials would improve the engineering knowledge base, for example understanding how to handle the problem of trace gas impurity that might arise from hydrogen storage via complex metal hydrides, for example NH_3 from ($2LiNH_2 + MgH_2$), or B_2H_6 from a future borohydride.

By comparison, much less work has been done unraveling the engineering issues for the "off-board reversible" materials, although systems based on these approaches have been analyzed for their potential to meet automotive hydrogen storage goals, as described in Chapter 11.

Overall, hydrogen storage methodology has advanced a great deal in the last 5 years. Hydrogen storage technology is sufficiently developed *right now* to support the creation of a hydrogen energy infrastructure, including use of hydrogen as the basis for electrical and shaft power in stationary applications, transportation and portable equipment. The Fuel Cell Mobile Light described in Chapter 2 is an example of performance-driven introduction of hydrogen PEM technology into the construction equipment marketplace, while hydrogen powered PEM fuel cell forklift trucks are proving to be ideal for efficient and pollution-free material handling applications. The H_2 ICE vehicles and hydrogen FCEVs shown in Chapter 2 have successfully used a range of hydrogen storage systems, ranging from LH_2 dewars, interstitial metal hydride tanks, to high-pressure hydrogen storage at 350 and 700 bar using composite-reinforced tanks, and storing from ~4–10 kg of hydrogen. The driving range of these vehicles has varied from ~150–430 miles, depending on design. All of these vehicles are high performance, high efficiency, near zero emissions, and can be economical and viable in the near term. They are "near-commercial-ready" for the early roll-out of the hydrogen-based transportation sector. The serial production of tens of thousands of hydrogen powered vehicles is planned world-wide for the 2015 time. The hydrogen codes and standards work described in Chapter 12 have been performed to fully support this deployment.

Although the driving range for vehicles constructed thus far can be as high as 430 miles (at 700 bar pressure), the more typical range is ~200–240 miles. This range, while acceptable for many situations, is an area for improvement and this is where the advanced solid-state storage methods hold promise. To increase the range for automotive applications, and to otherwise fully realize the potential of a hydrogen energy technology, new materials still need to be found with even better properties (capacity, thermodynamics, and kinetics) that will relieve the existing engineering challenges to their implementation. This is a perfectly acceptable, even exciting place to be. We can already see the benefits of using the currently available hydrogen technology pieces (hydrogen conversion devices, existing storage methods) in the start of a hydrogen energy infrastructure. At the same time, we can also see the path to a truly optimized hydrogen storage technology, where gravimetric density, volumetric density, kinetics of hydrogenation/dehydrogenation, energy efficiency, and convenience of use are at their most developed. My hope is that this book offers current and future scientists, engineers, and policy makers strong motivation and a solid foundation in the methods, materials, engineering, and applications of *Hydrogen Storage Technology*, so that even more progress can be made in the months and years ahead.

Before I close these remarks, I want to thank you, the reader, for your interest in *Hydrogen Storage Technology: Materials and Applications*, and for reading this book. I hope you enjoyed it, and found the content informative and useful. If you have any questions about the content presented here, feel free to contact me or the relevant Chapter author directly. All of the authors have their email addresses listed as part of their biographical information in the front of the book, and we want to hear from you if you need more information or explanation.

In closing, there are a number of people I would like to thank and acknowledge who helped me in one way or another with creating *Hydrogen Storage Technology: Materials and Applications*. First, I would like to thank the staff of Taylor & Francis for offering me the opportunity to serve the global hydrogen community as the Editor of this book. In particular, John Navas was there at the very beginning, offering advice and talking me down from the edge of the cliff on a number of occasions! During production, Judith Simon was extraordinarily patient with my many questions, and Scott Shamblin gets a lot of credit for the sweet cover design. Rachel Holt also helped to coordinate the early manuscript reviews. All of these folks at Taylor & Francis were highly professional and fun to work with.

I would like to thank Ned Stetson and the U.S. Department of Energy, Office of Energy Efficiency and Renewable Energy, as well as Sandia National Laboratories for supporting efforts on this book. It is important to note that the views expressed herein do not necessarily represent the views of the U.S. DOE or Sandia National Laboratories.

I would like to thank Jay Keller and Daniel Dedrick for serving as Associate Editors of Section I and Section III, respectively. They both helped to bring the best technical people to bear on their respective sections.

I also want to acknowledge the unique combination of passion and expertise that Jay Keller brings to anything he does. Jay was my manager at Sandia when the book project was conceived, and his early encouragement helped me enormously. Thank you, Jay.

I sincerely thank the many authors who contributed to *Hydrogen Storage Technology: Materials and Applications*. All of the authors put an enormous amount of talent, work, and care into their contributions, and with great understanding and collegiality accepted my edits. I'm very proud of the final product we have created together.

In the beginning of this project, I selected an "Advisory Board." Unfortunately, due to the book schedule, this group was not afforded the opportunity to review the manuscript. Hence they are innocent of any troubling content found herein! However, they did offer suggestions up-front about the organization of the book. I would like to thank the following who served in this advisory capacity: Bjorn Hauback (IFE), Bob Bowman (formerly JPL), Channing Ahn (Caltech), Craig Jensen (U. Hawaii), Greg Olson (formerly HRL), Jeff Long (U.C. Berkeley), Klaus Yvon (U. Geneva), Ned Stetson (U.S. DOE), Ping Chen (Dalian Institute for Chemical Physics), Scott Jorgensen (GM), Shin-ichi Orimo (U. Tohoku), and Tom Autrey (PNNL).

A review of the entire draft manuscript was performed by Klaus Yvon, Jeff Long, and Greg Olson. I want to thank these three in particular for taking time to read the page proofs and for giving early feedback that encouraged me we had achieved what we were striving for. Comments from Greg Olson were especially helpful, for which I am grateful. I also want to thank Brian Somerday and Tom Felter, both of Sandia National Laboratories, for giving me early feedback on the Editor's Introduction.

For some reason, the reference styles of the authors of Chapter 8 were hard to harmonize on the computer. Fortunately, the software abilities of Karen McWilliams at Sandia saved the day and we were able to get a readable version of Chapter 8 to the publisher. I also want to thank Martha Campiotti at Sandia for generating the Periodic Table which is found on the inside back cover of the book.

Finally, I want to thank my wife Nitcha. Nitcha, you have been wonderfully supportive and understanding during these many incredibly work-oriented months. I am grateful for your love, understanding, encouragement, kindness, and consideration, and all the wonderful and comforting Thai food. Please always stay 1000% Thai! Phom lak khun mak loy, krup!

Sincerely,

Lennie Klebanoff
Livermore, California

Index

A

AB-type intermetallic interstitial compounds, 113–115

AB$_2$-type intermetallic interstitial compounds, 117–122

AB$_5$-type intermetallic interstitial compounds, 115–116

Acid-catalyzed dehydrogenation, ammonia borane, 275–277

Aciplex, 53

Activated carbons, 78, 110, 227–228, 349, *See also* Adsorption materials; Carbon adsorbents
  AX-21, 221, 349–350, 361, 387, 397
  engineering assessments, 396–397
  heat exchange, 357–358
  performance objectives, 387
  surface excess values, 221, 349–350

Activation energy (E_a)
  aluminum hydride desorption, 249–250
  ball milling and improving hydrogen sorption/desorption kinetics, 185–187
  complex metal hydride hydrogen kinetics, 135
  fundamental hydrogen-water reaction, 31–32
  hydrogen sorption, 213
  nanoconfinement and, 187, 193

Additives, *See also* Catalysts
  AlH$_3$ desorption enhancement, 250
  ammonia borane dehydrogenation enhancement, 280–283
  ball milling and improving hydrogen sorption/desorption kinetics, 185–187
  effect on solid-state reaction kinetics, 200–201
  LiNH$_2$/Mg system, 161–166
  metal borohydride kinetics and, 152, 155
  thermal conductivity enhancement, 341–342
  thermodynamic destabilization effects, 173–175

Adsorption enthalpy, 215–217, 222, 228, 333, 396, *See also* Enthalpy
  isosteric enthalpy of adsorption, 215, 221, 228, 230, 396
  metal organic framework (MOF) compounds, 229
  optimal adsorption, 219–220

terminology, 215

volumetric heat generation, 356

Adsorption materials, 78–80, 110, 213–214, 428, 429, *See also* Activated carbons; Carbon adsorbents; Physisorption-based hydrogen storage; Porous media
  BOP components, 372
  carbon adsorbents, 223–228
  cryoadsorption, 110
  DOE technical targets, 397
  engineering assessments, 396–397
  future directions, 231–233
  gravimetric capacity considerations, 397
  intercalation compounds, 230–231
  interlayer expansion, 224–225
  metal organic framework (MOF) compounds, 78, 110, 196, 229–230, 349
  pressure and adsorption relationship, 78
  pressure/temperature and recoverable hydrogen, 350–351
  ranges of applicability, 400
  refueling thermodynamics, 357–358
  surface area, 78, 214, 216, 218, 222–230, 396–397
  surface excess values, 214–215, 220–221, 223–224
  system engineering design elements, 349–351, 397
  thermal management, 357–358
  insulation, 360–362
  zeolites, 228–229
  zeolite-templated carbons, 233

Aerogels, 191–198, 228

Airbus, 41

Alanates, 134, *See also* Metal alanates; *specific alanates*

Alane (AlH$_3$), *See* Aluminum hydride

Alkali intercalation compounds, 230–231

Alkali metal borohydrides, 148

Alkali metal carbonates, 47

Alkaline fuel cells (AFCs), 46

1-Alkyl(aryl)-3-methylimidazolinium N-bis(trifluoromethanesulfonyl)imidate ionic liquid salt, 316

Allowable stress, pressure vessel design, 367–368

Alpha (α) phase, 111
Aluminum alloys
    hydrogen compatibility, 424
    spacecraft hydrogen vessels, 72
Aluminum borohydride, 155
Aluminum foams, 78, 341
Aluminum hydride (AlH₃), 241–256
    aluminum metal rehydrogenation, 250–252
    electrochemical hydrogenation, 255–256
    organometallic approach, 252–255
    amine adducts, 252–254
    gravimetric and volumetric capacity, 241
    historical uses, 241
    hydrogen desorption kinetics, 247–250
    Lewis acid, 252
    structure and thermodynamics, 244–247
    synthesis, 242–244
    thermal decomposition, 244
    thermodynamic instability, 242, 247–248
Aluminum hydrides (alanates, AlH₄⁻), 134, *See
    also* Metal alanates
Aluminum liners, 68, 369
Aluminum oxide (AlO₄), 228
Aluminum powders, thermal conductivity
    enhancement, 341–342
Aluminum rehydrogenation (to AlH₃), 250–256
Aluminum trioxide (Al₂O₃), 198
American Institute of Aeronautics and
    Astronautics (AIAA), 369
American Society of Mechanical Engineers
    (ASME), 68
Amides, 134, 157, *See also* Lithium amide;
    Nitrogen-based hydrogen storage
    materials
Amine alane adducts, 252–254
Ammonia, 66, 160–161, 164
    borohydride compounds and, 201
    contamination target levels, 137
    hydrogen fuel cells and, 50–51
    metal salt adducts, 172
Ammonia borane (AB, NH₃BH₃), 81, 258
    computational studies, 269–273
    contaminant gas release, 84
    DADB isomer, 259–260, 263–272, 277–278,
        280, 286
    DOE technical targets, 387, 396
    endothermic hydrogen release, 84
    engineering assessments, 394–396
    performance, 396
    system descriptions, 394–396
    exothermic reactions, 273, 394
    free BH₃ and hydrogen release, 270

indoline-based autothermal systems, 313
Lewis acid, 269–270
metal amidoborane reactions, 280–283
nanoconfinement, 187–188, 278–280
NH₂ = BH₂ intermediate, 261, 267, 270,
    274–276, 281
outlook, 286–287
rehydrogenation, 284–286
solution-phase decomposition, 267–268
structure, 258–260
tailoring hydrogen release properties, 273
acid-catalyzed dehydrogenation, 275–279
chemical composition changes, 280–283
metal-mediated dehydrogenation, 273–275
nanoconfinement, 278–280
thermal decomposition in ionic liquids,
    277–278
thermal decomposition, 260–267, 277–278, 286
thermal management, 273
Ammonium chloride, ammonia borane
    rehydrogenation reaction, 284
Angola, 9
Anharmonic vibrations, 184
Apollo program, 71–72
Arc melting, 124
Arrhenius, Svante, 11
Arrhenius equation, 135
Astronomical applications, 73–74
Autoignition temperature, 33–34
Automatic shutoff valve, 407
Automotive hydrogen storage technical targets,
    See DOE light vehicle hydrogen storage
    technical targets
Automotive industry fuel cell development
    perspectives, 58–60
Autothermal hydrogen storage, 312–315
    coupled dehydrogenation reactions, 312–313
    dehydrogenation and selective partial
        oxidation, 314–315
Auxiliary power units (APUs), 46–47
Aviation-based hydrogen technology, 41
    liquid hydrogen, 72
    spacecraft systems, 69–72
AX-21, 221, 349–350, 361, 387, 397

B

Bahrain, 9
Balance-of-plant (BOP) components, 371–372,
    See also Heat exchangers; Thermal
    management
    chemical hydride system efficiency issues,
        396

electrical demands, 56–57
full-system performance criteria, 410
storage system cost and, 96
storage system volume and, 69
system descriptions and, 387
Ball milling
AlH$_3$, 250
improving hydrogen sorption/desorption
kinetics, 185–187
LiNH$_2$/Mg system, 169
lithium alanate and amide, 171
metal alanates, 135
metal borohydrides, 151, 152
mixed metal alanates, 139
Baseline performance, 411–412
Battery electric vehicles (BEVs), 17, 59, 406
energy carrier comparisons, 23–26
range-extended, 59
Battery technologies, 20, 430
Li, 157
Ni-MH, 74, 115–116, 118
range-extended, 59
BCC alloys, 123–126
Benzene, hydrogen fuel cells and, 51
Benzene/cyclohexane system, 81, 288, 289–290,
See also Cyclohexane/benzene reaction
system
Benzimidazoline compounds, 316
Beta (β) phase, 111
Biofuels, 17, 23–24
Biomass energy, 19, 23
BMW hydrogen internal combustion engine,
39–40, 72, 427
Boeing, 41
Borane-indoline systems, 312–313
Boranes, 81, *See also* Ammonia borane
Borazine, 187, 260, 267, 273, 284, 396
Borohydrides, 147–156, 429, *See also* Metal
borohydrides; *specific borohydrides*
ammonia compounds, 172
B$_{12}$H$_{12}$ systems, 175–179, 182, 201
destabilized compounds, 175–179
Boron-11 (^{11}B) NMR spectroscopy, 152f, 153, 155,
267–268, 273–275, 278, 280, 282–285
Boron-nitrogen heterocyclics, 312
Borosilicate, 148
Brake mean effective pressure (BMEP), 37
Brake thermal efficiency (BTE), 37–38
Breeder technology, 9, 10, 23
Bubble testing, 420
Buckyballs, 226
1-Butyl-3-methylimidazolium chloride ([BMIM]
[Cl]), 277–278, 387

C
Calcium alanate (Ca(AlH$_4$)$_2$), 141–142, 146–147
Calcium borohydride (Ca(BH$_4$)$_2$), 150, 154–155,
183, 201
Calcium borohydride ammonia, 172
Calcium hydride (CaH$_2$), 81, 110, 178–179
Callendar, G. S., 11–12, 15
Carbazoles, 81
Carbon adsorbents, 223–228, *See also* Activated
carbons
aerogels, 191–198, 228
fullerenes, 226–227
graphite, 224–226
pore sizes, 223
zeolite-templated, 233
Carbon aerogels, 191–198, 228
Carbonate fuel cells, 47
Carbon dioxide, synthetic hydrocarbon fuels
from, 23
Carbon dioxide emissions, xxviii–xxx, 10–15,
See also Greenhouse gas (GHG) emissions
contamination target levels, 137
hydrogen ICEs, 32–33
Carbon dioxide sequestration, 23
Carbon fibers, 68, 92, 342–343, 349, 369
Carbon-free energy technology, need for, *See*
Hydrogen-based energy technology,
need for
Carbon monoxide, 50, 137
Catalyst poisoning, 50–51
Catalysts, *See also* Additives; Platinum catalysts;
specific catalysts
ammonia borane dehydrogenation, 273–277
cost, 54, 56
fundamental hydrogen-water reaction, 32
hydrogen fuel cell, 45, 46, 49
alternatives and cost reduction, 56
stability, 54–56
iridium pincer complex, 273, 293, 309, 311
liquid organic hydrogen carrier reactions,
287–288
cyclohexane/benzene system, 289
decalin/naphthalene system, 292
methylcyclohexane/toluene system,
290–291
metal alanates, 134, 135, 144–145
metal hydride reactions and, 134, 135
NEDC/NEC cycle, 301–309
perhydrobenzofuran/dibenzofuran system,
311
self-catalyzing system, 169–170
stability, 54–56, 137

support materials, 55–56
voltage cycling effects, 52
Catalytic heaters, 32, 354, 362, 372, 373, 376
Centers of Excellence for hydrogen storage
materials, xxiv, 379, 387, 389, 428
Check valve, 407
Chemical Hydride Center of Excellence, xxiv
Chemical hydrides, 84, 187, 287, *See also*
Aluminum hydride; Ammonia borane;
Liquid organic hydrogen carriers;
Lithium alanate (LiAlH$_4$)
BOP components and efficiency issues, 396
engineering assessments, 394–396
materials properties and behavior, 394
performance, 387, 388*t*, 396
system descriptions, 394–396
ranges of applicability, 400
system engineering design elements,
394–396
Chemical species conservation, 366
Chrysler fuel cell program, 58
Chrysler Natrium, 82
CL-400 program, 41
Clausius-Clapeyron equation, 221
Coal, 40–44
Coal gasification, 42
Coal resources, 5, 6, 8, 11
Coatings, thermal conductivity enhancement,
340
Codes and standards, 378, 405–408
crash safety tests, 407
fueling station requirements, 406, 407, 414,
416
global harmonization, 406
historical perspective, 408–409
hydrogen gas storage, 68
inspection/recertification, 423–424
new facility requirements, 423
performance-based standards, 407, *See also*
DOE light vehicle hydrogen storage
technical targets
pressure vessels, 368–369
qualification testing, 409–421, *See also*
Performance qualification testing
risk management, 409–411
specific transport operations, 406
storage vessel leak rate, 407
stringency of new performance-based
requirements, 422–423
vehicle-based hydrogen system safety, 406
Cold system startup, 121, 354
Cold weather operation, 57–58

Combined heat and power (CHP) operation, 27,
43
Complex metal hydrides, 77–78, 110, 133–134,
429
alanates, 138–147, *See also* Metal alanates
amides, imides, nitrides, 157–169, *See also*
Nitrogen-based hydrogen storage
materials
ball milling and improving hydrogen
sorption/desorption kinetics, 185–187
borohydrides, 147–156, *See also* Metal
borohydrides
catalysts, 135
crystal structure prediction, 183–184
destabilized compounds, 172–179, 182, 190,
198
DOE technical targets, 137, 387
effective thermal conductivity, 335, 337
engineering assessments, 389–394
materials properties and behavior, 389–392
performance, 392–394
system descriptions, 392
engineering properties, *See* Engineering
properties
exothermic water reactions, 392
general hydrogen absorption/desorption
kinetics, 134–135
GM/Sandia system, 372–377, 430
gravimetric hydrogen capacities, 135, 136,
389, 392
hydrolysis reactions, 81
mixed-anion compounds, 169–172
nanoconfinement, 187–199
off-board reversible materials, 256–258, *See
also* Lithium alanate
on-board reversible materials, 134, 135
operating environment requirements, 349,
360
particle size and hydrogen storage
properties, 187–191
performance requirements, 135
ranges of applicability, 400
refueling, 351
reversibility, 110
self-catalyzing system, 169–170
simplified structural model, 183–184
stability and volatilization requirements, 137
system engineering design elements, 352,
392
theoretical prediction of materials, 179–185
thermal efficiency goals, 136
thermal management, 77–78, 362, 397

Compounded stresses test criteria, 410, 422

Comprehensive on-road stress conditions, 410–411

Compressed air energy storage (CAES), 20–21

Compressed Gas Association (CGA), 68

Compressed gas storage, 65, 67–69, 91, 92, 214, 427, *See also* Merchant hydrogen gas; Pressure vessels

  codes and standards, 68

  cost, 96, 97

  density, 109

  in fuel cell vehicles, 69

  geometric volume efficiency, 94

  historical perspective, 408–409

  limitations, 69

  metal hydride storage systems and, 392

  ranges of applicability, 399–400

  refueling thermodynamics, 100–103

  structural materials selection, 369

  system engineering design considerations, 352–354

  thermodynamic safety aspects, 105–106

Compressed natural gas (CNG), 408–409

Compression, 97

Compression ratio, brake thermal efficiency and, 37–38

Computational fluid dynamics (CFD) models, 363, 376, *See also* Computational modeling

Computational modeling, *See also* Density functional theory (DFT) methods

  ammonia borane hydrogen release properties, 269–273

  GM/Sandia complex metal alanate system, 373–376

  integrated vehicle systems, 387–388

  packed bed heat and mass transfer, 363–367

  energy conservation, 365–366

  heat exchanger design, 363

  mass conservation, 364

  momentum conservation, 364–365

  physical property models, 366–367

  species conservation, 366

  storage vessel design, 363–364

  physisorption systems, 222

  polyaromatic hydrocarbons, 297–299

  prediction of complex metal hydride reactions, 179–185

Condensed-phase hydrogen storage systems engineering, *See* Engineering assessments; Engineering properties; Hydrogen storage system engineering

Conformable storage vessel design, 352–353

Convective heat transfer coefficient, 358

Cooling systems, 45, 57, 355, 433, *See also* Heat exchangers; Thermal management

  gas diffusion layers, 49

  real fuel cell systems, 56–57

  solid hydrogen applications, 73

Coordination polymers, 229

Coordinatively unsaturated metal center (CUMC) compounds, 229–230

Copper

  magnesium hydride hydrogen storage properties and, 173

  thermal conductivity enhancing coatings, 340

  wire matrices, 341

Copper 1,3,5–benzenetricarboxylate ($Cu_3(btc)_2$), 196

Copper borohydride, 155

Coronene, 297–298

Corporate average fuel economy (CAFE) standards, 16

Cost

  barrier to hydrogen-powered vehicle technology, 136

  compression and liquefaction, 97

  delivery and dispensing, 97

  DOE light vehicle hydrogen storage goals, 378, 397

  energy carrier comparisons, 24

  hydrogen storage vessels, 95–97

  life-cycle, 97, 136

Coupled cluster (CCSD(T)) method, 222, 271

Crash safety tests, 407, 414

Critical pressure, 333, 338

Cryocompressed hydrogen storage, 73, 92, 95–97, 105, 109, 349, 361–362, 379, 428, *See also* Cryogenic hydrogen storage systems

Cryogenic hydrogen storage systems, 65–66, 69–74, 92, 348, *See also* Liquid hydrogen

  adsorption systems, 78

  cost, 96–97

  density, 93, 109

  DOE gravimetric capacity targets, 378

  dormancy, 98–100, 361–362

  evaporative loss, 70–71, 98

  handling risk, 106

  insulation, 359–362

  materials selection considerations, 349

  ranges of applicability, 400

  refueling thermodynamics, 103–104

  temperature and density behavior, 95

  thermodynamic safety aspects, 105–106

Cryogenic temperature effects, 349
Crystal structures
     ammonia borane, 258–260
     metal alanates, 139–142
     metal borohydrides, 149–150
     predicting based on electrostatic
          interactions, 183–184
Cyclodiborazane (CDB), 267
Cyclohexane, 297
Cyclohexane/benzene reaction system, 81, 288,
     289–290, 400
     electrochemical "virtual" hydrogen storage,
          316–318
     terphenyl-tercyclohexane system, 315–316
Cyclotriborazane (CTB), 267–268

D

DADB (diammoniate of diborane), 259–260,
     263–272, 277–278, 280, 286
Daimler, 58–59
Darcy-Brinkman equation, 365
Darcy's law, 365
Decalin, 297, 314
Decalin/naphthalene system, 81, 288, 292–297
     catalysts, 292–293
     catalytic dehydrogenation under
          nonequilibrium conditions, 293–296
     membrane reactors, 293
     reactor development, 296–297
Decrepitation, 368
Dehydrogenation-rehydrogenation cycles, *See
     also* Refueling
     life-cycle costs, 97, 136
     nanoscaffold stability, 187, 194
     performance qualification testing
          requirements, 413, 415–416
Densification, heat transport enhancement, 339
Density, packing, 332
Density functional theory (DFT) methods
     AlH_3 crystal structure prediction, 245
     limitations for physisorption systems, 222
     predicting dimensional effects on hydrogen
          storage properties, 188
     prediction of complex metal hydride
          reactions, 179–185
     Ti catalyzed metal alanates, 135
Density of hydrogen, 69, 92, *See also* Gravimetric
     hydrogen capacity; Volumetric hydrogen
     capacity
     compressed hydrogen, 109
     cryogenic hydrogen, 93, 109
     exergy, 97

liquid hydrogen, 70, 92, 93
     pressure-temperature relationship, 215
     pressure vessel considerations, 92–95
     refueling thermodynamics, 101–104
     in water, 70, 74, 109
Department of Energy (DOE), *See* U.S.
     Department of Energy
Desorption enthalpy, 309–312, *See also* Enthalpy
     ammonia borane, 394
     authothermal hydrogen storage, 312
     gas refueling, 353
     liquid organic hydrogen carrier selection
          recommendations, 311
     MgH_2 metal hydride systems, 154, 162, 173,
          175, 177, 184
     nanoconfinement and, 187–191, 196
     NEDC/NEC cycle, 301
     nitrogen/oxygen heterocyclic compounds,
          299–300, 310
     polyaromatic hydrocarbons, 297–298
Destabilized complex metal hydrides, 172–179,
     182, 198
     nanoconfinement systems, 190, 196
Deuteration studies
     ammonia borane decomposition, 263–264
     metal amidoborane reactions, 282–283
     metal hydrides, 140
     TiFe, 114
Dewar, James, 69
Diammoniate of diborane (DADB,
     $[NH_3BH_2NH_3]^+[BH_4]^-$), 259–260, 263–272,
     277–278, 280, 286
Dibenzofuran/perhydrodibenzofuran system,
     311
Diborane, 155, 193, 396
Diesel cycle, 32
Differential enthalpy of adsorption, 215, *See also*
     Adsorption enthalpy
Differential scanning calorimetry (DSC), 146,
     147, 164, 167, 246, 261–262, 272, 278, 281
Differential thermal analysis (DTA), 260–261
Diffusion and hydrogen storage chemical
     reactions, 135, 185–187, 199, 248, *See also*
     Permeability
     densification and heat transfer, 339
     gas diffusion layers, 49, 56
1,2-Dihydro-1,2-azaborine, 312
Dimethyl ether (DME), 257
Dimethylethylamine (DMEA), 252–255
Dodecahydrofluorene-fluorene-fluorenone
     system, 314–315
DOE light vehicle hydrogen storage technical
     targets, 136–138, 378–379, 387, 388t, 392

adsorption systems, 397
chemical hydride systems (AB), 396
cost, 378, 397
cycle lifetime, 136
gas adsorption terminology, 214
gravimetric capacity, 136, 378–379, 389
heat exchanger design, 389, 391
material reversibility, 136
metal hydride systems and, 127, 392–394
operating environments, 349
specific on-board hydrogen system targets, 137t
summary of assessments, 397–399
Dormancy, 95, 98–100, 361–362, 427
Driving range, 25, 69, 92, 97, 127, 413, 415, 429, 430
Durability (Hydraulic) Performance Test, 414

E

Efficiency of fossil fuel use, 15–17
Electricity production
combined heat and power, 27, 43
hydrogen gas turbine, 40–44
interstitial metal hydride applications, 122, 126
renewable resource flexibility problems, 318
solid oxide fuel cell, 46–47
UPS systems, 122
Electrification of transport sector, 24, 59
Electrochemical hydrogenation of alane, 255–256
Electrochemical "virtual" hydrogen storage, organic liquid carriers, 316–318
Electrolytic hydrogen production, 21, *See also* Hydrogen production
Electron concentration factor (*e/a*), 118
Electrostatic interactions
physisorption mechanisms, 222
predicting crystal structures, 183–184
Embrittlement, 349, 369, 411, 424
End-of-life strength criteria, 420, 422
Endothermic hydrogen release, 84
Energy carriers, 23–27
cost comparison, 24
refueling time, 24–25
Energy conservation, 365–366
Energy consumption, 4
Energy storage technologies (non-hydrogen-based), 20–22
Energy sustainability, 3
Engineering assessments, 385–387, *See also* Engineering properties

adsorbent-based systems, 396–397
chemical hydride systems, 394–396
complex hydride systems, 389–394
DOE technical targets, 387, 388t, *See also* DOE light vehicle hydrogen storage technical targets
drive cycle specifications, 389, 391t
"gravimetric capacity" issues, 386
integrated vehicle systems modeling, 387–389, 390f
performance objectives, 387
ranges of applicability, 399–400
status and outlook, 400
summary of assessments, 397–399
Engineering properties, 331–332, 337, 430, *See also* Engineering assessments; Hydrogen storage system engineering; Thermal management
complex metal hydride systems, 385–386, 389–394
enhancement strategies, 338–343
additives, 341–343
alloying, 340–341
coatings, 340
densification, 339
gas flow, 339–340
gas pressure, 338
particle size, 339
structures, 341
flow properties, 343–344
future outlook, 344
metal hydride volume expansion, 111–112, 115, 120, 124, 224–225, 332
packing density, 332
thermal properties, 333–338, *See also* Heat transfer; Thermal conductivity
Enthalpy (ΔH), 287, *See also* Adsorption enthalpy; Desorption enthalpy
additives and destabilization effects, 173–177
borohydride reactions, 143, 146, 154, 169, 172, 175, 177, 184
cyclohexane/benzene system, 289
decalin/naphthalene system, 292
general thermal management issues, 355, *See also* Thermal conductivity; Thermal management
harmonic vibration and free energy, 184
liquid organic hydrogen carrier reactions, 288
MCH-toluene-hydrogen cycle, 290
metal hydride reactions, 111–113, 134–135
nanoconfinement and, 187–191, 196, *See also* Nanoconfinement

optimal adsorption enthalpy, 219–220
prediction of complex metal hydride reactions, 180–184
refueling thermodynamics, 101, 216
van't Hoff expression, 112, 348
Entropy (ΔS), 219, 287, 348
cryogenic hydrogen, 99, 104
liquid organic hydrogen carrier reactions, 288
lithium borohydride system, 177
metal hydrides, 112, 134–135
van't Hoff expression, 112, 348
Environmental sustainability, 3
Equilibrium constant (*K*), 287–288
Equivalence ratio (φ), 33–34
hydrogen ICE power density and, 38–39
NO_x production and, 36–37
Ethanol, 23
Evaporative loss, 70–71, 98, 137, 361, *See also* Dormancy
Exergy, 97
Exothermic reactions, 372, *See also* Waste heat
ammonia borane, 273, 394
nanoconfinement and, 188
reversible adsorption process, 216
selective partial oxidation, 314
water and metal hydrides, 392
Expanded natural graphite (ENG) fibers, 343
Expansion, metal hydrides, 111–112, 115, 120, 124
Expansion energy, hydrogen gas, 105–106
Expected vehicle operations, qualification testing, 412–414, *See also* Performance qualification testing
External heat exchangers, 353, 355, 358, 359, 363, 369, 392, *See also* Heat exchangers
Extreme service durability requirements, 414–417, *See also* Performance qualification testing

F

Fire exposure, performance qualification criteria, 418
Fischer-Tropsch process, 23, 293
Fissionable material resources, 8–9, 10
Flammability limits of hydrogen, 34
Flemion, 53
Flow work, refueling thermodynamics, 101–104
Fluorene-dodecahydrofluorene system, 314–315
Ford hydrogen fuel cell program, 58
Ford hydrogen internal combustion engine, 39–40
Formic acid, 316

Fossil fuels
cost comparison with other energy carriers, 24
efficiency of use vs. net-zero, 15–17
resources, 5–9, 22
Fourier's law, 355
FreedomCAR, 136, 378, 392, 396, *See also* DOE light vehicle hydrogen storage technical targets
Frustrated Lewis pairs system, 316
Fuel cell electric vehicle (FCEV) fleet operations, 59
Fuel Cell Market Transformation project, 39
Fuel Cell Mobile Light project, 57–58, 69, 122, 432
Fuel cell systems, *See* Hydrogen fuel cell systems; Proton exchange membrane (PEM) fuel cells
Fuel economy standards for vehicles, 16–17
Fueling/defueling (driving), expected operations performance requirements, 413–416
Fueling station requirements, 406, 407, 414, 416, *See also* Codes and standards; Hydrogen fuel supply infrastructure
Fuel Partnership Program, 136
Fuel resource insecurity, xxvii–xxviii, 5–9
Fuel-to-air mixtures, 34
Fullerenes, 226–227
Full-system performance, 410
FutureGen program, 22–23, 42

G

Gas diffusion layer (GDL), 49, 56
Gas flow, *See also* Diffusion and hydrogen storage chemical reactions
momentum conservation in porous materials, 364–365
porous bed thermal conductivity enhancement, 339
Gasification of coal, 42
Gasification of hydrocarbon feedstock, 23
Gasoline
combustion properties, 33*t*
energy carrier cost comparisons, 24
energy equivalent density, 92
octane rating, 35
Gas turbine hydrogen internal combustion engine, 40–44
General Motors (GM)
Equinox, 59
fuel cell program, 58–59

metal hydride-based system, 59, 127, 372–377
vehicle development road map, 26
Geologic hydrogen storage, 21
Geometric volume efficiency, 94
Germany, hydrogen supply infrastructure, 59
Gibbs free energy change (ΔG), 135, 216, 287
    aluminum hydride reactions, 244–247
    catalyzed metal amide reactions, 135
    cyclohexane/benzene system, 288, 289
    decalin/naphthalene system, 292
    lithium alanate reactions, 143
    lithium amide borohydride system, 281
    lithium borohydride system, 257
    methylcyclohexane/toluene system, 290
    polyaromatic hydrocarbon systems, 300
    prediction of complex metal hydride
        reactions, 180
Gibbs surface excess, 220–221
Global climate change, 10–15, 17
Global energy demand projection, 3
Global harmonization of regulatory
    requirements, 406
Global warming temperature increases, 11–12
Gold (Au) catalyst, 316
Government regulation, 405–406, *See also* Codes
    and standards
Graphene slit-pore structure, 224–225
Graphite, 224–226
    expanded natural graphite (ENG) fibers,
        343
    intercalation compounds, 230–231
Gravimetric hydrogen capacity (or density),
    386
    adsorption systems, 397
    aluminum hydride, 241
    complex metal hydride systems, 135, 136,
        389, 392
    cyclohexane/benzene system, 289
    decalin/naphthalene system, 292
    DOE technical targets, 136, 378–379, 389
    graphite, 225–226
    MCH-toluene-hydrogen cycle, 290
    metal alanates, 143–144
    metal borohydrides, 147, 150, 154, 201
    nanoconfined metal hydrides, 196–198, 199
    NEDC/NEC cycle, 301
    problematic system-level issues, 386
    renewable energy infrastructure, 19
    "weight percentage" terminology, 214
Greenhouse gas (GHG) emissions, xxviii–xxx,
    3, 10–15, 81, *See also* Carbon dioxide
    emissions
Grid-scale energy storage options, 20–22

H

Hafnium borohydride, 155
Hammett (σ) parameter, 310
Harmonic vibrations, 184
Heat exchangers, 353, *See also* Cooling systems;
        Thermal management
    BOP components, 371–372
    computational modeling, 363–364
    DOE technical targets, 389, 391
    external, 353, 355, 358, 359, 363, 369, 392
    GM/Sandia complex metal alanate system,
        373, 376
    internal, 353, 355, 358, 363, 369
    physisorption systems, 358
    spacing and thermal conductivity, 391
    structural design, 369
    turbulent flow, 358
Heats of adsorption, *See* Adsorption enthalpy
Heat transfer, 355, 357–358, *See also* Engineering
        properties; Heat exchangers
    computational modeling, 363–364
    convective coefficient, 358
    densification and porosity, 339
    energy conservation, 365–366
    enhancement strategies, 338
        additives, 341–343
        alloying, 340–341
        coatings, 340
        densification, 339
        gas flow, 339–340
        gas pressure, 338
        particle size, 339
        structures, 341
    evaporative loss, 70–71, 98, *See also*
        Dormancy
    gas refueling, 353
    heat conduction, 355–357, *See also* Thermal
        conductivity
    insulation, 98, 99
    metal hydrides desorption/absorption, 78
    thermal properties of hydrogen storage
        beds, 333–338
    turbulent flow, 358
    volumetric heat generation, 356
Heat transfer fluids, 355, 376, 392, *See also* Heat
        exchangers
Henry's law, 215, 221, 228, 229
Hexabenzocoronene, 297
High-density polyethylene (HDPE), 369
High-pressure tanks, *See* Pressure vessels
HKUST-1, 196
Honda, 58–59

Hubbert's peaks and curves, 6, 14
Hybrid vehicles, 18, 59, 127, 351
Hydrazine, 285
Hydrocarbon-based hydrogen systems, *See* Liquid organic hydrogen carriers
Hydrocarbon contamination target levels, 137
Hydrocarbon feedstock gasification, 23
Hydrocarbon impurities, 51
Hydrogen, 65, 91–92
    commercial commodity, 427
    electrostatic interactions, 222
    orbital interactions, 222–223
Hydrogen, properties of, 67–68, 92, *See also* Engineering properties
    bond dissociation energy, 113
    combustion properties, 33–34
    density, 69, 70, 92, *See also* Density of hydrogen
    deviation from ideal gas, 92–93, 95
    flammability limits, 34
    lower flammability limit, 34, 409, 414, 419
    lower heating value, 67
    molar volume, 68
    non-toxic, 106
    quadrupole moment, 222
    size of molecular hydrogen, 223
    structural materials compatibility, 349, 369–371, 411, 424
Hydrogen adsorption, *See* Adsorption materials; Physisorption-based hydrogen storage
Hydrogen adsorption materials, *See* Adsorption materials
Hydrogen-air mixtures, 33–35
Hydrogen-based energy technology, need for, 3–4
    efficiency vs. net-zero, 15–17
    energy carrier comparisons, 23–27
    fuel resource insecurity, xxvii–xxviii, 5–9
    global climate change, xxviii–xxx, 10–15
    political energy insecurity, xxviii, 9–10
    renewable energy resources, 19–20
    standard of living for all, 4
    timescales for change, 17–19
Hydrogen bonding
    ammonia borane, 258
    fuel cell cathode water uptake, 56
    metal hydrides, 74, 110, 112, 118
Hydrogen combustion
    balance-of-plant components, 372
    catalytic heaters, 32, 354, 362, 372, 373, 376

complex metal hydride systems, 77, 397
    fuel cell heating, 354
Hydrogen conversion, fundamental chemical reaction, 31–32
Hydrogen desorption enthalpy, *See* Desorption enthalpy
Hydrogen fuel cell systems, 32, 44–47, 427, 430–431, *See also* Hydrogen-powered fuel cell vehicles; Proton exchange membrane (PEM) fuel cells
    alkaline, 46
    automotive OEM perspectives, 58–60
    balance-of-plant electrical demands, 56–57
    catalysts, 45, 49
    electric vehicle fleet operations, 59
    flow channel obstruction, 52
    future generations, 60–61
    hydrogen/oxygen reaction, 45
    infrastructure issues, 59
    molten carbonate, 47
    natural gas feedstock, 46, 47
    near-zero CO_2 emissions, 32
    PEM, 46, 47–50, *See also* Proton exchange membrane (PEM) fuel cells
    phosphoric acid, 46
    solid oxide, 46–47
    thermal efficiency, 45, 57
    types, 45–47
    waste heat elimination, 45, 57, *See also* Thermal management; Waste heat
Hydrogen fuel supply infrastructure, 59, 428, *See also* Refueling
    codes and standards, 406
    commercial prototype, 296–297
    fueling station requirements, 406, 407, 414, 416, *See also* Codes and standards
    liquid organic hydrogen carriers, 81, 319
    outlook, 429
    performance qualification testing requirements, 416
Hydrogen internal combustion engines (ICEs), 32, 427
    gas turbines, 40–44
    hydrogen combustion properties, 33–34
    knock, 35
    liquid hydrogen storage, 72
    near-zero CO_2 emissions, 32–33
    NO_x production, 32, 33, 35–37
    outlook, 430–431
    power density, 38–39
    preignition, 34–35

spark ignition engines, 32–40
thermal efficiency, 37–38
Hydrogen-powered fuel cell vehicles, *See also*
DOE light vehicle hydrogen storage
technical targets; Hydrogen fuel cell
systems; Proton exchange membrane
(PEM) fuel cells
barriers to use, 135–136
compressed hydrogen storage, 69
contamination target levels, 137–138
energy carrier comparisons, 24–27
sodium alanate system, 77*f*
supply requirements (300-mile driving
range), 69
Hydrogen-powered vehicle driving range, 25,
69, 92, 97, 127, 413, 415, 429, 430
Hydrogen production
electrolyzer, 21, 122
gasification, 40
nuclear power and, 23
steam reforming, 16, 46
Hydrogen sorption materials, *See* Adsorption
materials
Hydrogen storage beds, 331–332, *See also*
Engineering properties
Hydrogen storage capacity, *See* Gravimetric
hydrogen capacity; Volumetric hydrogen
capacity
Hydrogen Storage Engineering Center of
Excellence (HSECoE), 387, 389, 428
Hydrogen Storage Multi-Year Research
Development and Demonstration Plan
(MRDDP), 135
Hydrogen storage reactions, theoretical
prediction, 179–185
Hydrogen storage system engineering, 343–344,
379, 385–386, *See also* Engineering
assessments; Engineering properties
balance-of-plant components, 371–372
characteristic length, 355–357
future outlook, 379–381
general design concepts, 352–355
heat and mass transport computational
simulation, 363–367, *See also*
Computational modeling; Heat transfer
modular systems, 353–354, 373
performance and targets, 378–379, *See also*
DOE light vehicle hydrogen storage
technical targets
safety issues, 377–378

structural design, 367, 386
heat exchangers, 369
pressure vessels, 367–369, *See also* Pressure
vessels
structural materials selection, 369–370
system design elements
classic metal hydrides, 351
complex metal hydrides, 352–355, 392
operating environment, 348–349
sorption materials, 349–351
system example, GM/Sandia NaAlH$_4$ bed,
372–377, 432
system-level "gravimetric capacity" issues,
386
thermal management considerations,
355–363, *See also* Thermal management
Hydrogen storage systems, DOE technical
targets, *See* DOE light vehicle hydrogen
storage technical targets
Hydrogen storage systems, overview and
historical perspectives, 65–67
compressed gas, 67–69
cryogenic liquid hydrogen, 69–74
off-board reversible, 80–84, *See also* Off-
board reversible hydrogen storage
materials
on-board reversible, 74–80, *See also* On-board
reversible hydrogen storage materials
Hydrogen storage technology, codes and
standards, *See* Codes and standards
Hydrogen storage vessels, *See* Pressure vessels
Hydrogen Technology Advisory Committee
(HTAC), 22
Hydrolysis reactions, 81–84
Hydroperoxy radicals, 53
Hydropower energy storage, 20
Hysteresis, metal hydride absorption/
desorption, 114–115
Hyundai, 59

I

Imides, 157, 163
Indole-based systems, 300, 310–311
Indoline-borane systems, 312–313
Indoline/indole system, 300
Infrastructure, hydrogen supply, *See* Hydrogen
fuel supply infrastructure
Inspection/recertification requirements, 423–424
Insulation, 98, 99, 359–362
vacuum, 98, 361, 397

Integrated vehicle systems modeling, 387–389, 390*f*

Intercalation compounds, 230–231

Intergovernmental Panel on Climate Change (IPCC), xxx, 10

Internal combustion engines, *See* Hydrogen internal combustion engines

Internal heat exchangers, 353, 355, 358, 363, 369, *See also* Heat exchangers

International Organization for Standardization (ISO), 68, 369, 406

International vehicle standards, 406

Interstitial metal hydrides, 74–76, 109–110, 428
  AB_2-type, 117–122
  AB_5-type, 115–116
  absorption/desorption hysteresis, 114–115
  AB-type, 113–115
  advantages vs. compressed and liquid hydrogen states, 110
  applications, 74–76, 122, 126–128
  desorption plateau pressure, 121–122
  engineering assessments, 389–394
  enthalpy barriers for hydrogen release, 113
  entropy of formation, 112
  example system, 74
  hydrogen supply, 122
  mischmetals, 116
  operating condition requirements, 349
  refueling, 74, 351
  reversible materials, 110–113
  system engineering design elements, 351
  thermal management, 74
  Ti-Cr-Mn alloy, 351
  van't Hoff expression, 112
  volume expansion, 111–112, 115, 120, 124, 368
  volumetric and gravimetric hydrogen densities, 351, 428
  V-Ti-Cr-based BCC alloys, 123–125

Ionic conductivity, LiN_3, 157

Ionic liquids
  aluminum hydride electrochemical rehydrogenation, 255
  ammonia borane performance targets, 387
  ammonia borane thermal decomposition, 277–278
  lithium borohydride self-catalyzing reaction, 169
  methylimidiazolinium/imidate reversible hydrogenation, 316

Iran, 9

Iridium pincer complex [(POCOP)Ir(H)$_2$], 273, 293, 309, 311

Isosteric enthalpy of adsorption, 215, 221, 228, 230, 396

J

Japan hydrogen supply infrastructure, 59

Jet aircraft, 41

K

Knock, 35

L

Langmuir model, 217–220

Lanthanum-nickel ($LaNi_5$) powders, 340–341

Lanthanum nickel hydride ($LaNi_5H_6$), 110, 115–116

Leakage and permeation acceptance criteria, 407, 419–420

Leak and rupture risk, 409–410

Leak test, 420

Lewis acids
  alane, 252
  ammonia borane, 269–270
  ammonia borane dehydrogenation, 276–277
  frustrated Lewis pairs system, 316

Lewis base, 284

$Li_2Zr(BH_4)_6$, 155–156

$Li_4BN_3H_{10}$, 169–170

Libyan oil production, 9

Life-cycle cost, 97, 136

Light-duty vehicle hydrogen storage performance targets, *See* DOE light vehicle hydrogen storage technical targets

Liners, compressed gas pressure vessels, 68

Liquefaction, 97, 214

Liquid-film-type catalytic dehydrogenation, 294

Liquid hydrogen (LH_2), 65–66, 69–74, 92, *See also* Cryogenic hydrogen storage systems
  aviation applications, 41, 72
  critical point, 215
  density, 70, 92, 93, 109
  dormancy performance, *See* Dormancy
  evaporative loss, 70–71, 98, *See also* Dormancy
  handling risk, 106
  hydrogen adsorption storage systems, 78

ICEs, 72
  infrastructure issues, 428
  liquefaction and dispensing costs, 97
  mass fraction vs. capacity, 73
  properties, 68
  ranges of applicability, 400
  refueling thermodynamics, 103–104
  spacecraft systems, 69–72
  storage system cost, 96
Liquid organic hydrogen carriers (LOHC), 81,
    287, *See also specific systems*
  autothermal hydrogen storage, 312–315
  catalytic dehydrogenation under
    nonequilibrium conditions, 293–296
  catalytic hydrogenation/dehydrogenation
    basics, 287–288
  catalytic reactor design, 290–291
  "chemical hydride" designation, 287
  cyclohexane/benzene system, 81, 288,
    289–290, 316–318, 400
  decalin/naphthalene system, 81, 288, 292–297
  electrochemical "virtual" hydrogen storage,
    316–318
  formic acid, 316
  frustrated Lewis pairs system, 316
  ionic liquid salt, 316
  methylcyclohexane/toluene system, 81,
    290–292
  NEDC/NEC cycle, 301–309, 317–318
  nitrogen/oxygen heterocyclic compounds,
    299–300, 309–312
  outlook, 318–320
  perhydrodibenzofuran/dibenzofuran
    system, 311
  perhydro-phenanthrolene system, 309–310
  piperidine and octahydroindole systems,
    310–311
  polycyclic aromatic carriers, 297–300
  supply infrastructure, 81, 319
  terphenyl-tercyclohexane system, 315–316
Lithium alanate ($LiAlH_4$), 138, 256–258
  aluminum hydride synthesis, 242–243
  ammonia borane rehydrogenation reaction,
    284
  crystal structure, 139–140
  endothermic hydrogen release, 84
  hydrogen hydrogenation/dehydrogenation
    properties, 142–143
  mixed-anion system (Li-Al-N-H), 171–172
  off-board regeneration approaches, 257–258
  synthesis, 138

Lithium amide ($LiNH_2$)
  ammonia borane dehydrogenation reaction,
    280
  exothermic water reactions, 392
  Li-B-N-H system, 169–171
  Mg and, 161–166, 173, 182, 200
Lithium amide borohydride ($LiNH_2BH_3$),
    280–283
Lithium-based batteries, 157
Lithium borohydride ($LiBH_4$), 172
  crystal structure, 149
  destabilization, $LiBH_4/Mg_2NiH_4$ system,
    175–177
  harmonic vibration and free energy, 184
  hydrogen storage properties, 150–152, 175
  Li-B-N-H system, 169–171
  nanoconfinement and, 191–193, 198
  synthesis, 148
Lithium borohydride ammonia ($LiBH_4NH_3$),
    172
Lithium hydride (LiH), 110, 157–161, 174–175
Lithium magnesium alanate ($LiMg(AlH_4)_3$), 139,
    147
Lithium magnesium nitride (LiMgN), 166–169
Lithium nitride (Li_3N), 157–161
Lithium potassium aluminum hydride
    (LiK_2AlH_6), 139, 142
Lithium-silicon alloys, 174–175
Lithium sodium aluminum hydride
    ($LiNa_2AlH_6$), 139, 141
Lithium zinc borohydride ($Li_2Zn(BH_4)_4$), 201
Lockheed Skunk works, 41
Locomotive, fuel cell application, 69, 70*f*
Logistic growth curve, 6
Long-haul class 8 trucks, 24, 27
Lower flammability limit (LFL), 34, 409, 414, 419
Lower heating value (LHV), 67
2,6-Lutidine tris(pentafluorophenyl)borane, 316

M

Macropores, 223
Magnesium alanate ($Mg(AlH_4)_2$), 138
  crystal structure, 140
  ground-state structure prediction, 184
  hydrogen storage properties, 146
Magnesium borohydride ($Mg(BH_4)_2$), 148, 150,
    153–154, 201
Magnesium borohydride ammonia
    ($Mg(BH_4)_2(NH_3)_2$), 172
Magnesium copper ($MgCu_2$) hydride, 117–118

Magnesium hydride (MgH_2), 110, 172–173, 175, 182, 185, 193, 198, 200

Magnesium/$LiNH_2$ system, 161–166, 170–171, 173, 182, 200

Magnesium lithium nitride (LiMgN) system, 166–169

Magnesium nickel ($MgNi_2$) hydride, 117–118

Magnesium nickel hydride (Mg_2NiH_4)/LiBH system, 175–177

Magnesium zinc ($MgZn_2$) hydride, 117–118

Manganese borohydride, 155, 201

Manufactured synthetic hydrocarbon fuels, 23

Mass conservation, 364

Mass fractions, 214

MatLab model, 387, 389

Mazda fuel cell program, 59

Mazda hydrogen internal combustion engine, 39

Mechanical energy release, hydrogen gas, 105–106

Membrane degradation, PEM fuel cells, 49–50, 53–54

Membrane electrode assembly (MEA), 49

Mercedes Benz, 59

Merchant hydrogen gas, 68, 109
    interstitial metal hydride refueling, 74, 122
    refueling system design considerations, 352

Mesopores, 223

Mesoporous silica, 187–188, 194, 278–280

Metal alanates, 147, *See also* Lithium alanate; Sodium alanate; *other specific alanates*
    catalyst or additive effects, 134, 135, 144–145, 200
    crystal structures, 139–142
    hydrogen storage properties, 142–147
    intermediate species, 152
    mixed metal alanates, 139, 147
    nanoconfinement and, 194–196
    reacting species conservation, 366
    synthesis, 138–139
    thermal conductivity enhancement, 338–343, *See also* Thermal conductivity

Metal aluminum hydrides, *See* Metal alanates

Metal amidoboranes, 280–283

Metal borohydrides, 147–156, 429, *See also* Lithium borohydride; *specific borohydrides*
    amidoboranes and ammonia borane dehydrogenation reactions, 280–283
    ammonia borane decomposition and, 265–267
    catalyzed hydrolysis, 82–84
    crystal structures, 149–150, 183
    diversity of properties, 201

gravimetric and volumetric hydrogen densities, 147

hydrogen density, 154

hydrogen storage properties, 150–156, 201

intermediate species, 154

mixed-metal and mixed-cation, 155–156

nanoconfinement and, 191–193

synthesis, 148–149

transition metal borohydrides, 155

Metal Hydride Center of Excellence (MHCoE), xxiv, 215

Metal hydrides, 74–78, 110, 429, *See also* Chemical hydrides; Complex metal hydrides; Interstitial metal hydrides; *specific hydrides*
    alanates, *See* Metal alanates
    borohydrides, *See* Metal borohydrides
    bulk heat transfer modes, 333
    engineering assessments, 389–394, *See also* Complex metal hydrides; Engineering properties
    fundamental hydrogen reaction, 111
    gaseous hybrid application, 127, 351
    metal-hydrogen bonds, 110
    mixed-cation compounds, 141–142
    off-board regeneration methods, 250–258, *See also* Aluminum hydride; Lithium alanate
    operating environment requirements, 349
    packed bed flow properties, 343–344
    ranges of applicability, 400
    reversible materials, 110–113
    system engineering design elements, 351–352, 392
    thermal properties, 333–338, *See also* Thermal properties
    unsuitable metals for hydrogen storage, 113
    van't Hoff expression, 112, 134
    volume expansion, 111–112, 115, 120, 124, 368

Metal organic framework (MOF) compounds, 78, 110, 229–230, 349
    adsorption enthalpy, 229
    DOE gravimetric capacity targets, 378
    nanoconfinement systems, 196

Metals, hydrogen compatibility problems, 349, 369–371, 411, 424

Methane combustion properties, 33*t*, *See also* Natural gas

Methylcyclohexane (MCH), 289, 318, 319

Methylcyclohexane-toluene-hydrogen (MTH) cycle, 81, 290–291
    catalytic dehydrogenation, 295
    catalytic reactor design, 290–291

electrical energy storage, 292
vehicular applications, 291
Methylimidazolinium ionic liquid salt, 316
Micropores, 223
Microporous carbons, 223, 227–228, *See also*
Carbon adsorbents
Microporous silica, 291
Millennium Cell, 82–84
Mischmetals (Mm), 116
Mixed-anion complex metal hydrides, 169–172
Mixed-cation alanates, 141–142
Mixed-metal alanates, 139, 147
Mixed-metal borohydrides, 155–156
Modular system design, 353–354, 373
MOF-5, 229
MOF-74, 229
Moller-Plesset perturbation theory (MP2), 222
Molten carbonate fuel cell (MCFC), 47
Momentum conservation, gas flow through
porous materials, 364–365
Multilayer vacuum insulation (MVI), 362, 397

N

Nafion, 48, 53
Nanoconfinement, 187–199
ammonia borane system, 187–188, 278–280
gravimetric capacity considerations, 196–198,
199
lithium borohydride, 191–193
sodium alanate system, 194–196
Nanoscale particles, 187, *See also* Ball milling
agglomeration during cycling, 187
ball milling, *See* Ball milling
particle size and hydrogen desorption,
188–190
Naphthalene/decalin system, 81, 288, 292–297
NASA
Apollo missions, 71–72
pressure vessel codes and standards, 369
Space Shuttle, 69–70
Wide-Field Infrared Survey Explorer (WISE),
73–74
National Academy of Sciences (NAS), x, 10, 13
National Fire Protection Association (NFPA),
406
National Renewable Energy Laboratory
(NREL), 22
Natural gas, vii
compressed storage, 408–409
fuel cell feedstock, 46, 47
political energy insecurity, 9

resource depletion, 5
steam reforming, 16, 46
N-ethyldodecahydrocarbazole/N-
ethylcarbazole (NEDC/NEC) cycle,
301–309, 317–318
catalytic dehydrogenation, 305–309
catalytic hydrogenation, 301–305
dehydrogenation reactor development, 309
stereoisomers, 303–305, 306–307
Neutron diffraction, 139
Nickel catalyst systems
liquid organic hydrogen carrier reactions,
289
NEDC/NEC cycle, 305, 308
Nickel foams, 341
Nickel-lanthanum powders (LaNi$_5$), 340–341
Nickel metal hydride (Ni-MH) batteries, 74,
115–116, 118
Ni-MH batteries, 74, 115–116, 118
Niobium fluoride (NbF$_5$), 155
Nitrogen-based hydrogen storage materials, 157
ammonia formation and its effects, 51,
160–161, 164
borohydride compounds, 172
Li$_3$N, 157–161
LiMgN, 166–169
LiNH$_2$/Mg system, 161–166
mixed-anion complex metal hydride
materials, 169–172
Nitrogen dioxide (NO$_2$), 51
Nitrogen heterocyclic compounds, 299–300,
309–312
Noise, acoustic, 57–58
NO$_x$ production, 32, 33, 35–37, 43–44
Nuclear magnetic resonance (NMR)
spectroscopy
ammonia borane system, 260, 263, 265,
267–268, 273–275, 284–286
borohydride systems, 152–153
metal alanate systems, 144, 145
nanoconfined systems, 199
N-ethyldodecahydrocarbazole/N-
ethylcarbazole (NEDC/NEC) cycle,
302–303, 306
Nuclear power, 8–9, 23

O

Oceans and CO$_2$ absorption, 12–13
Octahydroindole systems, 310–311
Octane number, 35

Off-board reversible hydrogen storage
    materials, 66, 80–84, 110, 239–241,
    429–430, *See also* Ammonia borane;
    Chemical hydrides; Liquid organic
    hydrogen carriers
    alane regeneration, 250–255, *See also*
        Aluminum hydride
    aluminum hydride, 241–256
    ammonia borane, 284–286
    autothermal hydrogen storage, 312–315
    endothermic hydrogen release, 84
    engineering assessments, 394–396
    hydrolysis reactions, 81–84
    liquid organic hydrogen carriers, 287–320
Oil production insecurities, 9–10
Oil resources, 5, 6–8
On-board reversible hydrogen storage
    materials, 66, 74–80, 110–113, 127, 134,
        240, 429, *See also* Adsorption materials;
        Complex metal hydrides; Interstitial
        metal hydrides
    DOE light vehicle hydrogen storage goals,
        136
    DOE targets, 136–138, *See also* DOE light
        vehicle hydrogen storage technical
        targets
    system engineering, *See* Hydrogen storage
        system engineering
On-road stresses, 410–411, 414–415
Operating environments, 348–349, *See also*
    Pressure; Temperature
Orbital interactions and physisorption, 222–223
Ortho-to-para hydrogen interconversion, 69, 99
Otto cycle, 32
OV679, 122
OV694, 122
Oxygen contamination target levels, 137–138
Oxygen heterocyclic compounds, 299–300, 309

P

Packed bed engineering properties, 331–345,
    See also Engineering properties; Thermal
    conductivity
    computational modeling, 363–367
Packing density, 332, 394
Palladium (Pd) catalysts, 56, 289, 298, 301,
    305–308, 309, 311, 313, 314, 372, 376
Para-to-ortho hydrogen interconversion, 69, 99
Parking effects, performance qualification
    testing requirements, 99, 422

end-of-life strength, 420
    expected operation performance
        requirements, 414
    extreme service durability, 416–418
Particle size, thermal conductivity and, 337, 339
Particle size and hydrogen storage properties,
    187–191, *See also* Nanoconfinement;
    Nanoscale particles
PEM fuel cell, *See* Proton exchange membrane
    (PEM) fuel cells
Pentanethiol, 316
Performance objectives, *See* DOE light vehicle
    hydrogen storage technical targets
Performance qualification testing, 409–411
    acceptance criteria, 418
    end-of-life strength, 419–420, 422
    permeation and leakage, 419–420
    temperature specifications, 421
    additional requirements under discussion,
        423–424
    baseline performance, 411–412
    compounded stresses, 410, 422
    comprehensive on-road stress conditions,
        411, 414–415
    expected operation, 412–414
    fueling/defueling (driving), 413–414
    parking, 414
    extreme service durability, 414–418
    externally imposed stresses, 414–415
    fire, 418
    fueling/defueling (driving), 415–416
    parking, 416–418
    full-system performance, 410
    new facility requirements, 423
    protocols, 411
    stringency of new performance-based
        requirements, 422–423
    validation of qualification requirements, 421
Perhydro-4,7-phenanthrolene/4,7-
    phenanthrolene system, 309–310
Perhydrodibenzofuran/dibenzofuran system,
    311
Permeability, 332, *See also* Diffusion and
    hydrogen storage chemical reactions;
    Porous media
    flow properties and, 343–344
    GM/Sandia complex metal alanate system,
        375
    packing density and, 332
    parking and, 414, 416–418
    porosity, 334, 336, 339, 343, 365, 366–367

Permeation and leakage acceptance criteria, 419–420
Petroleum dependence of transport sector, 8, 23
Petroleum production insecurity, 9–10
Phenanthrolene/perhydro-phenanthrolene system, 309–310
Phosphoric acid fuel cells (PAFCs), 46
Physisorption-based hydrogen storage, 78–80, 213–234, 396–397, *See also* Adsorption materials
  absolute uptake and Langmuir model, 217–220
  BOP components, 372
  carbon adsorbents, 223–228, *See also* Activation energy
  engineering assessments, 396–397
  materials properties and behavior, 396
  performance, 397
  storage system description, 397
  future directions, 231–233
  gravimetric capacity, 397
  heats of adsorption, 215–216, *See also* Adsorption enthalpy
  importance of adsorption enthalpy, 219–220, 222
  mechanisms, 222
  electrostatic interactions, 222
  orbital interactions, 222–223
  size of molecular hydrogen, 223
  pressure-adsorption relationship, 215
  ranges of applicability, 400
  refueling time, 216
  surface excess values, 214–215, 220–221, 223–224, 349–350
  surface heterogeneity and, 217–218
  system engineering design elements, 349–351, 397
  terminology, 214–215
  thermal management, 357–358
  insulation, 360–362
Pinging, 35
Pipeline steels, 370
Piperidine systems, 310–311
Plass, Gilbert, 12
Plastic liners, 369
Platinum catalysts, 32
  autothermal systems, 313, 314
  cost, 54
  cyclohexane/benzene system, 317
  liquid organic hydrogen carrier reactions, 288, 289, 290, 293

NEDC/NEC cycle, 305
PEM fuel cell, 56
  stability, 54–56
  Pt/Al$_2$O$_3$, 290
  Pt-Re/Al$_2$O$_3$, 290
  terphenyl-tercyclohexane system, 315–316
  voltage cycling effects, 52
Pneumatic testing, 422
Political energy insecurity, xxviii, 9–10
Political sustainability, 3
Polyaminoboranes (PABs), 261, 263, 286
Polycyclic aromatic carriers, 297–300, 309–312
  nitrogen and oxygen heteroatomic substitution, 299–300
  polyaromatic hydrocarbons, 297–298
Polyfluorinated sulfonic acid (PFSA), 48, 53
Polyimidoboranes (PIBs), 261, 263, 286
Polytetrafluoroethylene (PTFE), 54
Pore sizes, carbon adsorbents, 223, *See also* Porosity
Porosity, 334, 336, 339, 343, 365, 366–367
Porous media, *See also* Adsorption materials; Carbon adsorbents; Permeability
  densification and heat transfer, 339
  mesoporous silica, 187–188, 194, 278–280
  microporous carbons, 223, 227–228
  microporous silica, 291
  momentum conservation, 364–365
Portable ranges of application, 399
Potassium alanate (KAlH$_4$), 138–139
  crystal structure, 140
  hydrogen storage properties, 145
Potassium amido borohydride (KNH$_2$BH$_3$), 281
Potassium hydride (KH), 164, 166, 200
Preignition, 34–35
Pressure
  adsorption materials system engineering considerations, 350–351
  adsorption systems, 78, 110
  aluminum-AlH$_3$ rehydrogenation, 250–252
  critical, 333, 338
  effective thermal conductivity and, 336–337
  exergy, 97
  hydrogen storage density and, 93–95
  Langmuir model and physisorption, 217–220
  metal hydride refueling, 351, 352
  metal hydrides and desorption plateau, 121–122
  parking effects, 414, 416–418
  performance qualification testing requirements, 413

refueling operating environments, 348–349
refueling thermodynamics, 101
regimes of effective thermal conductivity, 333
thermal conductivity enhancement, 338
venting, 99, 361
Pressure-activated pressure relief device (PRD), 407
Pressure-composition-temperature (PCT) curve, 111–112, 134
Pressure supercharging, hydrogen ICE, 39
Pressure vessels, 68–69, 91–92, 348, 407, 427–428, *See also* Compressed gas storage; Cryogenic hydrogen storage systems
    adsorption systems, 397
    allowable stress, 367–368
    capital cost, 95–97
    characteristic length, 355–357
    codes and standards, 368–369
    compression, 97
    computational modeling, 363–364
    conformable designs, 352–353
    density and, 92–95
    dormancy, *See* Dormancy
    geometric volume efficiency, 94
    historical perspective, 408–409
    hybrid metal hydride-gas system, 351
    internal volume as function of pressure, 93–95
    interstitial metal hydride applications, 127
    metal hydride storage systems and, 392
    modularity, 353–354
    optimum refueling pressure, 368
    pressure relief components, 407
    refueling thermodynamics, 100–104
    structural design, 367–369
    structural materials selection, 369–370
    system engineering design issues, 349, 352–354
    thermodynamic safety aspects, 105–106
    venting, 99, 361
    volumetric heat generation, 356
    wall strength, 94–95
2-Propanol, 318
Proton exchange membrane (PEM) fuel cells, 26, 46, 47–50, 427
    acoustic noise, 57–58
    automotive OEM perspectives, 58–60
    balance-of-plant electrical demands, 56–57
    catalyst poisons and impurity effects, 50–51
    cold weather operation, 57–58
    diagrams, 48*f*
    electrocatalyst, 49

factors affecting performance, 50
Fuel Cell Mobile Light project, 57–58, 69, 122, 432
future generations, 60–61
gas diffusion layer, 49, 56
locomotive application, 69, 70*f*
membrane degradation, 49–50, 53–54
thermal design issues, 333
thermal efficiency, 57
voltage cycling, 52
water freezing effects, 51–52
Prototype Electrostatic Ground State (PEGS) method, 184
Pulsed-spray reactor, 291, 296
Pumped hydro storage, 20
Pumps, chemical hydride system efficiency issues, 396
Pyrene, 297

Q

Quadrupole moment, 222
Qualification testing, *See* Performance qualification testing

R

Radiator, 372
Raman spectroscopy, 262, 265
Raney Ni catalysts, 305, 308
Rare earth-based-based AB_5-type intermetallic compounds, 118
Reciprocating internal combustion engine, 32
Refueling, 240, *See also* Hydrogen fuel supply infrastructure
    adsorption system thermal management issues, 357–358
    complex metal hydride systems, 351, 394
    cycle lifetime targets, 136
    DOE targets, 138, 389
    flow work, 101–104
    GM/Sandia complex metal alanate system, 373
    interstitial metal hydrides and, 74, 351
    liquid hydrogen infrastructure, 428
    operating environments, 348–349
    optimal pressure, 368
    outlook, 429
    performance qualification testing requirements
    expected operation, 414
    extreme service, 415
    physisorption considerations, 216

pressure vessel structural design, 368–369
pressure vessel thermodynamics, 100–104
storage vessel design issues, 352–354, *See also*
    Pressure vessels
system engineering, *See* Hydrogen storage
    system engineering
thermal design issues, 333, *See also* Thermal
    management
thermal management, 353, 355
vehicle driving range, 25, 69, 92, 97, 127, 413,
    415, 429, 430
Refueling time
    barriers to hydrogen-powered vehicle
        technology, 136
    DOE technical targets, 394, 396
    energy carrier comparisons, 24–25
    optimal pressure, 368
Renewable energy resources, 19–20
    energy storage technologies, 20–22
    flexibility problems for electricity
        generation/distribution, 318
    global energy supply fraction, 22
Resistance model of effective thermal
    conductivity, 334–335
Revelle, R., 13–14
Reynolds number, heat transfer and, 358
Rhodium (Rh) catalyst, 273, 288, 297
Risk management, 409–411
Russian natural gas supplies, 9
Ruthenium (Ru) catalysts, 56, 274–275, 288,
    301–303, 305, 311
Ruthenium black, 304, 305

S

Safety, *See also* Codes and standards;
        Performance qualification testing
    ambient hydrogen, 106
    crash safety tests, 414
    crash tests, 407
    fire exposure criteria, 418
    leak and rupture risk, 409–410
    pressure vessel thermodynamics, 105–106
    regulations, 406
    storage system engineering considerations,
        377–378
Saudi Arabia, 9
SBA-15, 187, 188
Scramjet, 44
Sequestration, 23
Shell-and-tube heat exchangers, 358, 359, 363
Short-haul class 8 trucks, 24, 27
Silica, mesoporous, 187–188, 194, 278–280

Silica, microporous, 291
Silicon-based compounds, 81
    lithium alloys, 174–175
    MgH_2 destabilization, 198
    zeolite adsorption materials, 228–229
Silicon dioxide (SiO_2), 198
SimuLink modeling, 387
Single-walled carbon nanotubes, 226
Sintered solids, effective thermal conductivities,
    337–338
Slit-pore structures, 224–225
Slurry-based hydrogen materials, 240–241, 394
Slush hydrogen, 73
Society of Automotive Engineers (SAE)
    international vehicle standards, 406
    J2578, 414
    J2579, 368, 409, 411, 412, 414, 418, 421, 422, 424,
        See also Performance qualification testing
    J2601, 421, 423
Sodium alanate ($NaAlH_4$), 77–78, 134, 138, 339,
        See also Complex metal hydrides
    alanate rehydrogenation methods, 255–256
    crystal structure, 140
    DOE technical targets, 387
    effective thermal conductivity, 335, 337
    exothermic water reactions, 392
    fuel-cell-powered vehicle design, 77*f*
    GM/Sandia system, 372–377, 432
    ground-state structure prediction, 184
    hydrogen storage properties, 142, 143–144
    nanoconfinement and, 194–196
    packed bed flow properties, 343–344
    reacting species conservation, 366
    refueling conditions, 351
    synthesis, 138
    thermal conductivity enhancement, 338–343,
        See also Thermal conductivity
    titanium and, 134, 200
Sodium amide borohydride ($NaNH_2BH_3$),
        280–281
Sodium borate ($NaBO_2$), 82
Sodium borohydride ($NaBH_4$)
    ammonia borane decomposition and, 265
    autothermal hydrogen storage, 312
    catalyzed hydrolysis, 82–84
    crystal structure, 150
    hydrogen storage properties, 153
    synthesis, 148
Sodium lithium aluminum hydride
        (Na_2LiAlH_6), 147
Sodium potassium aluminum hydride
        (NaK_2AlH_6), 139
Solar energy, 19

Solid oxide fuel cells (SOFCs), 46–47
Solid-solid interfaces, hydrogen sorption/
    desorption kinetics effects, 198–199
Solid-state hydrogen storage materials
    alanates, *See* Metal alanates
    engineering assessments, *See* Engineering
        assessments
    engineering properties, *See* Engineering
        properties
    hydrides, *See* Complex metal hydrides;
        Interstitial metal hydrides; Metal
        hydrides
    off-board reversible, *See* Off-board reversible
        hydrogen storage materials
    on-board reversible, *See* On-board reversible
        hydrogen storage materials
    system engineering, *See* Hydrogen storage
        system engineering
Sorption Center of Excellence, x, 215
Sorption enthalpy, *See* Adsorption enthalpy
Sorption materials, *See* Adsorption materials
Spacecraft applications, 69–72, 427
Spatial Infrared Imaging Telescope III (SPIRIT
        III), 73–74
Stainless steels, 92, 371, 373
Standards, *See* Codes and standards
Stationary storage, 399
    combined heat and power, 27, 43
    hydride systems, 76, 122, 126, 127
    pressure vessel suitability, 68
Steam reforming, 16
Steels
    cryogenic temperature effects, 349
    hydrogen compatibility (embrittlement), 349,
        370, 371, 424
Stress rupture, 416–418, 420, 421
Structural design, 386
    heat exchangers, 369
    materials selection, 369–370
    pressure vessels, 367–369, *See also* Pressure
        vessels
    conformable designs, 352–353
    materials selection, 369–370
Suess, H. E., 13–14
Sulfur dioxide (SO$_2$), 51, 137
"Suntan" project, 41
Superactivated carbon, 396–397, *See also*
        Activated carbons
Super Car Clean Car program, 18
Supercharging, 39
Supersonic combustion, 44

Surface area, 331
    adsorption systems, 78, 214, 216, 218, 222–230,
        396–397
    aerogels, 191, *See also* Carbon aerogels
    ball milling and, 185, *See also* Ball milling
    heat exchanger design, 358
    nanoparticles, 49, 54, 185, *See also* Nanoscale
        particles
    nanoscale complex metal hydrides, 185–187
    nanoscale scaffolds and, 199, 278, *See also*
        Nanoconfinement
    packed bed thermal conductivity
        enhancement, 339
Surface excess values, 214–215, 220–221,
        223–224, 349–350
Sustainable energy infrastructure, 3–4
    timescales for change, 18–19
Swelling, metal hydride volume expansion,
        111–112, 115, 120, 124, 332
Syngas, 42
Synthetic hydrocarbon fuels (synfuels), 23, 24
System engineering, *See* Hydrogen storage
        system engineering

T

Tap density, 332
Technology change timescales, 17–18
Telescopes, 73–74
Temperature, *See also* Thermal management
    adsorption materials system engineering
        considerations, 349–351
    cryoadsorption, 110
    cryogenic hydrogen storage systems, 95
    density pressure-temperature relationship,
        215
    material selection considerations, 349
    metal hydride refueling, 351, 352
    operating environments, 348–349
    performance qualification testing
        requirements, 413, 421
    end-of-life strength, 420
    refueling thermodynamics, 101–104
    storage vessel design considerations, 369
Temperature-activated pressure relief device
        (TPRD), 407
Temperature increases, climate change related,
        11–12
Terphenyl-tercyclohexane system, 315–316
Test facilities, new requirements for, 423
Tetrahydroaluminates, 134, *See also* Metal
        alanates

Tetrahydroborates, 147, *See also* Metal borohydrides

Tetrahydroboronates, 134, *See also* Metal borohydrides

Tetrahydrofuran (THF), 252, 255–256, 257

Tetrahydroquinoline, 300

Tetralin, 292

Theoretical prediction of hydrogen storage reactions, 179–185

Thermal conductivity, 333, 355–357, *See also* Heat transfer; Insulation; Thermal management
- energy conservation, 365–366
- enhancement strategies, 338, 353
  - additives, 341–343
  - alloying, 340–341
  - coatings, 340
  - densification, 339
  - gas flow, 339–340
  - gas pressure, 338
  - particle size, 339
  - structures, 341
- function of pressure and void fraction, 336–337
- heat exchanger spacing and, 391
- pressure regimes, 333
- resistance model for packed metal hydride bed, 334–335

Thermal efficiency, *See also* Waste heat
- complex metal hydride systems and, 136
- gas turbine hydrogen combustion, 43
- hydrogen fuel cell, 45, 57
- hydrogen ICE, 37–38, 45
- hydrogen-powered vehicle technology barrier, 136

Thermal management, *See also* Cooling systems; Heat exchangers; Heat transfer; Thermal conductivity
- ammonia borane hydrogen release, 273
- BOP components, 371–372
- catalytic heaters, 32, 354, 362, 372, 373, 376
- cold system startup, 354
- complex metal hydride systems, 77–78, 362, 397
- DOE technical targets, 389, 391
- enhancement strategies, 338–343
- fuel cell waste heat, 45, *See also* Waste heat
- gas storage systems, 353
- insulation, 98, 99, 359–362, 397
- interstitial metal hydrides and, 74
- physisorption considerations, 357–358
- refueling, 353, 355

Thermal properties, 333–338, *See also* Adsorption enthalpy; Desorption enthalpy; Enthalpy; Heat transfer; Thermal conductivity
- bulk modes of heat transfer, 333
- packing density and, 332

Thermodynamic analysis, *See also* Computational modeling; Enthalpy
- ammonia borane decomposition, 260–267
- compression and liquefaction, 97–98
- dormancy, 98–100
- exergy, 97
- Langmuir model and physisorption, 217–220
- metal alanates, 142–147
- metal borohydrides, 150–156
- operating environments, 348–349
- prediction of complex metal hydride reactions, 179–185
- pressure vessel refueling, 100–104
- pressure vessel safety, 105–106
- second law and fossil fuel use efficiency, 15
- van't Hoff expression, 112, 134, 166, 348, 366

Thermodynamic destabilization, 173–179

Thermogravimetric analysis, ammonia borane decomposition, 260–262

Thiophene liquid-carrier-based system, 316

Tin hydrides, 284

Titanium alloys, hydrogen compatibility problems, 369

Titanium borohydride, 155

Titanium catalysts or additives, *See also specific alloys or compounds*
- AlH_3 desorption enhancement, 250
- metal alanates, 134, 144–145, 146, 200

Titanium chloride ($TiCl_3$), 168
- $TiCl_3/Pd$, 155

Titanium chromide ($TiCr_2$), 118

Titanium-chromium-manganese alloy, 351

Titanium cobalt (TiCo) hydride, 113

Titanium iron (TiFe) hydride, 110, 113–115, 126

Titanium manganese hydride ($TiMn_2H_3$), 110, 118

Titanium nickel (TiNi) hydride, 113

Toluene/methylcyclohexane system, 81, 290–291

Tortuosity, 366–367

Total uptake, 215

Toyota
- fuel cell program, 58–59
- metal hydride-gaseous hybrid system, 127, 351

Toyota Prius, 18, 74, 127
Transition-metal-based intermetallic
    compounds, 118–122, 177
Transition metal borohydrides, 155
Transition metal hydrogen coordination
    complexes, 222–223
Transportable ranges of application, 399
Transportation sector
    electrification, 24
    energy carriers, 23–27
    fuel cell development perspectives, 58–60
    petroleum dependence, 8, 23
Trans-Siberian pipeline, 9
Triazinane, 299
Triethylamine (TEA) alane, 253
Triethylenediamine (TEDA), 252–253
Turbocharging, 39
Turbojet combustor, 41
Turbulent flow and heat transfer, 358
Tyndall, John, 10–11
Type I pressure vessels, 68
Type II pressure vessels, 68
Type III pressure vessels, 68, 69, 92, 351, 362,
    369
Type IV pressure vessels, 68, 69, 369, 392

U

Ukraine, 9
Ultimate recoverable resource (URR), 6–8
Uninterruptible power supply (UPS) systems,
    122
United Nations (UN) global technical
    regulation, 406
United Technologies Corporation (UTC), 58
Uranium resources, 8–9
U.S. Air Force, 41
U.S. Department of Energy (DOE), *See also* DOE
    light vehicle hydrogen storage technical
    targets
    Centers of Excellence for hydrogen storage
        materials, 379, 387, 428
    Energy Efficiency and Renewable Energy
        (EERE) targets, 50
    Fuel Cell Market Transformation project, 39
    Fuel Cell Mobile Light project, 57–58, 69, 122,
        432
    Hydrogen Storage Multi-Year Research
        Development and Demonstration Plan
        (MRDDP), 135

Hydrogen Technology Advisory Committee
    (HTAC), 22
U.S. energy consumption, 4

V

Vacuum insulation, 98, 361, 397
Vanadium/iron (V/Fe) catalyst, 314
Vanadium-titanium-chromium (V-Ti-Cr) solid
    solution BCC alloys, 123–126
Van der Waals force, 110, 216
Van't Hoff expression, 112, 134, 166, 348, 366
Vehicle crash tests, 407, 414
Vehicle driving range, 25, 69, 92, 97, 127, 413, 415,
    429, 430
Vehicle fuel economy standards, 16–17
Vehicle hydrogen storage system engineering
    assessments, *See* Engineering
    assessments
Vehicle hydrogen storage system performance
    targets, *See* DOE light vehicle hydrogen
    storage technical targets
Vehicle performance qualification testing, *See*
    Performance qualification testing
Vehicle technology change timescale, 18
Venting, 99, 361
Vibration-related free energy, 184
Vienna Ab initio Simulation Package (VASP), 183
Virtual hydrogen storage, 318
Volatilization, 137, *See also* Dormancy;
    Evaporative loss
Voltage cycling, 52
Volume expansion, metal hydride
    hydrogenation, 111–112, 115, 120, 124, 332,
    368
Volumetric hydrogen capacity (or density),
    109–110
    aluminum hydride, 241
    complex metal hydride systems, 135
    cyclohexane/benzene system, 289
    decalin/naphthalene system, 292
    DOE light vehicle hydrogen storage goals,
        378–379
    graphite, 226
    interstitial metal hydrides, 351, 428
    MCH-toluene-hydrogen cycle, 290
    metal borohydrides, 147
    nanoconfinement systems, 197–198
    NEDC/NEC cycle, 301
    renewable energy infrastructure, 19

W

Wall strength, pressure vessels, 94–95
Warm-up time, solid oxide fuel cell, 46–47
Waste heat, 45, 57, 354, *See also* Thermal
 management
 alane regeneration, 253
 balance-of-plant components, 372
 complex metal hydride systems, 77–78, 136,
 397
 cyclohexane/benzene system, 290
 interstitial metal hydrides and, 74, 122
Water, hydrogen storage mass density, 70, 74,
 109
Water freezing effects, 51–52
Water production, hydrogen-oxygen reaction,
 31–32
 freezing/thawing effects, 51–52
 hydrogen fuel cell, 47, 48, 56, 59
Water reactions, complex hydrides, 392
"Weight percentage" terminology, 214
Wells-to-wheels efficiency, 24, 97–98, 394
Wide-Field Infrared Survey Explorer (WISE),
 73–74
Wind power, 19–20, 21, 318

X

Xenon-129 (^{129}Xe) NMR, 278
X-ray diffraction (XRD), 139, *See also* Crystal
 structures
 ammonia borane, 259–260
 metal alanates, 139–142

Z

Zeolites, 228–229, 349
Zeolite-templated carbons, 233
Zero-carbon energy carrier system, xxviii,
 23–27, *See also* Hydrogen-based energy
 technology, need for
 efficiency vs. net-zero, 15–17
Zinc borohydride, 155
Zinc(II) borohydride, 148
Zirconium borohydride, 155
Zirconium cobalt (ZrCo) hydride, 113
Zirconium Cr_2 hydride, 118
Zirconium dioxide (ZrO_2), 198
Zirconium Mn_2 hydride, 118
Zirconium nickel hydride, 113
Zirconium V_2 hydride, 118

Periodic Table of the Elements

Key to Chart

```
      ┌──── Oxidation States
   1   +1
   H   -1  ──── Atomic Number / Element Symbol
 Hydrogen ──── Element Name
 1.00798  ──── 2009 Atomic Weight
```

1 IA	2 IIA	3 IIIB	4 IVB	5 VB	6 VIB	7 VIIB	8	9 VIIIB	10	11 IB	12 IIB	13 IIIA	14 IVA	15 VA	16 VIA	17 VIIA	18 VIIIA
1 +1 −1 **H** Hydrogen 1.00798																	2 0 **He** Helium 4.002600
3 +1 **Li** Lithium 6.9675	4 +2 **Be** Beryllium 9.01218											5 +3 **B** Boron 10.8135	6 +2 +4 −4 **C** Carbon 12.0106	7 +1 +2 +3 +4 +5 −3 **N** Nitrogen 14.00658	8 −2 **O** Oxygen 15.9994	9 −1 **F** Fluorine 18.9984	10 0 **Ne** Neon 20.1797
11 +1 **Na** Sodium 22.9897	12 +2 **Mg** Magnesium 24.3050											13 +3 **Al** Aluminum 26.9815	14 +2 +4 −4 **Si** Silicon 28.0850	15 +3 +5 −3 **P** Phosphorus 30.9738	16 +4 +6 −2 **S** Sulfur 32.0675	17 +1 +5 +7 −1 **Cl** Chlorine 35.451	18 0 **Ar** Argon 39.948
19 +1 **K** Potassium 39.0983	20 +2 **Ca** Calcium 40.078	21 +3 **Sc** Scandium 44.9559	22 +2 +3 +4 **Ti** Titanium 47.867	23 +2 +3 +4 +5 **V** Vanadium 50.9415	24 +2 +3 +6 **Cr** Chromium 51.9961	25 +2 +3 +4 +7 **Mn** Manganese 54.9380	26 +2 +3 **Fe** Iron 55.845	27 +2 +3 **Co** Cobalt 58.9332	28 +2 +3 **Ni** Nickel 58.6934	29 +1 +2 **Cu** Copper 63.546	30 +2 **Zn** Zinc 65.38	31 +3 **Ga** Gallium 69.723	32 +2 +4 **Ge** Germanium 72.63	33 +3 +5 −3 **As** Arsenic 74.9216	34 +4 +6 −2 **Se** Selenium 78.96	35 +1 +5 −1 **Br** Bromine 79.904	36 0 **Kr** Krypton 83.798
37 +1 **Rb** Rubidium 85.4678	38 +2 **Sr** Strontium 87.62	39 +3 **Y** Yttrium 88.9058	40 +4 **Zr** Zirconium 91.224	41 +3 +5 **Nb** Niobium 92.9064	42 +6 **Mo** Molybdenum 95.96	43 +4 +6 +7 **Tc** Technetium (98)	44 +3 **Ru** Ruthenium 101.07	45 +3 **Rh** Rhodium 102.905	46 +2 +4 **Pd** Palladium 106.42	47 +1 **Ag** Silver 107.868	48 +2 **Cd** Cadmium 112.411	49 +1 +3 **In** Indium 114.818	50 +2 +4 **Sn** Tin 118.710	51 +3 +5 −3 **Sb** Antimony 121.760	52 +4 +6 −2 **Te** Tellurium 127.60	53 +1 +5 +7 −1 **I** Iodine 126.904	54 0 **Xe** Xenon 131.293
55 +1 **Cs** Cesium 132.905	56 +2 **Ba** Barium 137.327	57–71 Lanthanide Series	72 +4 **Hf** Hafnium 178.49	73 +5 **Ta** Tantalum 180.948	74 +6 **W** Tungsten 183.84	75 +4 +6 +7 **Re** Rhenium 186.207	76 +3 +4 +6 +8 **Os** Osmium 190.23	77 +3 +4 **Ir** Iridium 192.217	78 +2 +4 **Pt** Platinum 195.078	79 +1 +3 **Au** Gold 196.966	80 +1 +2 **Hg** Mercury 200.59	81 +1 +3 **Tl** Thallium 204.383	82 +2 +4 **Pb** Lead 207.2	83 +3 +5 **Bi** Bismuth 208.980	84 +2 +4 **Po** Polonium (209)	85 **At** Astatine (210)	86 0 **Rn** Radon (222)
87 +1 **Fr** Francium (223)	88 +2 **Ra** Radium (226)	89–103 Actinide Series	104 **Rf** Rutherfordium (261)	105 **Db** Dubnium (262)	106 **Sg** Seaborgium (266)	107 **Bh** Bohrium (272)	108 **Hs** Hassium (277)	109 **Mt** Meitnerium (276)	110 **Ds** Darmstadtium (281)	111 **Rg** Roentgenium (280)	112 **Cn** Copernicium (285)	113 **Uut** Ununtrium (284)	114 **Uuq** Ununquadium (289)	115 **Uup** Ununpentium (288)	116 **Uuh** Ununhexium (293)	117 **Uus** Ununseptium (294)	118 **Uuo** Ununoctium (294)

Lanthanide Series

| 57 +3 **La** Lanthanum 138.905 | 58 +3 +4 **Ce** Cerium 140.116 | 59 +3 **Pr** Praseodymium 140.908 | 60 +3 **Nd** Neodymium 144.24 | 61 +3 **Pm** Promethium (145) | 62 +3 **Sm** Samarium 150.36 | 63 +2 +3 **Eu** Europium 151.964 | 64 +3 **Gd** Gadolinium 157.25 | 65 +3 **Tb** Terbium 158.925 | 66 +3 **Dy** Dysprosium 162.500 | 67 +3 **Ho** Holmium 164.930 | 68 +3 **Er** Erbium 167.259 | 69 +3 **Tm** Thulium 168.934 | 70 +2 +3 **Yb** Ytterbium 173.054 | 71 +3 **Lu** Lutetium 174.967 |

Actinide Series

| 89 +3 **Ac** Actinium (227) | 90 +4 **Th** Thorium 232.038 | 91 +4 +5 **Pa** Protactinium 231.036 | 92 +3 +4 +5 +6 **U** Uranium 238.029 | 93 +3 +4 +5 +6 **Np** Neptunium (237) | 94 +3 +4 +5 +6 **Pu** Plutonium (244) | 95 +3 +4 +5 +6 **Am** Americium (243) | 96 +3 **Cm** Curium (247) | 97 +3 +4 **Bk** Berkelium (247) | 98 +3 +4 **Cf** Californium (251) | 99 +3 **Es** Einsteinium (252) | 100 +3 **Fm** Fermium (257) | 101 +2 +3 **Md** Mendelevium (258) | 102 +2 +3 **No** Nobelium (259) | 103 +3 **Lr** Lawrencium (262) |

Atomic weight values are from Wieser, M.E., and Coplen, T.B., *Pure Appl. Chem.* **83**, 359 (2011). Weights for H, Li, B, C, N, O, Si, S and Cl are averages over a range of listed values. For radioactive elements that do not occur in nature, the mass number of the most stable isotope currently known is given in parentheses. Isotopes of elements 113 – 118 have been reported, but element names have not yet been adopted.